AN INTRODUCTION TO MARINE MAMMAL BIOLOGY AND CONSERVATION

E.C.M. PARSONS
George Mason University

with
AMY BAUER
George Mason University

DOMINIC McCAFFERTY
University of Glasgow

MARK PETER SIMMONDS
Whale and Dolphin Conservation Society

ANDREW J. WRIGHT
Aarhus University
George Mason University

JONES & BARTLETT
LEARNING

World Headquarters
Jones & Bartlett Learning
5 Wall Street
Burlington, MA 01803
978-443-5000
info@jblearning.com
www.jblearning.com

Jones & Bartlett Learning books and products are available through most bookstores and online booksellers. To contact Jones & Bartlett Learning directly, call 800-832-0034, fax 978-443-8000, or visit our website, www.jblearning.com.

> Substantial discounts on bulk quantities of Jones & Bartlett Learning publications are available to corporations, professional associations, and other qualified organizations. For details and specific discount information, contact the special sales department at Jones & Bartlett Learning via the above contact information or send an email to specialsales@jblearning.com.

Copyright © 2013 by Jones & Bartlett Learning, LLC, an Ascend Learning Company

All rights reserved. No part of the material protected by this copyright may be reproduced or utilized in any form, electronic or mechanical, including photocopying, recording, or by any information storage and retrieval system, without written permission from the copyright owner.

An Introduction to Marine Mammal Biology and Conservation is an independent publication and has not been authorized, sponsored, or otherwise approved by the owners of the trademarks or service marks referenced in this product.

Some images in this book feature models. These models do not necessarily endorse, represent, or participate in the activities represented in the images.

Production Credits
Chief Executive Officer: Ty Field
President: James Homer
SVP, Editor-in-Chief: Michael Johnson
SVP, Chief Marketing Officer: Alison M. Pendergast
Publisher: Cathleen Sether
Senior Acquisitions Editor: Erin O'Connor
Editorial Assistant: Rachel Isaacs
Production Editor: Leah Corrigan
Senior Marketing Manager: Andrea DeFronzo
V.P., Manufacturing and Inventory Control: Therese Connell
Composition: Achorn International, Inc.
Cover Design: Kristin E. Parker
Rights and Photo Research Associate: Lauren Miller
Cover Image: © NatalieJean/ShutterStock, Inc.
Printing and Binding: Courier Kendallville
Cover Printing: Courier Kendallville

Library of Congress Cataloging-in-Publication Data
    An introduction to marine mammal biology and conservation / E.C.M. Parsons.—
1st ed.
      p. cm.
    ISBN 978-0-7637-8344-0 (alk. paper)
    1. Marine mammals. 2. Marine mammals—Conservation. I. Title.
    QL713.2.P376 2012
    599.5—dc23
                                                                                2011044755
6048

Printed in the United States of America
16 15 14 13 12   10 9 8 7 6 5 4 3 2 1

To our families and friends.

Brief Contents

| Chapter 1 | What Is a Marine Mammal? | 1 |

PART I	**GENERAL BIOLOGY**	**13**
Chapter 2	Marine Mammal Evolution	15
Chapter 3	Marine Mammal Classification and Diversity	27
Chapter 4	Adaptations to a Marine Environment	43
Chapter 5	Underwater Sound	61

PART II	**ECOLOGY AND STATUS**	**79**
Chapter 6	Polar Bears	81
Chapter 7	Otters	91
Chapter 8	Sirenians	101
Chapter 9	Pinnipeds	115
Chapter 10	Mysticeti: The Baleen Whales	129
Chapter 11	Odontoceti: The Toothed Whales	141
Chapter 12	Delphinidae: The Oceanic Dolphins	167

PART III	**CONSERVATION**	**199**
Chapter 13	Humans and Marine Mammals	201
Chapter 14	Whaling and the International Whaling Commission	215
Chapter 15	Threats to Cetaceans	231
Chapter 16	Threats to Pinnipeds	253
Chapter 17	Marine Mammal Protection: Laws and Initiatives	271
Chapter 18	Marine Mammal Tourism	289

Appendix	307
Index	323
Photo Credits	343

Contents

Preface — xiii
Acknowledgments — xv
Author Profiles — xvii

1 What Is a Marine Mammal? — 1
Characteristics of Mammals — 1
Mammalian Evolution — 3
Defining a Marine Mammal — 3
Quasi-Marine Mammals — 5
Greater Bulldog Bat (*Noctilio leporinus*) — 5
Fishing (or Fish-Eating) Bat (*Myotis vivesi*) — 6
Arctic Fox (*Vulpes lagopus*) — 6
Exploring the Depths: International Union for Conservation of Nature — 6
Exploring the Depths: Sea Wolves and Sea Sheep? — 8
Exploring the Depths: Eurasian Otters: A Marine Mammal? — 8
Exploring the Depths: Steller's Sea Monkey: Were There More Species of Marine Mammals Historically? — 9
Why Are Marine Mammals Special? — 9
Selected References and Further Reading — 10

Part I
General Biology — 13

2 Marine Mammal Evolution — 15
The Changing Oceans — 15
Marine Carnivore Evolution — 15
Exploring the Depths: The Fossil Record — 17
Sirenian Evolution — 18
Exploring the Depths: What Are the Closest Modern Relatives of Whales? — 20
Exploring the Depths: The Earliest Aquatic Archaeocete — 21
Cetacean Evolution — 21
Exploring the Depths: Climate Changes in the Late Eocene and Oligocene — 23
Exploring the Depths: Basilosaurs — 24
Selected References and Further Reading — 24

3 Marine Mammal Classification and Diversity — 27
Marine Carnivores — 27
Polar Bears — 27
Otters — 28
Pinnipeds — 28
Exploring the Depths: California Sea Lion, Zalophus californianus — 30
Exploring the Depths: Walrus, Odobenus rosmarus — 31
Exploring the Depths: Recent Revisions in Phocinae Taxonomy — 32
Sirenians — 33
Cetaceans — 35
Baleen Whales (Suborder Mysticeti) — 35
Toothed Whales (Suborder Odontoceti) — 37
Exploring the Depths: Recent Genetic Studies on Toothed Whales — 39
Exploring the Depths: Value of Genetic Studies on Beaked Whales — 39
Selected References and Further Reading — 40

4 Adaptations to a Marine Environment — 43
Exploring the Depths: Ocean Zones — 43
Swimming — 45
Exploring the Depths: Blubber — 47
Exploring the Depths: Spermaceti Organ — 47
Thermoregulation — 48
Exploring the Depths: Surface Area to Volume Ratio — 49
Diving Physiology and Behavior — 52
Pressure Effects — 53
Osmoregulation — 55
Sensory Adaptations — 55
Exploring the Depths: Light in the Ocean — 56
Exploring the Depths: Sleep — 58
Selected References and Further Reading — 58

5 Underwater Sound — 61
Basics of Sound Physics — 61
Exploring the Depths: What Exactly Is the Decibel? — 63
Bioacoustics — 64
Exploring the Depths: Marine Mammal Sound Production — 64
Marine Mammal Hearing — 67
Exploring the Depths: Marine Mammal Ears — 67
    Cetacean Ears — 67
    Pinniped Ears — 68
    Other Marine Mammal Ears — 69
Echolocation — 69
Navigation — 70
Sound and Communication — 71
Intrasexual Selection — 71
Intersexual Selection — 71
Exploring the Depths: Song and Sexual Selection — 71
Exploring the Depths: Humpback Whale Song — 72
Mother–Calf Cohesion — 73
Group Cohesion — 73
Danger Avoidance — 73
Individual Recognition — 74
Exploring the Depths: Prey Stunning? — 74
Sound in the Marine Environment — 75
Exploring the Depths: What Is the Problem with Noise? — 75
Selected References and Further Reading — 76

Part II
Ecology and Status — 79

6 Polar Bears — 81
Aquatic Adaptations — 81
Distribution — 81
Arctic Adaptations — 83
Feeding Behavior and Ecology — 83
Reproduction — 84
Abundance and Status — 84
Hunting — 84
Pollution — 86
Climate Change — 86
Exploring the Depths: Sampling Contaminants in Polar Bears — 86
Disturbance — 87
Polar Bear Conservation — 88
International Polar Bear Agreement — 88
Exploring the Depths: Precautionary Principle — 88
Recent Changes in Polar Bear Status — 89
Exploring the Depths: Polar Bears in Captivity — 89
Selected References and Further Reading — 90

7 Otters — 91
Sea Otter — 91
Distribution — 92
Aquatic Adaptations — 92
Feeding Behavior and Ecology — 93
Exploring the Depths: Tool Use — 93
Reproduction — 94
Other Behavior — 94
Abundance and Status — 94
Exploring the Depths: Sea Otters and the Exxon Valdez Oil Spill — 96
Exploring the Depths: Sea Otter Declines and Killer Whales — 96
Marine Otter — 97
Distribution, Abundance, and Status — 97
Feeding Behavior and Ecology — 98
Reproduction — 98
Exploring the Depths: Eurasian Otter Rescue and Rehabilitation — 99
    Cubs — 99
    Juveniles — 99
    Adults — 99
Selected References and Further Reading — 100

8 Sirenians — 101
- **Distribution** — 102
- **Feeding Behavior and Ecology** — 104
- **Reproduction** — 105
- **Abundance and Status** — 105
  - *Exploring the Depths: Conservation of the Florida Manatee* — 106
  - *Exploring the Depths: Conservation of the Dugong* — 108
  - *Exploring the Depths: Extinction of Steller's Sea Cow* — 109
- **Selected References and Further Reading** — 111

9 Pinnipeds — 115
- **Distribution** — 115
  - Fur Seals and Sea Lions — 115
  - Walruses — 116
  - True Seals — 116
  - *Exploring the Depths: Guadalupe Fur Seal, Arctocephalus townsendi* — 116
  - *Exploring the Depths: Ross seal, Ommatophoca rossii* — 117
  - *Exploring the Depths: Lake Seals* — 118
  - *Exploring the Depths: Tagging Pinnipeds* — 120
- **Reproduction** — 121
  - *Exploring the Depths: Hooded Seal Mating Behavior* — 123
  - *Exploring the Depths: Nursing Behavior in Galápagos Fur Seals* — 124
  - *Exploring the Depths: Adult Molt* — 125
- **Abundance and Status** — 125
  - *Exploring the Depths: Caribbean Monk Seal, Monachus tropicalis* — 126
  - *Exploring the Depths: Japanese Sea Lion, Zalophus japonicus* — 126
- **Selected References and Further Reading** — 126

10 Mysticeti: The Baleen Whales — 129
- **Distribution** — 130
  - *Exploring the Depths: Migration* — 130
- **Feeding Behavior and Ecology** — 132
- **Reproduction** — 135
- **Abundance and Status** — 136
  - *Exploring the Depths: Antarctic Minke Whale Abundance* — 137
  - *Exploring the Depths: Public Perceptions of the Status of Baleen Whales* — 138
- **Selected References and Further Reading** — 138

11 Odontoceti: The Toothed Whales — 141
- **Sperm Whales** — 141
  - Distribution — 142
  - Reproduction — 142
  - Feeding Behavior and Ecology — 142
  - Abundance Status — 142
  - *Exploring the Depths: Pygmy and Dwarf Sperm Whales* — 143
- **Beaked Whales** — 144
  - *Exploring the Depths: Battling Beaked Whales* — 145
  - Distribution — 146
  - Deep Diving and Sonar — 146
  - Status — 147
- **River Dolphins** — 148
  - Indian River Dolphins — 148
  - Amazon River Dolphins — 149
  - Yangtze River Dolphin — 149
  - *Exploring the Depths: Franciscana or La Plata Dolphin, Pontoporia blainvillei* — 152
- **Beluga Whales** — 153
  - Distribution — 153
  - Abundance and Status — 153
  - *Exploring the Depths: Cook Inlet Beluga Whale* — 154
- **The Narwhal** — 154
  - *Exploring the Depths: The Narwhal's Tusk* — 155
- **Porpoises** — 156
  - *Exploring the Depths: Dall's Porpoise and Japanese Catches* — 156
  - *Exploring the Depths: Harbor Porpoise, Phocoena phocoena* — 158
  - *Exploring the Depths: A Particularly Perplexing Porpoise* — 160
- **Selected References and Further Reading** — 162

12 Delphinidae: The Oceanic Dolphins — 167
- **Killer Whales** — 167
  - Killer Whale Ecotypes — 168
  - *Exploring the Depths: Killer Whale Dorsal Fins* — 168
  - *Exploring the Depths: Killer Whale Behavior* — 169
  - Abundance Status — 170
  - *Exploring the Depths: Killer Whale Longevity* — 170
- **The "Blackfish"** — 171
  - Pilot Whales — 172
  - *Exploring the Depths: Strandings* — 172
- **"Blunt-Headed" Dolphins** — 172
  - Irrawaddy and Snubfin Dolphins — 172
  - Risso's Dolphin — 173

Right Whale Dolphins	**174**
"Beakless" and "Short-Beaked" Dolphins	**175**
Genus *Cephalorhynchus*	175
Exploring the Depths: Hector's Dolphin, Cephalorhynchus hectori	176
Genus *Lagenorhynchus*	177
Humpback Dolphins	**177**
Exploring the Depths: Climate Change and Conservation of Lagenorhynchus *Species*	179
Exploring the Depths: Critically Endangered Eastern Taiwan Strait Sousa chinensis *Population*	181
Long-Beaked Dolphins	**181**
Genera *Stenella* and *Steno*	181
Exploring the Depths: Rough-Toothed Dolphin, Steno bredanensis	184
Genus *Delphinus*	185
Exploring the Depths: Common Dolphin Bycatch in Europe	186
Exploring the Depths: Tucuxi and Costero, Genus Sotalia	188
Bottlenose Dolphins	**189**
Exploring the Depths: Sperm Competition	191
Exploring the Depths: Dolphin Intelligence	192
Exploring the Depths: Culture in Cetaceans	193
Selected References and Further Reading	**194**

PART III
CONSERVATION 199

13 Humans and Marine Mammals 201

Marine Mammals in Folklore and History	**201**
Exploring the Depths: Narwhals and Unicorns	206
Marine Mammal Hunts and Early Whaling	**207**
Exploring the Depths: Whales and Whaling in the United Kingdom	208
Commercial Whaling: The Early Years	208
Exploring the Depths: Life as a Whaler	210
Exploring the Depths: The Essex	211
Changing Attitudes Toward Marine Mammals	**212**
Exploring the Depths: Public Opinion, Whaling, and Seal Culls	213
Selected References and Further Reading	**213**

14 Whaling and the International Whaling Commission 215

History of Modern Whaling	**215**
Exploring the Depths: Japanese Commercial Whaling	217
International Convention on the Regulation of Whaling	**218**
Whaling Quotas and Bans	218
Whaling Moratorium	220
Norwegian Whaling	220
Japanese "Scientific" Whaling	220
Exploring the Depths: Structure of the IWC	220
Iceland's "Scientific" and Commercial Whaling	221
Exploring the Depths: Iceland and Whale Watching	222
Revised Management Procedure and Revised Management Scheme	222
Exploring the Depths: North Atlantic Marine Mammal Commission	223
Aboriginal Whaling	**223**
Exploring the Depths: Cetacean Meat Contamination	224
Exploring the Depths: IWC Conservation Committee	224
Exploring the Depths: Aboriginal or Indigenous Whaling	227
Exploring the Depths: Welfare and Whaling	228
Selected References and Further Reading	**229**

15 Threats to Cetaceans 231

Direct Takes of Small Cetaceans	**231**
Exploring the Depths: Cetacean Culls	233
Exploring the Depths: Live Takes	233
Fisheries and Cetaceans	**235**
Exploring the Depths: What Is Stress?	237
Exploring the Depths: U.S. Tuna Labeling Controversy	238
Ship Strikes	**238**
Pollution	**239**
Heavy Metals	239
Organohalogens	240
Harmful Algal Blooms	241
Marine Litter and Debris	241
Exploring the Depths: Butyltin	241
Exploring the Depths: Radioactive Discharges	242
Oil	242
Exploring the Depths: Polyaromatic Hydrocarbons	243
Sewage and Disease	243
Climate Change	**243**
Habitat Degradation	**245**
Selected References and Further Reading	**247**

16 Threats to Pinnipeds 253

Endangered and Threatened Pinniped Populations	**253**
Mediterranean Monk Seal, *Monachus monachus*	253
Hawaiian Monk Seal, *Monachus schauinslandi*	254
Steller Sea Lion, *Eumetopias jubatus*	255
Exploring the Depths: "Junk Food" Hypothesis	257
Exploring the Depths: Orca Predation Hypothesis	257
Galápagos Fur Seal, *Arctocephalus galapagoensis*	257
Australian Sea Lion, *Neophoca cinerea*	259

New Zealand Sea Lion, *Phocarctos hookeri*	260
Northern Fur Seal, *Callorhinus ursinus*	261
Caspian Seal, *Pusa caspica*	262
Exploring the Depths: Northern Elephant Seal, Mirounga angustirostris	263
Harbor Seal, *Phoca vitulina*	265
Exploring the Depths: "The Canadian Harp Seal Harvest	266
Selected References and Further Reading	**267**

17 Marine Mammal Protection: Laws and Initiatives — 271

International Union for Conservation of Nature	**271**
Exploring the Depths: United Nations Convention on the Law of the Sea	274
Convention on the International Trade in Endangered Species	**274**
Convention on Migratory Species	**275**
Exploring the Depths: ASCOBANS	276
Exploring the Depths: ACCOBAMS	276
U.S. Endangered Species Act	**276**
U.S. Marine Mammal Protection Act	**278**
Exploring the Depths: 1966 Fur Seal Act	279
Exploring the Depths: U.S. Marine Mammal Commission	279
Exploring the Depths: U.S. Magnuson-Stevens Act	280
Exploring the Depths: New Zealand Marine Mammals Protection Act	280
European Community Habitats Directive	**280**
Exploring the Depths: European Council Regulation on Bycatch	281
Exploring the Depths: Pinniped Conservation Law in the United Kingdom	282
Exploring the Depths: Legal Protections for Marine Mammals in the United Kingdom	283
Exploring the Depths: Public Opinion on Marine Mammal Protection in the United Kingdom	283
Exploring the Depths: Marine Protected Areas and Sanctuaries	284
Food For Thought: Science-Based Management?	**285**
Exploring the Depths: Domestic Laws Around the World	286
Food For Thought: Enforcement	**286**
Selected References and Further Reading	**286**

18 Marine Mammal Tourism — 289

Significance of Marine Mammal Tourism	**289**
Exploring the Depths: Whale Watching and Ecotourism	290
Exploring the Depths: Whale Watching Versus Whaling	291
Exploring the Depths: Pinniped Tourism	292
Exploring the Depths: Whale Watching Around the World	292
Who Watches Whales? The Nature of Marine Mammal Tourists	**294**
Negative Impacts of Marine Mammal Watching	**294**
Exploring the Depths: Negative Impacts of Pinniped Tourism	295
Managing Marine Mammal Tourism	**296**
Exploring the Depths: Sustainability Report Card	297
Exploring the Depths: Educational Potential of Marine Mammal Tourism	297
Exploring the Depths: Dolphinaria: Pros and Cons	298
Exploring the Depths: Keiko, the Whale From Free Willy	300
Solitary Sociable Dolphin Problem	**301**
Selected References and Further Reading	**302**

Appendix
Marine Mammal Research Techniques — 307

Line Transect Surveys	**307**
Exploring the Depths: Experiment Design	309
Mark-Recapture Analysis and Photo-Identification	**310**
Exploring the Depths: Photo-Identification	311
Exploring the Depths: Taking Marine Mammal Photographs	312
Land-Based Surveys	**313**
Exploring the Depths: Behavioral Observation Methods for Marine Mammals	315
Biopsy Darting	**315**
Exploring the Depths: Strandings Analysis	316
Exploring the Depths: A Brief Glimpse into Molecular Genetics	316
Tagging	**317**
Crittercams	**318**
Acoustic Techniques	**318**
Exploring the Depths: Public Sightings Schemes	319
Exploring the Depths: Welfare, Animal Research Ethics, and Invasive Studies	321
Selected References and Further Reading	**322**

Index — 323
Photo Credits — 343

Preface

The biology and conservation of marine mammals are topics on which there is a great deal of public interest, as evidenced by the vast number of TV documentaries and popular books on these animals. In universities, the topic of marine mammals attracts a great deal of student interest.

This textbook was initially developed from marine mammal course materials for undergraduate programs run for Glasgow, Leicester, London, and Stirling Universities in the U.K., and George Mason University in the U.S. The book is primarily aimed at providing an introduction to marine mammals to undergraduate students. We have suggested further texts at the end of each chapter for those students wishing for additional, more in-depth biological or conservation information on the topic. It is our hope that the text provides a useful reference for other courses that may incorporate marine mammals, e.g., animal behavior, conservation biology, environmental policy, marine biology, marine conservation, oceanography, vertebrate biology, or even wildlife tourism/ecotourism.

It was a deliberate intention that this book be multidisciplinary, containing information on marine mammal biology, history, physiology, law and policy, and social sciences, and even dipping into subjects as diverse as archeology, business, economics, and mythology where relevant. So hopefully even those students well versed in one field may find something new and of interest between the covers of this book.

We have also tackled several topics that are currently in the news (such as whaling, climate change, and underwater noise pollution), some of which are quite controversial. In dealing with these topics we have tried to show both sides. However, we quickly realized that giving equal weight in an argument to polarized sides of a debate can sometimes make fringe or extremist opinions appear more mainstream than they really are (for example, as seen in the debates over evolution and climate change). Therefore, when dealing with such issues we have ultimately tried to reflect the consensus opinions, or concerns, of scientists, marine mammal experts, and professional societies as appropriate.

Throughout the book there are Exploring the Depths discussion boxes on a variety of subjects written by guest experts in the field. While some of these textboxes are written by senior members of the marine mammal science and conservation community, where possible we have sought out younger experts, whose age is likely closer to the students reading this book, in a hope to inspire and encourage students to look at these young experts as role models.

We have written this book with a wider audience in mind than just college and university students, however, and we hope that it will appeal to marine conservationists, whether working for governmental, non-governmental, or local community-based organizations, those involved in marine mammal tourism (whale-watching staff and tourists), or simply anyone with an interest in marine mammals.

SELECTED REFERENCES AND FURTHER READING

As mentioned above, this is an introductory book. After reading this text, if you wish to investigate marine mammal biology or conservation in more detail, in addition to the references and suggested readings at the end of each chapter, here are some books that we would recommend:

Boyd, I.L., Bowen, W.D., & Iverson, S.J. (2010). *Marine Mammal Ecology and Conservation. A Handbook of Techniques.* Oxford University Press, Oxford.

Berta, A., Sumich, J.L., & Kovacs, K.M. (2006). *Marine Mammals: Evolutionary Biology, 2nd Edition.* Academic Press, New York.

Evans, P.G.H. & Raga, J.A. (2001). *Marine Mammals: Biology & Conservation.* Kluwer Academic, New York.

Hoezel, A.R. (2002). *Marine Mammal Biology: An Evolutionary Approach.* Wiley-Blackwell, Hoboken, NJ.

Jefferson, T.A., Weber, M.A., & Pitman, R. (2007). *Marine Mammals of the World: A Comprehensive Guide to Their Identification.* Academic Press, New York.

Perrin, W.F., Würsig, B., & Thewissen, J.G.M. (eds.) (2009). *Encyclopedia of Marine Mammals, 2nd Edition.* Academic Press, New York.

Reynolds, J.E. & Twiss, J.R. (eds.) (1999). *The Biology of Marine Mammals.* Smithsonian Press, Washington, DC.

Reynolds, J.E., Perrin, W.F., Reeves, R.R., Montgomery, S., & Ragen, T.J. (2005). *Marine Mammal Research: Conservation Beyond Crisis.* The Johns Hopkins University Press, Baltimore.

Simmonds, M.P. & Hutchinson, J.D. (eds.) (1996). *The Conservation of Whales and Dolphins: Science & Practice.* John Wiley & Sons, Chichester, UK.

Twiss, J.R. & Reeves, R.R. (eds.) (1999). *The Conservation and Management of Marine Mammals.* Smithsonian Press, Washington, DC.

Acknowledgments

Many people helped with providing information and editing this book, or simply by giving the authors helpful advice or support, including, on occasion, well-needed cups of tea/cakes/beer and a sympathetic ear. We would especially like to thank Isabel Baker, Bill Bauer, Chris Butler-Stroud, Lorelei Crerar, Beth Crossan, Megan Draheim, Sue Fisher, Fay Hoodock, Nicky Kemp, Eilidh McCafferty, Beth Miller, Jason O'Bryhim, Katheryn Patterson, Naomi Rose, Jakob Tougaard, Magnus Wahlberg, Leslie Walsh, Memnoch, and Bob.

We would like to thank Rupert Ormond, Andrew Campbell, and the University Marine Biological Station, Millport, University of London for providing the opportunity of developing an undergraduate/graduate course in Marine Mammal Biology, which helped to inspire this book.

We are very grateful to all of the experts who kindly contributed textboxes despite their busy schedules and many commitments.

We also thank Rachel Isaacs, Erin O'Connor, Molly Steinbach, Lauren Miller, and Leah Corrigan, our editorial, photo research, and production team at Jones & Bartlett Learning.

Author Profiles

E.C.M. Parsons

Chris Parsons has been involved in whale and dolphin research for over a decade and has conducted projects in South Africa, India, China, and the Caribbean as well as the UK. He has a BA/MA degree from Oxford University and a PhD from the University of Hong Kong. Dr. Parsons started teaching at George Mason University in 2003 and is an Associate Professor teaching various conservation and marine mammal biology classes. He is also currently the Undergraduate Director for marine biology, environmental science, and conservation biology.

Before moving to the U.S., Dr. Parsons was the Director of the Research and Education Departments of the Hebridean Whale and Dolphin Trust (HWDT), from 1998 until 2003. Prior to this, he was involved in research on Indo-Pacific humpback dolphins and finless porpoises in Hong Kong and China, which involved studies on the behavior and ecology of Hong Kong's cetaceans and marine pollution and its effects on marine life.

Dr. Parsons has been a member of the scientific committee of the International Whaling Commission since 1999. He was awarded a Fellowship by the Royal Geographical Society in 1997, won a Scottish Thistle Award in 2000 for his work in environmental tourism, and was acknowledged a young achiever in Scotland for his achievements in cetacean conservation by the Queen and the Duke of Edinburgh in 1999. In 2009, he was the Secretariat Director for the first International Marine Conservation Congress (IMCC)—the largest global marine conservation academic conference—in 2011 he was the program co-chair of the second IMCC held in 2011 and he is the chair of the third IMCC (to be held in 2014). He is currently serving a second term as the Marine Section President and a Governor of the Society for Conservation Biology. Dr. Parsons has published over 100 scientific papers and reports.

A. Bauer

Amy Bauer received an MPA from George Mason University's School of Public and International Affairs and currently is a PhD candidate studying marine conservation through the Environmental Science and Policy Department, also at George Mason. Amy's background includes working as a network security engineer and technical editing for the journals produced by the American Physiological Society. She currently resides in Northern Virginia.

D.J. McCafferty

Dominic McCafferty is a Senior Lecturer at the University of Glasgow, Scotland who specializes in pinniped and otter biology. He coordinates a field course in marine mammal biology at the Marine Biological Station, Millport on the Isle of Cumbrae and teaches on a range of continuing education, undergraduate, and postgraduate programs. Following a BSc (Hons) in Ecological Science and a PhD at the University of Edinburgh he worked for the British Antarctic Survey studying the diving behavior of Antarctic fur seals breeding on Bird Island, South Georgia. Since then he has undertaken research on gray seals, harbor seals, and Eurasian otters in Scotland. His current research is investigating the use of thermal imaging as a tool for monitoring animal welfare.

M.P. Simmonds

Mark Peter Simmonds has worked in the marine conservation and animal welfare field since the 1980s. For several years he was on the staff of Greenpeace International and, more recently, he was employed as a university lecturer. However, for the better part of the last two decades, Mark has worked full time for the Whale and Dolphin Conservation Society, where he is the International Director of Science overseeing work across a wide range of topics. He has been at the forefront of investigations into the impacts of human activities on marine wildlife, including studies into the effects of chemical and noise pollution and marine debris on marine mammals and the development of marine conservation policy, especially as it affects cetaceans. This includes nineteen years as part of the Scientific Committee of the International Whaling Commission (including in recent years as a member of the UK's official delegation to this body).

Mark is also involved in field research on cetaceans in UK waters, mainly on the trail of the elusive Risso's dolphin. He has also been the Chair of the UK's Marine Animal Rescue Coalition, which helps to coordinate the work of the UK's voluntary animal rescue organizations, since 1989. Mark has produced over 200 original papers and other contributions for scientific and popular periodicals and books. He jointly edited (with Judith Hutchinson) *The Conservation of Whales and Dolphins – Science and Practice* published by John Wiley and sons in 1996. He wrote *Whales and Dolphins of the World* first published by New Holland in 2004 (now available in a number of languages) and jointly edited (with Philippa Brakes) and part authored *Whales and Dolphins – Cognition, Culture, Conservation and Human Perceptions*, which was published in April 2011 by Earthscan.

A.J. Wright

Andrew Wright is currently at Aarhus University in Denmark and an Affiliate Professor at George Mason University in the United States. Currently investigating the diving and acoustic behavior of the Danish harbor porpoise, he continues to work on the science-policy boundary in the U.S. and abroad, as he has done for the last 8 years. Specifically he has been working with the management of cumulative impacts of noise and the effective communication of information between scientists and policy makers. This led to him becoming part of the Policy Committee for the Marine Section of the Society for Conservation Biology in 2010, the same year he was an invited participant at the scientific committee meeting of the International Whaling Commission. He was also elected to the Council of the European Cetacean Society in 2011, where he currently serves as the editor for the Society.

What Is a Marine Mammal?

CHAPTER 1

CHAPTER OUTLINE

- **Characteristics of Mammals**
- **Mammalian Evolution**
- **Defining a Marine Mammal**
- **Quasi-Marine Mammals**
  - Greater Bulldog Bat (*Noctilio leporinus*)
  - Fishing (or Fish-Eating) Bat (*Myotis vivesi*)
  - Arctic Fox (*Vulpes lagopus*)

Exploring the Depths: International Union for Conservation of Nature
Exploring the Depths: Sea Wolves and Sea Sheep?
Exploring the Depths: Eurasian Otters: A Marine Mammal?
Exploring the Depths: Steller's Sea Monkey: Were There More Species of Marine Mammals Historically?

- **Why Are Marine Mammals Special?**
- **Selected References and Further Reading**

Before we embark on this book about the biology and conservation of marine mammals, we should first define exactly what they are. Marine mammals are a diverse collection of species grouped together not because of a common evolutionary history, but because they inhabit marine environments. Nevertheless, they are all mammals and as such have many characteristics in common.

Characteristics of Mammals

Mammals (members of Class Mammalia) have several features. They all possess hair composed of keratin, a protein; three bones in their inner ear; and sweat glands. In females some of these sweat glands have been modified to become milk-producing mammary glands, or *mammae*, which give this group of animals their name. Most mammals have teats or nipples except for the monotremes (Subclass Prototheria, Order Monotremata), which exude milk directly from their pores to be lapped up by their young. The monotremes, which include the echidna, or spiny anteater, and the aquatic platypus (**Figure 1.1**) are also unusual in that they lay eggs, unlike the rest of the mammals (Subclass Theria), which produce live young. The monotremes also have a common cloaca, or orifice, through which they lay eggs, urinate, and defecate. These remaining (Theria) mammals are separated into the Metatheria, which include the marsupials, and the Eutheria, or placental mammals. All marine mammals are members of the Eutheria.

In addition to hair, sweat, and mammary glands, mammals also possess specialized teeth and, like birds, are endothermic (or homeothermic, commonly referred to as warm-blooded), metabolically generating their own heat and maintaining a constant internal body temperature. They also possess a four-chambered heart and their brain has a neocortex (the outer part of the cerebral cortex that is responsible for sensory perception, spatial awareness, and "higher functions," such as reasoning and language skills).

FIGURE 1.1 A member of the mammalian order Monotremata, the platypus.

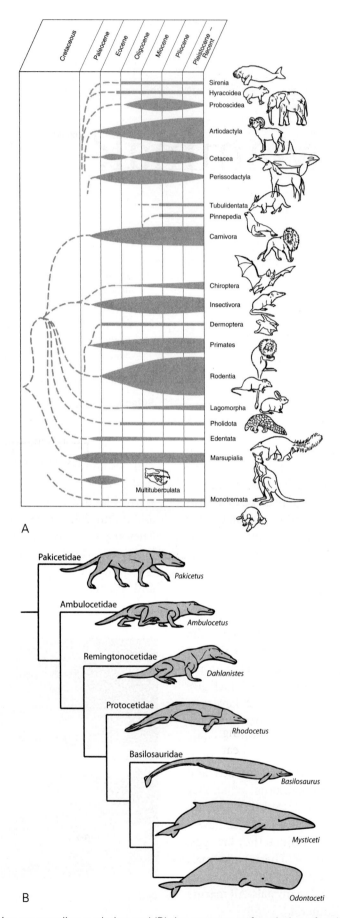

FIGURE 1.2 (A) A timeline showing mammalian evolution and (B) the sequence of evolution of various cetacean groups. Part A data from Gingerich, P.D., 1998. Vertebrates and evolution. *Evolution*, 52, 289–291. Part B adapted from Zimmer, C. *At the Water's Edge*. Simon & Schuster, 1998.

Mammalian Evolution

The evolution of the mammals really begins in the middle of the Permian period (299–251 million years ago) (**Figure 1.2**) when reptile-like therapsids were one of the main land-based predators. These creatures had several features in common, including similar skull structures and incisor teeth that were equal in size. A therapsid group that arose in the late Permian period was the cynodonts (**Figure 1.3**), animals with even more specialized teeth, including rear teeth that had crowns roughly resembling those of molar teeth. They also had a bony palate in the roof of the mouth that separated the mouth from their airways, so they could eat and breathe at the same time. Bones in their jaw also transformed, with two of the smaller jaw bones migrating to become the inner ear bones, presumably giving better hearing capabilities. It is hard to say to what extent or when these animals developed hair.

A mass extinction approximately 251 million years ago at the end of the Permian devastated terrestrial life, with approximately 70% of land vertebrate species going extinct. During the succeeding Triassic (251–199 million years ago) and Jurassic (199–145.5 million years ago) periods, the archosaurs (i.e., dinosaurs and crocodilians) overtook the therapsids as the dominant land predators and the proto-mammals appear to have shifted into a different ecological niche, becoming smaller, primarily nocturnal, and probably insectivorous, rodent-like creatures. This new niche may have led to further adaptations, such as hair to keep warm, improved hearing, and a higher metabolic rate. The first true mammals appeared in the early Jurassic, with the first marsupial, and possibly eutherians, appearing later in the fossil record at about 125 million years ago.

At the end of the Cretaceous period (145.5–65.5 million years ago) another extinction event wiped out many of the land and sea reptiles, including most of the (nonbird) dinosaurs. Mammals are then believed to have expanded and diversified, particularly in the Eocene (55.8–33.9 million years ago). Recent molecular studies confirm a peak in diversification at about 50 million years ago. However, they also suggest a (larger) peak in mammalian diversification between 100 and 85 million years ago, producing most current mammalian orders by around 75 million years ago, thus before the late Cretaceous extinction.

There are approximately 5,400 species of mammals ranging in size from the tiny Kitti's hog-nosed (or bumblebee) bat (*Craseonycteris thonglongyai*) and Etruscan shrew (*Suncus etruscus*), which weigh little more than a couple of grams, to the blue whale (*Balaenoptera musculus*), which is believed to be the largest animal that has ever lived (see Chapter 11). Mammals range from mouse-like rodents (Order Rodentia) and shrews (Order Soricomorpha) to the wolves and cats (Order Carnivora), even-toed hoofed mammals such as deer, cows, and hippos (Order Artiodactyla), odd-toed horses (Order Perissodactyla), the elephants (Order Proboscidea), the flying bats (Order Chiroptera), to the fully marine manatees and dugongs (Order Sirenia) and whales and dolphins (Order Cetacea), and to monkeys, apes, and humans (Order Primates) (**Figure 1.4**).

FIGURE 1.3 An artist's impression of how a cynodont could have appeared.

Domain Eukarya
    Kingdom Animalia
        Phylum Chordata
            Subphylum Vertebrata
                Superclass Tetrapoda
                    Class Mammalia

Defining a Marine Mammal

What is the definition of a marine mammal? In *Marine Mammals of the World, Systematics and Distribution* (1998), Dale Rice lists marine mammals as sirenians (manatees and dugongs), pinnipeds (seals, sea lions, and walruses), and cetaceans (whales, dolphins, and porpoises). He also notes several species of bats and carnivores that inhabit marine waters and are sometimes considered to be marine mammals, although he does not specifically define what a marine mammal is. These groups include the bulldog (*Noctilio leporinus*) and fishing bats (*Myotis vivesi*), the polar bear (*Ursus maritimus*), the Arctic fox (*Vulpes lagopus*), and several otters, including the sea otter (*Enhydra lutris*), marine otter (*Lontra felina*), and Eurasian otter (*Lutra lutra*). Reynolds et al. (1999, p.1) describe marine mammals as follows:

(a) "occupy or rely on aquatic, if not strictly marine, habitats";
(b) "have evolved similar anatomical features, including large body size, streamlined shape (compared to terrestrial relatives), insulation in the form of blubber and dense fur, and in most

FIGURE 1.4 Examples of the diverse range of animals contained in Class Mammalia. (A) Etruscan shrew; (B) hippopotamus; (C) giraffe; (D) elephant; (E) orangutan.

cases, a modified appendicular skeleton resulting in reduction in the size of appendages"; and
(c) "possess some similar physiological adaptations (e.g., for diving, thermoregulation, osmoregulation, communication, and orientation)."

They consider sirenians, pinnipeds, cetaceans, polar bears, and also marine otters and sea otters to be marine mammals but do not include species such as the North American river otter (*Lontra canadensis*) or the bulldog or fishing bats. They do note, however, that these animals can be dependent, at least in part, on the marine ecosystem. They then go on to say that this classification is influenced by the species listed under the U.S. Marine Mammal Protection Act (MMPA) of 1972, which states, "The term 'marine mammal' means any mammal which (a) is morphologically adapted to the marine environment (including sea

FIGURE 1.5 The polar bear, a species that is considered to be a marine mammal under the U.S. Marine Mammal Protection Act.

FIGURE 1.6 A sea otter eating crustaceans.

otters and members of the orders Sirenia, Pinnipedia and Cetacea), or (b) primarily inhabits the marine environment (such as the polar bear)" (Section 3 [6]). However, this legally based definition of a marine mammal does not specifically exclude bats or the Arctic fox, which may indeed primarily inhabit the marine environment.

Many marine mammal field guides use similar classifications, typically including polar bears and the sea and marine otters. Therefore, for the purposes of this textbook pinnipeds, sirenians, and cetaceans are the main marine mammal groups described, but we also consider that sea otters, marine otters (see Chapter 7), and polar bears (see Chapter 6) should be treated as marine mammals and devote entire chapters to them (**Figures 1.5** and **1.6**). However, we briefly discuss some of the quasi-marine mammals here.

Quasi-Marine Mammals

Greater Bulldog Bat (*Noctilio leporinus*)

Bulldog bats have large cheek pouches, in which they store food, giving them their name (**Figures 1.7A**). They are about 7 to 14 cm long and weigh up to 75 g, with reddish brown fur on their bodies. They have relatively large wings and unusually long feet and claws, which aid them in their fishing behavior. The species is found in forests and mangrove swamps in Central and South America from Mexico to Argentina, but they are also found in Trinidad and the Antilles.

They hunt both freshwater and saltwater fish (as many as 30 a night) using echolocation. The bat's large wings enable it to glide over the water when hunting for fish. They fly close to the water surface and trail their claws into the water. The claws and toes are laterally flattened, which reduces drag and allows them to slice through the water. After catching a fish with their feet, the bulldog bat transfers the fish up into its mouth. When the bat lands it chews up the fish, breaking it into large pieces. The pieces are stored in cheek pouches and chewed again before being swallowed.

Their diet exposes them to the threat of water pollution and makes them vulnerable to variations in water quality on prey abundance and distribution due to the changing climate, the alteration of freshwater influx as a result of human water usage, or increased turbidity (i.e., reduced clarity) associated with deforestation and other coastal and inshore activities. There are also reports that fish farmers in Guatemala kill this species, presumably due to their perceived impact on profits, but their population status is not known at this time.

FIGURE 1.7 Quasi-marine mammals. (A) a greater bulldog bat; (B) a fish-eating bat.

FIGURE 1.8 Map showing the current, known distribution of the fishing bat. Data from: IUCN Red List.

Fishing (or Fish-Eating) Bat (*Myotis vivesi*)

Like the greater bulldog bat above, the fishing bat (**Figure 1.7B**) has relatively large wings and long, laterally compressed claws that it uses to fish for crustaceans and small fish, in a similar method to that used by the bulldog bat. It has brown fur, with a paler underside, and weighs about 25 g. It has a very limited distribution, being found only on the islands and coastal areas on either side of the Sea of Cortez and on the central western coast of the Baja California peninsula (**Figure 1.8**). Because both the terrestrial and marine environments it inhabits are effectively arid, the bat has a modified urinary system, which is an adaptation to low water availability. This makes the fishing bat somewhat more adapted to a marine environment than the greater bulldog bat. The fishing bat has also been seen up to 8 km from the shore.

Because of its limited distribution, small population size, and risks of habitat destruction and degradation, the fishing bat is listed by the International Union for Conservation of Nature (IUCN) as "vulnerable" (see Chapter 17). Invasive species, such as rats and cats, are also thought to be a concern for this species.

Arctic Fox (*Vulpes lagopus*)

As its name suggests, this species of fox is found in the northern polar regions, from northern Scandinavia and Russia to northern Canada, Alaska, and even the islands of Svalbard, Iceland, and Greenland.

Exploring the Depths: International Union for Conservation of Nature

The IUCN includes governments as well as environmental and animal welfare nongovernmental organizations. A major activity of the IUCN is the collation of the Red List of Threatened Species. This list categorizes species of animals and plants according to their conservation status based on scientific information on species abundance trends, distribution, and threats. The IUCN Red List categories include, in order of species threat, "vulnerable," "endangered," and "critically endangered" (see Chapter 17).

FIGURE 1.9 An Arctic fox.

The Arctic fox is about 85 cm long and weighs 3 to 3.5 kg, with males being slightly larger (**Figure 1.9**). They are famous for their thick white coats of dense hair that provide camouflage in the Arctic snow and ice and insulation in one of the coldest regions on earth. During the summer their coat changes color to brown, blending in with the tundra. Pups are likewise born with a brown coat.

Further adaptations to a polar environment include fur on the pads of their paws to insulate against heat loss; a substantive fat layer, a typical feature of marine mammals; a countercurrent system (see Chapter 4) that reduces heat loss from their extremities; and rounded and small extremities (ears and muzzle), reducing the surface area through which they can lose heat to the environment (see Chapter 4 for an explanation of adaptations to cold temperatures). Indeed, many of their adaptations are similar to those of polar bears (see Chapter 6). Also similar to polar bears, the Arctic fox will walk on ice to find prey and may jump on ice and snow to break the crust and gain access to prey hiding beneath the snow. They may even eat seals and cetaceans by scavenging from carcasses of stranded animals or animals killed by polar bears.

Their main prey are lemmings (*Dicrostonyx richardsoni*), but when these are scarce up to two-thirds of their food can come from marine sources, such as scavenged seal carcasses. Changes in marine productivity can therefore greatly affect Arctic fox populations when lemming numbers are low.

The IUCN categorizes the Arctic fox as "Least Concern," meaning their survival as a species is reasonably secure. However, several populations are severely depleted and threatened, including the Scandinavian (Norway, Sweden, and Finland) population comprising an estimated 140 adult Arctic foxes. This is primarily the result of a long history of being hunted for their fur, despite the legal protections the species has in these countries. The population on and around Medny Island (Commander Islands, Russia) has also been severely depleted as the result of an outbreak of mange that stemmed from the introduction of ticks from dogs brought to the island by humans. The current population is only approximately 90 animals.

Historically, the Arctic fox faced competition from wolves (*Canis lupus*), but as wolf populations were depleted by humans this competition has decreased. Instead, they now compete with red foxes (*Vulpes vulpes*), whose ranges are expanding, perhaps also due to the decline in wolf populations. Climate change as well as reduced available

Exploring the Depths: Sea Wolves and Sea Sheep?

Although gray wolves (*Canis lupus*) are terrestrial mammals, there is a population of wolves on the coast of British Columbia, Canada and Alaska that occupy coastal habitats, with subpopulations inhabiting the outer islands of this region. These wolves have been found to have a diet that is predominantly (75%) marine based, consisting mainly of salmon and marine mammal carcasses (**Figure B1.1**). Their dependence on a marine diet makes them vulnerable to impact from marine threats such as oil spills and over-fishing of marine species. Should this population of "sea wolves" be considered marine mammals for conservation and management purposes?

An analogous case could be made for a breed of sheep (*Ovis aries*) that live on the remote island of North Ronaldsay in the Orkney Islands, Scotland (**Figure B1.2**). This semiferal flock lives almost entirely on seaweed for most of the year (except for the lambing period). They have been kept this way for almost 180 years when a dry-stone wall was built to exclude sheep from the center of the island. During this time the digestive physiology of this breed evolved to efficiently extract nutrients and sugars from *Laminaria* species. The behavior of the sheep has also adapted to the availability of seaweed with grazing linked to the tidal cycle and rumination generally occurring at high tide rather than at night as in other sheep.

FIGURE B1.1 A marine environment-inhabiting gray wolf from British Columbia.

FIGURE B1.2 A marine environment-inhabiting sheep from North Ronaldsay, Scotland.

Exploring the Depths: Eurasian Otters: A Marine Mammal?

The Eurasian otter (*Lutra lutra*) is widespread along Scotland's north and west coast (**Figure B1.3**). Its distribution is closely associated with coastal freshwater pools and streams because otters use fresh water to wash salt from their fur to maintain its insulation. For this reason coastal Eurasian otters are generally considered to be terrestrial rather than marine mammals. Based on their diet and behavior we should probably recognize these populations as marine ecotypes because they feed on small bottom-dwelling fish and crustaceans within the littoral and sublittoral zones. Their feeding behavior is mainly diurnal (in contrast to otters in fresh water, which are crepuscular (active during twilight) and nocturnal), which is an adaptation to optimize foraging success when prey are least active and easier to catch. Furthermore, DNA analysis of otters has revealed that some island populations such as those in the Orkney and Shetland Isles are genetically distinct from Scottish mainland populations.

Whether a species is marine or terrestrial has relevance for conservation and management. Scottish marine-dwelling otters have been excluded in discussions on marine biodiversity conservation and have not been included in national marine stranding recording schemes. Factors such as fisheries bycatch and marine pollution (i.e., threats in common with other marine species such as cetaceans and pinnipeds) do influence these populations, and therefore national agencies should include these mammals for conservation and management purposes.

FIGURE B1.3 A marine environment-inhabiting European otter in a Scottish sea loch.

Exploring the Depths: Steller's Sea Monkey: Were There More Species of Marine Mammals Historically?

Two species of marine mammal were discovered and first described to science in the 18th century by the naturalist Georg Wilhelm Steller: Steller's sea cow (*Hydrodamalis gigas*) and the Steller sea lion (*Eumetopias jubatus*) (see Chapters 8 and 16, respectively). However, two other marine animals that Steller describes in his writings remain a mystery to science. The first is the sea wolf, and Steller describes it as follows:

> There is another large marine mammal which resembles the whale but is smaller, with a proportionately much thinner circumference. The Russians call it "sea wolf"; the Itelmen call it plebun; on the Kamchatka River this animal is called tsheshshak. I have not been fortunate enough to see it during my stay here. Only its meat, cheeks [jaws], tongue, and intestines are used for food. The fat is used only as fuel in lamps and not eaten because, like mercury, as soon as it is ingested, it comes out again at the other end, and so the Itelmen only eat it when they are badly constipated or, just for fun, give it to others who do not know about it (Steller 2003, pp. 79–80).

The second is the sea monkey or sea ape. The famous 19th century biologist, Leonhard Stejneger, who conducted ground-breaking research on marine mammals and who also wrote a book on Georg Steller, assumed the animal Steller sighted was a sea lion (Stejneger 1936, p. 280). Steller describes the animal meticulously:

> "On August 10, we saw a very unusual and new animal, about which I shall write a short description since I watched it for two whole hours. The animal was about 2 ells long (1 ell = 1.378 ft). The head was like a dog's head, the ears pointed and erect, and on the upper and lower lips on both sides whiskers hung down which made him look like a Chinaman. The eyes were large. The body was longish, round, and fat, but gradually became thinner towards the tail; the skin was covered thickly with hair, gray on the back, russet white on the belly, but in the water it seemed to be entirely red and cow-colored. The tail, which was equipped with fins, was divided equally into two parts, the upper fin being two times as long as the lower one, just like on the sharks" (Steller 1988, p. 82).

Steller had his "cossack" shoot at it several times. The first shot missed. The second shot may have wounded the animal and it disappeared. "However, it was seen at various times in different parts of the sea" (Steller 1988, p. 83).

The naturalist Stejneger details an account by Tilesius that describes an animal which sounds remarkably like Steller's sea ape. Tilesius named it *Phoca mimica*, which is an older name for the Northern fur seal. Steller was an excellent observer, but it is very difficult to observe animals in the sea when one is really a land-trained naturalist. Stejneger also states that Steller had not seen a fur seal, either living or dead, when he made the detailed observation of the sea ape. It is, however, interesting to note that Steller makes no mention of the sea ape and the resemblance it had to a fur seal in his book *De Bestiis Marinis*. It seems as though an observer of Steller's caliber would have made that connection. Most scientists agree that Steller saw something but what it was remains a mystery.

Perhaps science will never know what Steller described in these few lines. It is important to note that Steller was a great observer, an extremely accomplished naturalist and wildlife expert, who recorded biological information with great precision and detail. Perhaps what he saw was a previously unknown species of marine mammal that did not survive until today and is now extinct!

—Contributing author, Lorelei Crerar, George Mason University.

habitat and prey species are expected to further increase competition between Arctic and red foxes.

Why Are Marine Mammals Special?

The question of why marine mammals are special has become an important one in recent years. Why are they often singled out for protection over, say, marine turtles? And why should commercial whaling not be allowed to resume in species that seem to have recovered in number?

First, marine mammals are unusual because air-breathing animals in the ocean are a rarity. They are also (for the most part) unique within mammals due to their different appearance and morphology. This is particularly true for cetaceans, which were considered to be fish throughout the world even late into the 19th century. In fact, an 1818 trial over taxation of fish oil in New York State (vs. the lack of taxation on whale oil) explored the taxonomy of the whale. Ultimately, the jury declared the whale to be a fish in line with popular (and biblically seated) opinion. Similarly, whales (like sturgeon) were declared to be royal fish in England in the early 14th century, a legal status that is still maintained today. They thus enjoy the various protections of being the property of the monarch (other European monarchies have also, at one time or another, made similar claims).

Marine mammals are also culturally important to humans. Any visitor to western Canada or the northwestern coast of the United States will see the images of killer whales made famous by the Nootkan peoples. In New Zealand marine mammals have an important cultural role, most famously perhaps (thanks to the movie *Whale Rider*) the Whangara people of eastern New Zealand, who tell of their legendary ancestor Paikea who rode on the back of a whale to New Zealand after his canoe sunk. Even in Europe marine mammals feature as important components of traditional cultures, for example, stories and songs of Selkies (shape-shifting seals who can turn into people) in Scotland. These are just a few examples of the place of marine mammals in the history, traditions, and cultures of humans.

Some species of marine mammals, particularly the bottlenose dolphin, have been shown to be highly intelligent. Common bottlenose dolphins are part of the small, but growing, number of animals that demonstrate a degree of self-awareness (see Chapter 12). They appear to have "names" or signature whistles (see Chapter 4), understand the linguistic rules of word order or syntax (see Chapter 12), and maximize the efficiency of their communication in a similar way to humans, with their most often used communicative elements being the shortest in their vocabulary. In fact, they are the first species demonstrated to follow this law of brevity, which is one of the basic rules that define all human languages.

Marine mammals are "keystone" species, meaning they are species that are essential for the proper functioning of an ecosystem and which if depleted or removed lead to a significant alteration, or even collapse, of that ecosystem. Top predators such as sharks and many marine mammals are known to be keystone species, with one of the best known examples being sea otters. When sea otters are removed from the kelp forest ecosystem, their sea urchin prey multiply, consuming kelp and denuding and destroying this marine "forest" ecosystem (see Chapter 7).

Finally, marine mammals are special because we believe them to be. Societies around the world have always been in awe of the whale or have depended on seals and sea lions, earning these animals a unique place in our collective hearts. The fact that tours specifically to see marine mammals, especially cetaceans, is a global industry worth over a billion dollars (see Chapter 18 on marine mammal tourism) indicates the extent of public interest and fascination for these animals. This is one reason marine mammals are treated as conservation "flagship" or "umbrella" species. A "flagship" species is a charismatic or iconic species with which people can identify, or be motivated by, to promote conservation in general (i.e., a rallying image). The Worldwide Fund for Nature (called the World Wildlife Fund [WWF] in the U.S.) uses the image of a panda as its conservation flagship species, but many other groups use marine mammals (e.g., the International Fund for Animal Welfare uses a seal pup in their logo).

The use of marine mammals as an "umbrella" species is a subtly different use of a charismatic or iconic animal. People may not be so motivated to conserve a rare benthic (seabed) habitat but may be more willing to support the conservation of marine mammals that use or rely on that habitat. Therefore, by conserving the umbrella species, collaterally other (less charismatic but no less important) species and habitats are protected.

Returning to the MMPA as our guide for what constitutes a marine mammal, we can see that the U.S. Congress subscribed to the view that marine mammals are special, stating, "marine mammals have proven themselves to be resources of great international significance, esthetic and recreational as well as economic, and it is the sense of the Congress that they should be protected and encouraged to develop to the greatest extent feasible."

SELECTED REFERENCES AND FURTHER READING

Angerbjörn, A., Tannerfeldt, M., Björvall, A., Ericson, M., From, J., & Noren, E. (1995). Dynamics of the Arctic fox population in Sweden. *Annales Zoologici Fennici* 32: 55–68.

Arita, H., & Ortega, J. (1998). The Middle American bat fauna: conservation in the Neotropical-Nearctic border. In: *Bat Biology and Conservation* (Ed. T. Kunz & P. Racey), pp. 295–308. Smithsonian Institution Press, Washington, DC.

Bininda-Emonds, O.R.P., Cardillo, M., Jones, K.E., MacPhee, R.D.E., Beck, R.M.D., Grenyer, R., Price, S. A., Vos Rutger, A., Gittleman, J.L. & Purvis, A. (2007). The delayed rise of present-day mammals. *Nature* 446: 507–512.

Burnett, D.G. (2007). *Trying Leviathan: The Nineteenth-Century New York Court Case That Put the Whale on Trial and Challenged the Order of Nature.* Princeton University Press, New Jersey.

Darimont, C.T., Paquet, P.C., & Reimchen, P.C. (2009). Landscape heterogeneity and marine subsidy generate extensive intrapopulation niche diversity in a large terrestrial vertebrate. *Journal of Animal Ecology* 78: 126–133.

Goltsman, M., Kruchenkova, E.P., & Macdonald, D.W. (1996). The Mednyi Arctic foxes: treating a population imperiled by disease. *Oryx* 30: 251–258.

Kruuk, H. (2006). *Otters: Ecology, Behaviour and Conservation.* Oxford University Press, Oxford.

McCafferty, D., & Parsons, E.C.M. (2011). Marine mammal ecotypes: implications for otter conservation and management. *Aquatic Mammals* 37: 205-207.

Moore, P.G. (2002). Mammals in intertidal and maritime ecosystems: interactions, impacts and implications. *Oceanography and Marine Biology* 40: 491–608.

Paterson, I.W., & Coleman, C.D. (1982). Activity patterns of seaweed-eating sheep on North Ronaldsay, Orkney. *Applied Animal Ethology* 8: 137–146.

Reynolds, J.E., Odell D.K., & Rommel, S.A. (1999). Marine mammals of the world. In: *Biology of Marine Mammals* (Ed. J.E. Reynolds & S.A. Rommel), pp. 1–14. Smithsonian Institution Press, Washington, DC.

Reynolds, J.E., Powell, J.A., & Taylor, C.R. (2009). Manatees, *Trichechus manatus, T. senegalensis* and *T. inunguis.* In: *Encyclopedia of Marine Mammals* (Ed. W.F. Perrin, B. Würsig & J.G.M. Thewissen), pp. 682–691. Academic Press, San Diego.

Rice, D.W. (1998). *Marine Mammals of the World. Systematics and Distribution.* The Society for Marine Mammalogy Special Publication 4. Society for Marine Mammalogy, Lawrence, KS.

Roth, J.D. (2003). Variability in marine resources affects arctic fox population dynamics. *Journal of Animal Ecology* 72: 668–676.

Scheffer, V. B. (1958). Seals, Sea Lions and Walruses: a Review of the Pinnipedia. Stanford University Press.

Stejneger, L. (1936). *Georg Wilhelm Steller: The Pioneer of Alaskan Natural History.* Harvard University Press, Cambridge, MA.

Steller, G. W. (Edited by O.W. Frost) (1988). *Journal of a Voyage with Bering, 1741–1742.* Stanford University Press, Stanford, CA.

Steller, G.W. (Edited by M.W. Falk) (2003). *Steller's History of Kamchatka: Collected Information Concerning the History of Kamchatka, Its Peoples, Their Manners, Names, Lifestyle, and Various Customary Practices.* (Translated by M. Engel & K. Willmore). University of Alaska Press, Fairbanks, AK.

Wilson, D.E., & Reeder, D.A.M. (2005). *Mammal Species of the World: A Taxonomic and Geographic Reference.* The Johns Hopkins University Press, Baltimore.

General Biology

PART

I

CHAPTER 2

Marine Mammal Evolution

CHAPTER OUTLINE

The Changing Oceans

Marine Carnivore Evolution

   Exploring the Depths: The Fossil Record

Sirenian Evolution

   Exploring the Depths: What Are the Closest Modern Relatives of Whales?

   Exploring the Depths: The Earliest Aquatic Archaeocete

Cetacean Evolution

   Exploring the Depths: Climate Changes in the Late Eocene and Oligocene

   Exploring the Depths: Basilosaurs

Selected References and Further Reading

The Changing Oceans

The evolutionary history of marine mammals begins in the Eocene (55.8 ± 0.2 million years ago); however, the Eocene world was much different from the world today (**Figure 2.1**; **Figure 2.2**). The beginning of the Eocene saw a period of warming when global temperatures increased by 6°C over 20,000 years, associated with rises in atmospheric carbon dioxide. This increase in temperature resulted in the melting of ice sheets and a rise in sea levels, both from the melted ice and because of the way water expands in volume as it becomes warmer. The climate of the polar regions was particularly affected due largely to a decline in albedo (reflection of solar energy) by the loss of ice and snow cover. This effect was so great, with Arctic Ocean temperatures reaching nearly 22°C, that subtropical marine species could survive in these colder regions. Similarly, temperate rainforests could be found on land near to both poles: the climate in polar regions probably more resembled that of Washington State than the icy climate of today. Global temperatures were so warm that palm trees extended to Alaska. The tropical regions had temperatures similar to those found there today, although the range of what we would recognize as tropical species extended much further north, covering most of the United States.

The continents were also substantially different in shape and location. North and South America had yet to fully join, with water from the Atlantic and Pacific Oceans connected through circulation in the region that would eventually become Panama. Antarctica and Australia were still connected at the beginning of the Eocene, although the land mass separated into two continents about 45 million years ago. At that point a change occurred in ocean circulation that resulted in Antarctica receiving less of the warm, equatorial waters and the continent becoming considerably cooler.

During this warm period mammalian groups, such as the even-toed ungulates (artiodactyls) and odd-toed ungulates (perissodactyls) and primates, began to appear and spread rapidly across the world. Hoofed predators called mesonychids became the top predators on the land while in the oceans the charcharinid sharks (the group of sharks containing the present-day great white shark, *Carcharodon carcharias*, and tiger shark, *Galeocerdo cuvier*) appeared. In the large, warm, shallow seas of the Eocene the first cetaceans and sirenians also appeared.

Marine Carnivore Evolution

Three of the accepted marine mammal groups belong to the Order Carnivora (see Chapter 3): polar bears, sea otters, and pinnipeds (seals, sea lions, and walruses). The most recently evolved of these groups are polar bears and sea otters.

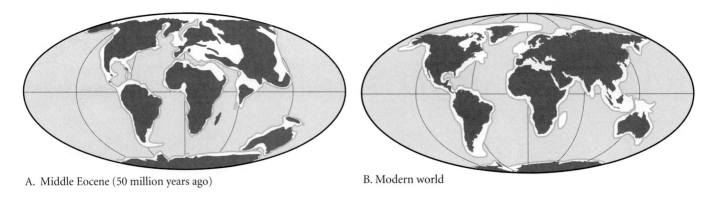

A. Middle Eocene (50 million years ago)

B. Modern world

FIGURE 2.1 Maps of (A) the Eocene, and (B) the modern world. Adapted from Dietz, R.S. and Holden, J.C., *Sci. Am.* 223 (1970): 30–41.

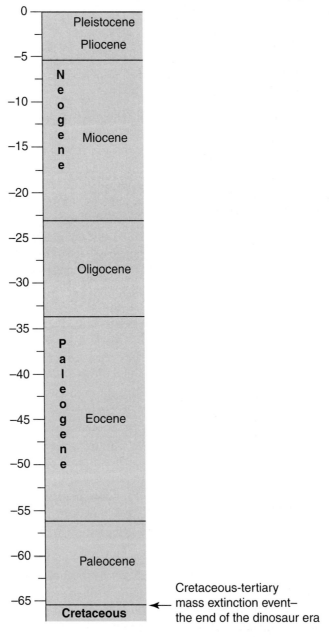

FIGURE 2.2 Geological timeline highlighting the Eocene.

Exploring the Depths: The Fossil Record

It is important to note that most of what we know about the early evolution of marine mammals is based on the fossil record, which means our knowledge is far from complete. Akin to looking at the past through a pinhole in a thick curtain, we are only able to view a tiny fraction of the animals that existed during this period. The environmental conditions that allow the remains of a dead creature to fossilize are very rare, and even then only hard body parts are preserved, except on extremely rare occasions. Many fossils are subsequently destroyed by geological, or even perhaps human, activities. Moreover, many remaining areas that might be rich in fossils are inaccessible (e.g., under water).

Sites where early marine mammal fossils have been found are few and far between. One of the most productive areas for early marine mammal fossils in recent years has been Pakistan. However, basing all of our knowledge of the evolution of marine mammals on one small area is a little like assessing the whole of human history by looking at just one archeological site. For example, you would get radically different impressions of human history and culture by looking at archeological finds from Pompeii compared with the ruins and Nazca lines of Peru (**Figure B2.1**).

FIGURE B2.1 (A) Nazca lines and (B) Pompeii.

Polar bears (*Ursus maritimus*) evolved in the Pleistocene (2.6 million to 10,000 years ago), a period of many ice ages and when up to 30% of the world's surface was covered in ice sheets. The oldest polar bear fossil is only dated at 100,000 years ago. However, genetic analyses have suggested that polar bears split from the brown (or grizzly) bears (*U. arctos*) between 1 and 1.5 million years ago. There was possibly a subspecies of, now extinct, giant polar bears, *U. m. tyrannus*, that existed in Pleistocene Britain.

The first otter (*Paralutra lorteii*) appeared in the Miocene approximately 20 million years ago. From the Miocene to early Pliocene (approximately 5 million years ago), there were two species of prehistoric sea otter: *Enhydritherium terraenovae*, reported from California and Florida, and *Enhydritherium lluecai*, reported from Spain. Modern sea otters (genus *Enhydra*) evolved during the beginning of the Pleistocene (approximately 2 million years ago). Remains have been found all along the west coast of North America, and the sea otter (*Enhydra lutris*) may have been sympatric (having an overlapping distribution, both in time and space) with a now extinct species of sea otter, *E. macrodonta*, remains of which have been found in California. Molecular data suggest the otter genera *Lutra* (which includes Eurasian otters, *L. lutra*) and *Aonyx* (African clawless, *A. capensis*, and Oriental small-clawed otters, *A. cinerea*) are more closely related to sea otters than other otter genera (see Chapter 3). Additionally, two other genera of extinct otters may have been marine in nature, *Enhydriodon* and *Enhydritherium*, both of which appeared in the late Miocene and from their morphology appear to be closely related to *Enhydra*. However, it is also possible that some or all otters of these two extinct genera may have lived in fresh water instead.

The evolutionary history of the pinnipeds is not terribly clear. Fossils of the otter-like *Potamotherium* date from the early Miocene, and this member of Family Mustelidae (weasels, badgers, and otters) has been suggested as an early ancestor of the pinnipeds. However, scientific opinion has, in general, been split between those who believe pinnipeds are more closely related to bears (Family Ursidae) and those who consider the true seals (Family Phocidae) to be closely related to the mustelids, whereas walruses and sea lions are more closely related to bears. Recent molecular data support the single ancestor idea, although the unclear divergences within Carnivora mean it is not obvious if the pinnipeds split from the Ursidae, the Mustelidae, or a common ursid–mustelid ancestor.

One of the earliest fossils of a pinniped-like species, *Puijila darwini*, dates to around 23 million years ago and was found in northern Canada, which could suggest the earliest pinnipeds dwelled in cold polar regions. The morphology (i.e., shape and structure) of this early pinniped fossil suggests the animal may have been most mobile on land, with heavy upright limbs. The morphology of another early pinniped from the same time period, *Enaliarctos* (Family Enaliarctidae), suggests this animal may have been more adapted to an aquatic life but was still relatively mobile on land (at least more mobile than modern pinnipeds). The structure of the teeth in this genus, with flesh-slicing rear teeth similar to modern dogs rather than piercing teeth of modern pinnipeds, may have meant these animals had to return to land to consume their prey, because chewing in a dog-like fashion would be difficult in the marine environment. Fossils of true seals start to appear in the fossil record from approximately 20 million years ago (e.g., *Prophoca* and *Leptophoca*).

Modern fur and sea lions (otariids) appear later in the fossil record, with the earliest otariid fossil (*Pithanotaria starri*), found in California, dating to 11 million years ago. Although the modern walrus (*Odobenus rosmarus*) is a relative newcomer (1 million years), the first fossil walruses (*Pelagiarctos thomasi*) appear 15 million years ago, although not all these prehistoric walruses had tusks or even enlarged canine teeth.

Sirenian Evolution

The Order Sirenia gains its name from the mythological sirens, beautiful sea creatures who drove sailors mad with their songs and lured ships onto rocks. The sirenians are currently divided into two families: Dugongidae (dugongs and the extinct Steller's sea cow) and Trichechidae (manatees).

The closest modern relatives of the Sirenia are believed to be elephants (Order Proboscidea) based on fossil and molecular evidence (**Figure 2.3**). The relationship can be seen easily if you look at the skeleton of a sirenian, especially the tusked dugong. Both have unusually osteosclerotic (dense) and pachyostotic (swollen) bones and large, ridged molar teeth in the back of their jaws.

Sirenia, elephants, and the now extinct Order Desmostylia (a group of probably aquatic herbivores that existed between 30 and 7 million years ago; **Figure 2.4**) are grouped together in the Tethytheria, a name that comes from the

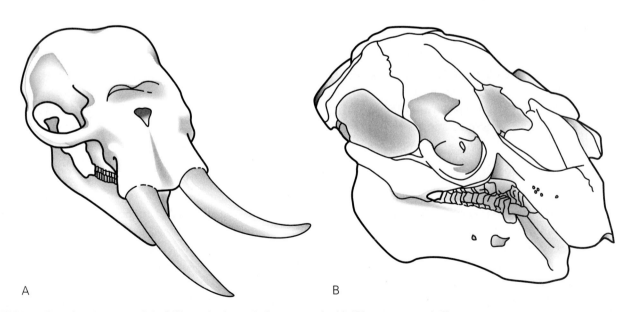

FIGURE 2.3 Drawing (not to scale) of (A) an elephant skull compared with (B) a manatee skull.

FIGURE 2.4 An artist's impression of a Desmostylian.

animals' assumed evolution around the shores of the ancient Tethys Sea.

The oldest known sirenian fossils are *Prorastomus* and *Pezosiren* found in early Eocene lagoon deposits in Jamaica, dated to approximately 50 million years old. These animals have well-developed legs and thus could have been terrestrial, but their thick, heavy bones suggest they may have been partly aquatic, similar to modern hippopotami. Successive sirenian fossils display a clear reduction in hind limbs, from the *Prorastomus* and *Pezosiren* which had legs, to *Protosiren* (Middle Eocene), which had reduced and weakened hind limbs, to *Eosiren* (Late Eocene), which probably had only small, vestigial external hind limbs (**Figure 2.5**).

The first members of the Dugongidae appear at the end of the Eocene. These fossils appear to be fully aquatic with features of a tail but no hind limbs, suggesting they had flattened tail flukes that could undulate up and down to propel the animals through the water. The now-extinct Steller's sea cow, *Hydrodamalis gigas*, and the dugong, *Dugong dugon*, are the most recent dugongines. The fossil record for manatees is less clear, although they probably also appeared in the late Eocene, probably before the Dugongidae. Miocene fossils imply they have long maintained their current distribution along the Atlantic coasts of North and South America and West Africa, although there are fossil remains of manatees from Europe too.

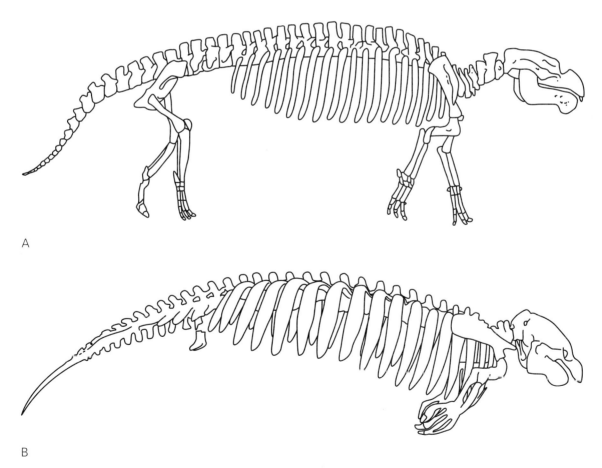

FIGURE 2.5 An artist's recreations of (A) *Pezosiren* and (B) *Eosiren* skeletons.

Exploring the Depths: What Are the Closest Modern Relatives of Whales?

Similarities in amino acid sequences in myoglobin and pancreatic ribonucleases suggest the cetaceans evolved from members of the Order Perissodactyla, the odd-toed hoofed animals, which include modern horses, tapirs, and rhinoceroses. However, amino acid sequences in a variety of proteins (including α-crystallin A, cytochrome *b*, and cytochrome *c*) and immunological comparisons suggest instead they may have evolved from the Order Artiodactyla, the even-toed hoofed animals, which include modern pigs, camels, hippopotami, deer, and cattle.

Cytochrome *c* and other protein and nucleotide sequences furthermore suggest members of Family Camelidae (camels and llamas) are cetaceans' closest relatives. However, other studies comparing mitochondrial DNA and other protein sequences suggested Suborder Ruminanta, which includes Family Cervidae (deer), Family Bovidae (cows and antelopes), and Family Giraffidae (giraffes) and Suborder Cetancodonta, which includes Family Hippopotamidae (hippos), are more closely related to cetaceans, with more recent genetic studies supporting the latter (hippos). There are anatomical clues to the ancestry of cetaceans too; for example, they have a multichambered stomach similar to ruminant ungulates (such as cows).

However, perhaps most convincing are recent studies revealing that hippos may have a similar auditory system to odontocetes (see Chapter 5) for hearing underwater. Hippos too have a body of fat in their jaws that may transmit sound to their inner ear through a thin part of their jaw.

The Suborder Archaeoceti is further split into several families: Pakicetidae, Ambulocetidae, Remingtonocetidae, Protocetidae, and Basilosauridae. The most "primitive" of the Archaeoceti belong to the Family Pakicetidae. As their name suggests, fossils of this earliest group of the Archaeoceti have primarily been obtained from modern-day Pakistan. This family of archaeocetes first appears in the fossil record approximately 53 million years ago, a time when the area that is now Pakistan was on the edges of a long shallow body of marine water called the Tethys Sea (**Figures B2.2** and **B2.3**). Originally, *Pakicetus* was assumed to have been seal-like in appearance, based on fragmentary fossil evidence. However, the subsequent discovery of more complete fossil remains have shown that *Pakicetus* was a terrestrial hoofed mammal approximately 2 meters long with its eyes and nose positioned high on its head. The bone structure and location of the ear in the skull of *Pakicetus* is similar to that found in cetaceans. The structure of the *Pakicetus* ankle bone is also more hippopotamus-like rather than mesonychid-like, adding to the evidence of an ancestral relationship between these animals. The teeth of *Pakicetus* are triangular, similar to sharks' teeth, and more closely akin to modern cetacean than other mammal teeth.

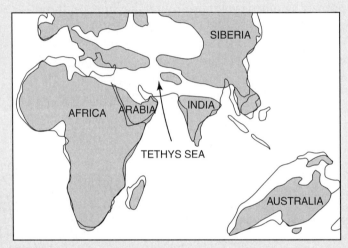

FIGURE B2.3 A close-up of Pakistan in the Eocene showing the Tethys Sea.

FIGURE B2.2 A drawing of *Pakicetus*. Courtesy of Carl Buell.

Exploring the Depths: The Earliest Aquatic Archaeocete

A fossil of a cat-sized, deer-like creature called *Indohyus* (Family Raoellidae), recently discovered by Dr. Hans Thewissen and his colleagues of Ohio University, was touted as being one of the best possibilities for the earliest direct ancestor of the cetaceans (**Figure B2.4**). This small, superficially raccoon-like animal had a cetacean-like ear structure, as mentioned above, but was also osteosclerotic, in that it possessed a heavy outer covering of their bones similar to modern-day hippopotami and other aquatic mammals. The researchers therefore suggested *Indohyus*' lifestyle was also aquatic. The fossils of *Indohyus* are, however, younger than *Pakicetus* and so it is less likely to be a direct ancestor of modern cetaceans.

FIGURE B2.4 An artist's impression of *Indohyus*. Courtesy of Carl Buell.

Cetacean Evolution

Whales, dolphins, and porpoises (Order Cetacea) are split into three suborders: the Mysticeti or baleen whales (see Chapters 3 and 10), the Odontoceti or toothed whales such as dolphins and porpoises (see Chapters 3, 11, and 12), and the Archaeoceti (the ancient whales). Cetacea are believed to be a monophyletic group (they all evolved from a single common mammalian ancestor species), with modern suborders descending from some members of the Archaeoceti.

The first archaeocete fossils appear in the Eocene, clearly showing that early cetaceans evolved from land mammals and became progressively more and more aquatic. It was initially believed that early archaeocete cetaceans split from a group of mammals called the mesonychids. These hoofed predators first evolved in North America in the Paleocene (approximately 60 million years ago) and spread through the northern hemisphere becoming the dominant predators on the planet for 20 million years. Many were wolf-like or hyena-like in build with large heads, powerful jaws, and hoofed toes. They also possessed triangular teeth similar in structure to cetacean teeth, hence the belief they were ancestors to the cetaceans. However, molecular studies (see Exploring the Depths: Basilosaurs) and more recent fossil evidence suggest cetaceans are more closely related to the hippopotamus and not the mesonychids (although the ancestral forms of hippopotami do not appear in the fossil record until well after the first archaeocete fossils at about 15 million years ago).

Fossils of *Ambulocetus natans* (Family Ambulocetidae), also found in Pakistan, appear about a million years after *Pakicetus* (50–49 million years ago). This 3-meter-long mammal had short legs, a long snout, and looked somewhat like a modern crocodile in shape and presumably lived a similar lifestyle. *Ambulocetus* had nostrils near the tip of the snout, possessed cetacean-like ear bones that would have aided hearing in the water, and had teeth similar to some modern cetaceans. From the structure of the bones it is possible to speculate they probably swam in a manner similar to modern otters, through undulation of the entire body. Members of Family Remingtonocetidae were similar in shape to *Ambulocetus* but appear to be more aquatic from their shape, with longer snouts and possibly more otter-like in appearance.

The bone structure of Protocetidae fossils suggests members of this family of archaeocetes were likely to have been semiaquatic swimmers. Because of the arrangement of the bones and muscle attachments in the tails of some protocetids, they might have used their tails to provide propulsion in the water, although many may have been more aquatic waders than swimmers. Members of the genus *Rodhocetus* were likely good swimmers (based on their tail morphology), although their large hind and forelimbs suggest they were amphibious or able to function both on land and in the water. The Protocetidae appear approximately 8 million years after the appearance of *Pakicetus*, and their nostrils were slightly further back from the tips of their snouts.

The first fully aquatic archaeocetes known were the basilosaurids (see Exploring the Depths: Basilosaurs). Although much more like modern cetaceans (apart from their serpentine length), they had relatively small (0.6 m) long vestigial hind limbs that, it has been suggested, may have helped keep *Basilosaurus* connected while mating.

Dorudon was a smaller (5 m or 16 ft) member of the Basilosauridae and was less serpentine, still possessing hind limbs, but probably more closely resembled modern cetaceans and may even be a direct ancestor (**Figure 2.6**). The Basilosauridae eventually became extinct in the late Eocene or early Oligocene (36 million years ago) possibly as the result of changing climate.

Moving to the evolution of the "modern" cetacean groups, the earliest mysticete whale (*Llanocetus denticrenatus*) appeared during the late Eocene (45 million years ago) and more mysticete species appeared in the early Oligocene. These early mysticetes possessed teeth. However, fossils of the early mysticete whale, *Aetiocetus weltoni*, show these animals also had grooves in the upper jaw similar to

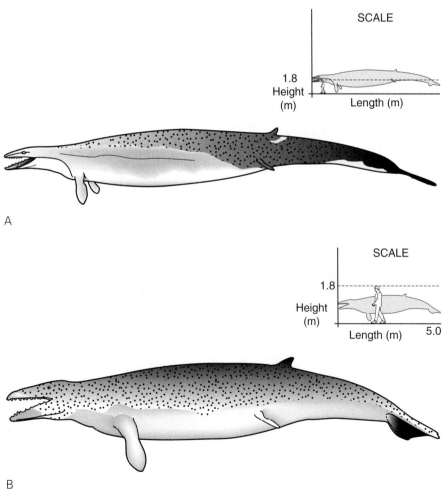

FIGURE 2.6 An artist's impression of (A) *Basilosaurus* and (B) *Dorudon*, with a size scale.

those caused by blood vessels that nourish baleen in modern mysticetes. Consequently, it is assumed this animal possessed baleen as well as teeth. In the late Oligocene mysticete whales with baleen and no teeth begin to appear.

Squalodon, an early odontocete, existed from the early to middle Oligocene to the middle Miocene (33–14 million years ago). Notably, the skulls of these cetaceans show possible evidence of a melon, the fatty structure important for echolocation and communication that forms the bump on the heads of toothed whales, dolphins, and porpoises. In the late Oligocene dolphin-like fossils began to appear. In particular, fossils of the Family Kentriodontidae, which have many similarities to modern dolphins and may have been an ancestral group, appeared during this time. It was not until the Miocene (23–5.3 million years ago) that most of the modern cetacean groups are present in the fossil record.

Exploring the Depths: Climate Changes in the Late Eocene and Oligocene

Continents continued to move, and South America and Australia both fully detached from Antarctica, allowing water to flow around the southern continent in what would become the Southern Ocean. This allowed Antarctica to cool substantially and ice sheets to reform. Ice sheet formation was so substantial that the ocean level dropped 55 meters over the period from 35.7 million years ago, at about the time the basilosaurids became extinct, to 33.5 million years ago. In addition, Africa and Europe moved closer together, leading to the shrinking and isolation of the Tethys Sea. There was a significant temperature drop at the end of the Eocene (approximately 8°C), largely due to oceanographic changes resulting from the movement of continents, so that the following Oligocene period (33.9–23 million years ago) was significantly cooler than the Eocene (**Figure B2.5**). This decrease in temperature was also accompanied by (and may have caused) the extinction of a large number of amphibian, invertebrate, and reptile species, including the basilosaurids.

The world continued to cool in the Miocene (23–5.3 million years ago), also becoming drier during this period. Antarctic ice sheets continued to build up, and sea levels dropped further during the Miocene, with the Mediterranean Sea even drying up at the end of this period. The Tethys Sea continued to shrink and eventually vanished as Africa collided with Europe. At this time India's collision with Asia resulted in extremely high mountain ranges, such as the Himalayas. Changes in the marine environment during this period include the appearance of kelp (which provides an important habitat for many marine mammals and their prey) and modern sharks.

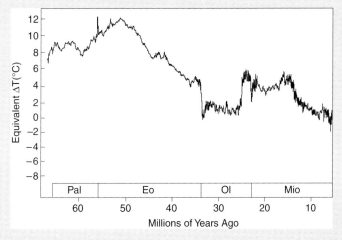

FIGURE B2.5 A historical record of the world climate based on isotope analysis showing the Paleocene, Eocene, Oligocene, and Miocene periods.

Exploring the Depths: Basilosaurs

Basilosaurus was the largest fossil whale known in the world during the Eocene. Specimens of this animal have been found in southeastern United States, Pakistan, Egypt, and Europe. *Basilosaurus* was about 16 meters long in life and lived in the warm subtropical seas that covered much of the northern subtropics during the Eocene, about 37 million years ago. *Basilosaurus* probably ate fish and squid, which it caught with its pointed front teeth and chewed with its large, triangular back teeth.

The name *Basilosaurus* was given to this animal by Dr. Richard Harlan in 1834 for a partial skeleton discovered in Louisiana. He mistook the long vertebrae for those of a reptile. Thinking he had discovered a fossil sea serpent, he called it *Basilosaurus*, which is derived from Latin meaning "king lizard." Despite this ill-conceived name for a mammal, the rules regarding names of animals state that the first name given to an animal must be retained, so we still call this early whale *Basilosaurus* today.

Basilosaurus is the namesake of the family Basilosauridae, which includes some of the earliest fully aquatic whales. The evidence of their fully aquatic lifestyle comes from their skeleton. First, basilosaurids have shortened necks compared with earlier whales. Second, the forelimbs of basilosaurids have several characteristics indicating they were used for swimming rather than walking: a broad, fan-shaped shoulder blade, an elbow with restricted movement only in one plane, and wrist bones with articular surfaces that do not allow the hands to be rotated under the body to support weight. Third, basilosaurids have a much higher number of lumbar vertebrae than earlier whales, making their backs long and providing more attachment sites for muscles for tail movement. Finally, the posterior tail vertebrae are flattened, a characteristic found only in modern mammals that have tail flukes, like modern whales.

Basilosaurus was unusual among basilosaurids in having very elongated trunk vertebrae, making it a very long animal, with a rather small head. More normally proportioned animals, such as *Dorudon*, known best from Egypt, or *Zygorhiza*, known best from the southeastern United States, provide good models for the ancestors of mysticetes and odontocetes, and thus all modern whales.

–Contributing author, Mark Uhen, George Mason University.

SELECTED REFERENCES AND FURTHER READING

Árnason, U., Gullberg, A., & Widegren, B. (1991). The complete nucleotide sequence of the mitochondrial DNA of the fin whale *Balaenoptera physalus*. *Journal of Molecular Evolution* 33: 556–568.

Árnason, U., Gullberg, A., Janke, A., Kullberg, M., Lehman, N., Petrov, E.A., & Väinölä, R. (2006). Pinniped phylogeny and a new hypothesis for their origin and dispersal. *Molecular Phylogenetics and Evolution* 41: 345–354.

Baba, M.L., Darga, L.L., Goodman, M., & Czelusnial, J. (1981). Evolution of cytochrome *c* investigated by the maximum parsimony method. *Journal of Molecular Evolution* 17: 197–213.

Beintema, J.J., & Lenstra, J.A. (1982). Evolution of mammalian pancreatic ribonucleases. In: *Macromolecular Sequences in Systematic and Evolutionary Biology* (Ed. M. Goodman), pp. 43–73. Plenum Press, New York.

Berta, A. (2009). Pinniped evolution. In: *Encyclopedia of Marine Mammals* (Ed. W.F. Perrin, B. Würsig & J.G.M. Thewissen), pp. 861–868. Academic Press, San Diego.

Berta, A., Ray C.E., & Wyss, A.R. (1989). Skeleton of the oldest known pinniped, *Enaliarctos mealsi*. *Science* 244: 60–62.

Berta, A., Sumich, J.L., & Kovacs, K.M. (2006). *Marine Mammals. Evolutionary Biology*. Academic Press, London.

Boisserie, J.R., Fabrice Lihoreau, F., & Brunet, M. (2005). The position of Hippopotamidae within Cetartiodactyla. *Proceedings of the National Academy of Sciences* 102: 1537–1541.

Boyden, A.A., & Gemeroy, D.G. (1950). The relative position of the Cetacea among the orders of mammals as indicated by precipitin tests. *Zoologica* 35: 145–151.

Czelusniak, J., Goodman, M., Koop, B.F., Tagle, D.A., Shoshani, J., Braunitzer, G., Kleinschmidt, T.K., de Jong, W.W., & Matsuda, G. (1990). Perspectives from amino acid and nucleotide sequences on cladistic relationships among higher taxa of Eutheria. In: *Current Mammalogy*, Vol. 2 (Ed. H.H. Genoways), pp. 545–572. Plenum Press, New York.

Deméré, T., McGowen, M., Berta, A., & Gatesy, J. (2008). Morphological and molecular evidence for a stepwise evolutionary transition from teeth to baleen in mysticete whales. *Systematic Biology* 57: 15–37.

Domning, D.P. (2001a). Evolution of the Sirenia and Desmostylia. In: *Secondary Adaptation to Life in Water* (Ed. J.M. Mazin & V. De Buffrenil), pp. 151–168. Verlag, Munich.

Domning, D.P. (2001b). The earliest known fully quadrupedal sirenian. *Nature* 413: 625–627.

Domning, D.P. (2009). Sirenian evolution. In: *Encyclopedia of Marine Mammals* (Ed. W.F. Perrin, B. Würsig & J.G.M. Thewissen), pp. 1016–1019. Academic Press, San Diego.

Domning, D.P., & Gingerich, P.D. (1994). *Protosiren smithae*, new species (Mammalia, Sirenia), from the late middle Eocene of Wadi Hitan, Egypt. *Contributions from the Museum of Paleontology, University of Michigan* 29: 69–87.

Domning, D.P., Ray, C.E., & McKenna, M.C. (1986). Two new Oligocene Desmostylians and a discussion of Tethytherian systematic. *Smithsonian Contributions to Paleobiology* 59: 1–56.

Fitzgerald, E.M.G. (2006). A bizarre new toothed mysticete (Cetacea) from Australia and the early evolution of baleen whales. *Proceedings of the Royal Society B: Biological Sciences* 273: 2955–2963.

Flynn, J.J., Finarelli, J.A., Zehr, S., Hsu, J., & Nedbal, M.A. (2005). Molecular phylogeny of the Carnivora (Mammalia): assessing the impact of increased sampling on resolving enigmatic relationships. *Systematic Biology* 54: 317–337.

Fordyce, R.E. (2009a). Cetacean evolution. In: *Encyclopedia of Marine Mammals* (Ed. W.F. Perrin, B. Würsig & J.G.M. Thewissen), pp. 201–207. Academic Press, San Diego.

Fordyce, R.E. (2009b). Cetacean fossil record. In: *Encyclopedia of Marine Mammals* (Ed. W.F. Perrin, B. Würsig & J.G.M. Thewissen), pp. 207–215. Academic Press, San Diego.

Gatesy, J. (1997). More DNA support for a Cetacea-Hippopotamidae clade: the blood-clotting protein gene gamma-fibrinogen. *Molecular Biology and Evolution* 14: 537–543.

Gatesy, J., Hayashi, C., Cronin, M. A., & Arctander, P. (1996). Evidence from milk casein genes that cetaceans are close relatives of hippopotamid artiodactyls. *Molecular Biology and Evolution* 13: 954–963.

Gheerbrant, E., Domning, D.P., & Tassy, P. (2005). Paenungulata (Sirenia, Proboscidea, Hydracoidea, and relatives). In: *The Rise of Placental Mammals* (Ed. K.D. Rose & J.D. Archibald), 84–105. John Hopkins Press, Baltimore.

Gingerich, P.D. (2003). Mammalian responses to climate change at the Paleocene-Eocene boundary: Polecat Bench record in the northern Bighorn Basin, Wyoming. In: *Causes and Consequences of Globally Warm Climates in the Early Paleogene: Boulder, Colorado* (Ed. Wing, S.L., Gingerich, P.D., Schmitz, B. & Thomas, E.), pp. 463–478. Geological Society of America Special Paper 369. Geological Society of America, Boulder, CO.

Gingerich, P.D., & Russell D.E. (1981). *Pakicetus inachus*, a new archaeocete (Mammalia, Cetacea) from the early-middle Eocene Kuldana Formation of Kohat (Pakistan). *Contributions from the Museum of Paleontology, University of Michigan* 25: 235–246.

Gingerich, P. D., Wells, N. A., Russel, D. E., & Shah, S. M. I. (1983). Origin of whales in epicontinental remnant seas: new evidence from the early Eocene of Pakistan. *Science* 220: 403–406.

Goodman, M., Czelusniak, J., & Beeber, J.E. (1985). Phylogeny of primates and other eutherian orders: a cladistic analysis using amino acid and nucleotide sequence data. *Cladistics* 1: 171–185.

Graur, D., & Higgins, D.G. (1994). Molecular evidence for the inclusion of cetaceans within the order Artiodactyla. *Molecular Biology and Evolution* 11: 357–364.

Hoelzel, R.A. (2002). *Marine Mammal Biology, An Evolutionary Approach*. Blackwell, Oxford.

Irwin, D.M., & Árnason, U. (1994). Cytochrome *b* gene of marine mammals: phylogeny and evolution. *Journal of Mammalian Evolution* 2: 37–55.

Irwin, D.M., Kocher, T.D., & Wilson, A.C. (1991). Evolution of the cytochrome *b* gene of mammals. *Journal of Molecular Evolution* 32: 128–144.

Kennett, J.P., & Stott, L.D. (1991). Abrupt deep-sea warming, palaeoceanographic changes and benthic extinctions at the end of the Palaeocene. *Nature* 353: 225–229.

Kurtén, B.J. (1968). *Pleistocene Mammals of Europe*. Weidenfeld & Nicholson, London.

Lavergne, A., Douzery, E., Strichler, T., Catzefiis, F.M., & Springer, M.S. (1996). Interordinal mammalian relationships: evidence for Paenungulate monophyly is provided by complete mitochondrial 12sRNA sequences. *Molecular Phylogenetics and Evolution* 6: 245–258.

Lento, G.M., Hickson, R.E., Chambers, G.K., & Penny, D. (1995). Use of spectral analysis to test hypotheses on the origin of pinnipeds. *Molecular Biology and Evolution* 12: 28–52.

McKenna, M.C. (1987). Molecular and morphological analysis of high-level mammalian interrelationships. In: *Molecules and Morphology in Evolution: Conflict or Compromise?* (Ed. C. Patterson), pp. 55–93. Cambridge University Press, Cambridge, UK.

Ozawa, T., Hayashi, S., & Mikhelson, V.M. (1997). Phylogenetic position of mammoth and Steller's sea cow within Tethytheria demonstrated by mitochondrial DNA sequences. *Journal of Molecular Evolution* 44: 406–413.

Rybczynski, N., Dawson, M.R & Tedford, R.H. (2009). A semi-aquatic Arctic mammalian carnivore from the Miocene epoch and origin of Pinnipedia. *Nature* 458: 1021–1024.

Sarich, V. M. (1993). Mammalian systematics: twenty-five years among their albumins and transferrins. In: *Mammalian Phylogeny* (Ed. F. S. Szalay, M. J. Novacek & M. C. McKenna), pp. 103–114. Springer, New York.

Sluijs, A., Schouten, S., Pagani, M., Woltering, M., Brinkhuis, H., Damsté, J.S.S., Dickens, G.R., Huber, M., Reichart, G.J., Stein, R., Matthiessen, J., Lourens, L.J., Pedentchouk, N., Backman, N., Moran, K., & the Expedition 302 Scientists. (2006). Subtropical Arctic Ocean temperatures during the Palaeocene/Eocene thermal maximum. *Nature* 441: 610–613.

Steeman, M. E. (2007). Cladistic analysis and a revised classification of fossil and recent mysticetes. *Zoological Journal of the Linnean Society* 150: 875–894.

Thewissen, J.G.M. (2007). Whales originated from aquatic artiodactyls in the Eocene epoch of India. *Nature* 450: 1190–1194.

Thewissen, J.G.M., & Domning, D.P. (1992). The role of Phenacodontids in the origins of modern orders of ungulate mammals. *Journal of Vertebrate Palaentology* 12: 494–504.

Thewissen, J.G.M., Hussain, S.T., & Arif, M. (1994). Fossil evidence for the origin of aquatic locomotion in archaeocete whales. *Science* 263: 210–212.

Thewissen, J.G.M., Madar, S.I., & Hussain, S.T. (1996). *Ambulocetus natans*, an Eocene cetacean (Mammalia) from Pakistan. *Courier Forschungsinstitut Senckenberg* 191: 1–86.

Thewissen, J.G.M., Williams, E.M., Roe L.J., & Hussain S.T. (2001). Skeletons of terrestrial cetaceans and the relationship of whales to artiodactyls. *Nature* 413: 277–281.

Ursing, B.M., & Árnason, U. (1998). Analyses of mitochondrial genomes strongly support a hippopotamus-whale clade. *Proceedings of the Royal Society B: Biological Sciences* 265: 2251–2255.

Yu, L., Qi, Q.W., Ryder, O.A., & Zhang, Y.P. (2004). Phylogeny of the bears (Ursidae) based on nuclear and mitochondrial genes. *Molecular Phylogenetics and Evolution* 32: 480–494.

Zachos, J.C., Dickens, G.R., & Zeebe, R.E. (2008). An early Cenozoic perspective on greenhouse warming and carbon-cycle dynamics. *Nature* 451: 279–283.

Zanazzi, A., Kohn, M.J., Macfadden, B.J., & Terry, D.O. (2007). Large temperature drop across the Eocene–Oligocene transition in central North America. *Nature* 445: 639–642.

Marine Mammal Classification and Diversity

CHAPTER 3

CHAPTER OUTLINE

Marine Carnivores
 Polar Bears
 Otters
 Pinnipeds
  Fur Seals and Sea Lions (Family Otariidae)
  True Seals (Family Phocidae)
  Walruses (Family Odobenidae)
  Exploring the Depths: California Sea Lion, Zalophus californianus
  Exploring the Depths: Walrus, Odobenus rosmarus
  Exploring the Depths: Recent Revisions in Phocinae Taxonomy

Sirenians

Cetaceans
 Baleen Whales (Suborder Mysticeti)
 Toothed Whales (Suborder Odontoceti)
 Exploring the Depths: Recent Genetic Studies on Toothed Whales
 Exploring the Depths: Value of Genetic Studies on Beaked Whales

Selected References and Further Reading

Taxonomy (from the Greek *taxis*, meaning "order," and *nomos*, meaning "science") is the branch of science in which living organisms are classified into groups. In this chapter we describe the various marine mammal species and their taxonomic groups.

Marine Carnivores

The name "carnivore" comes from *carnis* (flesh) and *vorare* (to devour). The Carnivora are split into two suborders, Feliformia (cat-like carnivores) and Caniformia (dog-like carnivores). The marine carnivores all belong to the latter suborder. All fully marine carnivores are members of the infraorder Arctoidea, which also includes superfamily Ursoidea (and within that family Ursidae [bears]), superfamily Musteloidea (which contains the family Mustelidae [weasels, minks, wolverines, badgers, and otters]), and superfamily Pinnipedia (seals and sea lions).

Order Carnivora
 Suborder Feliformia (cat-like carnivores)
 Suborder Caniformia (dog-like carnivores)
  Family Canidae: dogs and wolves
  Infraorder Arctoidea
   Superfamily Ursoidea
    Family Ursidae: bears
     Ursus maritimus: polar bear
   Superfamily Musteloidea
    Family Ailuridae: red panda
    Family Mephitidae: skunks
    Family Mustelidae: weasels and otters
    Family Procyonidae: raccoons
   Superfamily Pinnipedia

Polar Bears

Family Ursidae, the bears, includes the giant panda, *Ailuropoda melanoleuca*, and seven species of bear, including the polar bear, *Ursus maritimus* (the Latin name means "the maritime bear" or "bear that lives by the sea"). Until recently, two subspecies of polar bears were in use: *U. m. maritimus*, which was distributed in the Arctic circle to the north of the Atlantic including northeastern Canada, and *U. m. marinus*, which is found in the Arctic to the north of the Pacific. However, splitting living polar bears into two subspecies is probably not warranted. A larger, now extinct, subspecies of polar bear, *U. m. tyrannus*, existed in the Pleistocene (see Chapter 2).

As mentioned in Chapter 2, the polar bear's closest relative is the brown bear (*U. arctos*). In fact, polar bears are so genetically similar to grizzly bears (a Canadian and North American subspecies of brown bear, *U. arctos horribilis*) that there have been hybrids (**Figure 3.1**). Several of these hybridizations occurred in zoos, but in 2006 a strange-looking wild Canadian bear was confirmed as a polar–grizzly bear hybrid by genetic analyses. As grizzly bears and polar bears come into closer contact because of warming in the Arctic and grizzly bear habitat expanding northward (see Chapter 6), more polar–grizzly bear hybrids may be seen in the future.

FIGURE 3.1 A polar bear/grizzly bear hybrid.

Otters

Otters all belong to the Mustelidae family and are related to minks, badgers, weasels, and wolverines. Although some species inhabit marine environments (e.g., the Eurasian otter, *Lutra lutra* [see Chapter 1]), only the sea otter, *Enhydra lutris*, and the marine otter, *Lontra felina*, are currently considered to be marine mammals (**Figure 3.2**). Molecular data suggest the otter genera *Lutra* and *Aonyx* may be more closely related to sea otters than the other otter groups.

Order Carnivora
    Suborder Caniformia (dog-like carnivores)
        Superfamily Musteloidea
            Family Mustelidae: weasels and otters
                Aonyx capensis: African clawless otter
                Aonyx cinerea: Oriental small-clawed otter
                Enhydra lutris: sea otter
                Lutra lutra: Eurasian otter
                Lutra sumatrana: hairy-nosed otter
                Hydrictis maculicollis: spotted-necked otter
                Lutrogale perspicillata: smooth-coated otter
                Lontra canadensis: northern river otter
                Lontra provocax: southern river otter
                Lontra longicaudis: neotropical river otter
                Lontra felina: marine otter
                Pteronura brasiliensis: giant otter

Pinnipeds

Pinnipeds (seals, sea lions, and walruses) derive their name from the Latin *pinna* (for wing) and *ped* (meaning foot). The pinnipeds comprise over a quarter of marine mammal species, with an estimated 50 million pinnipeds worldwide. The pinnipeds are split into three main groups: the fur seals and sea lions, the walruses, and the "true" seals.

A

B

FIGURE 3.2 (A) A Chilean marine otter and (B) a sea otter.

Order Carnivora
    Suborder Caniformia (dog-like carnivores)
        Infraorder Arctoidea
            Superfamily Pinnipedia
                Family Enaliarctidae (extinct)
                Family Odobenidae: walrus
                Family Otariidae: sea lions and fur seals
                    Subfamily Arctocephalinae: fur seals
                    Subfamily Otariinae: sea lions
                Family Phocidae: true seals
                    Subfamily Phocinae: northern seals
                    Subfamily Monachinae: southern seals

Fur Seals and Sea Lions (Family Otariidae)

There are currently nine recognized species of fur seals (**Table 3.1**) and seven species of sea lions, one of which is extinct (**Table 3.2**). All otariids have several features in common. One of the most obvious is that otariids have obvious external ears (pinnae), which true seals lack. They also have hinged hind flippers, which allow them to pull their flippers forward, and a pelvic structure that allows them to put more weight onto their "hips." These features allow fur seals and sea lions to waddle on land, which may not seem terribly graceful but allows them to be more maneuverable in a terrestrial environment than phocid seals. The otariids also use their fore flippers like beating wings to "fly" through the water, with their hind flippers being used like rudders. The coat (pelage) of fur seals and sea lions is dense and contains a dense underfur, which is sparse or lacking in phocids. Male otariids have a noticeable thick "furry" mane.

Of these fur seals, only two are found in the United States, the northern fur seal (see Chapter 16) and the Guadalupe fur seal (see Exploring the Depths: Guadalupe Fur Seals in Chapter 9); none are found in Europe. There are seven recognized species of sea lion, one of which (the Japanese sea lion) is extinct (see Exploring the Depths: Japanese Sea Lions in Chapter 9). No sea lions occur in European waters, but two sea lion species, the California sea lion and Steller's sea lion (see Chapter 16), are found in the United States.

Walruses (Family Odobenidae)

The name "walrus" probably comes from the Nordic word *hrossvalr,* meaning "horse-whale." The scientific name *Odobenus* comes from the Greek words for tooth and walk, *odous* and *baino,* respectively, because of walruses' habit of using their tusks to pull themselves onto ice floes. There is

TABLE 3.1 List of Recognized Fur Seal Species

northern fur seal	*Callorhinus ursinus*
Guadalupe fur seal	*Arctocephalus townsendi*
Galápagos fur seal	*Arctocephalus galapagoensis*
Juan Fernández fur seal	*Arctocephalus philippii*
South American fur seal	*Arctocephalus australis*
New Zealand fur seal	*Arctocephalus forsteri*
Subantarctic fur seal	*Arctocephalus tropicalis*
Antarctic fur seal	*Arctocephalus gazella*
brown (or cape) fur seal	
The brown fur seal is generally divided into two subspecies:	
South African fur seal	*Arctocephalus pusillus pusillus*
Australian fur seal	*Arctocephalus pusillus doriferus*

TABLE 3.2 List of Recognized Sea Lion Species

Japanese sea lion	*Zalophus japonicus*
California sea lion	*Zalophus californianus*
Galápagos sea lion	*Zalophus wollebaeki*
Steller sea lion	*Eumetopias jubatus*
South American sea lion	*Otaria flavescens*
Australian sea lion	*Neophoca cinerea*
New Zealand sea lion	*Phocarctos hookeri*

Exploring the Depths: California Sea Lion, *Zalophus californianus*

Probably the most commonly seen sea lion species in the United States, the California sea lion (**Figure B3.1**), is found along the Pacific coast of Mexico, Canada, and the United States to southern Alaska (**Figure B3.2**). This species breeds between May and July on sandy California beaches. They feed on fish and cephalopods (squid), some of which are commercially harvested by humans, resulting in entanglement in fishing gear and discord with fishermen, who have shot and killed animals because of this perceived conflict with fisheries.

Sea lion occupation of beaches, as well as their tendency to haul out on boats and marinas, has led them to come into further conflict with humans, with claims of damage to property and conflict over access to bathing beaches. Possibly the most famous clash between California sea lions and humans occurred in San Francisco, where sea lions began to haul out on the K dock of Pier 39 in the late summer of 1989. This handful of animals quickly rose to over 400 by March 1990 when the dock was officially closed and set aside for the sea lions, despite calls from boat owners for the city to remove the "nuisance" animals. The highest official count to date was on September 3, 2001, when an astonishing 1,139 sea lions were recorded. Although the sea lions originally vacated the dock every year over the height of summer (roughly June to August) in favor of the breeding sites, smaller numbers of mostly younger animals now haul out at the location throughout the year. The sea lions have now become a highly lucrative tourist attraction.

It has always been normal for California sea lions to be in the bay, especially during the winter months when herring provide an abundant food source. However, it is not known why the sea lions moved from their traditional haul out on Seal Rock (where numbers have declined) to Pier 39. One attractive idea is that K dock provides a safer place to rest, as predators (great white sharks and orcas) do not enter the bay. Regardless, the California sea lions of Pier 39 have now been joined by (presumably) a single male Steller's sea lion, who can be found there from late summer to early fall and also occasionally in the winter. He originally appeared in August 1993 and seems not to mind sharing with his smaller companions.

FIGURE B3.1 California sea lions at Pier 39 in San Francisco.

FIGURE B3.2 Map showing the current, known distribution of the California sea lion. Data from: IUCN Red List.

---

currently only one species of living walrus, which is split into two recognized subspecies:

Odobenus rosmarus rosmarus: Atlantic walrus
Odobenus rosmarus divergens: Pacific walrus

Some scientists also recognize a northern Russia subspecies, *O. rosmarus laptevi*, or the Laptev Sea walrus.

Although closely related to fur seals and sea lions, the walrus has many unique characteristics. Walruses have hinged back flippers that allow some mobility on land, similar to sea lions and fur seals. However, they lack external ear pinnae like the true seals. Their characteristic tusks are enlarged canine teeth that can reach up to 1 meter long and are found in both males and females, although they tend to be slightly larger and more robust in males. As well as being used to aid haul out, tusks are also used during fighting bouts.

True Seals (Family Phocidae)

The true seals, or phocids, have several characteristics in common. They have no external ears, just openings in the sides of their heads with no pinnae. Their hind flippers are not hinged, meaning their flippers drag behind them on land and do not support their body weight, resulting in less mobility than other pinnipeds. They use their hind flippers for propulsion (moving side to side, rather than up and down). Their fore flippers are used to direct themselves, rather than for propulsion, and the claws at the end of their

Exploring the Depths: Walrus, *Odobenus rosmarus*

Walruses are usually found in shallow (Arctic) coastal waters and pack ice (**Figure B3.3**). They are the most gregarious of the pinnipeds and can be found huddled together in large groups at haul-out sites. They eat benthic invertebrates (clams, worms, snails, and shrimp) and slow-swimming fish, using their sensitive vibrissae (whiskers) to find food in seabed sediments. Walruses can suck the flesh out of a clam using their powerful tongue and mouth muscles. Some walruses have been known to even eat seals and small whales.

Humans have hunted walruses for thousands of years, with some walrus populations being decimated in the 19th and 20th centuries. Aboriginal hunts still occur today, as well as illegal poaching. Climate change is also a concern for these Arctic animals because melting ice sheets reduce available habitat (walruses rest on floating ice floes) and warming waters alter Arctic ecosystems. Surveys in the Arctic have estimated fewer and fewer walruses in some regions being studied, and there are concerns that this species may be in decline (**Figure B3.4**).

FIGURE B3.3 A walrus.

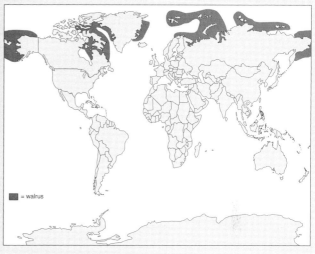

FIGURE B3.4 Map showing the current, known distribution of the walrus. Data from: IUCN Red List.

flippers help these seals to get a grip on rocky shores, sand, or ice and to pull themselves along on land. The true seals are split into two subfamilies:

Northern seals (Phocinae): ten species (**Table 3.3**)
Southern seals (Monachinae): nine species (**Table 3.4**)

Although called "northern" and "southern," some species of southern seals actually occur in the northern hemisphere, for example, monk seals and the northern elephant seal.

Many of the Arctic-dwelling northern seal species (ringed seal, ribbon seal, spotted seal, harp seal, bearded seal, and hooded seal; **Figure 3.3**) come into U.S. waters, particularly Alaskan waters (**Figure 3.4**). Likewise, many of these Arctic seals may visit northern European waters. Although it is an Arctic species, the hooded seal is known to travel considerable distances and has been reported as far south as the Caribbean. Harbor seals (generally known as common seals in the United Kingdom and Europe) and gray seals can be

TABLE 3.3 List of Recognized Northern Seal (Phocinae) Species

bearded seal	*Erignathus barbatus*
harbor (or common) seal	*Phoca vitulina*
spotted (or largha) seal	*Phoca largha*
ringed seal	*Phoca hispida*
Caspian seal	*Pusa caspica*
Baikal seal	*Pusa sibirica*
gray (or grey) seal	*Halichoerus grypus*
ribbon seal	*Histriophoca fasciata*
harp seal	*Pagophilus groenlandicus*
hooded seal	*Cystophora cristata*

TABLE 3.4 List of Recognized Southern Seal (Monachinae) Species

Mediterranean monk seal	*Monachus monachus*
Caribbean monk seal (extinct)	*Monachus tropicalis*
Hawaiian monk seal	*Monachus schauinslandi*
northern elephant seal	*Mirounga angustirostris*
southern elephant seal	*Mirounga leonina*
weddell seal	*Leptonychotes weddellii*
ross seal	*Ommatophoca rossii*
crabeater seal	*Lobodon carcinophaga*
leopard seal	*Hydrurga leptonyx*

A

B

C

D

FIGURE 3.3 (A) A bearded seal, (B) a harp seal, (C) a gray seal, and (D) a ribbon seal.

Exploring the Depths: Recent Revisions in Phocinae Taxonomy

In a comprehensive review of marine mammal taxonomy published in 1998, marine mammal scientist Dale Rice suggested the ringed seal should be called *Pusa hispida*, instead of the more widely recognized *Phoca hispida*. However, more recent work suggests the ringed seal should remain in the genus *Phoca*. Similarly, molecular data indicate the genus *Halichoerus* should also be merged with *Phoca*, bringing the gray seal into the genus. Despite these studies, management agencies are lacking somewhat behind the science. Although the U.S. National Marine Fisheries Service lists ringed seals as *Phoca hispida*, they are listed as *Pusa hispida* by the International Union for Conservation of Nature on their Red List. Furthermore, both agencies continue to list the gray seal as *Halichoerus grypus*.

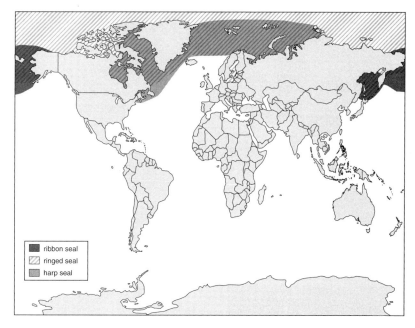

FIGURE 3.4 Map showing the current, known distribution of northern seal species. Data from: IUCN Red List.

commonly found in coastal areas in the northeastern United States and also in the United Kingdom, which possesses the largest populations of these two species in Europe.

One species of monk seal, the Caribbean monk seal, became extinct in the 1950s (see Chapter 9). Most southern seals are found in Antarctica or the sub-Antarctic islands. One species of monk seal, the Mediterranean monk seal (see Chapter 16), can be found, as their name suggests, in the Mediterranean, with some colonies on the Atlantic islands near Madeira. Two species of southern seal are found in the United States: the Hawaiian monk seal (see Chapter 16) and the northern elephant seal (see Chapter 16) (**Figure 3.5** and **3.6**).

Sirenians

Living sirenians are split into the dugongs and manatees. The dugong, a Tagalong (Philippino) word meaning "Lady of the Sea," is in the same family as the now extinct Steller's sea cow. The name manatee comes from *manati*, the Taino (a pre-Columbian indigenous group in the Caribbean) word for breast, a reference to the way in which manatees nurse their offspring by holding them to their mammary glands reminiscent of a nursing human female. Manatees differ from the dugongs in that the manatees have rounded spatula-shaped tails as opposed to the crescent-shaped, cetacean-like tails of dugongs. Dugongs have tusks, whereas manatees do not. Moreover, manatees also have teeth that move in a serial fashion towards the front of the animal's mouth, falling out when the tooth is worn down.

There are currently three recognized species of manatee: the Amazonian, West African, and West Indian. The West Indian manatee is split into two subspecies: the Antillean manatee (found in the Caribbean) and the Florida manatee (found along the U.S. coast especially, as the name suggests, Florida).

Warm periods in the Quaternary may have allowed manatees to move northward into the United States from the Caribbean. Cooler winters along the northern Gulf of Mexico and the Straits of Florida then, possibly, allowed the evolution of endemic North American forms (*Trichechus manatus bakerorum* and the living Florida manatee, *T. m. latirostris*) in genetic isolation. Subsequent oscillations between warm and cool conditions then moved the ranges of all the resulting subspecies north and south, respectively, allowing some gene flow and causing periodic extinctions of various manatees in North America. More details on the distribution and biology of the Sirenians are found in Chapter 8.

Order Sirenia
    Family Protosirenidae: extinct
    Family Prorastomidae: extinct
    Family Dugongidae
        Subfamily Hydrodamalinae
            Hydrodamalis gigas: Steller's sea cow (extinct)
        Subfamily Dugonginae
            Dugong dugon: dugong
    Family Trichechidae: manatees
        Trichechus manatus: West Indian manatee
            Trichechus manatus latirostris: Florida manatee
            Trichechus manatus manatus: Antillean manatee
        Trichechus senegalensis: West African manatee
        Trichechus inunguis: Amazonian manatee

FIGURE 3.5 (A) A northern elephant seal, (B) a Hawaiian monk seal, (C) a crabeater seal, and (D) a leopard seal.

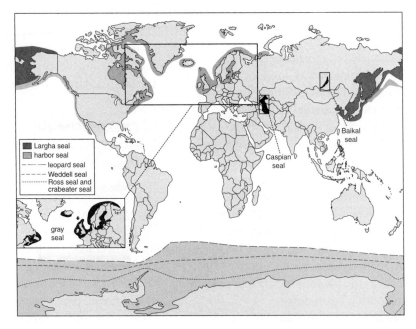

FIGURE 3.6 Map showing the current, known distribution of southern seal species. Data from: IUCN Red List.

Cetaceans

The whales, dolphins, and porpoises belong to the Order Cetacea, which is derived from the Greek word *cetos*, or *ketos*, meaning large sea creature. There are nearly 90 recognized species in the Order Cetacea (see Chapters 10–12). The order is divided into three suborders:

Mysticeti: baleen whales (see Chapter 10)
Odontoceti: toothed whales, dolphins, and porpoises (see Chapters 11 and 12)
Archaeoceti: prehistoric whales (see Chapter 2)

The terms "whale," "dolphin," and "porpoise" have varying degrees of taxonomic distinction. The word "whale," derived from the Anglo-Saxon word *hwael*, has no real taxonomic meaning but historically was instead applied to any large cetacean. Thus, large dolphins and other odontocetes or baleen whales are often referred to as whales, such as the killer whale. Currently, the term loosely refers to all the mysticetes, the odontocete families *Physeteridae* and *Kogiidae* (the sperm whales), and *Ziphiidae* (the beaked whales).

Likewise, the term "porpoise," derived from the Latin word *porcopiscus* (*porcus*, pig and *piscus*, fish), was until quite recently used to refer to smaller dolphins, including the bottlenose porpoise (now the bottlenose dolphin). The term is now linked more solidly with the odontocete family *Phocoenidae*. Finally, the word "dolphin" originates from the ancient Greek word *delphís* or *delphys*, for womb, derived from their ability to bear live young. Taxonomically, this term is particular to the family *Delphinidae* but is generally used to refer to all the remaining odontocete families.

Baleen Whales (Suborder Mysticeti)

The name mysticete means mustached whales, a reference to the whiskery baleen plates that hang down from the upper jaw of these whales. Baleen whales all possess hundreds of plates of baleen instead of teeth. The baleen is made of the protein keratin, just like human fingernails. The fringed baleen plates are used by whales to trap engulfed prey and filter out water when feeding (see Chapter 10). In addition, all baleen whales have two nostrils, differing from the odontocete cetaceans, which have just one. The baleen whales are split into four families. The Balaenidae, or right whales, have large arch-shaped mouths and long baleen plates (**Figure 3.7** and **3.8**) used for "skim feeding" plankton, especially copepods (**Figure 3.9**), which are microscopic crustaceans (see Chapter 10). They also have thick blubber layers and lack a dorsal fin on their backs and pleats or grooves on the undersides of their throats.

The family Neobalaenidae only consists of one species, the pygmy right whale, *Caperea marginata*. This whale looks like a much smaller (<6.5 m), slimmer right whale. The pygmy right whale, similar to its larger cousins, also feeds on copepods.

Family Eschrichtiidae also has only one species, the gray whale, *Eschrichtius robustus*. Their scientific name was given to honor zoologist Daniel Eschricht. The gray whale is currently found exclusively in the Pacific; an Atlantic gray whale population was hunted to extinction in the 1600s to 1700s. Gray whales are distinguishable from other baleen whales by having blunt, rounded heads and a lower jaw of approximately the same width as the upper jaw. They also have two to five shallow grooves on the undersides of their throats. They lack dorsal fins, instead possessing a bumpy ridge on their backs. It has recently been suggested by genetic analyses that gray whales do not belong in their own family but may be genetically more closely related to humpback and fin whales. Therefore, the classification below may change with further genetic studies.

FIGURE 3.7 A North Atlantic right whale.

FIGURE 3.8 The skeleton of a North Atlantic right whale can be seen hanging from the ceiling in the background. A rorqual skeleton is in the foreground for comparative purposes.

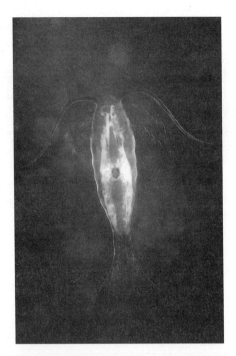

FIGURE 3.9 A copepod, a major prey item of right whales.

FIGURE 3.10 A humpback whale — a rorqual — showing its throat grooves.

Family Balaenopteridae, or the rorquals (from the Norwegian *röyrkval* or furrow whale), all have a similar long, tapered shape and possess ventral pleats on the underside of their throats (**Figure 3.10**). These pleats concertina out when the whale engulfs prey species, a little like the skin pouch on the underside of a pelican's bill. The rorquals also all possess small sickle-shaped dorsal fins, similar to the curved dorsal fins of many dolphin species. Although humpback whales, Subfamily Megapterinae, are very distinctive, with their long white flippers and distinctive tail flukes, many of the Subfamily Balaenopterinae rorqual whales look very similar to each other and can be very difficult to tell apart

at sea. For example, the Bryde's whale (pronounced *broo-dahs* whale) and the sei whale are virtually identical and often can only be told apart because Bryde's whales have three ridges on their snouts and sei whales only have one. Genetic studies have made the situation even more complex as, for example, minke whales were just one species but have now been split into two (the northern minke whale and Antarctic minke whale). There may even be a third species, the dwarf minke whale. Moreover, the Bryde's whale may be not one but two or perhaps three (or more) different species (Bryde's whale, Eden's whale, and possibly also Omura's whale). These animals superficially look identical in appearance but seem to reach different sizes as adults and have distinct differences at a genetic level. Morphological examination of more specimens is needed to investigate the exact taxonomy in the "Bryde's whale complex." However, many of the taxonomic issues surrounding the complex will not be resolved until further analysis, including genetic testing, is permitted on the *Balaenoptera edeni* holotype specimen (the first described specimen) in Calcutta.

Because of genetic and morphological differences and differences in distribution, many rorqual species have also been split into subspecies, which are summarized below. Not all scientists recognize these new species or subspecies, however, and identification and classification of many baleen whale species can thus be difficult and controversial. These classification issues in turn make the estimation of numbers of animals in a particular species or population difficult, which is undoubtedly leading to flawed abundance estimates and management actions, such as quotas for animals that can be taken during whaling activities (see Chapter 14).

Suborder Mysticeti
    Family Balaenidae: right and bowhead whales
        Balaena mysticetus: bowhead whale
        Eubalaena glacialis: North Atlantic right whale
        Eubalaena japonica : North Pacific right whale
        Eubalaena australis: southern right whale
    Family Neobalaenidae: pygmy right whale
        Caperea marginata: pygmy right whale
    Family Eschrichtiidae: gray whale
        Eschrichtius robustus: gray whale
    Family Balaenopteridae: rorquals
        Subfamily Megapterinae
            Megaptera novaeangliae: humpback whale
        Subfamily Balaenopterinae
            Balaenoptera acutorostrata: northern minke whale*

*Species that have been recently reclassified and split or are in dispute.

Balaenoptera bonaerensis: Antarctic minke whale*
Balaenoptera borealis: sei whale
Balaenoptera brydei: Bryde's whale*
Balaenoptera edeni: dwarf Bryde's or Eden's whale*
Balaenoptera omurai: Omura's whale
Balaenoptera musculus: blue whale
Balaenoptera physalus: fin whale

Toothed Whales (Suborder Odontoceti)

As the name suggests, the toothed whales (the scientific name odontoceti means literally "toothed whale") have enamel-covered teeth, similar to other mammals. Some species in this group may not have any teeth (for example, female beaked whales do not) or may only have a few (male beaked whales or narwhals, for example). The teeth may be strangely shaped, whereas others may have scores of near identical teeth. As mentioned above, the term "whale" has no specific taxonomic meaning, and the odontoceti include dolphins and porpoises as well as sperm and beaked whales. The ecology and biology of most of the toothed whale species are discussed in Chapter 11 and 12.

Family Delphinidae is the largest of the odontoceti cetaceans, and their biology is described in Chapter 12. Despite their name the river dolphins are actually not in the family Delphinidae but are in their own families because of substantial genetic and morphological differences.

Although some people still refer to dolphins as porpoises (particularly in the United States), the members of the family Phocoenidae have distinct morphological differences from the dolphins, specifically they have flat, incisor-like teeth, whereas dolphins have pointed, canine-like teeth. Porpoises also tend to be relatively small, with rounded heads and high metabolic rates (but some dolphins also are small and have rounded heads, so these are not unique characteristics, such as tooth shape).

Suborder Odontoceti
    Family Physeteridae: sperm whale
        Physeter macrocephalus: sperm whale
    Family Kogiidae: pygmy sperm whales
        Kogia breviceps: pygmy sperm whale
        Kogia sima: dwarf sperm whale
    Family Ziphiidae: beaked whales
        Subfamily Ziphiinae
            Ziphius cavirostris: Cuvier's beaked whale
            Berardius arnuxii: Arnoux's beaked whale
            Berardius bairdii: Baird's beaked whale (or North Pacific bottlenose whale)
            Tasmacetus shepherdi: Shepherd's (or Tasman) beaked whale
        Subfamily Hyperoodontidae
            Hyperoodon ampullatus: northern bottlenose whale
            Hyperoodon planifrons: southern bottlenose whale
            Indopacetus pacificus: Longman's (or Indo-Pacific) beaked whale
            Mesoplodon densirostris: dense (Blainville's) beaked whale
            Mesoplodon grayi: Gray's beaked whale
            Mesoplodon ginkgodens: ginko-toothed beaked whale
            Mesoplodon hectori: Hector's beaked whale
            Mesoplodon carlhubbsi: Hubb's beaked whale
            Mesoplodon peruvianus: lesser (or pygmy) beaked whale
            Mesoplodon bidens: Sowerby's beaked whale
            Mesoplodon europaeus: Gervais' beaked whale
            Mesoplodon mirus: True's beaked whale
            Mesoplodon layardii: strap-toothed (or Layard's beaked) whale
            Mesoplodon bowdoini: Andrew's beaked whale
            Mesoplodon stejnegeri: Stejneger's beaked whale
            Mesoplodon perrini: Perrin's beaked whale
            *Mesoplodon traversii**: spade-toothed (or Bahamonde's beaked) whale
    Family Platanistidae: Indian river dolphin
        Platanista minor: Indus river dolphin
        Platanista gangetica: Ganges river dolphin
    Family Iniidae: Amazon river dolphin
        Inia geoffrensis: Amazon river dolphin
            Inia geoffrensis geoffrensis: Amazon river dolphin
            Inia geoffrensis humboldtiana: Orinoco river dolphin
            Inia geoffrensis boliviensis: Rio Madeira river dolphin
    Family Lipotidae: Chinese river dolphin
        Lipotes vexillifer: Baiji or Yangtze river dolphin
    Family Pontoporiidae: La Plata dolphin
        Pontoporia blainvillei: La Plata dolphin
    Family Monodontidae: beluga whales and narwhal
        Delphinapterus leucas: beluga whale
        Monodon monoceros: narwhal
    Family Delphinidae: dolphins
        Orcinus orca: killer whale
        Pseudorca crassidens: false killer whale
        Feresa attenuata: pygmy killer whale
        Peponocephala electra: melon-headed whale
        Globicephala melas: long-finned pilot whale

*Species that have been recently reclassified and split or are in dispute.

*was previously *M. bahamondi*.

Globicephala macrorhyncus: short-finned pilot whale
Lissodelphis borealis: northern right whale dolphin
Lissodelphis peronii: southern right whale dolphin
Cephalorhynchus commersonii: Commerson's dolphin
Cephalorhynchus eutropia: Chilean dolphin
Cephalorhynchus heavisidii: Heaviside's dolphin
Cephalorhynchus hectori: Hector's dolphin
    Cephalorhynchus hectori hectori: Hector's dolphin
    Cephalorhynchus hectori maui: Maui's dolphin
Grampus griseus: Risso's dolphins
Orcaella brevirostris: Irrawaddy dolphin
Orcaella heinsohni: snubfin dolphin
Sotalia fluviatilis: tucuxi
Sotalia guianensis: costero (or estuarine dolphin, or marine tucuxi)
Lagenodelphis hosei: Fraser's dolphin
Lagenorhynchus albirostris: white-beaked dolphin
Lagenorhynchus acutus: Atlantic white-sided dolphin
Lagenorhynchus obliquidens: Pacific white-sided dolphin
Lagenorhynchus australis: Peale's (or black-chinned) dolphin
Lagenorhynchus cruciger: hourglass dolphin
Lagenorhynchus obscurus: dusky dolphin
    Lagenorhynchus obscurus fitzroyi: South American dusky dolphin
    Lagenorhynchus obscurus obscurus: South African dusky dolphin
    Lagenorhynchus obscurus (as yet unnamed subspecies): New Zealand dusky dolphin
Sousa teuszii: Atlantic humpback dolphin
Sousa chinensis: Indo-Pacific humpback dolphin
    Sousa chinensis plumbea: Indian humpback dolphin
    Sousa chinensis chinensis: Pacific humpback dolphin
Steno bredanensis: rough-toothed dolphin
Stenella attenuata: pantropical spotted dolphin
    Stenella attenuata graffmani: Eastern Pacific coastal spotted dolphin
Stenella frontalis: Atlantic spotted dolphin
Stenella clymene: clymene dolphin
Stenella coeruleoalba: striped dolphin
Stenella longirostris: spinner dolphin
    Stenella longirostris centroamericana: Central American spinner dolphin
    Stenella longirostris longirostris: cosmopolitan spinner dolphin
    Stenella longirostris orientalis: tropical Pacific spinner dolphin
Delphinus delphis: short-beaked common dolphin
Delphinus capensis: long-beaked common dolphin
Tursiops truncatus: common bottlenose dolphin
Tursiops aduncus: Indo-Pacific bottlenose dolphin
Family Phocoenidae: porpoises
    Phocoena dioptrica: spectacled porpoise
    Neophocaena phocaenoides: Indo-Pacific finless porpoise
    Neophocaena asiaeorientalis asiaeorientalis: Yangtze finless porpoise
    Neophocaena asiaeorientalis sunameri: East Asian finless porpoise
    Neophocaena phocaenoides sunameri: Chinese finless porpoise
    Phocoenoides dalli: Dall's porpoise
        Phocoenoides dalli dalli: North Pacific Dall's porpoise
        Phocoenoides dalli truei: West Pacific Dall's porpoise
    Phocoena phocoena: harbor porpoise
        Phocoena phocoena phocoena: North Atlantic harbor porpoise
        Phocoena phocoena (subspecies?): Western North Pacific harbor porpoise
        Phocoena phocoena vomerina: Eastern North Pacific harbor porpoise
    Phocoena sinus: Vaquita or Gulf of California harbor porpoise
    Phocoena spinipinnis: Burmeister's porpoise

*Researchers have recently suggested a new bottlenose dolphin species, *Tursiops australis*: burrunan dolphin.

Exploring the Depths: Recent Genetic Studies on Toothed Whales

Various dolphin populations, despite looking very similar morphologically, have been recently analyzed genetically and have been found to be so genetically different as to constitute separate species (**Table B3.1**). There are likely to be more species splits in the near future. For example, genetic studies suggest that dwarf sperm whales (*Kogia sima*) and Indo-Pacific humpback dolphins may both consist of two separate species.

Changes are also expected in bottlenose dolphins, which are currently split into two species: the common bottlenose dolphin and the Indo-Pacific bottlenose dolphin. However, several researchers suggest three or more species of bottlenose dolphins. For example, Australia researchers suggested that a bottlenose dolphin population there is genetically different enough to warrant species status (the burrunan dolphin, *Tursiops australis*). Also, there is a small, apparently isolated and highly endangered population in the Black Sea that may represent a distinct subspecies, if not a full species, although animals of an intermediate size do inhabit the Mediterranean. Another possibility is found in the coastal and offshore bottlenose dolphins that live off the east coast of the United States. These forms have different ecologies (one is found in shallow near-shore waters, the other in deep oceanic waters), are different sizes (the offshore form is larger), and have slightly different morphological characteristics. In this case molecular studies have also demonstrated that these forms are more genetically different from each other that some other pairs of species within the family, such as *Delphinus capensis* and *D. delphis*. Furthermore, the Indo-Pacific bottlenose dolphin may be so genetically different from the other bottlenoses that it may even belong in a different genus, with data suggesting it may be more closely related to *Stenella* spp. dolphins.

To complicate matters further, some genetic studies indicate the classification of toothed whales may be even more complicated than previously thought. Molecular studies of cytochrome *b* from Peale's dolphin (*Lagenorhynchus australis*) and the hourglass dolphin (*L. cruciger*) suggest these two dolphins do not belong genetically with the other *Lagenorhynchus* dolphins they morphologically resemble but should in fact be in the genus *Cephalorhynchus*. Although a reclassification of these species needs more evidence to be definitive, this study and those noted above suggest that current odontocete taxonomy may need substantial review as more molecular and genetic studies are undertaken and we get a clearer view of how these various species are related to each other.

TABLE B3.1 New Species Designations

Previous Species	New Species
Bottlenose dolphin—*Tursiops truncatus*	Common bottlenose dolphin—*Tursiops truncatus* Indo-Pacific bottlenose dolphin—*Tursiops aduncus*
Irrawaddy dolphin—*Orcaella brevirostris*	Irrawaddy dolphin—*Orcaella brevirostris* Snubfin dolphin—*Orcaella heinsohni*
Tucuxi—*Sotalia fluviatilis*	Tucuxi—*Sotalia fluviatilis* Costero—*Sotalia guianensis*
Finless porpoise—*Neophocaena phocaenoides*	Indo-Pacific finless porpoise—*Neophocaena phocaenoides* Narrow-ridged finless porpoise—*Neophocaena asiaeorientalis*

Exploring the Depths: Value of Genetic Studies on Beaked Whales

Although conflicting with morphological studies in many odontocetes, genetic studies have managed to substantially reduce confusion over the classification of beaked whales. Many species of beaked whale are rarely (and only very briefly) observed, and stranding specimens are often hard to find. Many of these whales also look very similar, being distinguished primarily by the position and shape of their teeth (**Figure B3.5**). However, only males have these identifiable teeth, and thus skulls of females are especially difficult to identify. Graduate student Merel Dalebout investigated museum specimens of beaked whales around the world using the latest genetic techniques and discovered that many beaked whale skull specimens had been misidentified (for example, many new specimens of Longman's beaked whale were discovered), species names were given incorrectly (Bahamonde's beaked whale, *Mesoplodon bahamondi*, should really be called the spade-toothed whale, *M. traversii*), or because of genetic differences animals were assigned to the wrong genus (e.g., Longman's beaked whale should be *Indopacetus pacificus*, not *M. pacificus*). Most strikingly, a mislabeled beak whale specimen was actually found to be a completely new, and previously undescribed, beaked whale species, which was named Perrin's beaked whale (*M. perrini*) after the famous cetacean biologist William "Bill" Perrin (**Figure B3. 6**).

Exploring the Depths: Value of Genetic Studies on Beaked Whales (continued)

FIGURE B3.5 A selection of beaked whale skulls kept at the Smithsonian Institution.

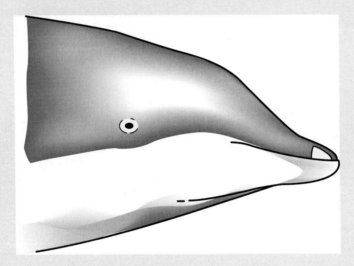

FIGURE B3.6 An artist's depiction of the newly discovered Perrin's beaked whale.

SELECTED REFERENCES AND FURTHER READING

Árnason, U., & Gullberg, A. (1996). Cytochrome *b* nucleotide sequences and the identification of five primary lineages of extant cetaceans. *Molecular Biology & Evolution* 13: 407–417.

Árnason, U., Gullberg A., & Widegren, B. (1993). Cetacean mitochondrial DNA control region: sequences of all extant baleen whales and two sperm whale species. *Molecular Biology and Evolution* 10: 960–970.

Arnold, P., Marsh, H., & Heinsohn, G. (1987). The occurrence of two forms of minke whales in east Australian waters with a description of external characters and skeleton of the diminutive and dwarf form. *Scientific Reports of the Whales Research Institute, Tokyo* 38: 1–46.

Beasley, I., Robertson, K.M., & Arnold, P. (2005). Description of a new dolphin, the Australian snubfin dolphin *Orcaella heinsohni* sp. n. (Cetacea, Delphinidae). *Marine Mammal Science* 21: 365–400.

Best, P.B. (1985). External characters of southern minke whales and the existence of a diminutive form. *Scientific Reports of the Whales Research Institute, Tokyo* 36: 1–33.

Caballero, S., Trujillo, F., Vianna, J.A., Barrios-Garrido, H., Montiel, M.G., Beltrán-Pedreros, S., Marmontel, M., Santos, M.C., Rossi-Santos, M., Santos, F.R., & Baker, C.S. (2007). Taxonomic status of the genus *Sotalia*: species level ranking for "Tucuxi" (*Sotalia fluviatilis*) and "Costero" (*Sotalia guianensis*) dolphins. *Marine Mammal Science* 23: 358–386.

Charlton-Robb, K., Gershwin, L., Thompson, R., Austin, J., Owen, K., & McKechnie, S. (2011). A new dolphin species, the Burrunan dolphin *Tursiops australis* sp. nov., endemic to southern Australian coastal waters. PLOS One 6 (9) e24047: 1–17.

Dalebout, M.L., Baker, C.S., Mead, J.G., Cockcroft, V.G., & Yamada T.K. (2004). A comprehensive and validated molecular taxonomy of beaked whales, Family Ziphiidae. *Journal of Heredity* 95: 459–473.

Dalebout, M.L., Mead, J.G., Baker, C.S., Baker, A.N., Best, P.B. & Van Helden, A.L. (2002). A new species of beaked whale *Mesoplodon perrini* sp. n. (cetacean Ziphiidae) discovered through phylogenetic analyses of mitochondrial DNA sequences. *Marine Mammal Science* 8: 577–608.

Dalebout, M.L., Ross, G.J.B., Baker, C.S., Anderson, R.C., Best, P.B., Cockcroft, V.G., Hinsz, H.L., Peddemors, V., & Pitman, R.L. (2003). Appearance, distribution and genetic distinctiveness of Longman's beaked whales, *Indopacetus pacificus*. *Marine Mammal Science* 19: 421–461.

Davis C.S., Delisle, I., Stirling, I., Siniff, D.B., & Strobeck, C. (2004). A phylogeny of the extant Phocidae inferred from complete mitochondrial DNA coding regions. *Molecular Phylogenetics and Evolution* 33: 363–377.

Dizon, A.E., Lux, C.A., LeDuc, R.G., Urban, J., Henshaw, M., Baker, C.S., Cipriano, F., & Brownell, R.L. (1996). Molecular phylogeny of the Bryde's/Sei whale complex: separate species status for the pygmy Bryde's form? Paper SC/48/27 presented to the Scientific Committee at the 48th Meeting of the International Whaling Commission, June 1996, Aberdeen, Scotland.

Domning, D.P. (2005). Fossil sirenia of the West Atlantic and Caribbean region. VII. Pleistocene *Trichechus manatus* Linnaeus, 1758. *Journal of Vertebrate Paleontology* 25: 685–701.

Hale, P.T., Barreto, A.S., & Ross, G.J.B. (2000). Comparative morphology and distribution of the *aduncus* and *truncatus* forms of bottlenose dolphin *Tursiops* in the Indian and Western Pacific Oceans. *Aquatic Mammals* 26: 101–110.

Heyning, J.E., & Perrin, W.F. (1994). Evidence for two species of common dolphins (genus *Delphinus*) from the eastern North Pacific. *Natural History Museum of Los Angeles County, Contributions in Science* 442: 1–35.

Hoezel, R. (2001). *Marine Mammal Biology: An Evolutionary Approach*. Blackwell Science, Oxford.

International Whaling Commission. (2001). Annex U. Report of the Working Group on Nomenclature. *Journal of Cetacean Research and Management* 3(Suppl.): 363–365.

Jefferson, T.A., & Van Waerebeek, K. (2002). The taxonomic status of the nominal dolphins species *Delphinus tropicalis* Van Bree, 1971. *Marine Mammal Science* 18: 787–818.

Jefferson, T.A., & Wang, J.Y. (2011). Revision of the taxonomy of finless porpoises (genus *Neophocaena*): The existence of two species. *Journal of Marine Animals and Their Ecology* 4: 3–16.

Kasuya, T. (1999). Finless porpoise *Neophocaena phocaenoides*. In: *Handbook of Marine Mammals*. Vol. 6 (Ed. S.H. Ridgway & R. Harrison), pp. 411–442. Academic Press, San Diego.

Kingston, S.E., & Rosel, P.E. (2004). Genetic differentiation among recently diverged Delphinid taxa determined using AFLP markers. *Journal of Heredity* 95: 1–10.

Koepfli, K.P., Deere, K.A., Slater, G.J., Begg, K., Grassman, L., Lucherini, M., Vernon, G., & Wayne, R.K. (2008). Multigene phylogeny of the Mustelidae: resolving relationships, tempo and biogeographic history of a mammalian adaptive radiation. *BMC Biology* 6: 4–5.

Koepfli, K.P., & Wayne, R.K. (1998). Phylogenetic relationships of otters (Carnivora: Mustelidae) based on mitochondrial cytochrome *b* sequences. *Journal of Zoology, London* 246: 401–416.

Koepfli, K.P., & Wayne, R.K. (2003). Type I STS markers are more informative than cytochrome in phylogenetic reconstruction of the mustelidae (Mammalia: Carnivora). *Systematic Biology* 52: 571–593.

LeDuc, R.G., Perrin, W.F., & Dizon, A.E. (1999). Phylogenetic relationships among the delphinid cetaceans based on full cytochrome *b* sequences. *Marine Mammal Science* 15: 619–648.

Lento, G.M. Haddon, M., Chambers, G.K., & Baker, C.S. (1997). Genetic variation of Southern Hemisphere fur seals (*Arctocephalus* spp.): investigation of population structure and species identity. *Journal of Heredity* 88: 202–208.

Mallet, J. (2008). Hybridization, ecological races and the nature of species: empirical evidence for the ease of speciation. *Philosophical Transactions of the Royal Society B* 363: 2971–2986.

May-Collado, L., & Agnarsson, I. (2006). Cytochrome *b* and Bayesian inference of whale phylogeny. *Molecular Phylogenetics and Evolution* 38: 344–354.

Natoli, A., Peddemors, V.M., & Hoelzel, A.R. (2004). Population structure and speciation in the genus *Tursiops* based on microsatellite and mitochondrial DNA analyses. *Journal of Evolutionary Biology* 17: 363–375.

Nowak, R.M., Heyning J.E., Reeves, R.R., & Stewart, B.S. (2003). *Walker's Marine Mammals of the World*. Johns Hopkins University Press, Baltimore and London.

Parsons, E.C.M., & Wang J.Y. (1998). A review of finless porpoises (*Neophocaena phocaenoides*) from the South China Sea. In: *The Marine Biology of the South China Sea*

3 (Ed. B. Morton), pp. 287–306. Hong Kong University Press, Hong Kong.

Rice, D.W. (1998). *Marine Mammals of the World. Systematics and Distribution*. Special Publication No. 4. The Society for Marine Mammalogy, Lawrence, KS.

Sasaki, T., Nikaido, M., Hamilton, H., Goto, M., Kato, H., Kanda, N., Pastene, L.A., Cao, Y., Fordyce, R.E., Hasegawa, M., & Okada, N. (2005). Mitochondrial phylogenetics and evolution of mysticete whales. *Systematic Biology* 54: 77–90.

Van Helden, A.L., Baker, A.N., Dalebout, M.L., Reyes, J.C., Waerebeek, K., & Baker, C.S. (2002). Resurrection of *Mesoplodon traversii* (Gray, 1874), senior synonym of *M. bahamondi* Reyes, Van Waerebeek, Ctirdenas and Ytiiiez, 1995 (Cetacea: Ziphiidae). *Marine Mammal Science* 18: 609–621.

Wada, S., Kobayashi, T., & Numachi, K.I. (1991). Genetic variation and differentiation of mitochondrial DNA in minke whales. *Reports of the International Whaling Commission* (Special Issue) 13: 203–215.

Wada, S., & Numachi, K.I. (1991). Allozyme analyses of genetic differentiation among the populations and species in *Balaenoptera*. *Reports of the International Whaling Commission* (Special Issue) 13: 125–154.

Wada, S., Oishi, M., & Yamada, T.K. (2003). A newly discovered species of living baleen whale. *Nature* 426: 278–281.

Wang, J.Y., Chou, L.S., & White, B.N. (1999). Mitochondrial DNA analysis of sympatric morphotypes of bottlenose dolphins (genus: *Tursiops*) in Chinese waters. *Molecular Ecology* 8: 1603–1612.

Wang, J.Y., Chou, L.S., & White, B.N. (2000a). Differences in the external morphology of two sympatric species of bottlenose dolphins (Genus *Tursiops*) in the waters of China. *Journal of Mammalogy* 81: 1159–1165.

Wang, J.Y., Chou, L.S., & White, B.N. (2000b). Osteological differences between two sympatric forms of bottlenose dolphins (genus *Tursiops*) in Chinese waters. *Journal of Zoology* 252: 147–162.

Weber, D.S., Stewart, B.S., & Lehman, N. (2004). Genetic consequences of a severe population bottleneck in the Guadalupe fur seal (*Arctocephalus townsendi*). *Journal of Heredity* 95:144–153.

Adaptations to a Marine Environment

CHAPTER 4

CHAPTER OUTLINE

Exploring the Depths: Ocean Zones

Swimming

Exploring the Depths: Blubber
Exploring the Depths: Spermaceti Organ

Thermoregulation

Exploring the Depths: Surface Area to Volume Ratio

Diving Physiology and Behavior

Pressure Effects

Osmoregulation

Sensory Adaptations

Exploring the Depths: Light in the Ocean
Exploring the Depths: Sleep

Selected References and Further Reading

Marine mammals can be found in a wide range of habitats, from freshwater rivers, lakes, and lagoons (e.g., Amazonian manatees and Amazon, Indus, and Ganges river dolphins) to estuaries and intertidal zones (e.g., Indo-Pacific humpback dolphins and common bottlenose dolphins) to the continental shelf and the deep waters of the bathyl zone (e.g., sperm whales and beaked whales). In these habitats marine mammals can be in water a few feet deep to depths of over 2 km, the latter in the case of the deep-diving sperm whale. They can also be found in waters that range in temperature from 40°C or higher for Indo-Pacific humpback dolphins or dugongs in the Arabian Gulf to −1.8°C for seals and baleen whales in the Antarctic. As a result of living in these extreme environments, marine mammals have a variety of specialized adaptations. But what distinguishes them as a particularly special group of mammals are their adaptations for living in a fluid environment.

Exploring the Depths: Ocean Zones

Oceanographers divide the marine environment into a series of zones, depending on their depth and location (**Figure B4.1**). The two main zones are the pelagic zone, or the open water environment, and the benthic zone, or the seabed environment. Other designations include the neritic zone, which encompasses coastal waters and marine waters above the continental shelf, and the oceanic zone, which entails the open ocean away from the influence of land. The closest part of the neritic zone to land is the littoral zone, also called the intertidal zone (i.e., the area between high and low tide), which is usually covered and uncovered by seawater twice a day. The sublittoral or subtidal zone is the constantly submerged area up to the edge of the continental shelf. The continental shelf is the flat, relatively shallow area extending from the shore to a depth of 20 to 500 m (average 130 m) at the edge of the shelf (the shelf break). The continental shelf

(continues)

Exploring the Depths: Ocean Zones (continued)

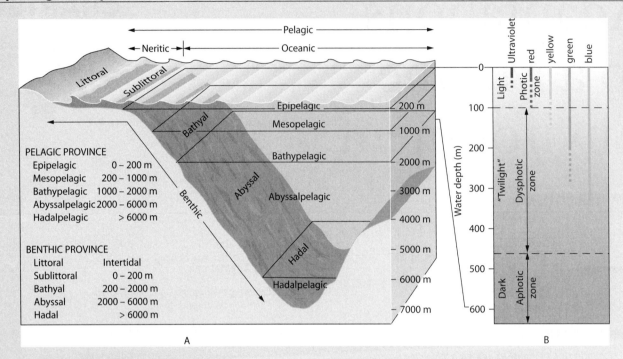

FIGURE B4.1 (A) Marine zones and (B) light zones.

is on average 65 km wide. Over the edge of the continental shelf leads to the steeper continental slope and then to the deep waters of the bathyl zone. Below 4 km the bathyl zone becomes the abyssal zone, an area that is too deep for even the deepest diving whales. At the base of the continental slope the seabed levels out and becomes the abyssal plains. These sediment-covered plains range from depths of 4 to 6 km and comprise more of the world's surface than the total area of land. Occasionally rising above the abyssal plains are seamounts, steep-sided volcanoes that can sometimes break the surface to form islands. Many, however, are submarine, including guyots, which are flat-topped seamounts **(Figure B4.2)**. Seamounts are often associated with upwellings of nutrients and high levels of productivity and thus large amounts of marine life, including marine mammals.

Two other important zones for biological species are the photic and aphotic zones. The photic zone is the depth of water in which there is sufficient light for photosynthesis to occur (sometimes referred to as the epipelagic zone); this zone usually extends to a depth of 100 m to 200 m in typical conditions (although the amount of suspended matter, or turbidity, in the water column can substantially reduce the distance to which light penetrates). The water below this photosynthesis limit is referred to as the dysphotic or aphotic zone (Figure B4.1). From about 200 to 1,000 m is the mesopelagic zone. In this zone, although photosynthesis cannot occur, there is a minute amount of light available (less than 1% of the light at the surface), and so marine species in this zone often have large or specialized eyes to detect these minute amounts of light. Below the mesopelagic zone no surface sunlight penetrates whatsoever, and the waters of these depths are in permanent darkness. If there is any light, it is produced by marine species through bioluminescent organs.

FIGURE B4.2 A flat-topped seamount or "guyot."

Swimming

Marine mammals have a variety of adaptations to maneuver and survive in a liquid environment. As air-breathing mammals they need to come to the surface but typically feed beneath the surface. Their adaptations to stay underwater are described in more detail below (see Diving Physiology and Behavior), but this diving behavior is aided by having nostrils or blowholes that remain closed underwater (in the case of sirenians, pinnipeds, and cetaceans). Pinnipeds and sirenians have to poke their heads and nostrils above water to breathe, but with nostrils (or blowholes) on top of their heads, cetaceans do not have to bring their snouts above the surface and they are therefore able to exchange respiratory gases at the surface more quickly and efficiently.

All marine mammals have modified forelimbs, ranging from webbed feet in the polar bears and sea otters to more flat, paddle-like flippers in sirenians, pinnipeds, and cetaceans (**Figure 4.1**). These forelimbs have relatively high surface areas, which aid mobility in the water.

Polar bears, sea otters, and pinnipeds have hind limbs that are webbed to some extent. Sirenians and cetaceans lack hind limbs altogether (although some larger whale species retain minute fragments of vestigial pelvic and thigh bones). Instead of rear limbs the sirenians and cetaceans possess dorsal-ventrally (i.e., horizontally) flat-ended tail flukes, which are largely composed of cartilage and which extend as flanges on either side of the end of the tail vertebrae. These animals undulate their tails in an up-and-down motion (as opposed to the side-to-side motions of fish tails). To aid with this undulation of the vertebrae sirenians, pinnipeds, and cetaceans have reduced, or totally lack, zygapophyses, processes on the vertebrae that help to lock together vertebral disks in land animals (**Figure 4.2**). Because they lack these processes, the backbones of these marine mammals are more flexible. However, their backbones are also weaker, should the marine mammals ever be outside of the supporting medium of water and on the land.

The eyes of cetaceans lack tear ducts. Marine mammals typically have thick corneas and produce viscous mucus

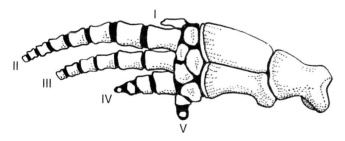

FIGURE 4.1 A dorsal view of the left forelimb of the bottlenose dolphin. Digit numbers are shown.

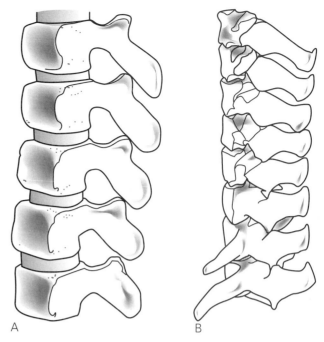

FIGURE 4.2 A portion of the spine of (A) a sperm whale, lacking zygapophyses and for comparison (B) a horse, showing these processes that help to lock vertebral disks in land animals.

to further protect their eyes from abrasion as water, containing particles and sediment, flows past their eyes. Other adaptations seen in marine mammal eyes to deal with living in a marine environment are discussed below (see Sensory Adaptations).

The main adaptations of marine mammals to their environment, however, relate to hydrodynamics. In other words, as marine mammals pass through water, their body surface produces drag, the main resisting force that acts to slow down the forward motion of the animal. To counteract this drag many marine mammals have a pointed, spindle-like, or tear-drop shape, which aids fluid flow over the surface of the animal when they are swimming (i.e., they have a hydrodynamic shape).

Although shape is most important, friction at the skin surface also increases drag. A simple way to reduce drag is to reduce surface area. Evolution has therefore favored large body size, which has a lower surface area to volume ratio. Sirenians, cetaceans, and to a certain degree pinnipeds also have reduced surface hair to the extent that many cetacean species have no hair at all (although some species may have vestigial facial bristles on young animals). The skin of cetaceans is also, in general, very smooth, and in some species this surface layer of skin (epidermis) is replaced every 2 hours, which keeps their body surface smooth and reduces friction.

The blubber layer of marine mammals also helps to improve their hydrodynamics, smoothing out their body shape and filling in contours that might otherwise "catch"

FIGURE 4.3 (A) A beluga whale and a 747 and (B) a spinner dolphin and a fighter jet.

flowing water and create resistance (e.g., the necks of cetaceans). Projections from the body surface can also increase drag; therefore, marine mammals can typically flatten their limbs against the sides of their body (except when being used to control position), and gonads, genitals, and mammary glands are internal. As a result of all these adaptations the shape of a marine mammal generally indicates the speed at which they travel. For example, a slower moving, but maneuverable Risso's dolphin or beluga whale has a blunt head, similar in shape to aircraft such as a 747 jumbo jet. In contrast, faster moving cetaceans, such as spinner dolphins, have a more tapered snout, similar to that of a jet fighter (**Figure 4.3**).

Marine mammals, particularly faster species, have to deal with problems of roll, yaw, and pitch, much in the same way as airplanes must tackle similar control issues as they fly through the air. These are movements around the longitudinal axis in roll, up and down in pitch, and side to side in yaw (**Figure 4.4**). To maintain stability marine mammals have outstretched limbs, for example, broad tail flukes and a flattened peduncle (or portion of a dolphin's tail stock

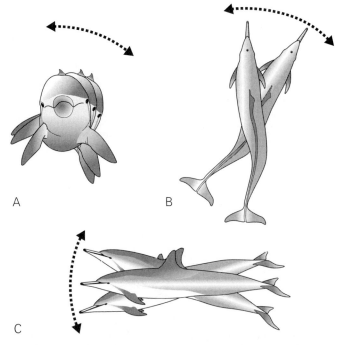

FIGURE 4.4 A diagram showing (A) roll, (B) yaw, and (C) pitch movements.

Exploring the Depths: Blubber

Pinnipeds, sirenians, and cetaceans all possess a blubber layer, which may vary considerably in thickness between species. For example, in bowhead whales blubber layers may be up to 50 cm thick, whereas the blubber of an adult harbor porpoise may only be 2 to 3 cm thick. Blubber differs from fat or adipose tissues in mammals (including polar bears) in that blubber is in a more defined layer, it is strongly attached to the underlying muscle layer, the structure is maintained by a network of collagen fibers, and the layer is vascularized (interlaced with blood vessels) (**Figure B4.3**). The insulating properties of blubber depend on how much lipid (fats and oils) the blubber layer contains, which may vary among species or even in an individual between seasons or during different life history stages.

FIGURE B4.3 A cross-section of blubber.

just in front of the tail flukes) that acts like a "keel." Stiffened forelimbs and rigid dorsal fins also give fast-moving spinner dolphins stability.

Adaptations for buoyancy are important for marine mammals. The buoyancy of a marine mammal is the upward force that keeps it afloat (or at least reduces its rate of descent) and is influenced by the density of the animal. The buoyancy is also increased by the modified forelimbs or flippers that have a cross-section that increases lift, and these can be moved to act like hydroplanes, similar to the pectoral fins in fish.

Although marine mammals have lungs and many other air-filled sinuses, many of these compress or even collapse relatively quickly with depth. Furthermore, marine mammals lack gas-filled swim bladders that many teleost (bony) fish possess, which provide them with buoyancy. Instead, marine mammals have abundant blubber layers that provide buoyancy, because these fat-filled tissues are typically less dense than seawater. To further increase buoyancy some marine mammals have porous or fat-filled bones; for example, many mysticete whales have fat-filled vertebrae. To aid orientation in the water and to

Exploring the Depths: Spermaceti Organ

The spermaceti organ can extend for up to one-third of the length of an adult male sperm whale (and comprise one-fourth of the mass of the animal). It is composed of two parts: the upper spermaceti sack and the lower "junk," (**Figure B4.4**) both of which have a high oil content, although the oil in the junk is denser and compartmentalized. A network of blood vessels weaves through the spermaceti organ.

The organ gets its name because whalers once thought the oil resembled and smelled like human semen and so assumed it was a large gonad or reproductive organ. It is probable that the spermaceti organ plays a role in directing and focusing the echolocation clicks and calls of sperm whales. It also may be a secondary sexual characteristic, akin to antlers in deer by which females can assess the

(continues)

Exploring the Depths: Spermaceti Organ (*continued*)

fitness of males, as males have extremely large spermaceti organs in relation to those of females. Another suggestion is that the spermaceti organs may be used by males during aggressive attacks, by using their larger heads as battering rams. It is also possible the spermaceti organ is used for a combination of all the above.

A particularly interesting idea is that the spermaceti organs are buoyancy control devices. The proposal is that sperm whales allow the waxy material in the spermaceti organ to cool down (by reducing blood flow to the organ) before they dive, permitting the spermaceti to begin to solidify and become denser. This would make the whale's head heavier and negatively buoyant and thus help assist in descent. When the whale wishes to surface, metabolic heat and blood flow to the spermaceti organ may warm the oils, causing them to liquefy, become less dense, and increase the whale's buoyancy, thus aiding the whale in surfacing. Although empirical evidence to date supports the idea that the spermaceti organ is primarily used in sound production from echolocation, it does not preclude the possibility that the organ may have multiple functions.

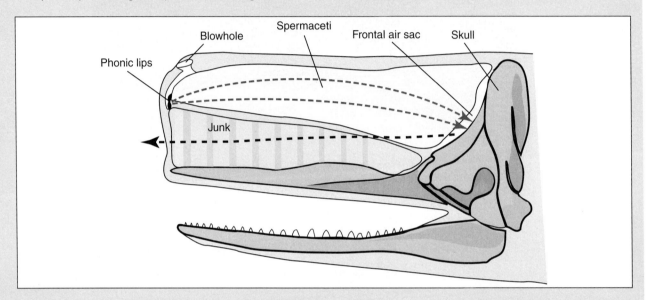

FIGURE B4.4 Cross section of a sperm whale head showing the spermaceti organ. Adapted from Wahlberg, P.T., M. and Mohl, B. *Behav Ecol Sociobiol*, 53 (2002): 31–41.

reduce problems with roll, many species also have denser bones on their ventral side (e.g., the ribs and sternum). This means their dorsal sides have lighter porous or fat-filled vertebrae and their ventral sides have denser, less porous ribs. As a result their ventral sides have a tendency to sink, whereas the dorsal sides have a tendency to orient toward the surface.

Thermoregulation

Any organism, in terms of regulating its internal body temperature, can be either poikilothermic (Greek *poikilos*, changeable, and *therm*, heat), or "cold-blooded," in that their internal body temperature and thus metabolic rate largely varies with and is controlled at the surface by the external environment (ectothermic) or homeothermic (Greek *homoios*, same), or "warm-blooded," meaning its internal body temperature is maintained at a constant temperature, usually between 37 and 40°C by internal metabolic heat production (endothermic).

The marine environment offers temperatures that are relatively more stable than those on land because water has a high specific heat capacity, which means substantial amounts of energy need to be absorbed or lost before the temperature of water changes. Therefore, poikilothermic species tend to be more common in the ocean than on land. However, the metabolic rates of poikilotherms change by a factor of 2 for every 10°C temperature change. Thus the rate at which metabolic reactions occur for poikilotherms is substantially reduced as temperatures drop. This gives homeothermic species a big advantage over poikilothermic prey or competitors in cooler water temperatures. That marine mammals are homeothermic explains, in part, why they are so abundant and successful in polar and subpolar regions. However, being homeothermic does have a cost, because these species require more energy to support a high metabolic rate, which means they must consume relatively more food to survive.

The thermal conductivity of water is 25 times greater than air. Although water temperatures in the marine environment, as noted above, are relatively stable compared with those of the terrestrial environment, marine mammals do have difficulties, nonetheless, with heat loss to the surrounding environment. This is especially true for the many marine mammals that occupy cool waters, where the temperature gradient between their internal body temperature and the surrounding seawater is high, and thus heat loss is greater. Heat loss (W) can be determined from

$$W = A \times C \times (T_{body} - T_{water})$$

where A is surface area (m^2), C is thermal conductance (Wm^{-2}°C^{-1}), and ($T_{body} - T_{water}$) is the gradient between body and water temperature (°C).

Thus, heat loss is greatest for animals with a high surface area (relative to their volume), a body surface that conducts heat well, and a large difference between their internal body temperature and the surrounding water temperature. As a result marine mammals are able to minimize heat loss by increasing body size, reducing conductance (i.e., increasing insulation), or, in some cases, avoiding low water temperatures. For example, sea otters experience heat loss because they are relatively small and their surface to volume ratio is high (see Exploring the Depths: Surface Area to Volume Ratio). One way to decrease surface area is to have small appendages. Allen's rule, an ecological observation made by Joel Allen in 1877, states that endotherms from colder climates usually have shorter limbs (or appendages) than the equivalent animals from warmer climates because this reduces their relative surface area. Sea otters have relatively blunt muzzles and shorter tails than their otter counterparts in warmer climates.

Although sea otters are one of the smallest of the marine mammals, they can weigh up to 45 kg (100 lb) and be over 1.5 m in length, making them the heaviest and largest mustelid species. Therefore, they have a relatively smaller surface area to volume ratio than other otters. This is an example of another ecological rule: Bergmann's rule (after Christian Bergmann, who published the rule in 1847) states that animals within a specific group tend to get larger as the latitude increases (and decreases in temperature). For example, Arctic-dwelling polar bears and sea otters are the largest species of bear and otters, respectively.

Heat loss is also minimized by reducing the conductance (increasing insulation) of the exterior surface. Sea otters possess a thick fur layer and have one of the densest furs of all mammals, with up to one million hairs per square inch (150,000 hairs/square cm) (**Figure 4.5**). This fur layer traps air (air has a lower conductance than water) and therefore insulates the otter well. However, when a sea otter dives, water pressure forces air out of their fur and they lose much of their insulation. Salt from seawater also clogs their

Exploring the Depths: Surface Area to Volume Ratio

The principle of surface area to volume ratio is an important one in biology. In general, larger organisms have a relatively smaller ratio of surface area to volume from which to absorb or lose heat through or to lose or gain fluids. The simplest way to visualize this is by taking a series of cubes. A cube with each side 1 cm long has a volume of 1 cm^3 (1 cm × 1 cm × 1 cm) and a surface area of 6 cm^2 (6 sides × 1 cm × 1 cm). Its surface area to volume ratio is 6 cm^2:1 cm^3. A cube with each side 2 cm long has a volume of 8 cm^3 (2 cm × 2 cm × 2 cm) and a surface area of 24 cm^2 (6 sides × 2 cm × 2 cm). Its surface area to volume ratio is 24 cm^2:8 cm^3 or 3:1. A cube with each side 3 cm long has a volume of 27 cm^3 (3 cm × 3 cm × 3 cm) and a surface area of 54 cm^2 (6 sides × 3 cm × 3 cm). Its surface area to volume ratio is 54 cm^2:27 cm^3 or 2:1. So as an object becomes larger, its ratio of surface area to volume becomes smaller (**Figure B4.5**).

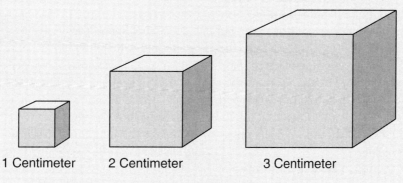

FIGURE B4.5 Drawing of 1, 2, and 3 centimeter cubes.

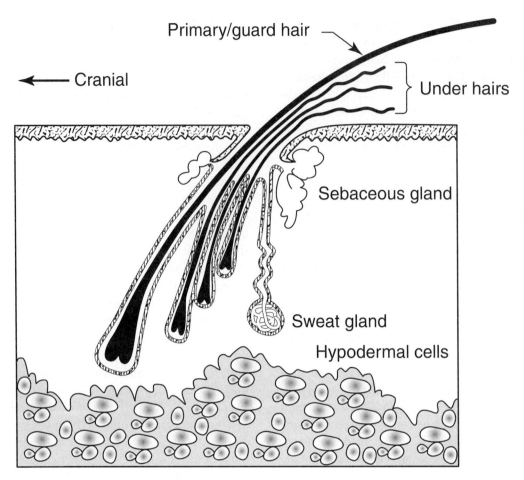

FIGURE 4.5 A diagram of the hair structure of a fur seal showing the position of primary and insulating under hairs. Modified from Bonner, W. N. *The Natural History of Seals*. Christopher Helm, London.

fur, making hairs stick together. Thus, much of a sea otter's behavior is dedicated to grooming, cleaning fur, and "fluffing it up" so air can get between the fur hairs.

Polar bears face a similar problem; although their fur is an effective insulator on land, when they enter the water their fur becomes waterlogged and loses much of its insulating properties. However, polar bears have a substantive layer of adipose, or fatty, tissue. This helps insulate them even when their fur is wet. A thick layer of fat does have some disadvantages on land. For example, if a polar bear exerts itself in warmer weather, it may actually be in danger of overheating because of an inability to lose heat from the body surface. To counteract this, erector muscles at the base of hairs relax, making hairs lie flat instead of standing up straight on end, thus reducing the thickness of the trapped air layer in the fur. The danger of overheating is not totally avoided, however, as similar rapid alteration of the thickness of a fat layer is not possible.

Because fur relies on trapping a layer of air to keep an animal warm, the more highly adapted a marine mammal is to the aquatic environment, the less fur or hair it possesses and the more it relies on a fat layer, or specifically a blubber layer, for insulation.

When marine mammals are first born, their blubber layers tend to be thin. This in combination with a relatively high surface area to volume ratio means that young marine mammals can be very susceptible to heat loss. Many large cetaceans migrate to warmer waters to breed (for more discussion on migration see Chapter 11). Seals and sea lions must breed on land or ice. Many seals are born with a thick neonatal coat of fur, or lanugo, which is effective only because they spend their early days entirely out of the water. The lanugo helps to keep seal pups warm until their blubber layer is thick enough to adequately insulate them (**Figure 4.6**). If this fur gets wet, it no longer insulates. Thus, the lanugo is shed before the pup starts swimming. In some species, such as the harbor seal, the lanugo is actually molted within the uterus as an adaptation for immediate aquatic existence. However, if a young seal pup is forced into seawater before its blubber layer is fully developed, there is a risk of hypothermia (lowered body core temperature) and even death. Adult pinnipeds also have a fur layer, or pelage. In fur seals

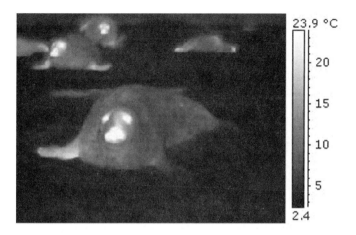

FIGURE 4.6 A thermal image of adult gray seals. It shows relatively high heat loss from the head compared to the cooler well-insulated body.

and sea lions the fur layer is thick and consists of underhair and protective guard hairs (diagram), but in phocid seals hair fibers in the adult coat are considerably shorter, more akin to a military "buzz" or "crew" cut hairstyle than the fluffy lanugo of most pups.

Marine mammals also decrease the rate of heat loss by reducing the difference between the temperature of the surrounding water and that of its exposed tissue. Thus, to reduce heat loss from appendages (which inherently have a relatively high surface area to lose heat through) such as flippers, marine mammals have a countercurrent blood system in their appendages (**Figure 4.7**). Put simply, blood vessels carrying blood away from the core of the animal and into the appendages run parallel and right next to blood vessels returning blood to the core from the appendages. As a result heat is able to transfer from the warm outgoing vessel into the adjacent cool incoming vessel. As heat is transferred to this partner vessel, the blood reaching the extremity is closer in temperature to that of the surrounding seawater. This means there is less of a temperature difference, or gradient, and so less heat is lost to the surroundings. Similarly, the cool blood in the vessel heading back to the body is warmed via heat transferred from the neighboring outgoing vessel. This means that by the time the blood reenters the body, it has returned to near-body temperature (and the marine mammal does not suffer a decrease in core body temperature).

Sea otters use yet another way to maintain their body temperature despite high rates of heat loss: they simply produce more body heat by increasing their metabolism. As a result their resting metabolic rate is on average 2.4 times greater than that of similarly sized terrestrial animals. However, this high metabolic rate means they need a very high daily energy intake (see Chapter 7).

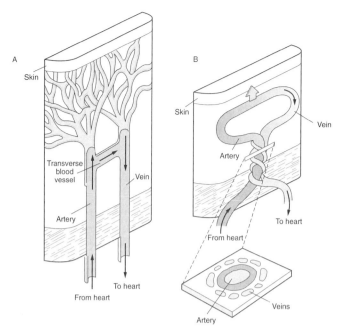

FIGURE 4.7 Diagrams of countercurrent blood circulation systems.

In addition to these physiological adaptations, marine mammals may also have behavioral adaptations for thermoregulation. For example, Antarctic seals in the winter winds may be exposed to temperatures as low as −40°C. Although the seawater is warmer at −1.8°C, heat loss is still high due to the high thermal conductivity of seawater. Instead, seals are likely to benefit from the active generation of heat as a byproduct of swimming rather than reducing their overall heat loss in water. In cold conditions animals may also huddle together. Pinnipeds in particular exhibit this behavior when hauled out, sharing body heat and gaining shelter from cold winds. Other animals, such as manatees, migrate to warmer waters (see Chapter 8). Manatees have also learned to congregate around the warm water outflows from power plants and other industrial facilities for thermal comfort.

When adult seals come onto land to breed or molt or when cetaceans are in warm oceanic regions, they could overheat because of their thick blubber layers. To avoid overheating, special blood vessels passing through the blubber layer can shunt warm blood to the skin surface to increase heat loss. In some species with pigmented skins this can be difficult to observe, but in parts of the skin lacking pigmentation (e.g., the underside of Indo-Pacific bottlenose dolphins) or in species that lack pigmentation in general (Amazon river dolphins, some Indo-Pacific humpback dolphins, and walruses) the skin surface can become bright pink. Weddell seals on pack ice in Antarctica can shunt so much heat through their skin they can actually start to "steam" and the ice below them can melt (producing something akin to a seal-shaped "snow angel"). Many cetacean species are also thought to be able to flush their dorsal fins with blood and use this large surface area with thin blubber layer as a "thermal window" to vent heat. Elephant seals when they get too hot throw moist gravel onto their backs. The water evaporates, absorbing heat and cooling the seal, in much the same way as we cool from sweating. Seals also pump blood to the surface of their outstretched flippers in warm conditions, thereby increasing their exposed surface area for heat loss by radiation and convection. Fur seals and sea lions wave or fan their flippers on hot days, moving cold air over their bodies and increasing convective heat loss. Finally, many pinnipeds seek shade or take to the water in warm conditions to thermoregulate.

Diving Physiology and Behavior

Feeding on prey underwater means that marine mammals must spend a good portion of their lives diving, which presents a variety of problems for air-breathing creatures. Marine mammals have thus adapted in various ways to overcome these issues. One of the basic problems for diving mammals is access to and storage of oxygen. Oxygen is required for aerobic respiration, the chemical reaction in which glucose is broken down to produce energy that "powers" essential biochemical reactions within cells.

$$C_6H_{12}O_6 + 6O_2 \rightarrow 6CO_2 + 6H_2O + \text{energy}$$
glucose   oxygen   carbon dioxide   water

Species like sea otters have much larger lungs than terrestrial animals of a similar size, allowing them access to more oxygen for their tissues while diving. Large lungs also have the advantage of providing more buoyancy, although conversely it means it is harder to dive to depth. Consequently, it is perhaps not surprising that many marine mammals have smaller lungs than would be expected based on their size (**Figure 4.8**). However, oxygen can also be stored in the blood (bound to hemoglobin in red blood cells) or in muscle tissue (bound to myoglobin), both of which have higher levels of stored oxygen in cetaceans and pinnipeds than in terrestrial mammals (**Figure 4.9**). Myoglobin is, in fact, better than hemoglobin for binding and thus storing oxygen, and so marine mammals typically have much higher levels of myoglobin in their muscle tissue than do other terrestrial mammals.

Cetaceans have some additional adaptations that increase their ability to store oxygen within their tissues. They have a large spleen, which can store additional red blood cells that can be released during dives. They also (as do sirenians) possess well-developed, intricate blood capillary networks, the *retia mirabilia* (Latin for "wonderful nets"; **Figure 4.10**), which effectively grant them a greater blood volume and aid in redistributing blood during diving cycles, as discussed below.

Marine mammals may also reduce their consumption of oxygen during a dive. Blood flow may be reduced to "nonessential" organs, such as the intestines or muscle tissue, whereas the flow of blood to organs that must have a supply of oxygen, such as the brain, continues with the aid of retia mirabilia. Even with this diverted blood flow, cetacean brains can also continue to function at oxygen levels that would render a human unconscious.

The heart requires oxygen when beating, so another way to reduce oxygen consumption is by reducing the heart rate, or bradycardia. In ringed seals, for example, heart rate

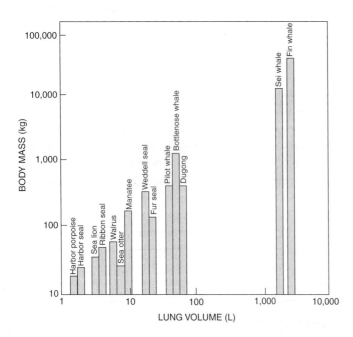

FIGURE 4.8 The relationship between lung volume and body mass in marine mammals. Data from Kooyman (1973).

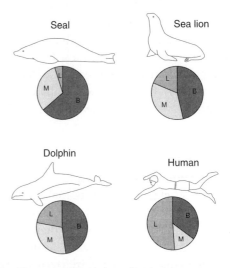

FIGURE 4.9 The proportion of total oxygen stores contained in lungs (L), muscle (M), and blood (B) in different marine mammals compared to a human.

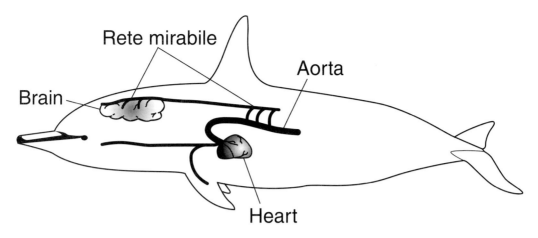

FIGURE 4.10 Drawing showing the location of the rete mirabile.

decreases from nearly 150 beats per minute at the surface to fewer than 10 beats per minute during a dive, with an associated decrease in oxygen consumption. Normally, such a slow rate of blood flow might lead to circulatory disorders, such as congestion within the blood vessels. However, marine mammal blood often has a lower viscosity ("stickiness") that allows it to continue moving, despite this.

The metabolic rate in other organs may also slow down, by up to 50%, which reduces overall oxygen consumption. Cold temperatures at depth may play a part in slowing down the metabolic rate as muscle tissue and internal organs may drop by 1.5°C in temperature during a dive. This reduction in temperature may be, in part, an effect of blood being diverted away from muscle and internal organs during a dive, as noted above.

In this low oxygen environment, tissues, especially muscle, can still produce energy. Lactic acid is produced by anaerobic (without oxygen) respiration. High levels of lactic acid can be toxic for cells, but the muscles of pinnipeds and cetaceans can tolerate a high level of lactic acid buildup by storing the chemical as less toxic and less problematic lactate. It should be noted, however, that a tolerance for anaerobic respiration at levels much higher than in nondiving mammals does not mean marine mammals can do away with aerobic respiration altogether. Aerobic respiration produces 18 times the energy of anaerobic respiration and thus is a major source of cellular energy. Moreover, nerve tissues only respire aerobically, presumably due to the toxic effects the lactic acid would have before it could be transported away. Similarly, the divers must still repay their oxygen debt by removing lactic acid when they are able to breathe once again. Fortunately, this can be done quickly, as marine mammals, and especially cetaceans, can exchange a much higher proportion of air in the lungs on each breath than terrestrial mammals can. For example, a human exchanges about 17% of the lung volume on each breath, whereas a bottlenose dolphin exchanges more than 80%. Marine mammals also tend to have nasal openings higher on their snouts or on top of their heads to make breathing more efficient at the surface.

$$C_6H_{12}O_6 \rightarrow 2C_3H_6O_3 + \text{energy}$$
$$\text{glucose} \qquad\qquad \text{lactic acid}$$

Pressure Effects

Water pressure increases by 1 atmosphere every 10 m in depth (or 1 kg/cm² or 15 lb/in²). Therefore, organisms are exposed to a pressure twice that of atmospheric pressure at 10 m, pressure three times atmospheric pressure at 20 m, and pressure 101 times atmospheric pressure at 1 km. This pressure can have many effects, and marine mammals have many adaptations that overcome or reduce the effects of this pressure. One of the most infamous problems associated with water pressure in air-breathing, diving species (including human scuba divers) is decompression sickness, or "the bends."

At higher pressures (e.g., greater depth) it becomes easier for gases to dissolve in a liquid. This is the result of a physical effect referred to as Henry's gas law. Air contains approximately 70% nitrogen, which is effectively inert and normally plays little role in respiratory processes. However, when air in the lungs is under increased pressure, more nitrogen dissolves in the wet linings of the lungs and can be carried in blood plasma. When pressure decreases (for example, when surfacing), body fluids can hold less gas in solution and nitrogen is released. If pressure is reduced slowly, this gas can be released into the lungs. But if the pressure is released too quickly, nitrogen can come out of solution in blood vessels as bubbles of gas. These bubbles can become lodged in blood capillaries, and when these bubbles form in nervous tissue and joints they can cause crippling pain. Scuba divers who surface too quickly from depth can get this decompression sickness, or "the bends," so-called because divers who succumb are "bent over" in agony. In severe cases paralysis and death can result.

So how do marine mammals avoid the bends? Some marine mammal species, such as polar bears, sea otters, and manatees, do not dive to a depth or for a duration in which nitrogen absorption becomes a problem. For pinnipeds and cetaceans, many of the problems caused by decompression can be prevented by reducing exposure of high-pressure gases to the blood. One way to do this is by having a smaller lung volume, as mentioned above. Seals reduce the air in their lungs further by exhaling before diving (although sea lions, manatees, and cetaceans inhale before diving). Cetaceans and pinnipeds have extremely flexible rib cages that compress as water pressure increases. This compression squeezes the lungs and alveoli (air sacs), which collapse (**Figure 4.11**). Gaseous exchange is almost completely prevented as a result, although the cartilage lining of the air passages leading from the alveoli keep these airways open so air can flow out of the lungs (and are not pinched shut as the result of pressure) trapping small pockets of high-pressurized gases in the lungs. However, most of the air is forced instead into the upper parts of the air passages (the trachea or windpipe), which are thickened and impermeable to gases.

When marine mammals move toward the surface, transfer of nitrogen from blood into lungs is rapid, and some nitrogen is absorbed into mucus lining the airways. Because of the diversion of blood flow away from muscles when diving, there is less risk of a blockage in blood capillaries. Moreover, it has been suggested that the retia mirabilia could act as a filter, trapping and filtering out nitrogen bubbles from the bloodstream. It was thought that pinnipeds and cetaceans never got the bends. However, lesions similar to those caused by decompression sickness in humans have been found in several cetacean species, in particular beaked whales and sperm whales. (See Chapters 5 and 11 for discussions on decompression sickness–like effects observed in beaked whales, which have been linked to underwater noise impacts.)

Another consequence of absorbing higher levels of dissolved nitrogen while under pressure is nitrogen narcosis. Dissolved nitrogen causes a thinning in myelin sheaths (nerve coverings) and other effects. In human scuba divers the symptoms of nitrogen narcosis are similar to drunkenness. An adaptation of marine mammals to reduce nitrogen absorption not only limits the potential effects of decompression sickness but also serves to reduce nitrogen narcosis.

In human scuba divers even oxygen becomes a problem when the partial pressure becomes too high (generally at a depth of 55 m or below), with the onset of "oxygen toxicity." This affects the central nervous system, initially causing nausea and changes in vision but eventually leading to seizures and death. Experiments on pinnipeds placed in hyperbaric (high pressure) chambers in the 1970s found that marine mammals can suffer oxygen toxicity. However, in the wild, compression of the lungs while diving not only reduces the effects of nitrogen absorption but also prevents the absorption of oxygen at high pressures.

A further effect of pressure is high pressure nervous syndrome, which results from pressure on internal air spaces. In humans this can cause tremors, dizziness, and nausea at depths greater than 15 m. However, cetacean lungs collapse when diving, as noted above, and both pinnipeds and cetaceans lack facial sinuses (air spaces in the skull). Furthermore, blood vessel volume in the middle ear increases during dives, in both cetaceans and pinnipeds, thus filling that air space.

Additionally, it has been suggested that deep-diving cetaceans, such as beaked whales, may suffer from a state of "hyperexcitability" as the result of the effect of high pressure on the nervous system. This might make beaked whales

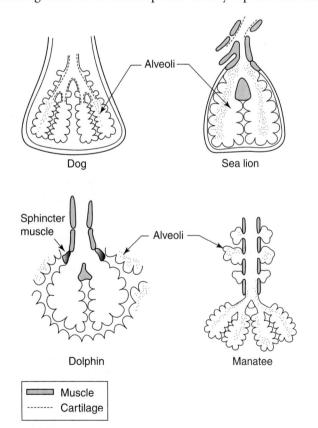

FIGURE 4.11 A diagram of the respiratory bronchioles and alveoli in the lungs of a dog in comparison to those seen in marine mammals. Note the additional muscle used to control respiratory gas exchange in diving mammals.

more likely to startle and behave in an extreme fashion, for example, beaching themselves when exposed to underwater sonar systems (see Chapter 11).

Because of these adaptations some pinnipeds can dive to depths of over 1.5 km; northern elephant seals currently hold the pinniped diving record at a depth of just over 1.6 km. Of the cetaceans with recorded diving depths, narwhals are known to reach 1 km in depth and beaked whales can dive to over 2.9 km. Sperm whales are also deep divers (to depths of 2 km or more) and may be able to dive even deeper, as carcasses have been found entangled in seabed cable at a depth of 3 km.

Osmoregulation

Molecules of a liquid or gas tend to move from an area of high concentration to low concentration through a process known as diffusion. When water diffuses through a semipermeable membrane or selective membrane (a membrane that allows some substances through and not others, such as a cell membrane) from an area with lower dissolved substances to that of higher dissolved substances (i.e., higher concentration of water to a lower concentration of water), this process is called osmosis.

Marine mammals, like other organisms, must maintain a reasonably balanced internal environment (homeostasis). Thus, to maintain a balance of salt and water concentrations within the cells of marine organisms, the balancing of water and dissolved materials gained and lost by cells (or osmoregulation) is important. In seawater, marine mammals are surrounded by a medium that has a higher level of dissolved substances (or solutes) and relatively lower concentration of water molecules than the cells in their bodies. Therefore, there is a tendency for marine mammals to lose water from their bodies to the surrounding environment by osmosis (they are hypo-osmotic). To replace this lost water sirenians need to find and drink fresh water (e.g., rivers). Cetaceans extract water from ingested prey, produce a certain amount of water as a byproduct of metabolizing fats, and drink a certain amount of seawater, from which they extract salts. Marine mammals have highly specialized kidneys that are globular and resemble a bunch of grapes. These "grapes," or reniculi, have a high surface area through which they can effectively filter marine mammal blood, extracting salts (which helps elevate the solute content of marine mammal cells). These efficient kidneys also reduce the amount of water lost during urination; marine mammals produce highly concentrated urine.

In freshwater, river, and some estuarine species marine mammals have the opposite problem. The solute content of their cells is higher than their freshwater surroundings and there is a higher relative concentration of water outside their bodies than within their tissues, so they take on water by osmosis (they are hyperosmotic). To remove this excess water species such as river dolphins urinate copiously.

Sensory Adaptations

Because of lower levels of light in the marine environment compared with the terrestrial environment, marine mammals tend to have large eyes that superficially resemble those of nocturnal land mammals. There are, however, some substantive differences in their structure. Terrestrial mammal eyes work by refracting light so it is focused at the back of the eye onto the retina. Most of this focusing is done by the cornea with small adjustments made by the lens. This refraction by the cornea is principally because the cornea is a curved surface that separates air from the fluid-filled interior of the eye and the difference in refractive index between these two media is high. However, seawater has a refractive index similar to the contents of the eye, so on immersion in water the refractive power of the cornea is lost. To compensate for this loss pinnipeds and cetaceans have large spherical lenses in their eyes, similar in structure to the lenses of fish. It is the lens that does all the focusing of light onto the retina. Sea otters have lenses similar to land mammals, although the front of the lens has a curved bump and the iris is attached to the lens, which allows the otter to change the shape of the lens, giving it better vision when underwater. Sirenians have lenses that are more spherical in appearance than terrestrial mammals but not completely spherical like the cetaceans and pinnipeds (**Figure 4.12**).

When cetaceans and pinnipeds put their heads above water, the situation changes, and light would not be properly focused if the cornea now started to function. Marine mammal eyes overcome this in two ways. First, their cornea is relatively flat so that even in air it has very little refractive power. Second, they can constrict their pupils so in effect the eye becomes more like a pinhole camera that bypasses the optical system of both the lens and cornea. This means objects at all distances are brought to focus on the retina, but the image produced in the eye is very dim.

Similar to nocturnal species, pinnipeds and cetaceans have large numbers of rod cells in the retina (cells specialized to detect low light levels) but have fewer cone cells (cells sensitive to higher light levels and can give rise to color vision). Thus, color discrimination in marine mammals is poorer than in humans. When one considers that several wavelengths of light (e.g., red) only penetrate the most shallow of ocean waters, having an ability to perceive these wavelengths would only be of limited use anyway. Another adaptation similar to nocturnal mammals and seen in marine mammals (albeit not sirenians) is a *tapetum lucidum* a reflective layer of protein crystals behind

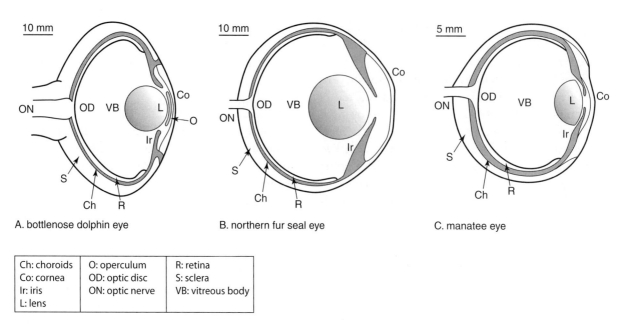

FIGURE 4.12 A diagram showing the eye structure of (A) a bottlenose dolphin, (B) a northern fur seal, and (C) a manatee. Modified from Mass, A.M. and Supin, A.Y. (2009).

Ch: choroids	O: operculum	R: retina
Co: cornea	OD: optic disc	S: sclera
Ir: iris	ON: optic nerve	VB: vitreous body
L: lens		

the retina, which means that each parcel of light that passes through the retina is bounced back through it, further increasing the sensitivity of marine mammal eyes to low light conditions.

Most marine mammals, cetaceans in particular, also have an acute sense of touch, possessing skin with many sensitive nerve endings. This may explain why cetaceans spend so much time in tactile contact, touching and caressing (spinner dolphins may spend up to 30% of their time in tactile contact with other group members). Several species, particularly sirenians and some pinnipeds, have very sensitive muzzles, with vibrissae or sensory whiskers they can use to detect food items, by direct contact, or by detecting vibrations in the water caused by nearby prey.

In contrast, cetaceans appear to have a less developed sense of smell (olfaction), or indeed taste (gustation) compared with terrestrial mammals. Sea otters, however, have large complex nasal cavities and may use smell to detect the reproductive state of females when out of the water. It is notable that sea otters lack scent glands that are characteristic of other mustelids. Polar bears have extremely acute senses of smell in air, which they use to find their seal prey from distances of several miles. Pinnipeds also have an acute sense of smell on land, and females of several species use scent to locate and identify their pups when they return to breeding beaches after a feeding bout. The olfactory system is poorly adapted for communication in water, and this sense declines with increasing aquatic specialization.

Exploring the Depths: Light in the Ocean

Light in the marine environment is very different from the terrestrial environment. Sunlight is absorbed quite quickly as it moves deeper in the water column. This absorption varies by wavelength, with longer wavelengths of light absorbed first, so that little red light penetrates deeper than 10 m depth in the ocean. Blue and green penetrate deeper, but only 10% of blue light penetrates to 100 m or more. This effect can be observed by scuba divers if they cut themselves. At the surface their blood looks red, at 20 m it looks green, but at 40 m it looks more purple-black in color.

Turbidity (the amount of suspended matter in the sea) further reduces the depth that light penetrates while also increasing the absorption of blue light (**Figure B4.6**). For example, in a turbid river estuary only 1% of surface levels of green light may penetrate to 30 m. Because of this loss of sunlight, as noted above, photosynthesis is not possible in marine waters below 100 m (and sometimes at much shallower depths). The zone of ocean waters in which there is sufficient light for photosynthesis is called the photic zone. Below this are the dark waters of the aphotic zone.

Some light penetrates the aphotic zone, but only a tiny percentage of what is found at surface levels. No light, however, penetrates deeper than 1 km. This "twilight" region is called the mesopelagic zone, where animals typically have special adaptations to perceive the minute amounts of light present, produce their own light (bioluminescence), or use other senses such as detecting sounds and vibrations to find prey.

Exploring the Depths: Light in the Ocean (*continued*)

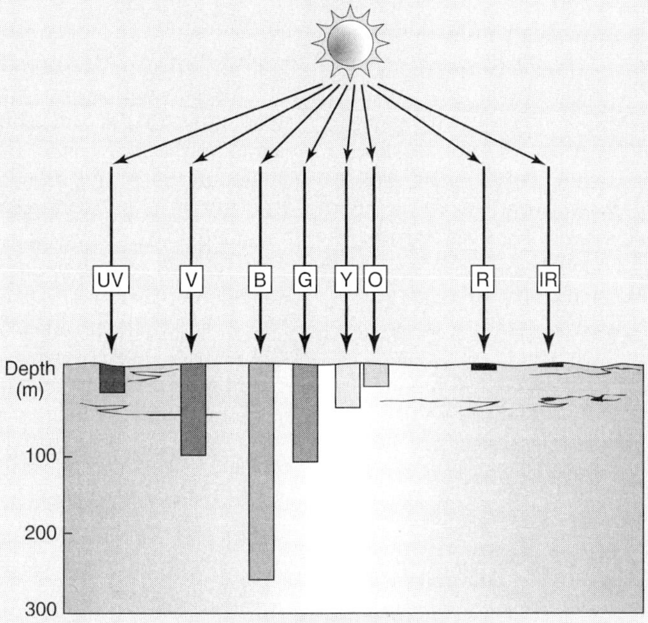

FIGURE B4.6 Chart of light penetration in the ocean.

In addition, all groups of marine mammals cannot smell in water because they close their nasal apertures or blowholes when underwater. Therefore, for marine mammals chemoreception is a matter of taste rather than smell. Experiments with cetaceans and pinnipeds have shown they can taste and are able to discriminate between different salt concentrations and other simple tastes but with different sensitivities. Most cetacean species appear to lack taste buds on their tongues, which gives their tongues a smoother appearance than terrestrial species (**Figure 4.13**). It is likely that sea otters and polar bears have a sense of taste equivalent to closely related terrestrial species.

Therefore, because marine mammals do not appear to have particularly highly developed senses of smell and taste underwater and as light is so limited in the ocean, many species, especially cetaceans, use hearing instead of sight as their primary sense to gain extra information about objects remote from them and to communicate with each other. This is reflected in the ratio of the number of fibers in the auditory nerve in comparison with the optic nerve, which is two to three times greater in cetaceans as in terrestrial mammals. It is not only that hearing is constrained less than the other senses underwater, but it is actually enhanced because sound travels further and faster in water than in air. The nature of sound and communication in the ocean is described in detail in Chapter 5.

FIGURE 4.13 A killer whale showing its tongue.

Exploring the Depths: Sleep

Cetaceans are voluntary breathers such that, unlike humans and most other mammals, they must be conscious to open their blowholes. A few species of cetaceans sleep as we would recognize it. Right whales, being very buoyant with a blowhole high out of the water on a dorsal ridge, appear to sleep for short periods. Occasionally, human whale watchers have been able to swim or kayak up to a sleeping right whale, and this sleeping behavior may be the root of folkloric stories of humans climbing onto rocks or small islands only to find they were in fact clambering onto whales that then sank beneath the surface.

Sperm whales with their ability to adjust their buoyancy and adaptations for diving appear to be able to rest, possibly sleeping, while submerged. Sperm whales are sometimes sighted floating immobile in the water, oriented vertically. When sleeping these whales would be less aware of their environment, and it is possible that high rates of boat collisions with right whales and sperm whales in some parts of the world might be in part due to resting whales being less able to react to oncoming boat traffic. Pinnipeds largely sleep like terrestrial mammals while on land, but they can also sleep at sea. Sometimes pinnipeds can be observed floating with their muzzles pointing straight up in the air—a behavior called "bottling"—when they appear to be sleeping.

However, in dolphins an unusual type of sleeping behavior has been observed. The animals effectively allow one half of their brain to "sleep" while the other half remains awake, and so the dolphins still consciously swim and breathe. This may in part explain why dolphins' brains are so large and complex, as certain parts of the brain may have to be duplicated.

SELECTED REFERENCES AND FURTHER READING

Clarke, M.R. (1970). The function of the spermaceti organ in the sperm whale. *Nature* 228: 873–874.

Clarke, M.R. (1978a). Structure and proportions of the spermaceti organ in the sperm whale. *Journal of the Marine Biological Association of the United Kingdom* 58: 1–17.

Clarke, M.R. (1978b). Physical properties of spermaceti oil in the sperm whale. *Journal of the Marine Biological Association of the United Kingdom* 58: 19–26.

Cranford, T.W. (1999). The sperm whale's nose: sexual selection on a grand scale? *Marine Mammal Science* 15: 1133–1157.

Davis, R.W., Williams, T.M., Thomas, J.A., Kastelein, R.A., & Cornell L.H. (1998). The effects of oil contamination and cleaning on sea otters (*Enhydra lutris*). II. Metabolism, thermoregulation, and behavior. *Canadian Journal of Zoology* 66: 2782–2790.

DeLong, R.L., & Stewart, B.S. (1991). Diving patterns of northern elephant seal bulls. *Marine Mammal Science* 7: 369–384.

Elsner, R. (1999). Living in water: solutions to physiological problems. In: *Biology of Marine Mammals* (Ed. by J.E. Reynolds & S.A. Rommel), pp. 73–116. Smithsonian Institution Press, Washington, DC.

Fernández, A., Arbelo, M., Deaville, R., Patterson, I.A.P., Castro, P., Baker, J.R., Degollada, E., Ross, H.M., Herráez, P., Pocknell, A.M., Rodríguez, E., Howie, F.E., Espinosa, A., Reid, R.J., Jaber, J.R., Martin, V., Cunningham, A.A., & Jepson, P.D. (2004). Pathology: whales, sonar and decompression sickness. *Nature* 428: 1–2.

Fernández, A., Edwards, J.F., Rodriguez, F., Espinosa de los Morteros, A., Herraez, P., Casstro, P., Jaber, J.R., Martin, V., & Arbelo, M. (2005). "Gas and fat embolic syndrome" involving a mass stranding of beaked whales (Family Ziphiidae) exposed to anthropogenic sonar signals. *Veterinary Pathology* 42: 446–457.

Gnone, G., Moriconi, T., & Gambini, G. (2006). Sleep behaviour: activity and sleep in dolphins. *Nature* 441: 10–11.

Goley, P.D. (1999). Behavioral aspects of sleep in Pacific whitesided dolphins (*Lagenorhynchus obliquidens*, Gill 1865). *Marine Mammal Science* 15: 1054–1064.

Heide-Jörgensen, M. P., & Dietz, R. (1995). Some characteristics of narwhal, *Monodon monoceros*, diving behaviour in Baffin Bay. *Canadian Journal of Zoology* 73: 2120–2132.

Hooker, S.K., & Baird, R.W. (1999). Deep-diving behaviour of the northern bottlenose whale, *Hyperoodon ampullatus* (Cetacea: Ziphiidae). *Proceedings of the Royal Society B* 266: 671–676.

Hooker, S.K., Baird, R.W., & Fahlman, A. (2009). Could beaked whales get the bends? Effect of diving behaviour and physiology on modelled gas exchange for three species: *Ziphius cavirostris, Mesoplodon densirostris* and *Hyperoodon ampullatus*. *Respiratory Physiology & Neurobiology* 167: 235–246.

Kerem, D., Kooyman, G., Schroeder, J.P., Wright, J.J., & Drabek, C.M. (1972). Hyperbaric oxygen-induced seizure in a marine mammal, the seal. *American Journal of Physiology* 222: 1322–1325.

Kooyman, G.L. (2009). Diving physiology. In: *Encyclopedia of Marine Mammals*, 2nd Edition. (Ed. W.F. Perrin, B. Würsig, & J.G.M. Thewissen), pp. 327–332. Academic Press, San Diego.

Lyamin, O. I., Manger, P.R., Sam, H., Ridgway, S.H., Mukhametov, L.M., & Siegel, J.M. (2008). Cetacean sleep:

an unusual form of mammalian sleep. *Neuroscience and Biobehavioral Reviews* 32: 1451–1484.

Lyamin, O. I., Mukhametov, L.M., & Siegel, J.M. (2004). Relationship between sleep and eye state in cetaceans and pinnipeds. *Archives italiennes de Biologie* 142: 557–568.

Lyamin, O.I., Pryaslova, J., Kosenko, P., & Siegel, J.M. (2007). Behavioral aspects of sleep in bottlenose dolphin mothers and their calves. *Physiology and Behavior* 92: 4725–4733.

Lyamin, O.I., Pryaslova, J., Lance, V., & Siegel, J.M. (2005). Animal behaviour: continuous activity in cetaceans after birth. *Nature* 435: 1177.

Lyamin, O.I., Pryaslova, J., Lance, V., & Siegel, J.M. (2006). Sleep behaviour: sleep in continuously active dolphins; activity and sleep in dolphins (Reply). *Nature* 441: E11.

Mass, A.M., & Supin, A.Y. (2009). Vision. In: *Encyclopedia of Marine Mammals*, 2nd Edition. (Ed. W.F. Perrin, B. Würsig, & J.G.M. Thewissen), pp. 1200–1211. Academic Press, San Diego

Mitchell, E.D. (2005). What causes lesions in sperm whale bones? *Science* 308: 631.

Moore, M.J., & Early, G.A. (2004). Cumulative sperm whale bone damage and the bends. *Science* 306: 2215.

Pabst, D.A., Ballantyne, C., & Merte, H. (2009). How do marine mammals avoid freezing to death? Do they ever feel cold? *Scientific American* 301(2): 80.

Sekiguchi, Y., Arai, K., & Kohshima, S. (2006). Sleep behaviour: sleep in continuously active dolphins; activity and sleep in dolphins. *Nature* 441: 9–10.

Stewart, B. (2009). Diving behavior. In: *Encyclopedia of Marine Mammals*, 2nd Edition. (Ed. W.F. Perrin, B. Würsig, & J.G.M. Thewissen), pp. 321–327. Academic Press, San Diego.

Talpalar, A.E., & Grossman, Y. (2005). Sonar versus whales: noise may disrupt neural activity in deep-diving cetaceans. *Undersea & Hyperbaric Medicine* 32: 135–139.

Thewissen, J.G.M. (2009). Sensory biology: overview. In: *Encyclopedia of Marine Mammals*, 2nd Edition. (Ed. W.F. Perrin, B. Würsig, & J.G.M. Thewissen), pp. 1003–1005. Academic Press, San Diego.

Thomas, J.A., & Kastelein, R.A. (1990). *Sensory Abilities of Cetaceans: Laboratory and Field Evidence.* Plenum, New York.

Tyack, P.L., Johnson, M., Soto, N.A., Sturlese, A., & Madsen, P.T. (2006). Extreme diving of beaked whales. *Journal of Experimental Biology* 209: 4238–4253.

Wartzok, D., & Ketten, D.R. (2009). Marine mammal sensory systems In: *Biology of Marine Mammals* (Ed. J.E. Reynolds & S.A. Rommel), pp. 117–175. Smithsonian Institution Press, Washington, DC.

Watkins, W.A., Daher, M.A., Fristrup, K.M., Howald, T.J., & Notarbartolo-di-Sciara, G. (1993). Sperm whales tagged with transponders and tracked underwater by sonar. *Marine Mammal Science* 9: 55–67.

Wickham, L.L., Elsner, R., White, F.C., & Cornell L.H. (1989). Blood viscosity in phocid seals: possible adaptations to diving. *Journal of Comparative Physiology B* 159: 153–158.

CHAPTER 5

Underwater Sound

CHAPTER OUTLINE

Basics of Sound Physics
   Exploring the Depths: What Exactly Is the Decibel?

Bioacoustics
   Exploring the Depths: Marine Mammal Sound Production

Marine Mammal Hearing
   Exploring the Depths: Marine Mammal Ears
      Cetacean Ears
      Pinniped Ears
      Other Marine Mammal Ears

Echolocation

Navigation

Sound and Communication

   Intrasexual Selection
   Intersexual Selection

   Exploring the Depths: Song and Sexual Selection

   Exploring the Depths: Humpback Whale Song

   Mother–Calf Cohesion
   Group Cohesion
   Danger Avoidance
   Individual Recognition

   Exploring the Depths: Prey Stunning?

Sound in the Marine Environment
   Exploring the Depths: What is the Problem with Noise?

Selected References and Further Reading

Sound travels faster in water (around 1,500 m/s) than in air (about 330 m/s), because sound travels faster in a medium that is less elastic, "stiffer" (a factor related to the strength of bonds between molecules), and less compressible, and water is much less compressible and elastic than air. Sounds can also travel substantial distances in water because of low attenuation of sound in water when compared with in air. Attenuation is where there is a decrease of sound intensity as sound energy is "lost" by being transformed into other types of energy as the sound wave travels. As mentioned in Chapter 4, light in the ocean decreases with depth, and factors such as plankton concentration and suspended sediments increase the turbidity (i.e., muddiness) of water and further reduce the ability to see in marine waters. Therefore, many marine mammals use sound and hearing instead of sight as their primary sense (through echolocation in odontocetes) to navigate and to communicate underwater. Toothed whales can detect a wider frequency range of sound than humans can, with upper auditory limits roughly five times higher than in humans. Some cetaceans produce sounds that can be heard over hundreds of kilometers.

Basics of Sound Physics

In water, propagating (i.e., traveling) sound waves consist of alternating compressions and rarefactions (i.e., extensions or decompressions) between the water molecules. Waves essentially consist of alternating compressive pulses of high and low pressure (discussed later in this section). The amount of pressure is one way of defining the amplitude (or size) of the sound wave. This is generally represented by the vertical size of the signal on the page (**Figure 5.1**). However, this can be measured by:

1. Measuring the largest "height" of the sound wave peak-to-peak;
2. Measuring the largest "height" of the sound wave zero to the highest peak (known as peak pressure); or

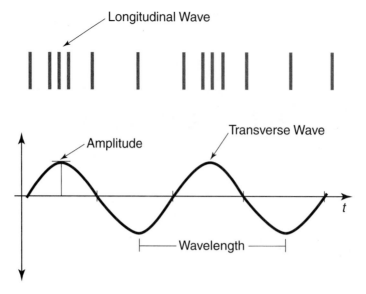

FIGURE 5.1 A diagram of a horizontal wave showing amplitude and wavelength over time (t). Transverse waves occur when a disturbance creates oscillations perpendicular to the direction of propagation. Longitudinal waves occur when oscillations are parallel to propagation.

3. Measuring the "average height" of the sound wave over a specified time period (which is, unfortunately, often not reported in much of the marine mammal literature), which is known as root mean square (or "rms") pressure.

Transient sounds are often measured differently by their highest pressure level, or peak level. The wavelength (the distance between the peak of one wave to the peak of the next) of sound in water can be calculated from the equation:

$$\lambda = c/f$$

where λ is wavelength (m), c is the speed of sound in water (m/s), and f is the frequency (how many waves are produced in 1 second, measured in Hz).

Musical notes are made of different frequencies of sound, which are commonly referred to as their "pitch." For example, an A above middle C has a frequency of 440 Hz, but an A an octave higher has a frequency of 880 Hz. A sound that consists of only one frequency is said to be tonal. Sounds consisting of pure tones (i.e., a single frequency) are very rare in nature and typically can only be produced by electronic equipment. Even the best musical instruments produce a range of frequencies when a note is played, rather than a single tonal frequency. Sound that incorporates such a small range of frequencies is said to be narrowband. Conversely, a sound that consists of a wide range of frequencies is said to be broadband. A sound that changes in frequency with time, such as a police siren, is said to be frequency modulated. Finally, the range of frequencies in a sound is known as the spectral content of the sound. The frequency spectrum is often presented visually through use of a spectrogram, and is usually determined by applying an algorithm known as a fast Fourier transform (FFT) to a recording. Different species of marine mammals produce sounds of different spectral content and often use different frequencies for different functions. A selection of sound spectra for odontocete cetaceans is shown in **Table 5.1**.

The absolute intensity (or "power") of sound, or the average amount of energy passing in a specified direction through a unit area per unit time, is measured in Watts/m². Sound intensity measured in this way can be compared between media. For example, the energy in sound underwater can be compared with the energy in sound in air. The level of sound is often expressed using the decibel scale (see equation below), which is itself a comparison between the intensity of a sound and that of a reference value. One further complication arises because scientists studying sound in water have traditionally used a different standard reference intensity from those working with sound in the air. Put simply, this means that decibel data on sound sources measured in air have to be recalculated when considering levels in water, and thus effects and measurements of airborne noise are not directly equivalent to waterborne sound. This makes comparisons between the level of sounds in air and water somewhat more complicated than first appears.

The human ear can detect sound intensities in air from as low as 1 pW/m² (one picowatt or 1 million millionth of a watt per meter square) with sound levels becoming painful at 10 W/m². In judging relative loudness, our hearing system (from ear to brain) translates sound intensities logarithmically. To reflect this process acousticians use a logarithmic scale to measure relative sound intensity known as the decibel (dB) scale:

$$\text{Sound intensity (dB)} = 10 \log \left[\frac{I}{I_0} \right]$$

where I_0 is the reference intensity (e.g., 1 pW/m) and I is the intensity of the signal (W/m²).

TABLE 5.1 **The Range of Sound Frequencies Used by Cetaceans.**

		Spectrum (kHz)
Beluga whale	Whistles	0.26–20
	Pulsed tones	0.4–12
	Misc. vocalizations	0.5–16
	Echolocation clicks	40–60, 100–120
Common dolphin	Whistles	2–18
	Chirps	8–14
	Barks	<0.5–3
	Clicks	0.2–150
Long-finned pilot whale	Whistles	1–8
	Clicks	1–18
	Echolocation clicks	6–11
Killer whale	Whistles	1.5–18
	Clicks	0.1–35
	Screams	2
	Pulsed calls	0.5–25
	Echolocation clicks	12–25

The decibel value should be followed with the notation relating to the reference values, such as "… dB re: 1 µPa," although this convention is not always followed in older literature and some less helpful modern papers. In the example, the notation indicates the sound level in terms of decibels is relative to the reference ("re") value of the intensity of sound at the pressure level of 1 micropascal, the standard reference value for sound in water. However, the standard reference pressure in air is 20 µPa. As noted above, there are also differences between the measurement of sound in water and air because of differing physical properties of the two media. As a result, while the absolute intensity (in Watts per unit area) may be the same, the relative intensity on the decibel scale typically differs between sound in water and sound in air by 61.5 dB (35.5 dB from the properties of the media and 26 dB due to the typical reference values). Consequently, a sound in water would need a relative intensity of 201.5 dB re: 20 µPa to have the same effective "loudness" as a sound in air with a relative intensity of 140 dB re: 20 µPa. Consequently, it is important to be exceedingly careful about comparing the intensity of sound in water with sound in air.

Sounds can be of (relatively) short duration or transient, typically having obvious starting and stopping points. Alternatively they can be of longer duration or continuous, with no obvious starting and stopping point. A possible subcategory of transient sounds are impulsive sounds, which are very short duration, broadband sounds, such as an underwater explosion. In contrast, nonimpulsive transients (such as those produced by sonar systems) and continuous sounds (such as those produced by the ocean-wide low-frequency hum of global marine shipping activity) tend to have much more limited frequencies.

The levels of continuous sounds are generally described in terms of their intensity at a root mean square pressure (explained above). Intensity in transient sounds is usually measured differently by determining their peak (or peak-to-peak) pressure level, relative to a sound at the reference pressure level. As a result, there may be large differences between these two measurements, even though they might both be expressed in terms of decibels! Put simply, it is generally inadvisable to try to compare sounds measured

Exploring the Depths: What Exactly is the Decibel?

Many people find it difficult to understand the dB (deciBel), as well as units in acoustics in general. This has the consequence that not everything one reads in books and papers is correct, causing more confusion and creating a feedback loop of misunderstanding. The most important thing to understand is that **the dB is not a unit**. It is a relative measure, much like percentages. Most people find it clear that, for example, 5% income tax cannot in any meaningful way be compared with 5% alcohol content. Just remember that exactly the same applies to dB.

–Contributing author, Jakob Tougaard, Aarhus University.

in these two different ways to avoid comparing "apples with oranges."

Sound generally moves out from a source evenly in all directions (if not focused or impeded) in a spherical manner. As it does so, the total energy contained within the sound is spread over an increasingly large area, leading to the reduction in the pressure at any specific point on the surface of the expanding sphere. This reduction in source level due to distance is known as geometric spreading loss. (It should be noted that energy isn't actually "lost" as a result of this. It is simply spread over a greater area, like spreading a lump of butter over a piece of toast). Additional losses also result from absorption, where some of the sound energy is transformed into heat during transmission. Higher frequency sounds are more susceptible to absorption, while little absorption occurring for very low frequency sounds. Combined, the loss of energy to spreading and absorption is known as the transmission loss. Additionally, the ocean environment is full of things that can impede the propagation of sound.

Sound, like other waves, is reflected or refracted when it comes in contact with a medium with different physical properties. For underwater sound this could be the surface of the ocean, the seabed, or the boundaries between layers in nonhomogeneous (i.e., not fully mixed) waters, such as when fresh water remains above saltwater in an estuary or warm, less dense surface seawater sits above cooler, more dense seawater. Sound is reflected when the sound waves bounce off such a surface, like waves of light bouncing off a mirror. Accordingly, sound can bounce off the seabed or the ocean surface. In contrast, refraction entails the "bending" of a wave, slightly changing its direction, as it passes through substances of differing densities and thus also different transmission properties. This occurs with light too, such as when it passes through thick glass or from air into water, perhaps causing a distortion of the image. Thus, sound often "bends" as it passes from warmer surface water to denser bottom water. As a general (although not exclusive) rule, sound reflects at boundaries where the speed of sound changes greatly and refracts at boundaries separating media with similar speeds of sound.

In the open ocean, an unfocused sound produced at depth could initially radiate away from the animal spherically. Spreading loss (in terms of decibels in whatever form lost at a specific point on the sphere) in these conditions is calculated as follows:

$$\text{Spreading loss (spherical)} = 20 \log \left[\frac{R}{R_0}\right]$$

where R is the range (m) and R_0 is the standard reference range (often 1 m from the source). Therefore, a sound at any point on the sphere will be 6 dB (allowing a consistent amplitude measurement and reference value) lower at a distance twice as far away from a source and be 20 dB ten times as far away. Of course, absorption will reduce this further, especially for higher frequency sounds where it will become greater than spreading loss after just a few hundred meters. However, this is highly dependent upon the frequency of the sound and the exact density of the transmission media, which in turn can be affected by temperature and other conditions. (Urick, 1983 offers the relevant equations for those who really want to know.)

Sound can also be produced in a focused manner (as is most often the case) or constrained within a layer, such as between the surface and the seafloor, or some other reflective layer. This is especially true in "shallow" waters (shallow here being relative to wavelength). As a result sound often travels in more of a cylindrical fashion and spreading loss is calculated using a different equation:

$$\text{Cylindrical spreading loss (dB)} = 20 \log R_1 + 10 \log \left[\frac{R}{R_0}\right]$$

between the starting point of the sound at distance R_1 and distance R (nb R_1 is the distance in meters when spherical spreading stops and cylindrical spreading starts). Under these conditions, sound pressures may decrease by only 3 dB with a doubling in distance and by 10 dB if distances increase by a factor of 10. This means that higher frequency sound can, under certain conditions, effectively travel twice as far in shallow waters as it would in deeper waters and retain the same amount of energy. However, sounds at lower frequencies begin to interact with the seabed, dramatically increasing losses and thus cannot propagate effectively in shallow waters at all.

Bioacoustics

Exploring the Depths: Marine Mammal Sound Production

In 1966 Jarvis Bastian noted that the "sensory communication between individuals is an integral part of any social mode of adaptation." Indeed, marine mammals not only produce sound for the purpose of communication but, like many other marine animals, also generate sounds for hunting and navigating. In fact, marine mammals, particularly cetaceans, have extremely highly developed acoustic sensory systems.

Just like humans, and most other mammals, polar bears, sea otters, and pinnipeds produce at least the vast majority of their vocal sounds by vibrating their vocal folds in the larynx. These sounds may resonate in the sinuses and be modified by the movement of the mouth. However, the cetacean, and especially the odontocete, larynx has some substantial

Exploring the Depths: Marine Mammal Sound Production (*continued*)

physical differences, including a lack of vocal chords of the type found in terrestrial mammals. Furthermore, cetaceans routinely emit sounds underwater without letting out air; thus, they use a closed vocalization system quite different from that normally used by terrestrial mammals, as described above.

Consequently, heated debate began in the 1960s about exactly how cetaceans produced sound, and the scientific literature was totally divided. Some scientists located the sound-producing mechanism among the various pneumatic chambers found in the upper part of the nasal passages just below the blowhole, whereas others identified the larynx as the source. Several chose to straddle the fence and championed a dual system, suggesting whistles were generated in the larynx and echolocation clicks in the nasal sacs.

Over time, the nasal passage location theory grew in popularity as the role of the "melon" (the globular fatty organ that gives odontocetes their domed foreheads) in sound production was unraveled. Sound produced by any source radiates out spherically (or cylindrically), but to be useful for echolocation (see Echolocation section) it would need to be focused into a beam. Researcher Ken Norris proposed that the melon could act as an "acoustic lens" with various asymmetrical tissues and air spaces acting as reflecting surfaces. Later work confirmed that the core of the melon is made of less dense fat, with a slower rate of sound transmission, than the outer layers. Consequently, as sound passes through this area to the outer layers of denser fats, it is refracted (i.e., the sound waves are deflected to the side, or bent), narrowing and focusing the echolocation beam, a little like a magnifying glass can be used to focus a beam of sunlight. More recently, a high-density tissue encapsulating the posterior region of the melon, called the connective tissue theca, has been proposed as a reflective surface to help focus the sound energy forward. It is also likely that the sound source itself is somewhat directional as well.

Eventually, a generalized theory of click production across all odontocetes came together. Immediately below the blowhole is an air sac, called the ventral vestibular air sac. Immediately below the slit-like opening to this sac (and typically right behind the melon) are structures that have now become known as the phonic or phonetic lips. (Before their function was determined, they were called the "museau de singe," or "monkey lips," because of their appearance) (**Figure B5.1**). These lips are valves made of ridges in the anterior and posterior walls of the nasal passages or, in sperm whales (which have only one set), they consist of denser connective tissue. To produce clicks air is pushed through these phonic lips at pressure, vibrating not only the air (which actually appears not to contribute to the sound that leaves the head of the animal at all), but also the lips and surrounding tissue. The sound is then transferred through the tissues and associated fat bodies into the melon for eventual transmission into the water column. The rate of repetition can be controlled by the air pressure and the tension across the vibrating portion of the phonetic lips. The air used to produce the sound is captured in the ventral vestibular air sac and recycled into the nasal cavity as needed.

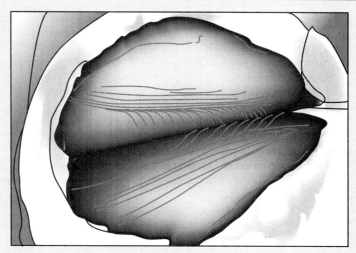

FIGURE B5.1 A drawing showing the phonic lips of a sperm whale with the anterior tissue cut away (the white area).

The dominant frequency of any sound partly depends on the size of the sound source. The phonic lips are almost symmetrical in porpoises and some have suggested that this could produce the observed narrowband echolocation clicks when used in unison. Similarly, the right phonic lips are twice the size of the left in most delphinids, which might explain the production of two different types of signals containing two dominant frequency peaks (for example, clicks for echolocation and whistles for communication). However, this is the subject of another heated debate as other scientists believe only one pair of phonic lips is used in any odontocete and the signal structure results from the physical structures in the sound production system.

This discussion is redundant in the sperm whales, in which only the right pair of phonic lips can be recognized. However, this is not the only odd characteristic of the nasal anatomy of sperm whales. The blowhole in the physeterids is situated in front of the "spermaceti organ" (thought to be an extremely hypertrophied part of the right phonic lips complex) and the "junk" (homologous to the melon in other odontocetes) in these species (**Figure B5.2**). This led Bertel Møhl and colleagues to propose the now-accepted theory (based on the earlier ideas of Ken Norris and George Harvey) that the air sacs that partly enclose the anterior (front) of the spermaceti organ send the sound produced by the phonic lips backwards at first. The sound travels through the spermaceti organ until it is reflected by the air sacs at the posterior end. It then moves forward through the junk and out into the marine environment in a highly focused beam. Some of the energy also moves directly out from the whale's head into the environment, whereas more rebounds back and forth inside the spermaceti, uniquely generating multiple pulses within a single click (**Figure B5.3**).

Although the origin and mechanism of click production has been largely settled in odontocetes, the site of production of tonal sounds, such as whistles and moans in both odontocetes and mysticetes, has yet to be unequivocally

Exploring the Depths: Marine Mammal Sound Production (*continued*)

FIGURE B5.2 Photos showing that the blowhole in (A) sperm whales is situated in front of the melon (also called the spermaceti organ) and in dolphins (B) it is situated behind the melon.

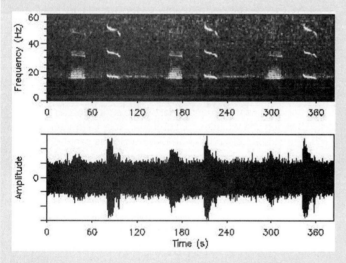

FIGURE B5.3 A spectrogram image of sperm whale clicks.

lungs offer a big enough air space to reach the extreme low-frequency sounds produced by blue whales. Additionally, sound production systems relying on resonance of air spaces would produce different frequencies at different pressures at different depths, but this runs contrary to observations that mysticete signals appear to be consistent in frequency, irrespective of depth. Therefore, the search for a site of sound production continues. Little is known about sound production anatomy in sirenians.

determined. Evidence is mounting in support of whistle production in odontocetes also occurring at the phonic lips. However, as the larynx of mysticetes is in many ways more similar to that of terrestrial animals than the odontocetes' larynx, many scientists believe their vocal folds are the source of their low-frequency calls. The issue here is that although sound produced by the vocal folds may indeed resonate in the laryngeal air sac and be transferred to the water without distortion through the throat pleats of the rorquals, only the

–Contributing author, Elly Roland, George Mason University.

–Contributing author, Andrew Wright, Aarhus University.

Marine Mammal Hearing

Marine mammals have a broad spectrum of hearing abilities. Collectively, marine mammals have functional hearing ranges spanning from probably below 10 Hz up to as high as 200 kHz in some animals (**Figure 5.2**). Mysticetes are thought to hear across frequencies ranging from (less than) 10 Hz to 20 to 30 kHz, with excellent infrasonic hearing. Moving up in frequency, otariids are thought to hear best a little below 10 kHz in air and with peak sensitivity between 15 and 30 kHz in the water. Phocid hearing is believed to be similar to otariid hearing underwater, with greatest sensitivity at frequencies of 10 to 30 kHz in the water. However, phocid hearing in air is thought to peak at generally lower frequencies than otariids, between 3 and 10 kHz. Manatees seem to be most sensitive to sound at frequencies from 16 to 18 kHz, meaning their hearing overlaps with that of pinnipeds. Finally, odontocetes as a group hear over the widest range of frequencies, from 200 Hz to 180 kHz and even 200 kHz in some cases.

The acoustic properties of tissue is very similar to that of water. Thus, animals in the water do not need to channel sound waves in toward the inner ear, as terrestrial mammals do. Sound waves in marine mammals travel through the body tissue and cause vibrations of the skull and ear bones directly. The problem, therefore, for marine mammals is isolating the inner ear from the rest of the body and skull so that directional hearing can function properly, as well as for echolocating odontocetes to avoid damaging their own hearing with every pulse. Data on sea otter and polar bear hearing are limited, so the discussion below focuses on cetaceans and pinnipeds.

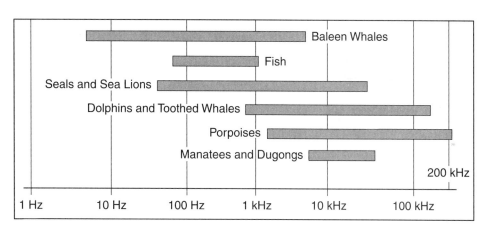

FIGURE 5.2 A diagram showing frequency ranges of marine mammal sound. Modified from B. Southall, NMFS/NOAA.

Exploring the Depths: Marine Mammal Ears

Cetacean Ears

During their evolution cetaceans have completely lost their external pinnae (ear flaps). In addition, cetaceans have no air-filled external ear canals, which might be problematic for diving animals facing rapid changes in pressure. Instead, the vestigial external ears of cetaceans are completely blocked by wax (in mysticetes), or closed completely (in odontocetes). Odontocetes are known to have developed an alternative primary path of sound transmission to the middle ear (**Figure B5.4**). In this case sounds enter the ears very efficiently through transfer from the lower jaw. The lower jawbones of odontocetes are quite thin in specific areas and are associated with fatty tissues that, in combination, act as an acoustic window channeling sound to the middle and inner ears (the auditory bullae). The auditory bullae (bony shells or sheaths that surround the ear bones of cetaceans) are also positioned such that they are perfectly aligned with the fatty channel from the jaw. It is not known yet how mysticetes transmit sound from the ocean to their auditory bullae, although other paths of sound transmission have been proposed in various odontocetes (including through the upper jaw and through the gullet between the lower jawbones).

Although the auditory bullae are made of dense bone in all mammals, these are especially dense in cetaceans. More importantly, two of the bones most involved in hearing, the tympanic and periotic bones, have become separated from the rest of the skull, forming structures known as the tympanic bullae. In odontocetes, the bullae are completely isolated from

Exploring the Depths: Marine Mammal Ears (continued)

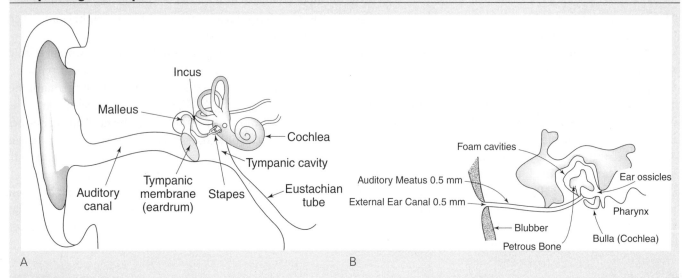

FIGURE B5.4 (A) The human ear and (B) a dolphin ear.

the rest of the skull by ligaments, mucosa, and air sacs. This is thought to facilitate directional hearing, but it may also reduce the transmission of self-produced sound directly to the ears. (How directional hearing works on land and underwater is discussed later in this section.) In mysticetes the isolation from the skull is not complete, providing some hard tissue as well as soft tissue sound channels to the inner ear.

In all mammals the structure in the mammalian cochlea, inside the inner ear, that determines the ranges of frequencies ultimately processed by the brain is known as the organ of Corti. The organ of Corti contains the basilar membrane, the properties of which determine the frequencies to which an ear is sensitive. In odontocetes the basilar membrane is exceptionally stiff, allowing them to perceive and process the high, ultrasonic frequencies used in echolocation. The stiffness of the basilar membrane is a trait that is also found in bats. In mysticetes, the basilar membrane is less stiff and more adapted to respond to low and infrasonic frequencies. Cetaceans also have a very high number of nerves associated with the inner ear. Their auditory ganglion cell counts range from 68,000 (harbor porpoises) to 160,000 (fin whales) compared with around 32,000 in humans. In odontocetes this level of "neural investment" has been assumed to be needed for processing echolocation, but this may not be the case given that similar levels are also found in mysticetes.

One final adaptation in cetacean ears to hearing underwater can be found in how the sound is transmitted across the middle ear. In humans the bones involved (the ossicles) amplify the pressures on the eardrum to improve hearing sensitivity. However, pressures are much greater at depth, meaning that the changes in pressure are much smaller relative to the overall pressure. Accordingly, the prevailing theory is that the differences in the ossicles and their arrangement in cetaceans have arisen to amplify and transfer, through a lever system, the motion of particles in the sound wave (known as the particle motion) to the inner ear, rather than the pressure itself.

Pinniped Ears

Unlike cetaceans, pinnipeds need functional hearing in two very different media. However, they are only able to optimize for hearing in either air or water, although it is possible that a single hearing system could adapt to hear reasonably well in both media.

When on land pinniped ears work in the same way as those of terrestrial mammals in that sound is received by the outer ear, channeled up to the middle ear, transferred to the inner ear, and converted to neurological impulses. The only difference here between phocids and otariids is that the otariids have not completely lost their external pinnae, which may assist them in localizing aerial sound sources. All pinnipeds, except for elephant seals, have good hearing in air, although not necessarily across a wide range of frequencies.

In water, the differences between phocid and otariid hearing become more apparent. Otariids have few anatomical differences from the standard terrestrial carnivore auditory system, apparently representing a trade-off in favor of hearing in air for these animals. In contrast, phocids have completely lost their external pinnae and have a ring of muscle that allows them to close the entrance to the ear canal when in water. The inner ears of phocids are also adapted for underwater hearing with ossicles that are 10 times larger than those of terrestrial mammals with similar skull size. This is thought to improve bone conduction of sounds by phocid ears underwater, although the large ossicles limit the ability of these animals to hear higher frequencies in air. Thus, phocids are constrained with lower peak hearing frequencies (3–10 kHz) in air than in water (10–30 kHz).

Phocids also exhibit two further adaptations to hearing underwater. Firstly, it seems that, like odontocetes, they have developed a more direct path for sound to enter the inner ear underwater. However, in phocids underwater sound enters through a specialized portion of the skull just below the ear canal. Secondly, there is cavernous tissue (tissue that can be

Exploring the Depths: Marine Mammal Ears (*continued*)

filled with blood, such as is also present in the erectile tissue of the penis) in the phocid external ear canal and middle ear. This is thought to resolve the issue of pressure equalization during diving.

Other Marine Mammal Ears

Little research has been done on the hearing of polar bears and sea otters. Polar bears actually do not spend a lot of time underwater, making it unlikely that their ears would be highly adapted for underwater hearing. Similarly, the anatomy of sea otter ears is not substantially different from that of terrestrial carnivores, which may represent a lack of adaptation to hearing in the marine environment. However, sea otters do fold their external ears down when diving, which may mean the ear canal is closed and does not fill with water.

—Contributing authors, Elly Roland and Andrew Wright, George Mason University and Aarhus University.

Echolocation

Echolocation is the ability by which animals can produce sounds (generally, but not exclusively, at high frequencies) and detect the echoes of these sounds that are reflected from objects in the environment. This allows echolocators to determine some of the physical features of their surroundings. In the terrestrial environment microchiropteran bats are the most noteworthy animals using echolocation. Of the marine mammals only odontocetes are definitely known to echolocate.

As mentioned above, the echolocation clicks (and also their other sounds) produced by odontocetes are formed in the complex system of air sacs and valves that sits below the blowhole and is connected to the animal's windpipe. These high-frequency clicks pass through, and are focused by, the melon (**Figure 5.3**). The sounds are ultimately emitted in a beam ahead of the animal. Objects in the environment reflect these clicks; the cetacean detects the echoes and is able to interpret them.

Echolocation allows odontocetes to detect objects only a few centimeters in size at distances greater than 10 m. They can distinguish small differences in the composition of their targets due to differences in density, which influence transmission and reflective properties. The optimum frequency of echolocation clicks used by odontocetes depends on the size of the object to be detected. If an object is smaller than the wavelength of the echolocation clicks, it becomes difficult to detect. Therefore, the size of preferred prey of cetaceans determines frequency of echolocation clicks. For example, to detect a 1-cm-circumference solid object a cetacean would need an echolocation click of a certain frequency, which we can calculate using the formula $f = c/\lambda$. Solving this equation for a minimum wavelength of 1 cm gives us $f = 1,500/0.01 = 150,000$, or 150 kHz. A cetacean would therefore need echolocation clicks ranging up to 150 kHz to detect the 1-cm-prey item. However, this is a very simplified explanation. Animals can detect targets that are much smaller than the equation suggests,

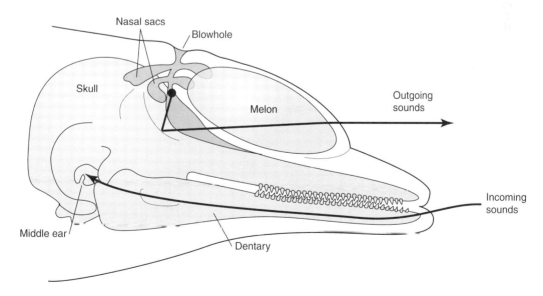

FIGURE 5.3 Cross-section of a dolphin's head showing the transmission of echolocation clicks to the water through the melon and the hearing of the returning echos through the jaw. Adapted from Thomas, J. and Kastelein, R., eds. *Sensory Abilities of Cetaceans: Laboratory and Field Evidence.* Plenum Press, 1990.

provided they are close enough. But you should now be able to understand why higher frequencies allow the detection of smaller objects. If higher frequency echolocation clicks have much finer resolution, however, should not all odontocetes just produce high-frequency clicks? Why do they produce lower frequency sounds? It is because high frequencies attenuate (lose and dissipate energy) more quickly and therefore do not travel as far as low-frequency sounds. Thus, there is a compromise between the distance at which prey can be found and the smallest size of prey that can be detected. As a result, the sound frequency that odontocetes use for echolocation clicks can (although not always) give us an indication of the sizes of their prey items.

However, in an area with a lot of background noise, odontocetes have been found to modify the frequency of their echolocation clicks, so their clicks are not obscured or masked by the noise in the environment. The problem is that, while such frequency changes may allow the clicks to better stand out from background noise, it also changes the range or resolution of the clicks. Either may reduce the efficiency with which a cetacean can find food or acoustically "see" in the marine environment.

Navigation

Mysticete cetaceans are known to produce low-frequency calls of a high source level (i.e., sounds that are typically loud but low). As mentioned above, the higher the frequency, the more energy that is dissipated into water with distance. Consequently, the lower the frequency, the further

sound can travel. It has been estimated that a 20-Hz call from a fin whale could potentially be detected hundreds of kilometers away.

An oceanographic effect is also sometimes used by mysticetes that allows the sounds they produce to travel even further. In areas where stratification occurs (i.e., there are water layers of differing temperatures and densities), sound waves tend to get trapped within these layers, reflecting and refracting at the layer boundaries. The sound waves become concentrated and channeled over great distances, in effect similar to light traveling through fiber optics. This effect is particularly heightened for low-frequency sounds due to their very low attenuation (such as at 15–30 Hz, the spectrum of many mysticete whale calls).

The depth of water where the effects of pressure, temperature, and water salinity combine to produce the best conditions for transmission of sound is known as the SOFAR (SOund Fixing And Ranging) channel. This layer of water was discovered in the 1940s and has been used since by submarines to assist transmission of sonar and communication signals.

It has been proposed that mysticetes could use their low-frequency calls for navigation, like a long-distance form of echolocation, bouncing the sounds back from large physical structures (such as the seabed, islands, submerged seamounts, the continental shelf, or ice edge) instead of food items (**Figure 5.4**). This, combined with passive listening to the acoustic scene (known as "auditory scene analyses"), may allow whales to navigate over long migrations across entire ocean basins.

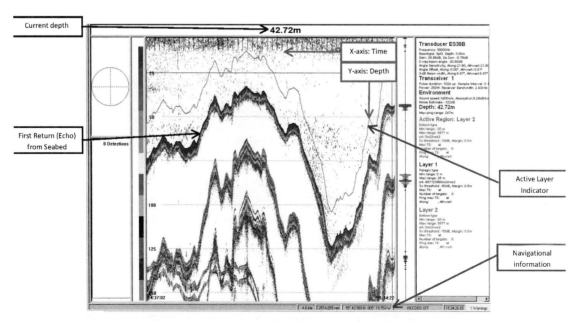

FIGURE 5.4 An echogram of water column and seabed taken at the Isle of Arran, Scotland from R.V. *Aora*, University Marine Biological Station, Millport.

Sound and Communication

In addition to using sound for finding food and navigation, sound can also be used for communication (which can be defined as the transmission of information from one individual to another). Within marine mammals sound-based communication can have a variety of functions:

- Intrasexual (male–male or female–female) display and selection
- Intersexual (male–female) display and selection
- Mother–calf cohesion
- Group cohesion
- Alerts and danger avoidance
- Individual recognition

Intrasexual Selection

Sound can be used in marine mammals to compete against other members of the same gender, to establish dominance hierarchies and also to establish territories. A good example is the elephant seal roar, where the males' trunk-like proboscis resonates and amplifies their calls. It has been suggested that humpback whale song might also be used for intrasexual selection because only male humpback whales sing. This song appears to maintain distances between whales, and aggressive interactions have been observed between singers and other males. Similar behaviors, with respect to spacing of whales, have been observed in calling northern minke whales.

Intersexual Selection

Many vertebrate species use visual displays as a means of displaying their fitness, via their ability to maintain a costly display such as a rack of antlers or the spectacular tail of a male peacock. Males can also use vocal calls as a means of demonstrating their fitness to females; for example, the volume of roar produced by red deer (*Cervus elaphus*) males

---

Exploring the Depths: Song and Sexual Selection

In his book, *The Descent of Man and Selection in Relation to Sex*, Charles Darwin defined sexual selection as the "struggle between the individuals of one sex, generally the males, for the possession of the other sex." Sexual selection involves competition and aggression between members of the same sex to gain access to mates (intrasexual selection) or selection between sexes based on some sort of mate choice (i.e., intersexual selection). Darwin (1871, p. 87) explained it as follows:

> The sexual struggle is of two kinds: in the one it is between the individuals of the same sex, generally the males, in order to drive away or kill their rivals, the females remaining passive; while in the other, the struggle is likewise between the individuals of the same sex, in order to excite or charm those of the opposite sex, generally the females, which no longer remain passive, but select the more agreeable partners.

Examples of sexual selection in pinnipeds include male elephant seals fighting each other to maintain their dominance over females (intrasexual selection) (**Figure B5.5**). Male hooded seals blow a bright red membrane out of their noses, which they wobble as a display to females. Females choosing males on the basis of this display are an example of intersexual selection. Some displays might incorporate both inter- and intrasexual selection, simultaneously advertising fitness and strength to both males and females.

FIGURE B5.5 (A) Two male elephant seals fighting and (B) two male hooded seals displaying.

is linked to their size and fitness. Fitness is the group of evolutionary characteristics of an individual that make it best adapted to the environment. "Fit" is not used in the sense of physical stamina, such as would be found in an athlete, but derives from best "fitted" to the environment.

Choosing a male based on his roar allows female deer to choose the best male possible to father her offspring, increasing the likelihood she will have healthy offspring, including larger, more attractive males that would, in turn, increase her potential for a large number of second generation offspring (which we would call grandchildren in humans). It has been suggested that the tusks of narwhals, which are usually only present in males, might be a display for females akin to that of a peacock's tail. Similarly, female humpback whales may assess the fitness of males via their song. It has been suggested that song length is an indicator of male breath-holding ability and therefore male fitness. But other researchers have suggested that other whale song might demonstrate "mental" fitness of male whales (i.e., the ability to innovate or remember), which might be an important ability for humpback whales (see Exploring the Depths: Humpback Whale Song).

Exploring the Depths: Humpback Whale Song

One of the most interesting behaviors of humpback whales (**Figure B5.6**) is their singing. This, arguably most famous, whale behavior was only discovered by scientists in the early 1970s. Only male humpback whales sing, and although song is heard sporadically in summer feeding grounds, whales primarily sing in, or when approaching, winter breeding grounds.

It has been suggested that whale song is a means by which males stake out "territories" (or perhaps "maritories" as they are in the sea and not on land) because playbacks of whale song seem to affect the spacing of whales. However, singing males frequently join with females before initiating mating behavior, so singing may be part of a courtship display or in addition to it.

A few researchers have suggested that whale song is a form of sonar that allows males to find females, but others have argued that characteristics of the song and whale behavior do not support this hypothesis. Because songs are sung louder at night, it has also been suggested that singing may influence courtship associations in the daytime. It has even been suggested that songs are a record of patterns of association, and possibly cooperation, between males. Most scientists agree, however, that humpback whale song

FIGURE B5.6 Two humpback whales.

Exploring the Depths: Humpback Whale Song (*continued*)

plays a role in intrasexual or intersexual selection or possibly both.

A typical humpback whale song lasts from 10 to 20 minutes and is repeated for hours at a time. The whale song changes continuously (evolves) throughout the breeding season. However, the song is exactly the same at the beginning of one breeding season as it was at the end of the previous breeding season, so that the state of the song appears to be "remembered" between seasons. This would be explained if certain sections of the brain related to song learning and production, but not memory, were to atrophy (i.e., shrink in size) out of season, as has been recently suggested. Within 5 years the songs have changed and evolved so much that it is completely different from the original song.

Each breeding population of humpbacks sings a different song, although neighboring populations share some song characteristics, but there is a little overlap. Scientists observing one particular event reported that when outsider males joined a breeding population other males in the population copied the outsiders' songs. This copying is called song matching and is a phenomenon also seen in several bird species. One hypothesis for changing the nature of humpback song is that it allows the singer of the changed song to stand out from the other males, giving an advantage, which in turn other males attempt to gain by copying and matching the new song, and so the song keeps changing and evolving.

In many species of mammals females choose mates as the result of a display or signal that indicates the fitness of a male. By choosing a male with a high level of fitness to mate with, the chances are reasonably good that the male's high fitness level will be passed on to the next generation, giving offspring greater chances of survival and in turn better reproductive success for the female. Humpback whale song may indicate good physical condition (i.e., breath-holding ability), mental fitness (e.g., as an indication of memory and learning abilities, which could be important for remembering migration routes and finding feeding locations), or perhaps some other aspect of fitness. It has also been hypothesized that by an individual changing and singing new components of a song, it displays to females a male's cognitive or innovate ability. The innovative male may then become "popular" with females, resulting in other males copying the innovative singer to increase their chances of mating, which would again lead to an evolution in the song structure of the population.

Mother–Calf Cohesion

One of the most important social bonds in marine mammals is that of mother and her offspring. For some species of cetaceans a calf may stay with its mother for up to a decade learning important life skills such as foraging and social behavior. Therefore, effective communication is required between the mother and her offspring (**Figure 5.5**). Bottlenose and other dolphins communicate with their offspring with unique whistles (known as contact calls or signature whistles). If a mother and calf are separated, they both start whistling at high rates. Calves will also whistle to bring the mother closer. A study on Indo-Pacific humpback dolphins in Australia found that mother–calf pairs significantly increased whistling rates when disturbed by boat traffic, possibly to increase cohesion of these pairs when faced by a possible threat. Similarly, recent work in harbor porpoises indicates that mothers use particular click patterns to call their calves closer.

Group Cohesion

Many odontocete cetaceans, and even some mysticetes such as humpback whales, forage in groups. Group predation allows cetaceans to catch larger prey than they would individually (e.g., killer whales attacking baleen whales) and increases success rate (e.g., dolphins schooling shoals of herring). To effectively coordinate these activities, good group communication is required. For example, fish-feeding specialist killer whales in the eastern north Pacific form large social groups. These groups possess distinct calls that are unique to group members, and killer whales produce different call types according to their behavior (i.e., foraging or resting), with certain specialized calls used almost exclusively during specific activities, such as schooling fish.

Danger Avoidance

Many vertebrates produce alarm calls to warn of danger. In some species these calls provide information as to the type of threat so the animals can engage in an appropriate response. For example, vervet monkeys (*Chlorocebus pygerythrus*) produce different calls whether a leopard, a snake, or an eagle has been spotted, which leads to the monkeys adopting different avoidance behaviors.

Several studies have shown increases in certain odontocete calls believed to be "alarm" calls in the presence of a particular threat. For example, specific calls may be produced by dolphins in response to boat traffic approaching

FIGURE 5.5 Dolphin mother and calf.

a group, and this seems to lead to changes in dolphin behavior. However, whether cetaceans produce specific calls for specific dangers is unknown.

Individual Recognition

Most otariid species, as well as some smaller phocids, leave their pups on beaches while they go to sea to forage (see Chapter 9). When the mother returns she must be able to find her pup on an often very crowded breeding beach. Studies have found that both pup and mother are able to identify each other by unique characteristics and variations in their calls. In addition to recognizing offspring or relatives, individual recognition can also play an important role in social animals in coordinating behaviors such as foraging and repelling competitors or predators.

In 1965, Melba and David Caldwell reported that dolphins produced whistles that were unique to individual animals. They suggested these "signature" whistles (now often referred to more generally as "contact calls") allowed dolphins to distinguish individuals and closely related animals from others. Researchers in Florida found bottlenose dolphins to be more likely to turn toward a whistle being played that resembled the whistle of a close relative, which indicates these whistles do provide some information on relationship to the receiver.

These whistles are not simply instinctive. Although newborn calves can produce a whistle sound, the unique signature whistles seem to develop at 3 months to 1 year old, although there is some variation. Further studies on Florida dolphins have found whistle similarities between mothers and offspring, and these similarities between mother and calf whistles are stronger in males, perhaps to help avoid inbreeding. These whistles act in ways similar to names; one component might be equivalent to a surname, partly indicating the matrilineal lineage of the dolphin calf (i.e., the identity of the calf's mother and family), with also an individual unique component, perhaps akin to a first name. Bottlenose dolphins in Florida (when in isolation) have been recorded mimicking the signature whistle of a close companion, adding to the idea that it might act like a "name." It may also play other functions, however, such as an alarm call, or a call to aid group cohesion, or a combination of all these as well as other functions.

> **Exploring the Depths: Prey Stunning?**
>
> Another possible use of sound by cetaceans may be to debilitate and stun prey. One hypothesis proposes that cetaceans use intense sound sources in foraging. It has been shown experimentally that acoustic pulses can be used to stun fish. Experiments have been conducted, for example, in which intense sound waves were directed at fish bearing air-filled swim bladders. The experiment resulted in the fish suffering rapid pulsations of their swim bladders and cellular damage. In another experiment reverberation of microscopic bubbles in seawater led to tissue damage in the lateral line of fish (a sensory structure that detects vibrations in water) or around the gills. Some cetaceans are certainly capable of producing pulses loud enough to cause such stunning effects. Whether they actually do, however, has yet to be proven either by experiments in the laboratory or by observing the behavior of wild cetaceans and their prey. Prey stunning could also be achieved not by vocal sound production but by powerful tail slaps that species such as killer whales use to capture prey.

Sound in the Marine Environment

The oceans are noisy places. Natural phenomena in the marine environment that produce noise include waves, surf, wind, lightning, precipitation, animals (fish, shrimp, and marine mammals), earthquakes, and subsea volcanoes.

Noise from earthquakes dominates underwater noise frequencies below 100 Hz, but rain and wave sounds dominate higher frequencies. Marine mammals have evolved alongside these underwater noises. Since the industrial revolution, however, humans have introduced sometimes substantial amounts of additional noise with increasing regularity. Research at Point Sur, California compared underwater sound levels from 1963 to 1965 and 1994 to 2001. A 10-dB increase in noise levels occurred over 33 years, that is, a 10-fold increase in the level of noise attributed to human activities.

Shipping is one of the main sources of human-made sound in the marine environment. Although there are louder single sources, the huge number of commercial vessels active around the world each produces low-frequency noise and can propagate over large distances. To give an example of the potential for underwater noise from boat traffic, the operating noise from a supertanker can be detected by hydrophones up to 400 km away. In fact, much of the above-mentioned increase in noise levels since the 1960s has been attributed to shipping.

Other sources include air guns used in oil and gas exploration, fish finders and depth sounders, sound sources used in oceanographic research, predator-deterrent devices (including pingers and seal scrammers, which are intended to warn cetaceans away from fishing gear and to repel seals from fish farm cages, respectively), dredging, trawl fishing, coastal construction, oceanic wind farms, bridge traffic, aircraft (to some extent), and military activities.

Exploring the Depths: What Is the Problem with Noise?

Intuitively, we would probably expect the problems caused by marine noise pollution for marine mammals would be similar to those we experience from noise in our environment. To some extent this is correct. Marine mammals can be startled, disturbed, and displaced by noise. We might also expect chronic (i.e., repeated or continuous) disturbance to potentially have health effects. Such things are very difficult to investigate for marine mammals, and there are also some differences between marine and terrestrial "noise pollution."

For cetaceans hearing is their primary sense; consequently, anything that interferes with their ability to hear their environment is likely to have negative consequences for them. Over the last few decades there has been a substantial increase in background noise in the seas. This primarily comes from shipping and, along with other noise, may cause "masking"—the obscuring of sounds of interest (e.g., the sound of potential breeding partners, prey, or calves) by interfering with the propagation of the sounds animals produce.

In the late 1980s and early 1990s a series of unusual stranding events affecting rarely seen beaked whales in the Spanish Canary Islands in the North Atlantic alerted the world to a new and unexpected problem (see Deep Diving and Sonar in Chapter 11). The strandings seemed to correlate with naval exercises being conducted offshore. At first it seemed likely that the disturbance from the exercises might be causing the whales to strand by driving them ashore. However, later investigations into similar events in the Canary Islands, on the Spanish coast, and in the Bahamas included detailed autopsies and showed something else may have been involved. Some of the bodies had unusual internal wounds called embolisms, which are holes in tissues probably caused when gases held in blood and other tissues were released from solution. This is very similar to the condition known as the "bends," a condition suffered by divers who come to the surface too quickly (see Chapter 4). It was previously thought that whales would be immune to such an effect, as they were expected to have evolved physiological or behavioral mechanisms to avoid it. However, the current theory is that these mechanisms, in particular ones relating to typical diving behavior, are disrupted when the whales become startled or are threatened by loud noises (see Deep Diving and Sonar in Chapter 11). Another compatible theory is that the noise could have direct effects on the tissues of exposed animals. The precise mechanism that caused these whales to die remains controversial and hotly debated, but there is a strong correlation between the deployment of certain naval sonar and these events (see Deep Diving and Sonar in Chapter 11).

The effects of marine noise pollution more generally are also far from straightforward. It is generally accepted that behavioral changes can occur (including flight and displacement) and that beaked whale strandings can be induced, as discussed above. Temporary and permanent hearing damage has been seen in response to noise exposure in captive animals but has yet to be recorded in the wild. Similarly, it can be expected that wild marine mammals undergo stress responses when exposed to at least certain noise sources. Thus, the potential exists for them to be chronically stressed by prolonged and/or repeated exposure to noise alone or in conjunction with other anthropogenic threats, with all the associated consequences for health and reproduction. These and many other effects have yet to be observed in wild marine mammals in response to noise. However, this does not mean they are not happening.

More and more noise is being introduced into the oceans both deliberately and incidentally. The latest generation of naval sonar is louder and lower in frequency than previous generations (designed to seek increasingly stealthy enemy submarines), and prospecting for and monitoring underwater fossil

Exploring the Depths: What Is the Problem With Noise? (*continued*)

fuels uses loud powerful noise. We are also increasingly building in the sea. This includes large areas of marine wind farms in which the placement of each turbine usually requires "pile driving" (a process involving a massive hammer that drives the foundations into the seabed, generating considerable noise).

Unfortunately, many humans remain oblivious to the increased marine noise because we can hear very little of it thanks to the air–water interface that substantially hinders the transmission of such sounds into our lives.

SELECTED REFERENCES AND FURTHER READING

Andrew, R.K., Howe, B.M., Mercer, J.A., & Dzieciuch, M.A. (2002). Ocean ambient sound: comparing the 1960s with the 1990s for a receiver off the California Coast. *Acoustics Research Letters Online* 3: 65–70.

Aroyan. J.L., Cranford, T.W., Kent, J., & Norris, K.S. (1992). Computer modeling of acoustic beam formation in *Delphinus delphis*. *Journal of the Acoustical Society of America* 92: 2539–2545.

Aroyan, J.L., Cranford, J.K., & Norris, K.S. (1990). Super-computer modeling of delphinid sonar beam formation. *Journal of the Acoustical Society of America*, 8(Suppl. 1): S4.

Au, W.W.L. (1993). *The Sonar of Dolphins*. Springer-Verlag, New York.

Au, W.W.L., Frankel, A., Helweg, D.A., & Cato D.H. (2001). Against the humpback whale sonar hypothesis. *IEEE Journal of Oceanic Engineering* 26: 295–300.

Au, W.W.L., Mobley, J., Burgess, W.C., Lammers, M.O., & Nachtigall, P.E. (2000). Seasonal and diurnal trends of chorusing humpback whales wintering in waters off western Maui. *Marine Mammal Science* 16: 530–544.

Bastian, J. (1966). The transmission of arbitrary environmental information between bottlenose dolphins. In: *Les Systems Sonars Animaux: Biologie at Bionique* (Animal Sonar Systems: Biology and Bionics). Vol. 2 (Ed. R.-G. Busnel), pp. 803–873. Laboratoire de Physiologie Acoustique, Jouy-en-Josas, France.

Busnel, R.-G. (1966). Information in the human whistled language and sea mammal whistling. In: *Whales, Dolphins, and Porpoises* (Ed. K.S. Norris), pp. 544–568. University of California Press, Berkeley and Los Angeles.

Caldwell, M.C., & Caldwell D.K. (1965). Individualized whistle contours in bottlenosed dolphins, *Tursiops truncatus*. *Nature* 207: 434–435.

Caldwell, M.C., Caldwell, D.K., & Tyack, P.L. (1990). Review of the signature-whistle hypothesis for the Atlantic bottlenose dolphin. In: *The Bottlenose Dolphin* (Ed. S. Leatherwood & R. R. Reeves), pp. 199–234. Academic Press, New York.

Charrier, I., & Harcourt, R.G. (2006). Individual vocal identity in mother and pup Australian sea lions (*Neophoca cinerea*). *Journal of Mammalogy* 87: 929–938.

Chu, K. (1988). Dive times and ventilation patterns of singing humpback whales (*Megaptera novaengliae*). *Canadian Journal of Zoology* 66: 1322–1327.

Chu, K., & Harcourt, P. (1986). Behavioral correlations with aberrant patterns in humpback whale songs. *Behavioral Ecology and Sociobiology* 19: 309–312.

Clapham, P.J., & Palsboll, P.J. (1997). Molecular analysis of paternity shows promiscuous mating in female humpback whales (*Megaptera novaengliae*, Borowski). *Proceedings of the Royal Society London B* 264: 95–98.

Clark, C.W. (1993). Bioacoustics of baleen whales: from infrasonics to complex songs. *Journal of the Acoustical Society of America* 94: 1830.

Clarke, M.R. (1978). Structure and proportions of the spermaceti organ in the sperm whale. *Journal of the Marine Biological Association UK* 58: 1–17

Cook, M.L.H., Sayigh, L.S., Blum, J. E., & Wells, R. S. (2004). Signature-whistle production in undisturbed free-ranging bottlenose dolphins (*Tursiops truncatus*). *Proceedings of the Royal Society B* 271: 1043–1049.

Craig, A.S., Herman, L.M., & Pack A.A. (2002). Male mate choice and male male competition coexist in the humpback whale (*Megaptera novaengliae*). *Canadian Journal of Zoology* 80: 745–755.

Cranford, T.W. (1999). The sperm whale's nose: sexual selection on a grand scale? *Marine Mammal Science* 15: 1133–1157.

Cranford, T.W. (2000). In search of impulse sound sources in odontocetes. In: *Hearing by Whales and Dolphins* (Ed. W.W.L. Au, A.N. Popper, & R.R. Fay), pp. 109–155. Springer-Verlag, New York.

Cranford, T.W., Amundin, M., & Norris, K.S. (1996). Functional morphology and homology in the odontocete nasal complex: implications for sound generation. *Journal of Morphology* 228: 223–285.

Cranford, T.W., Van Bonn, W.G., Chaplin, M.S., Carr, J.A., Kamolnick, T.A., Carder, D.A., & Ridgway, S.H. (1997). Visualizing dolphin sonar signal generation using high-speed video endoscopy. *Journal of the Acoustical Society of America* 102: 3123.

Darwin, C. (1871). *The Descent of Man and Selection in Relation to Sex*. John Murray, London.

Dolman, S., Parsons, E.C.M., & Wright, A.J. (2011). Cetaceans and military sonar: a need for better management. *Marine Pollution Bulletin*: in review.

Dormer, K. J. (1979). Mechanism of sound production and air recycling in delphinids: Cineradiographic evidence. *Journal of the Acoustical Society of America* 65: 229–239.

Evans, W.E., & Maderson, P.F.A. (1973). Mechanisms of sound production in delphinid cetaceans: a review and some anatomical considerations. *American Zoology* 13: 1205–1213.

Ford, J.K.B. (1989). Acoustic behavior of resident killer whales (*Orcinus orca*) off Vancouver Island, British Columbia. *Canadian Journal of Zoology* 67: 727–745.

Ford, J.K.B. (1991). Vocal traditions among resident killer whales (*Orcinus orca*) in coastal waters of British Columbia. *Canadian Journal of Zoology* 69: 1454–1483.

Frazer, L.N., & Mercado, E. (2000). A sonar model for humpback whale song. *IEEE Journal of Oceanic Engineering* 25: 160–182.

Gedamke, J., Costa, D., Dunstan, A., and O'Neil, F. (2001). Do minke whales sing? Analysis of discrete categories in a repetitive breeding season sound sequence. In: *Abstracts of the 14th Biennial Conference on the Biology of Marine Mammals, Vancouver, Canada, December 2001*. p. 80. Society of Marine Mammalogy, Vancouver, Canada.

Glockner, D.A. (1983). Determining the sex of humpback whale (*Megaptera novaeangliae*) in their natural environment. In: *Communication and Behavior of Whales* (Ed. R. Payne), pp. 447–464. Westview Press, Boulder, CO.

Goold, J., & Coates, R. (2001). Acoustic monitoring of marine wildlife. Seiche Technical Education. Retrieved from http://www.seiche.com/4.%20Courses/Courses/Monitoring/MonitoringWindow.html on 1 Aug 2011.

Hatch, L., & Wright, A.J. (2007). A brief review of anthropogenic sound in the oceans. *International Journal of Comparative Psychology* 20: 121–133.

Helweg, D.A., Cato, D.H., Jenkins, P.F., Garrigue, C., & McCauley, R.D. (1998). Geographic variation in South Pacific humpback whale songs. *Behaviour* 135: 1–27.

Helweg, D.A., Frankel, A.S., Mobley, J.R., & Herman, L.M. (1992). Humpback whale song: our current understanding. In: *Marine Mammal Sensory Systems* (Ed. J.A. Thomas, R. Kastelein, & A.Y. Supin), pp. 459–483. Plenum Press, New York.

Hemila, S., Nummela, S. Berta, A., & Reuter, T. (2006). High-frequency hearing in phocid and otariid pinnipeds: an interpretation based on inertial and cochlear constraints (L). *Journal of the Acoustical Society of America* 120: 3463–3466.

Janik, V.M. (2000). Whistle matching in wild bottlenose dolphins (*Tursiops truncatus*). *Science* 289: 1355–1357.

Janik, V.M., & Slater, P.J.B. (1998). Context-specific use suggests that bottlenose dolphin signature whistles are cohesion calls. *Animal Behavior* 56: 829–838.

Janik, V., Sayigh, L.S., & Wells, R.S. (2006). Signature whistle shape conveys identity information to bottlenose dolphins. *Proceedings of the National Academy of Sciences* 103: 8293–8297.

Kamminga, C., & Van Der Ree, A.F. (1976). Discrimination of solid and hollow spheres by *Tursiops truncatus* (Montagu). *Aquatic Mammals* 4: 1–9.

Kellogg, W.N. (1961). *Porpoises and Sonar*. University of Chicago Press, Chicago.

Ketten, D.R. (1994). Functional analysis of whale ears: adaptations for underwater hearing. *IEEE Proceedings in Underwater Acoustics* 1: 264–270.

Ketten, D.R. (1997). Structure and function in whale ears. *Bioacoustics* 8: 103–135.

King, S., Sayigh, L.S., Wells, R.S., Fellner, W. & Janik, V. (2011). What's in a name? Why do bottlenose dolphins copy each other's signature whistles? In: *Abstracts. 19th Biennial Conference on the Biology of Marine Mammals, Tampa, Florida, 2011*. p. 158–159. Society for Marine Mammalogy, Tampa, FL.

MacKay, R.S., & Pegg, J. (1988). Debilitation of prey by intense sounds. *Marine Mammal Science* 4: 356–359.

Madsen, P.T., Payne, R., Kristiansen, N.U., Wahlberg, M., Kerr, I., & Mohl, B. (2002). Sperm whale sound production studied with ultrasound time/depth-recording tags. *Journal of Experimental Biology* 208: 1899–1906.

Mattila, D.K., Guinee, L.N., & Mayo, C.A. (1987). Humpback whale songs on a North Atlantic feeding ground. *Journal of Mammalogy* 68: 880–883.

McCowan, B., & Reiss, D. (2001). The fallacy of "signature whistles" in bottlenose dolphins: a comparative perspective of "signature information" in animal vocalizations. *Animal Behaviour* 62: 1151–1162.

McGregor, P.K., & Krebs, J.R. (1982). Mating and song types in the great tit. *Nature* 297: 60–61.

McSweeny, D.J., Chu, K.C., Dolphin, W.F., & Guinee, L.N. (1989). North Pacific humpback whale songs: a comparison of southeast Alaskan feeding ground songs and Hawaiian wintering ground songs. *Marine Mammal Science* 5: 116–138.

Miller, D., & Williams, A. (1983). Further investigations of ATP release from human erythrocytes exposed to ultrasonically activated gas-filled pores. *Ultrasound in Medicine and Biology* 9: 303.

Mobley, J.R., Herman, L.M., & Frankel, A.S. (1988). Responses of wintering humpback whales (*Megaptera novaeangliae*) to playback recordings of winter and summer vocalisations and of synthetic sound. *Behavioural Ecology and Sociobiology* 23: 211–223.

Mohl, B. (2001). Sound transmission in the nose of the sperm whale, *Physeter catodon*. A post mortem study. *Journal of Comparative Physiology, Part A* 187: 335–340.

Noad, M.J., Cato, D.H., Bryden, M.M., Jenner, M.N., Curt, K., & Jenner, S. (2000). Cultural revolution in whale songs. *Nature* 408: 537.

Norris, K.S. (1964). Some problems of echolocation in cetaceans. In: *Marine Bio-Acoustics* (Ed. W.N. Tavolga), pp. 317–336. Pergamon Press, Oxford, UK.

Norris, K.S., & Møhl, B. (1983). Can odontocetes debilitate prey with sound? *American Naturalist* 122: 85–104.

Nummela, S., Thewissen, J.G.M., Bajpai, S., Hussain, T., & Kumar, K. (2007). Sound transmission in archaic and modern whales: anatomical adaptations for underwater hearing. *Anatomical Record* 290: 716–733.

Parsons, E.C.M., & Dolman, S. (2003a). The use of sound by cetaceans. In: *Oceans of Noise* (Ed. M. Simmonds, S. Dolman, & L. Weilgart), pp. 44–52. Whale and Dolphin Conservation Society, Chippenham, UK.

Parsons, E.C.M., & Dolman, S. (2003b). Noise as a problem for cetaceans. In: *Oceans of Noise* (Ed. M. Simmonds, S. Dolman, & L. Weilgart), pp. 53–58. Whale and Dolphin Conservation Society, Chippenham, UK.

Parsons, E.C.M., Dolman, S., Jasny, M., Rose, N.A., Simmonds, M.P., & Wright, A.J. (2009). A critique of the UK's JNCC Seismic Survey Guidelines for minimizing acoustic disturbance to marine mammals: best practice? *Marine Pollution Bulletin* 58: 643–651.

Parsons, E.C.M., Dolman, S., Wright, A.J., Rose, N.A., & Burns, W.C.G. (2008). Navy sonar and cetaceans: just how much does the gun need to smoke before we act? *Marine Pollution Bulletin* 56: 1248–1257.

Parsons, E.C.M., Swift, R., & Dolman, S. (2003). Sources of noise. In: *Oceans of Noise* (Ed. M. Simmonds, S. Dolman, & L. Weilgart), pp. 24–43. Whale and Dolphin Conservation Society, Chippenham, UK.

Parsons, E.C.M., Wright, A.J., & Gore, M. (2008). The nature of humpback whale (*Megaptera novaeangliae*) song. *Journal of Marine Animals and Their Ecology* 1: 22–31.

Payne. K. (1999). The progressively changing songs of humpback whales: a window on the creative process in a wild animal. In: *The Origins of Music* (Ed. N.L. Wallin, B. Merker, & S. Brown), pp. 135–150. MIT Press, Cambridge, MA.

Payne, R., & McVay, S. (1971). Songs of humpback whales. *Science* 173: 587–597.

Payne, K., Tyack, P., & Payne, R. (1983). Progressive changes in the song of humpback whales (*Megaptera novaeangliae*): a detailed analysis of two seasons in Hawaii. In: *Communication and Behavior of Whales* (Ed. R. Payne), pp. 9–57. Westview Press, Boulder, CO.

Purves, P. E., & Pilleri, G.E. (1983). *Echolocation in Whales and Dolphins*. Academic Press, London.

Richardson, W.J., Greene, C.R., Malme, C.I., & Thomson, D.H. (1995). *Marine Mammals and Noise*. Academic Press, San Diego.

Sayigh, L.S., Tyack, P.L., Wells, R.S., Scott, M.D., & Irvine, A.B. (1995). Sex differences in signature whistle production of free-ranging bottlenose dolphins, *Tursiops truncatus*. *Behavioral Ecology and Sociobiology* 36: 171–177.

Sayigh, L.S., Tyack, P.L., Wells, R.S., Solow, A.R., Scott, M.D., & Irvine, A.B. (1999). Individual recognition in wild bottlenose dolphins. *Animal Behaviour* 57: 41–50.

Scarpaci, C., Bigger, S.W., Corkeron, P.J., & Nugegoda, D. (2000). Bottlenose dolphins (*Tursiops truncatus*) increase whistling in the presence of "swim-with-dolphin" tour operations. *Journal of Cetacean Research and Management* 2: 183–185.

Simmonds, M.P., & Lopez-Jurado, L.F. (1991). Whales and the military. *Nature* 351: 448.

Spiesberger, J.L., & Fristrup, K.M. (1990). Passive localization of calling animals and sensing of their acoustic environment using acoustic tomography. *American Naturalist* 135: 107–153.

Tavolga W.N. (Ed.) (1964). *Marine Bio-Acoustics*. Pergamon Press, Oxford, UK.

Tyack, P. (1981). Interactions between singing Hawaiian humpback whales and conspecifics nearby. *Behavioral Ecology and Sociobiology* 8: 105–116.

Tyack, P. (1983). Differential responses of humpback whales, *Megaptera novaeangliae*, to playback of song or social sounds. *Behavioral Ecology and Sociobiology* 13: 49–55.

Tyack, P. (1986). Whistle repertoires of two bottlenosed dolphins, *Tursiops truncatus*: mimicry of signature whistles? *Behavioral Ecology and Sociobiology* 18: 251–257.

Tyack, P. (1997). Studying how cetaceans use sound to explore their environment. *Perspectives in Ethology* 12: 251–297.

Tyack, P.L., & Miller, E.H. (2002). Vocal anatomy, acoustic communication and echolocation. In: *Marine Mammal Biology: an Evolutionary Approach* (Ed. A.R. Hoelzel), pp. 142–184. Blackwell Science, Oxford, UK.

Tyack, P., & Payne, R. (1983). Progressive changes in the songs of humpback whales (*Megaptera novaeangliae*): a detailed analysis of two seasons in Hawaii. In: *Communication and Behavior of Whales* (Ed. R. Payne), pp. 9–57. Westview Press, Boulder, CO.

Urick, R.J. (1983). *Principles of Underwater Sound*. 3rd Ed. McGraw-Hill, New York.

Van Parijs, S.M., & Corkeron, P.J. (2001). Boat traffic affects the acoustic behaviour of Pacific humpback dolphins, *Sousa chinensis*. *Journal of the Marine Biological Association UK* 81: 533–538.

Wenz, G.M. (1962). Acoustic ambient noise in the ocean: spectra and sources. *Journal of the Acoustical Society of America* 34: 1936–1956.

Whitehead, H. (2002). Culture in whales and dolphins. In: *Encyclopedia of Marine Mammals*. 2nd Ed. (Ed. W.F. Perrin, B. Würsig, & J.G.M. Thewissen), pp. 292–294. Academic Press, New York.

Winn, H.E., Bischoff, W.L., & Taruski, A.G. (1973). Cytological sexing of cetaceans. *Marine Biology* 23: 343–346.

Zagaeski, M. (1987). Some observations on the prey stunning hypothesis. *Marine Mammal Science* 3: 275–279.

Part II

Ecology and Status

CHAPTER 6

Polar Bears

CHAPTER OUTLINE

Aquatic Adaptations

Distribution

Arctic Adaptations

Feeding Behavior and Ecology

Reproduction

Abundance and Status

    Hunting
    Pollution
    Climate Change

Exploring the Depths: Sampling Contaminants in Polar Bears

Disturbance

Polar Bear Conservation

    International Polar Bear Agreement

    Exploring the Depths: Precautionary Principle

    Recent Changes in Polar Bear Status

    Exploring the Depths: Polar Bears in Captivity

Selected References and Further Reading

Polar bears are considered to be marine mammals under the U.S. Marine Mammal Protection Act (MMPA). They are considered to be marine mammals because they:

- typically are found on marine pack ice (up to 1,300 km from shore);
- consume (predominantly marine) prey in the ice;
- are able to swim large distances and have adaptations that allow them to do so; and
- depend on the marine environment for their continued existence.

Polar bears are larger than the brown bear (*Ursus arctos*), although slightly more slender in profile, with an elongated head and neck. Polar bears are sexually dimorphic; males and females differ substantially in size and weight. Males can be up to 3 m in length and weigh up to 800 kg; females are smaller, up to 2 m in length and approximately 300 kg in weight. However, female polar bears have a life expectancy of 25 to 30 years, whereas the life expectancy of males is slightly lower, at approximately 20 years.

Aquatic Adaptations

Polar bears have been recorded swimming, in extreme cases, hundreds of kilometers, at speeds of up to 10 km per hour. To aid in swimming they possess partially webbed front paws, which have a large surface area. Their eyes have an additional nictitating membrane to protect them in harsh Arctic conditions. This membrane also helps the bears see underwater (**Figure 6.1**). Their thick (10-cm) adipose (fat) layer provides them both with buoyancy and insulation in the water, where the insulating effects of their fur are reduced (see Chapter 4).

Distribution

Polar bears are distributed throughout the Arctic (from Greek *arktos*, for bear), although approximately two-thirds of the world population is found in North America and neighboring Greenland (**Figure 6.2**). Small numbers of polar bears can be found on permanent pack ice in the mid-Arctic region. However, most bears inhabit annual pack ice (i.e.,

FIGURE 6.1 A polar bear swimming underwater.

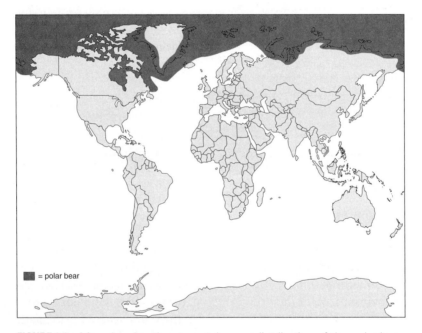

FIGURE 6.2 Map showing the current, known distribution of the polar bear. Data from: IUCN Red List.

ice that breaks up and refreezes once a year). This pack ice is found over continental shelves and around coastlines of the Arctic Ocean.

The distribution of polar bears is related to abundance of their main prey species, ringed seals (*Pusa hispida*). In areas where annual ice completely melts in the summer, polar bears may stay on land and fast during the warm season. Generally, polar bears are solitary, but they can congregate in larger numbers around substantial sources of food, such as a stranded whale carcass.

Arctic Adaptations

Arctic temperatures can drop to as low as −45°C for days or weeks, with an average midwinter temperature of −34°C. To survive these harsh conditions, polar bears have a number of adaptations to survive the cold and also to cope with snow-covered and icy terrain.

One obvious adaptation to their habitat is their white fur for camouflage when they are stalking prey. The hairs of their fur are not, however, colored white but are actually translucent. Some scientists have suggested that these hairs can also act like optic fibers, channeling sunlight energy (and heat) to the polar bear's skin. However, measurements on the structure of polar bear hair have now ruled this out. The skin of a polar bear is actually black, which may aid heat absorption and/or protect against ultraviolet radiation.

In fact, polar bears have two layers of fur, a dense undercoat and longer guard hairs (up to 15 cm long), that insulate them while on land. They also have a thick fat layer that adds insulation, particularly while in the water (see Chapter 4). The large size of adult polar bears means they have a relatively small surface area through which to lose heat and their relatively small ears and tail (which is an example of Allen's rule) further reduces their small surface area (see Chapter 4).

In contrast, the feet of polar bears have a large surface area (approximately 35 cm in diameter). They are also covered in fur (**Figure 6.3**). The large surface area distributes their weight across the surface of the snow, similar to snowshoes, while the fur reduces conduction to the cold ground.

Polar bears also have various behavioral adaptations to reduce heat loss. During storms and harsh weather conditions they dig holes in snow banks as shelter. In these shelters they curl up into a ball, reducing their exposed surface area, as well as cover their nose with their paws, all to reduce heat loss. Polar bears are actually so effective at retaining heat that, even in cold weather, they can quickly overheat when exercising vigorously.

Feeding Behavior and Ecology

The diet of polar bears primarily consists of ringed seals, but they also hunt bearded seals (*Erignathus barbatus*), harp seals (*Pagophilus groenlandicus*), and hooded seals (*Cystophora cristata*). They have also been reported to hunt walruses (*Odobenus rosmarus*), beluga whales (*Delphinapterus leucas*), Arctic cod (*Arctogadus glacialis*), shellfish, terrestrial mammals (from reindeer to small rodents), and seabirds. On occasion they have even been reported to be cannibalistic. Polar bears obtain most of their food during the spring and summer, preying on seal pups (which may have as much as a 50% fat content). Their digestive system is very effective at digesting and storing lipids in this high fat diet (98% efficient).

The most common method of hunting used by polar bears is to lie alongside a breathing hole in the ice waiting for a seal to surface (ringed seals keep holes open in the ice with their teeth), or they may patrol areas of broken ice with channels to the ocean (where seals swim between the ice) and polynyas (areas where tide/currents break up ice) and similar locations where there is a gap in (or thin) ice cover. In spring, polar bears hunt on stable sea ice covered in snowdrifts, searching for buried ringed seal dens by smell (**Figure 6.4**). They can detect such dens even at depths up to 1 m below the snow surface. They can also smell a seal breathing hole from around 1 km away and have been reported to detect a seal carcass as far away as 20 km.

Polar bears can hunt through the depths of winter, except for pregnant females, which exhibit a form of hibernation (see Reproduction). If food sources are low (7 to

FIGURE 6.3 Underside of polar bear feet.

FIGURE 6.4 Polar bears attacking a snow den of a ringed seal.

10 days without food), they can also go into another mild hibernation (often called "walking hibernation"), from which they can very quickly awaken. They are the only bear species able to hibernate (in any form) directly in response to a lack of food.

Reproduction

Female polar bears become sexually mature at 4 to 5 years of age, and their reproductive peak is between 8 and 18 years. Mating occurs from April to June, coinciding with the peak seal pupping season. A male may mate with one or more females, but will defend the female against other rival males, in an example of a polygynous mating system. Females have induced ovulation; they must mate multiple times to ovulate. After the egg is fertilized, there is a delayed implantation onto the uterine wall for several months, until October/November. At this time, the female excavates a maternal den from the snow, where she will give birth during the winter. Between mating and denning, pregnant females begin to build up fat stores and may double in weight. In the den the female goes into a mild form of hibernation. In this state she becomes bradycardic (decreased heart rate), but her metabolism and body temperature do not drop as much as in the hibernation of other bears.

One to three cubs are born in December or January, weighing less than 1 kg. Approximately two-thirds of cubs are twins. They are nursed by their mother on fat-rich milk until March and April (although sometimes in February) when they achieve a weight of 10 to 15 kg and emerge from the buried snow den. After a period of acclimatization for the cubs, and again at the peak of the seal pupping season, the female treks to the feeding grounds to replenish her body reserves, with the cubs following along (**Figure 6.5**).

FIGURE 6.5 A polar bear mother and cubs.

Cubs are weaned at 2 (low arctic) to 3 (high arctic) years old. During these early years the mother teaches the cub hunting skills. Males have been known to attack cubs, and mothers can be very aggressive in protecting their cubs; for example, they have even been seen trying to attack helicopters. Interestingly, female polar bears have been recorded "adopting" cubs that were not their own.

Abundance and Status

There are approximately 19 recognized populations of polar bears, with each population varying in size from hundreds of animals to thousands. There are currently an estimated 22,000 to 25,000 polar bears worldwide. The 2005 meeting of the International Union for Conservation of Nature (IUCN) Polar Bear Specialist Group determined that, of the 19 polar bear populations, there was insufficient information to make a determination on several populations. Despite this, the Group expressed belief that two populations were increasing, five were stable and five were in decline. However, at the 2009 meeting it was determined that only one population was increasing, three were stable, and eight were in decline.

Mortality in polar bear cubs can be high because they cannot survive independently of their mothers. As subadults, the biggest problem is starvation, as young bears without an established territory have to compete with adult bears. The younger animals may consequently be relegated to more marginal territories, which in recent years may mean areas of thin and melting ice. The natural mortality rate of adult bears is low (<5% per year), and attacks by other bears or human activities pose the largest cause of mortality for these animals. In recent years, three of the greatest conservation issues for polar bears have been hunting, pollution, and climate change. Disturbances due to industrial activity in the Arctic and tourism are also causes of concern.

Hunting

There has been a long history of hunting polar bears throughout their range. The Inuit of northern Canada have, for example, hunted polar bears for centuries, using their fur for clothing, their fat for fuel, and their meat for food. Europeans have also hunted polar bears, primarily for their fur in Norway and Russia, since the Middle Ages. Hunting of polar bears was once unregulated, and large numbers were harvested in the 1960s and 1970s. In 1973 the International Agreement on the Conservation of Polar Bears and their Habitat (see section on Polar Bear Conservation) was signed in Oslo and largely brought an end to unregulated hunting.

Today, polar bears can only be hunted for aboriginal subsistence use by native people in the Arctic and for

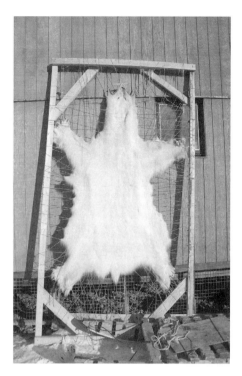

FIGURE 6.6 A polar bear skin mounted for drying and preservation. Trophy and sport hunting of polar bears is rising in popularity.

trophy or sport hunting. No hunting of either type is permitted in Norway. Aboriginal subsistence hunting occurs in Russia, Canada, Greenland, and the United States (in Alaska). Sport or trophy hunting is permitted in Canada under a quota system, through which non-natives and non-Canadian citizens can take polar bears legally (**Figure 6.6**). This type of hunting was relatively uncommon until the 1980s, when it began to rise. It has been argued that such sport hunting provides important income for indigenous populations and the fees involved in the hunts subsidize conservation efforts. However, the issue is controversial and environmental and animal welfare groups have argued that the majority of the income from sport hunting tours goes to non-native outfitters and trip organizers rather than to native communities. Similarly, they also argue that the gain in conservation funding is outweighed by the loss of large adult (mostly male) polar bears.

Under the U.S. MMPA the importation of polar bear products (such as body parts for trophies and skins hunted in Canada) was banned from 1972 to 1994. An amendment in 1994 allowed the import of polar bear products by sport hunters for personal use into the United States; this amendment is the only part of the MMPA that allows importation of marine mammal products that are not for educational (public display), scientific research purposes or for the enhancement (conservation) of the species. Between 1994 and 2008 about 900 polar bear trophies were imported into the United States from Canada, when this loophole was effectively closed and the polar bear was listed under the U.S. Endangered Species Act (ESA; see Recent Changes in Polar Bear Status).

In some countries the monitoring of polar bear kills is arguably effective (e.g., Norway and the United States). However, this is not so in other areas (e.g., Russia and Greenland). Of the declining polar bear populations, the Baffin Bay population (shared between Canada and Greenland) is primarily declining because of overharvesting (and climate change effects; **Figure 6.7**). Somewhat controversially, in 2005 the harvest quota for polar bears in the Canadian part of this region (Nunavut) was increased, despite the declining Baffin Bay population. Also in 2005, the Greenland authorities introduced a quota for polar bear harvests, some management controls on harvesting methods, and documentation requirements. Until that point hunting of polar bears by indigenous people in Greenland was effectively unlimited because no quota system existed. This legal change also introduced provisions to allow sport hunting of polar bears in Greenland. The Chukchi Sea population (shared between the United States and Russia) is also in decline partly due to illegal hunting. Regardless of the legal regime, polar bears can also be killed if done so in the defense of human life or property.

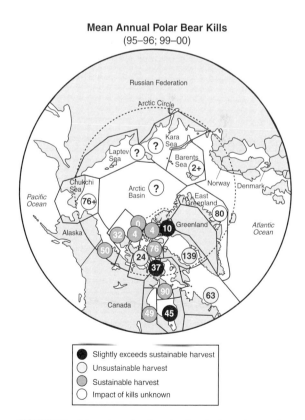

FIGURE 6.7 Map showing levels of polar bear harvests. Data from: IUCN Polar Bear Specialist Group.

Pollution

Arctic marine mammals have been reported to accumulate high concentrations of pollutants in their blubber (for example, organic pesticides such as DDT). Polar bears prey on these marine mammals, so they accumulate even higher concentrations of pollutants in their tissues. In some areas there is concern about the effects of these pollutants on the health of polar bears. In the United States (Alaska), organic contaminant concentrations have been reported in polar bears that were nearly five times higher than in their seal prey. There has been particular concern expressed about the impacts of these contaminants on the reproductive system of females, as many act as hormone mimics and disruptors. Instances of pseudohermaphroditism (a female possessing structures resembling penises) have been found in Svalbard, Norway, and it has been suggested that this may be due to high levels of contaminants in these bears. To explore the various possible problems, scientists are exploring new ways to investigate contaminant loads, such as by using polar bear hair (see Exploring the Depths: Sampling Contaminants in Polar Bears).

Climate Change

Climate change is the issue considered by the IUCN Polar Bear Specialist Group to be the foremost threat to polar bears, specifically the impacts of melting annual ice in terms of the loss of polar bear habitat (and that of their prey). Arctic seasonal ice has recently been melting earlier in the year, and the extent of pack ice in the summer has receded. The years 2007 through 2009 were the 3 lowest years for summer ice coverage since satellite recordings began in 1979 (**Figure 6.8**).

This melting ice disrupts polar bear migratory routes and has stranded many animals on land, meaning they are unable to reach their preferred feeding grounds. Polar bear

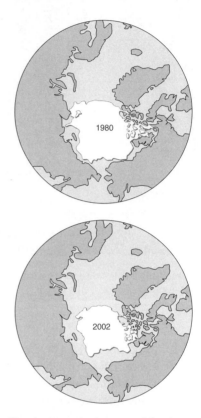

FIGURE 6.8 The decrease in the area of Arctic sea ice between 1980 and 2002.

Exploring the Depths: Sampling Contaminants in Polar Bears

Normally samples of tissues such as liver, kidney and blood have been used when studying the general physiological state of polar bears. However, as the polar bear is a protected species it is often hard to acquire these tissue samples, which can only be obtained by major invasive procedures (which in zoo animals could be an operation, but in the wild this usually means they are only available from a dead bear).

Scientists have therefore recently turned their attention towards the non-invasive method of analyzing polar bear hair, which is much easier to acquire, and can be obtained from both live and dead animals. It is even possible to sample repeatedly from the same animal at different points in time. In addition, hair samples are a great deal easier to work with logistically than, for example, blood or liver samples, as they require nothing more than a sharp tool and a plastic bag. They also have no particular handling or storage requirements on the way from the field to the laboratory.

Hair samples can be used in genetic studies, as well as for analyzing mercury, persistent organic pollutants, and hormones. This way it can be used as a non-invasive indicator for monitoring whether the various compounds in the bear exceed the general guideline values in wildlife management. An added benefit regarding hair is that it does not reflect short-term fluctuations in biological measures. For instance, in hormones, it can express the extent to which the animal is chronically stressed, rather than just responding to a short-term threat, such as capture. There are, at the moment, several ongoing projects studying polar bear hair, in order to determine how much more this non-invasive method can tell us about the physiological state of the polar bear.

—Contributing author, Thea Østergaard Bechshøft, Aarhus University.

populations may already be experiencing the impact of lost feeding opportunities. The Hudson Bay population of polar bears has declined by 22% (between 1987 and 2004), whereas the average weight of females has declined by 20% over roughly the same period, presumably due to a lack of prey. Additionally, increasing numbers of females have had to den on land rather than on the ice, where there is increased competition for denning sites and less than ideal habitat for the bears. The proportion of females denning on ice has decreased by over one-third: between 1985 and 1994, 62% of female polar bears in northern Alaska denned on the ice versus 37% in 1998 to 2004.

Researchers have also reported increasing numbers of polar bears in open waters and there have been reports of notable numbers of young polar bears having drowned, possibly after being caught in open water after premature melting of ice. Despite adults being competent swimmers, the distances involved in these stranded animals may have been too great for the young polar bears to swim.

The diminishing ice coverage and shifting distribution of polar bears is also forcing them into closer proximity to human settlements. This is leading to more conflicts with humans as polar bears raid garbage, damage properties, injure pets, and threaten the safety of humans. In Churchill, Manitoba, Canada the authorities have instigated a "polar bear jail." Bears that become a nuisance or a possible threat to humans are tranquilized and then placed into the jail, before being translocated to an area away from humans. This increase in proximity to humans has led some people to get the false impression that polar bear numbers are increasing, which is not the case. It simply reflects a recent shift in bear distribution.

Disturbance

Many locations in the Arctic are rich in natural resources, especially oil and gas (**Figure 6.9**). The exploration for and extraction of these resources involve construction that removes habitat, causes pollution, and creates noise and other disturbances that could directly affect polar bear populations.

Another form of disturbance is tourism. Trips to watch polar bears are a growing industry, especially in Churchill, Manitoba, Canada (**Figure 6.10**). There are increasing concerns about disturbance of bears by these activities. However, tourism does confer a non-consumptive economic value to polar bears and is an alternative to hunting, so the advantages and disadvantages must be weighed carefully.

FIGURE 6.9 Oil and gas development in the Arctic.

FIGURE 6.10 Polar bear tourism in Churchill, Canada.

Polar Bear Conservation

In the United States polar bears are afforded protection under the MMPA. A subsistence harvest by Alaskan indigenous people is allowed under the Act, as noted previously, and an 1994 amendment to the MMPA allowed the importing of sport-hunted polar bear trophies from some Canadian populations, although the 2008 listing of the polar bear under the ESA once again made this importation illegal. The MMPA and ESA also protect polar bears from incidental "takes," which include death, injury, and disturbance. Activities that occur in polar bear habitats that could potentially disturb them require a "take authorization" permit from the U.S. government.

International Polar Bear Agreement

In 1965 a meeting was held in Fairbanks, Alaska to discuss concerns about the conservation status of polar bears. This meeting was attended by representatives from all the polar bear range states, namely the United States, Canada, Norway, the Soviet Union, and Denmark (for Greenland). At this meeting it was recognized that:

- polar bears are an international circumpolar resource;
- each polar bear range state should take whatever steps are necessary to conserve the polar bear until the results of research could be applied (an application of what is now known as "the precautionary principle");
- polar bear cubs and females accompanied by cubs are especially vulnerable and should be protected throughout the year;
- each polar bear range state should conduct research programs on polar bears within their territory (as best they could);
- there should be a free exchange of information on polar bears; and
- further meetings were necessary (and proceedings of the meetings should be published).

However, it was not until May 1973 that the Agreement on the Conservation of Polar Bears and their Habitat was signed in Oslo, Norway. This agreement included prohibitions on unregulated sport hunting and various hunting methods, such as from helicopters. Initially, the Agreement

---

Exploring the Depths: Precautionary Principle

The precautionary principle is often hotly debated (especially with regards to the inclusion of any consideration of financial costs) and many definitions for the term exist. However, the precautionary principle was formally defined by international agreement as part of the 1982 United Nations World Charter for Nature and later incorporated into the 1992 Rio Declaration (on Environment and Development), the result of the so-called Earth Summit meeting. Principle 15 of the Rio Declaration described the precautionary principle as follows:

> Where there are threats of serious or irreversible damage, lack of full scientific certainty shall not be used as a reason for postponing cost-effective measures to prevent environmental degradation.

Furthermore, the European Commission (European Commission Communication on the Precautionary Principle, 2nd February 2000) has directed the following:

> "The precautionary principle applies where scientific evidence is insufficient, inconclusive or uncertain and preliminary scientific evaluation indicates that there are reasonable grounds for concern that the potentially dangerous effects on the environment, human, animal or plant health may be inconsistent with the high level of protection chosen by the EU."

Therefore European Union members are obliged to enact conservation measures proactively whenever there might be a risk to animals, including marine mammals, even if definitive scientific data are lacking. However, many other nations do not abide by the precautionary principle, but instead shift the "burden of proof" and insist that those wanting to enact protections have to prove definitively that something has an impact to the environment (and/or a species) before any action is taken. Unfortunately, for the case of marine mammals definitive proof may simply be too difficult, costly, and time-consuming to gather.

underwent a 5-year trial period, but it was eventually made indefinite in January 1981.

In addition to this major agreement there are also several other regional agreements and treaties between nations with respect to polar bear management and conservation, such as the 1988 and 2000 Inuvialuit–Inupiat Agreements (between the United States and Canada), which set a voluntary quota for subsistence harvests of bears in the southern Beaufort Sea. In October 2000 the Agreement on the Conservation and Management of the Alaska-Chukotka Polar Bear Population was also signed. In addition, there are Norwegian–Russian and Greenland–Canadian agreements.

Recent Changes in Polar Bear Status

In 1996 polar bears were listed as "conservation dependent" by the IUCN (see Chapter 17 for details). In 2006 they were considered to be under greater threat and their listing was changed to "vulnerable." This assessment was made based on a suspected population reduction of more than 30% within three generations (45 years) because of the following factors:

- a decrease in area of occupancy and extent of occurrence;
- a decline in habitat quality; and
- the potential for legal and illegal overharvesting.

In the United States, on February 16, 2005, the Center for Biological Diversity and Greenpeace USA won lawsuits petitioning the U.S. Department of the Interior (the government agency responsible for polar bear conservation) to list polar bears as "threatened" under the ESA. According to the ESA a "threatened" species is one that is likely to become an endangered species within the "foreseeable future" throughout all or a significant portion of its range (for details of the ESA see Chapter 17). On December 27, 2007 the U.S. Secretary of the Interior, Dirk Kempthorne, announced the Department's intent to list the polar bear as "threatened" under the ESA. Subsequently, public hearings to discuss the issue were to be conducted in Anchorage and Barrow, Alaska and in Washington, DC. The legal deadline for the listing was supposed to be in January 2008; however, the listing was not finalized until May 15, 2008, several months beyond the deadline. There was much criticism of this delay. Specifically, critics claimed that it was a tactic by the George W. Bush administration to allow the leasing of oil and gas activities in polar bear habitat in the Chukchi Sea prior to increases in their protection. However, the government responded that it needed additional time to fully consider the scientific data on the polar bear.

Exploring the Depths: Polar Bears in Captivity

Possibly the first recorded polar bear in a zoo was a white bear owned by Pharaoh Ptolemy II, nearly 2,300 years ago. Romans also captured polar bears and the poet Calpurnius wrote that he saw seals with bears, presumably polar bears, "pitted against them" in the arena of Emperor Nero.

The Norse hunted polar bears, usually killing the mother for her pelt and capturing the cubs. The kings of Norway frequently kept pet polar bears, dating from King Harold the Fairhaired in 880 AD. In 1252 a polar bear was given to King Henry III by the King of Norway. The bear was kept in the Tower of London and, according to legend, would be allowed out on a leash to go fishing in the River Thames.

Many zoos around the world currently hold polar bears (**Figure B6.1**). Many of these enclosures historically were not cooled and were little more than concrete pits. However, modern enclosures are typically chilled, with pools and landscaping. Polar bears in captivity are very prone to stereotypical pacing, a behavior associated with stress in many wide-ranging predatory mammals. In 1995, the Calgary Zoo tried to reduce polar bear pacing by feeding bears antidepressant medicine. More recently, many zoos try to reduce stereotypical pacing by providing activities to the bears, such as introducing objects for them to manipulate or gravel pits for them to dig in. However, there are also bears kept in far worse conditions. In 2002 six polar bears were seized by U.S. government officials from the Suarez Brothers Circus in Puerto Rico after complaints by animal welfare organizations and for violations under the MMPA.

FIGURE B6.1 Polar bears in a zoo.

SELECTED REFERENCES AND FURTHER READING

Aars, J., Lunn, N.J., & Derocher, A.E. (2006). *Proceedings of the 14th Working Meeting of the IUCN/SSC Polar Bear Specialist Group*. IUCN, Gland, Switzerland.

Andersen, M., Lie, E., Derocher, A.E., Belikov, S.E., Bernhoft, A., Boltunov, A.N., Garner, G.W., Skaare, J.U., & Wiig, Ø. (2001). Geographic variation of PCB congeners in polar bears (*Ursus maritimus*) from Svalbard east to the Chukchi Sea. *Polar Biology* 24: 231–238.

Bechshøft, T.Ø., Sonne, C., Dietz, R., Born, E.W., Novak, M.A., Henchey, E., & Meyer, J.S. (2011). Cortisol levels in hair of East Greenland polar bears. *Science of the Total Environment* 409: 831–834.

Bernhoft, A., Wiig, Ø., & Skaare, J.U. (1997). Organochlorines in polar bears (*Ursus maritimus*) at Svalbard. *Environmental Pollution* 95: 2159–2175.

Best, R.C. (1985). Digestibility of ringed seals by the polar bear. *Canadian Journal of Zoology* 63: 1033–1036.

Clubb, R., & Mason, G.J. (2004). Pacing polar bears and stoical sheep: testing ecological and evolutionary hypotheses about animal welfare. *Animal Welfare* 13: S33–S40.

Clubb, R., & Mason, G.J. (2007). Natural behavioural biology as a risk factor in carnivore welfare: how analysing species differences could help zoos improve enclosures. *Applied Animal Behaviour* 102: 303–328.

Derocher, A.E., Andersen, M., & Wiig, Ø. (2005). Sexual dimorphism of polar bears. *Journal of Mammalogy* 86: 895–901.

Derocher, A.E., Lunn, N.J., & Stirling, I. (2004). Polar bears in a warming climate. *Integrative and Comparative Biology* 44: 163–176.

Derocher, A.E., & Stirling, I. (1990). Observations of aggregating behaviour in adult male polar bears (*Ursus maritimus*). *Canadian Journal of Zoology* 68: 1390–1394.

Derocher, A.E., & Stirling, I. (1994). Age-specific reproductive performance of female polar bears (*Ursus maritimus*). *Journal of Zoology* 234: 527–536.

Fischbach, A.S., Amstrup, S.C., & Douglas, D.C. (2007). Landward and eastward shift of Alaskan polar bear denning associated with recent sea ice changes. *Polar Biology* 30: 1395–1405.

Haave, M., Ropstad, E., Derocher, A.E., Lie, E., Dahl, E., Wiig, Ø., Skaare, J.U., & Jenssen, B.M. (2003). Polychlorinated biphenyls and reproductive hormones in female polar bears at Svalbard. *Environmental Health Perspectives* 111: 431–436.

Jennison, G. (1922). Polar bears at Rome. Calpurnius Siculus, Ecl. VII. 65–6. *Classical Review* 36(3/4): 73.

Koon, D.W. (1998). Is polar bear hair fiber optic? *Applied Optics* 37: 3198–3200.

Kucklick, J.R., Struntz, W.D.J., Becker, P.R., York, G.W., O'Hara, T.M., & Bohonowych, J.E. (2002). Persistent organochlorine pollutants in ringed seals and polar bears collected from northern Alaska. *Science of the Total Environment* 287: 45–59.

Monnett, C., & Gleason, J.S. (2006). Observations of mortality associated with extended open-water swimming by polar bears in the Alaskan Beaufort Sea. *Polar Biology* 29: 681–687.

Oleson, T.J. (1950). Polar bears in the middle ages. *Canadian Historical Review* 31: 47–55.

Øritsland, N.A., Lentfer, J.W., & Ronald, K. (1974). Radiative surface temperatures of polar bear. *Journal of Mammalogy* 55: 459–461.

Ramsay, M.A., & Hobson, K.A. (1991). Polar bears make little use of terrestrial food webs: evidence from stable-carbon isotope analysis. *Oecologia* 86: 598–600.

Regehr, E.V., Hunter, C.M., Caswell, H., Amstrup, S.C., & Stirling, I. (2007). *Polar Bears in the Southern Beaufort Sea. I. Survival and Breeding in Relation to Sea Ice Conditions, 2001–2006*. U.S. Geological Survey, Reston, Virginia.

Regehr, E.V., Lunn, N.J., Amstrup, N.C., & Stirling, I. (2007). Effects of earlier sea ice breakup on survival and population size of polar bears in western Hudson Bay. *Journal of Wildlife Management* 71: 2673–2683.

Rode, K., Amstrup, S., & Regehr, E. (2010). Reduced body size and cub recruitment in polar bears associated with sea ice decline. *Ecological Applications* 20: 768–782.

Shepherdson, D.J., Carlstead, K.C., & Wielebnowski, N. (2004). Cross institutional assessment of stress responses in zoo animals using longitudinal monitoring of faecal corticoids and behaviour. *Animal Welfare* 13: S105–S113.

Stirling, I. (1999). *Polar Bears*. University of Michigan Press, Ann Arbor, MI.

Stirling, I., & Derocher, A.E. (1993). Possible impacts of climatic warming on polar bears. *Arctic* 46: 240–245.

Stirling, I., & Derocher, A.E. (2007). Melting under pressure: the real scoop on climate warming and polar bears. *Wildlife Professional* 1(3): 24–43.

Stirling, I., & Parkinson, C.L. (2006). Possible effects of climate warming on selected populations of polar bears (*Ursus maritimus*) in the Canadian Arctic. *Arctic* 59: 261–275.

Verreault, J., Muir, D.C.G., Norstrom, R.J., & Stirling, I. (2005). Chlorinated hydrocarbon contaminants and metabolites in polar bears (*Ursus maritimus*) from Alaska, Canada, East Greenland, and Svalbard: 1996–2002. *Science of the Total Environment* 351/2: 369–390.

Wiig, Ø., Derocher, A.E., Cronin, M.M., & Skaare, J.U. (1998). Female pseudohermaphrodite polar bears at Svalbard. *Journal of Wildlife Diseases* 34: 792–796.

Otters

CHAPTER 7

CHAPTER OUTLINE

Sea Otter

Distribution
Aquatic Adaptations
Feeding Behavior and Ecology

Exploring the Depths: Tool Use

Reproduction
Other Behavior
Abundance and Status

Exploring the Depths: Sea Otters and the Exxon Valdez *Oil Spill*
Exploring the Depths: Sea Otter Declines and Killer Whales

Marine Otter

Distribution, Abundance, and Status
Feeding Behavior and Ecology
Reproduction

Exploring the Depths: Eurasian Otter Rescue and Rehabilitation

   Cubs
   Juveniles
   Adults

Selected References and Further Reading

Two species of otter exclusively inhabit the marine environment and are generally considered to be marine mammals: the sea otter (*Enhydra lutris*) and the marine otter (*Lontra felina*).

Sea Otter

Sea otters look superficially similar to other otters but have rounder heads and thicker fur (**Figure 7.1**). In addition, their muzzles are rather blunt, and they have relatively shorter tails than other otters, which may be another example of "Allen's rule" (see Chapter 4). Sea otters often swim and rest on their backs, which other otters do infrequently. Sea otters have the greatest body mass of any otter, with males growing up to 1.5 m and weighing 45 kg and females growing up to 1.4 m and weighing over 32 kg. The large body size

FIGURE 7.1 A sea otter feeding.

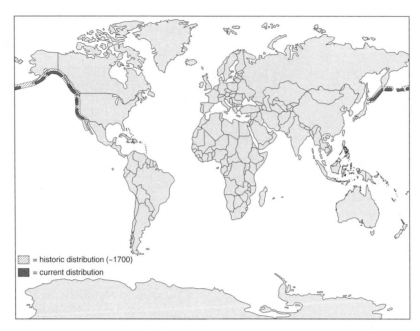

FIGURE 7.2 Map showing the current and historic distribution of the sea otter. Data from: U.S. Fish and Wildlife Service.

of sea otters appears to be another example of "Bergmann's rule" within groups of marine mammals (see Chapter 4).

Distribution

Sea otters are distributed around the northern rim of the Pacific Ocean from northern Japan in the west through the Aleutian Islands, Alaska, Canada, and the U.S. western coast to California (**Figure 7.2**). There are possibly three subspecies:

- *Enhydra lutris lutris*, extending from Japan to Kamchatka, Russia
- *Enhydra lutris kenyoni*, extending from the Aleutian Islands to mainland Alaska
- *Enhydra lutris nereis*, extending from northern California to Baja

Sea otters are found in near-shore coastal waters and are often associated with kelp forests and are believed to be a keystone species in the ecology of these habitats (see below).

Aquatic Adaptations

Sea otters have flexible spines and swim with an undulating motion, with their flattened tails acting like paddles and webbed toes with wide rear feet providing propulsion. They can swim up to 2.5 m/sec (9 km/h), although they typically swim at a speed closer to 0.5 m/sec (1.8 km/h). Their thick vibrissae (whiskers) allow them to detect vibrations in the watery environment, similar to seals and walruses.

Because of their relatively small size they have a relatively high rate of heat loss in cold water. Sea otters lack a thick insulating blubber or fat layer, and it is their dense fur (>100,000 hairs per cm^2) that provides most insulation. However, wet fur is a poor insulator, especially when compressed. To keep fur in good condition and to prevent it from clumping due to salt penetration, sea otters can spend 10% of their time grooming each day.

To generate sufficient metabolic heat to keep warm, sea otters have relatively higher metabolic rates than would be expected for mammals of their size. To fuel this higher metabolic rate, they have to consume 25 to 38% of their body mass per day, which on average is greater than the food intake of a similar-sized terrestrial mammal. Sea otters, especially cubs, may therefore be susceptible to starvation if prey are scarce.

Sea otters have proportionally larger lungs compared with a similar-sized mammal (two and a half times larger than typical). Large lungs provide buoyancy, for floating at the sea surface, and are their main store of oxygen when they dive, although oxygen bound to hemoglobin provides an additional oxygen store. They have a short, wide trachea that allows them to exchange gases from their lungs rapidly when they surface, and cartilage in their lung arterioles ensures that these narrow airways do not constrict and collapse under pressure when the otter is diving (**Figure 7.3**).

Sea otters on average dive to 5 to 35 m to feed but have been recorded diving to 100 m. This is relatively shallow

FIGURE 7.3 A sea otter diving underwater.

FIGURE 7.4 A sea otter breaking open shellfish with a rock "anvil".

compared with other marine mammals, such as pinnipeds but is nonetheless deeper than is possible for a human recreational scuba diver. Sea otter dives last on average 30 to 60 seconds (to a maximum of approximately 260 seconds).

Feeding Behavior and Ecology

Otters search for food on or near the seabed, using their sensitive vibrissae to locate items in murky waters. They use their forepaws to search for food items, turn over rocks, and catch fish. Their diet consists of clams and other shellfish, such as snails, abalone, urchins, and crabs. To help carry food, sea otters have folds of skin under their arms in which they can tuck food items. Often, they break open shellfish using a stone as a tool (see Exploring the Depths: Tool Use). While floating on their back they place a flat stone on their belly, hold the prey in both paws and smash open the prey item using the stone as an anvil (**Figure 7.4**).

Sea otters have been recorded to feed on more than 150 prey species throughout their range. An individual's diet is dominated by only a few important species and may be influenced by habitat and season. Individual otters have preferences as to the prey items they consume. As these preferences seem to be matrilineal (passed down the female line), it is likely that cubs learn the skills to handle and consume these specific prey items from their mothers. Because of their high daily energy requirements and their predation on major herbivores of kelp, sea otters are considered to be keystone species in kelp forests. If sea otters are removed, kelp consumers, specifically sea urchins, increase in numbers, which leads to kelp "deforestation." The loss of the kelp forest habitat in turn leads to declines in fish and other species that depend on the kelp habitat. Thus, sea otters can have a major impact on the ecology of a coastline.

Exploring the Depths: Tool Use

Tool use is considered to require special cognitive abilities because it involves causally relating two or more objects that are external to one's body. One of the earliest explicit definitions of tool use focuses on abstract properties of behavior, including functionality and goals, and states that "Tool use is the use of an external object as a functional extension of mouth or beak, hand or claw, in the attainment of an immediate goal" (Van Lawick-Goodall, 1970, p. 195). However, a more recent definition has served the field of animal cognition for over 25 years, "the external employment of an unattached environmental object to alter more efficiently the form, position, or condition of another object, another organism, or the user itself when the user holds or carried the tool during or just prior to use and is responsible for the proper and effective orientation of the tool" (Beck, 1980, p. 10).

However, Beck's definition does not distinguish some dependent properties necessary or causal properties of tool use. For example, in Beck's definition object manipulation is central to tool use, whereas the unattached property is not as critical. This property arguably stands for the degree of control the animal can exert in manipulating an object as a tool; however, bottlenose dolphins (*Tursiops truncatus*) have been observed to use sponges as tools to protect their rostrums while foraging. In this example the sponge is neither a functional extension of the mouth or hand nor does it alter the condition of the tool user. Therefore, the use of sponges by dolphins does not

Exploring the Depths: Tool Use (*continued*)

comply with Beck's definition of tool use. It is for this reason that St. Amant and Horton (2008, p. 1203) redefined tool use as follows:

> The exertion of control over a freely manipulable external object (the tool) with the goal of (1) altering the physical properties of another object, substance, surface or medium (the target, which may be the tool user or another organism) via a dynamic mechanical interaction or (2) mediating the flow of information between the user and the environment or other organisms in the environment.

One of the most famous examples of tool use in marine mammals is the use of rocks by sea otters. Sea otters native to northern and eastern parts of the Pacific Ocean prey upon sea urchins, mollusks (specifically bivalves), crustaceans, and some large fish. Sea otters dive to the seafloor and carefully select a flat rock that they use to hammer open bivalves such as abalone, mussels, and clams at a rate of 45 hits in 15 seconds. These animals are successfully able to open the shells of their prey in two to three dives.

Humans were previously believed to be superior and above all other primate species in intelligence because they were "tool-using apes." However, upon the discovery that chimpanzees used tools the renowned paleontologist/primatologist, Louis Leakey, famously said, "Now we must redefine tool, redefine man, or accept chimpanzees as humans."

—Contributing author, Katheryn Patterson, George Mason University.

Reproduction

Female sea otters become sexually mature at about 3 years of age. Males defend a large territory that encompasses several smaller female territories (polygynous mating system). Males may locate females by detecting their waterborne scent; males have large and convoluted nasal cavities that increase their ability to "smell." When mating, a male mounts the female and may bite her nose to hold on, which can lead to injury and scarring for the female. Males have been known to be so aggressive when mating they have killed their mates. After a 4- to 6-month gestation period, involving a period of delayed implantation of the embryo, females give birth to a single cub, weighing approximately 1.8 kg. Cubs are born throughout the year but mostly in May and June (Alaska) and peaking either in January through March or August through September (California). Cubs are normally weaned after 6 months.

Other Behavior

Sea otters are often seen singly but can form groups (referred to as rafts) of 50 (in California) to 2,000 animals (in Alaska). They feed in shallow coastal waters, especially in kelp beds. When resting, they may wrap kelp around their body to anchor themselves against waves and currents (**Figure 7.5**). When resting they often hold their head out of the water, with their chins resting upon their chests, as well as their front and back paws, which minimizes heat loss from their appendages. Their large lungs provide them with buoyancy to allow them to float easily when resting. They have been observed holding their forepaws over their eyes during the day, perhaps when sleeping. When awake they spend much of their time foraging and grooming.

Abundance and Status

When European colonizers first discovered sea otters in the 1700s, they ranged from Baja California along the west coast of the United States into Alaska and around the Pacific to the eastern coast of Russia and down into Japanese waters. In the early 1800s, however, sea otters were hunted to near extinction because of their luxuriant fur. They were thought to have gone extinct in California; however, a raft of about 300 was unexpectedly discovered in 1938, although

FIGURE 7.5 A resting sea otter wrapped in kelp.

biologists had first noted their presence in 1915 but kept it secret. Since that time the population in California has slowly grown to number about 2,200, which is a substantial increase, but this is only about 10% of their estimated historic population size. Sea otter numbers in Alaska were also depleted but were recovering in the 20th century. However, in the 1990s sea otters in the Aleutian Islands declined by 80 to 90%; the cause of this decline is not known. Much of the sea otter's former range in Asia has also been reoccupied.

A major cause of sea otter mortality is entanglement in fishing gear, particularly gill nets. Pollution is also a problem, and high levels of organic contaminants (polychlorinated biphenyls, or PCBs) have been found in sea otters in the Aleutian Islands, with the contamination probably coming from military bases in the area. The *Exxon Valdez* oil spill in March 1989 led to a high mortality of sea otters in Prince William Sound, Alaska (see Exploring the Depths: Sea Otters and the *Exxon Valdez* Oil Spill).

As with many marine mammal species, aboriginal harvest of sea otters is permitted, with no restrictions on numbers taken from some populations. In addition to this permitted hunting, poaching for pelts and illegal culling of otters (due to perceived competition for shellfish) are other causes of mortality. It has been suggested that killer whale predation may be a significant source of mortality for sea otters in the Aleutian Islands, although this hypothesis is somewhat controversial (see Exploring the Depths: Sea Otter Declines and Killer Whales).

Sea otters are classified as "endangered" by the International Union for Conservation of Nature (IUCN). Sea otters are protected under the U.S. Marine Mammal Protection Act and the U.S. Endangered Species Act. Californian sea otters are listed as "threatened" under the Endangered Species Act (for details on laws affecting marine mammals, see Chapter 17). As a result of this listing the U.S. Fish & Wildlife Service adopted a recovery plan for Californian sea otters in 1981.

The U.S. Congress enacted Public Law 99-625 (PL 625) to increase the Californian sea otter population, and text was also inserted to protect shell fisheries. In 1987 sea otters were reintroduced to San Nicolas Island; however, somewhat contrarily, PL 625 enacted a "no-otter zone" and fishery interests demanded that otters be removed from this zone. Other measures to help promote sea otter populations include rehabilitation programs for stranded and orphaned otters, such as the one run by the Monterey Bay Aquarium in California.

Exploring the Depths: Sea Otters and the *Exxon Valdez* Oil Spill

On March 24, 1989 the tanker *Exxon Valdez* spilled 10.9 million gallons of crude oil into Prince William Sound, Alaska. This was the largest oil spill in U.S. history and is currently the 35th largest single spill in the world. By June 20 the oil covered 28,500 km². In total, 40% of the oil stranded on beaches in Prince William Sound; 1,900 km of the coast were oiled (900 km of which were heavily oiled) but up to 5,221 km of shoreline received oil contamination. Three years later some 20% of the original oil spill still remained on beaches in 1992, and oil was found in sediments nearly a decade later.

The area where the spill occurred was important sea otter habitat, and an estimated 3,500 to 5,500 otters were oiled (**Figure B7.1**). A minimum number of 751 sea otters were estimated to have died as a result of the spill, but mortality from the spill could be as high as 2,650. Actions were taken to try to clean and rehabilitate these oiled otters, which was a major undertaking. The estimated cost of the rehabilitation program was $51,000 (U.S.) per sea otter. Despite the efforts of the U.S. government, the recovery of the otter population from the spill was much slower than expected, and reduced survival rates were recorded in sea otters a decade after the spill. It is likely that toxic effects from oil in the marine ecosystem have had a long-term detrimental effect on the health of the otter population.

FIGURE B7.1 A dead otter covered in oil.

Exploring the Depths: Sea Otter Declines and Killer Whales

In southwestern Alaska populations of sea otters have seriously declined (as well as Steller sea lions, northern fur seals, harbor seals, and seabirds) in the past three decades, some populations by as much as 80 to 90%. Such a widespread pattern indicates the collapse of an entire ecosystem, and researchers, conservation groups, and wildlife agencies have expressed serious concerns.

Some scientists have developed a hypothesis that the sea otter (and other species) declines are due to killer whales "fishing down the food chain," that is, eating smaller prey than they had in the past. Transient orcas (see Chapter 12) that eat marine mammals are the focus of this hypothesis rather than resident orcas, which eat fish. The hypothesis presupposes that transient orcas were forced to abandon a previously favored food source (i.e., large whales). These researchers point out, correctly, that historic Pacific whaling devastated fin, gray, and humpback whale populations. Then they speculate that orcas, deprived of the whales, ate through the populations of mid-sized sea lions and fur seals and finally had to focus on the relatively small sea otters for their food supply.

However, although there have been occasional sightings of orcas killing sea lions and parts of sea lions and sea otters have been found occasionally in the stomachs of stranded killer whales (among other prey items), actual sightings of orcas eating sea otters are extremely rare. The scientists built their hypothesis on 10 instances of orcas harassing or attacking otters during their studies in the 1990s, behavior that had not been previously observed, but the killer whales were not actually observed eating these otters. The researchers then performed a theoretical exercise, calculating how many sea otters an average orca could eat, and concluded that orcas could be responsible for the otters' decline. Their hypothesis was based on this theoretical exercise rather than on empirical observations of high levels of otter consumption.

Moreover, several biologists have pointed out that the hypothesis is not supported by convincing evidence that high levels of orca predation have indeed caused these population crashes. Indeed, there is no direct evidence that orcas ever preferred to eat large whales. North Pacific minke whales were not historically targeted by whalers and remained abundant

Exploring the Depths: Sea Otter Declines and Killer Whales (*continued*)

throughout the 20th century, as did Dall's porpoises (which are commonly eaten by orcas). Gray whale populations were recovering during the time orcas supposedly had no large whales to eat and are now seasonally abundant, and fin whales are increasingly sighted off western Alaska.

The hypothesis is further compromised because populations of pinnipeds are flourishing in other regions inhabited by transient orcas. In particular, orca researchers found that in southeast Alaska, transients were much more common than residents, yet the pinniped populations there were either stable or increasing, and so the feeding down the food web phenomenon does not appear to be happening in areas where transients are common. Indeed, about three-fourths of the orcas identified in the eastern North Pacific are fish-eating residents and not those who eat marine mammals at all.

Unfortunately, many members of the public (e.g., a display at the Monterey Bay Aquarium calls the hypothesis "an airtight case") and journalists have latched onto this speculative hypothesis as if it were fact. Some journalists have even typecast the orcas as villains; for example, a leading U.S. newspaper, *The Washington Post*, had a headline that read "From Ocean Icons to Prime Suspects: Orcas Devastate Seal, Otter Populations." *The Washington Post* article was also full of hyperbole, saying that Alaskan orcas were on a "species-threatening rampage," where they were "wiping out" sea lions and otters. Such a framing of the situation has led to members of the public and even local politicians calling for culls of orcas to protect sea otter and sea lion populations. Furthermore, articles have frequently failed to include the perspective of large-whale biologists, many of whom reject the notion that orcas ever preferred to hunt large whales.

Unfortunately, articles such as the above rarely mention the negative impacts of regional overfishing, pollution, or climate change, which might be a more logical explanation for the decline of the sea otter. Several politicians and special interest group leaders have been only too eager to embrace the hypothesis and blame orcas for declines in marine mammal populations to avoid closer scrutiny into other possible causes, such as poor fishery management or global warming.

It is very dangerous for the media to report conjecture as fact and for scientists not to clarify the difference between theoretical, unsubstantiated hypotheses and tested and verified scientific research. Many orca populations are beset by human pressures such as pollution and overfishing, whereas some orca stocks are threatened with extinction. By deflecting the focus off current human-caused hazards facing the Pacific Northwest ecosystems, examination of, and action on, more likely causes of these sea otter and sea lion declines may be delayed while this more sensational and politically convenient "explanation" takes center stage.

–Contributing author, Naomi Rose, Humane Society International.

Marine Otter

The marine otter looks similar to an American river otter (*Lontra canadensis*) or Eurasian otter (*Lutra lutra*). Marine otters are considerably smaller than sea otters, weighing 3 to 6 kg, and grow to approximately 1 m in length. It is dark brown on its head and back, with a cream underside and a black nose (**Figure 7.6**). The tail is flattened but is thinner and more tapered than that of the sea otter. Unlike American river and Eurasian otters the marine otter does not appear to depend so heavily on freshwater streams/pools for drinking and for washing salt from its fur. The fur is also rougher and coarser than that of those species, which may allow it to maintain its insulation properties in the marine environment.

Distribution, Abundance, and Status

The original distribution of the marine otter is along the Pacific coast of South America from Peru, down the length of Chile to Cape Horn and Isla de Los Estados in Argentina (**Figure 7.7**). It has now become locally extinct (or extirpated) throughout its range and is found mainly on exposed rocky shores and offshore islands. In the past the marine otter was heavily hunted. The current estimated population size is less than 1,000 individuals, and population has declined due to habitat loss, persecution due to perceived competition with fisheries, accidental death, and hunting for their valuable pelt. The marine otter is classified as "endangered" by the IUCN, and in Argentina, Chile, and Peru the marine

FIGURE 7.6 A Chilean marine otter.

otter is a protected species. IUCN predicts future population decreases of 50% over the next 30 years unless conservation measures are enforced.

Feeding Behavior and Ecology

Relatively little is known about the behavior of the marine otter. Usually one or two animals are observed, but occasionally they are seen in groups of three or more. They feed mainly on crustaceans (crabs, shrimp), fish, and mollusks from the littoral and sublittoral zone and sometimes enter rivers to feed on freshwater species. Along the Chilean coast up to 25 species of prey have been recorded in the diet. Diet is studied by identifying undigested prey in spraints (feces) that are deposited on conspicuous rocks and entrances to holts or by visual observations of prey caught. Fish species taken may overlap with the South American sea lion, which is sympatric with marine otters.

Home ranges typically extend 30 m inland to 150 m offshore in exposed areas of coastline, with abundant marine algae that provides suitable prey habitat. Marine otters forage alone and dive to depths of 30 to 40 m for on average of 30 seconds (occasionally more than 1 minute) to locate prey. Killer whales are the main natural predator of marine otters. Adults may be fed on by sharks, and birds of prey have been recorded taking cubs on land.

Reproduction

When on land the marine otter lives in holts (otter dens) beneath large rocks or in sea caves, frequently in small family groups consisting of more than one adult and several cubs. Breeding may occur January through March, and two to four cubs are born after a gestation period of 60 to 70 days. More than one adult has been seen to bring back prey to feed young, which has been interpreted as indicating either a monogamous breeding system or adult offspring from previous litters cooperating with females to feed young. However, further research is required to establish the nature of social organization within marine otters.

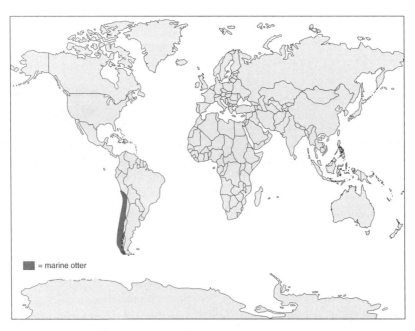

FIGURE 7.7 Map showing the current, known distribution of the marine otter. Data from: IUCN Red List.

Exploring the Depths: Eurasian Otter Rescue and Rehabilitation

Otter rescue and rehabilitation should not be undertaken by anyone who is not familiar with the species or who is not willing to ensure the animal remains wild so it can survive once released. Eurasian otter cubs stay with their mothers for 12 to 15 months, so they must be released at this age. This makes it even more important to keep them away from human contact, because an imprinted or tame animal cannot be released. Casualties can be broadly divided into three groups: cubs, juveniles, and adults, each with different needs.

Cubs

Cubs are born at more or less any time of the year and so can appear in a rehabilitation center during any month. Most cubs arrive at around 10 to 12 weeks of age, the time they first venture out of the holt. Usually, it is not known how they have been orphaned, but if possible it is important to wait and give the mother a chance to return.

When they arrive at the center they are usually cold, wet, and dehydrated, and it is therefore vital to give the usual first aid treatment of warmth and liquids. Individuals are weighed and checked for any signs of injury or illness.

At the International Otter Survival Fund, the cub-rearing program has three parts: indoors, nursery pens, and croft pens. All new cubs are first kept indoors in a specially designed warm cub unit. If the cub is very young, it is given milk substitute (because otters cannot tolerate lactose). Older cubs may start feeding on filleted or pureed fish. They then progress on to whole fish.

It is always better to rear two cubs together even if they are not related so they develop "normal" otter behavior. Young animals are gradually introduced to outside nursery pens with a small pool and sleeping area that can be heated. Here they can be monitored regularly.

The third stage is when they are transferred to the croft pens, which are 25-m^2 areas of natural moorland with a natural pool and sleeping box. They are fed and checked only once a day, and in this way the cubs are totally wild when it comes to release.

Choosing the release site must be done with care. It is generally accepted policy that cubs must be released in the same geographical area in which they were found unless there are problems.

Juveniles

If an adult otter is injured it will often seek somewhere quiet, like a shed, where it can rest. If an animal appears uninjured it is likely to be a juvenile that has become separated from its mother too early and is finding it difficult to adapt. In these cases the juvenile is kept until its normal dispersal age, keeping human contact to a minimum, and then released into the area where it was found.

Both juveniles and adults have to be treated with care when being caught. Sometimes it is possible to trap them using a large cat trap with a spring door. However, otters do learn quickly, and if they spring the trap but manage to escape, it is unlikely they will go in the trap again. A grasper is used to restrain the animal, but the loop of the grasper must be positioned behind the front legs because it will slip off if only placed around the neck. It is also better for the animal to be held in this way rather than around the neck.

Adults

Any adult that comes into a center obviously has a serious problem that needs treatment by a vet. Road traffic accidents are the most common problem, followed by bites due to intraspecific fighting. Bite wounds are often on the face and/or genital area, and can often be extremely serious. Sometimes an animal may appear to have no obvious wounds because they are hidden in the thick fur, and this reinforces the need for a good veterinary examination to ensure that such wounds are not missed. Adults must be released to the same area (within reason!) because they already have an established home range.

–Contributing author, Grace Yoxon, International Otter Survival Fund.

SELECTED REFERENCES AND FURTHER READING

Beck, B.B. (1980). *Animal Tool Behavior: The Use and Manufacture of Tools.* Garland STPM Press, New York.

Bentley-Condit, V.K., & Smith, E.O. (2010). Animal tool use: current definitions and an updated comprehensive catalog. *Behaviour* 147: 185–221.

Bodkin, J.L., Ballachey, B.E., Dean, T.A., Fukuyama, A.K., Jewett, S.C., McDonald, L., Monson, D., O'Clair, C.E., & VanBlaricom, G.R. (2002). Sea otter population status and the process of recovery from the 1989 *Exxon Valdez* oil spill. *Marine Ecology Progress series* 241: 237–253.

Caro, T.M., & Hauser, M.D. (1992). Is there teaching in nonhuman animals? *Quarterly Review of Biology* 67: 151–174.

Costa, D.P., & Kooyman, G.L. (1982). Oxygen consumption, thermoregulation, and the effect of fur oiling and washing on the sea otter, *Enhydra lutris. Canadian Journal of Zoology* 60: 2761–2767.

Dean, T.A., Bodkin, J.L., Fukuyama, A.K., Jewett, S.C., Monson, D.H., O'Clair, C.E., & VanBlaricom, G.R. (2002). Sea otter (*Enhydra lutris*) perspective: mechanisms of impact and potential recovery of nearshore vertebrate predators following the 1989 *Exxon Valdez* oil spill. *Marine Ecology Progress series* 241: 255–270.

Estes, J.A. (1990). Growth and equilibrium in sea otter populations. *Journal of Animal Ecology* 59: 385–401.

Estes, J.A., Riedman, M.L., Staedler, M.M., Tinker M.T., & Lyon, B.E. (2003). Individual variation in prey selection by sea otters: patterns, causes and implications *Journal of Animal Ecology* 72: 144–155.

Estes, J.A., Smith, N.S., & Palmisano, J.F. (1978). Sea otter predation and community organization in the western Aleutian Islands, Alaska. *Ecology* 59: 822–833.

Estes, J.A., Tinker, M.T., Williams, T.M., & Doak, D.F. (1998). Killer whale predation on sea otters linking oceanic and nearshore ecosystems. *Science* 282: 473–476.

Fisher, E.M. (1939). Habits of the southern sea otter. *Journal of Mammalogy* 20: 21–36.

Hall, K.R.L., & Schaller, G.B. (1964). Tool-using behavior of the California sea otter. *Journal of Mammalogy* 45: 287–298.

Hatfield, B.B., Marks, D., Tinker, M.T., Nolan, K., & Peirce, J. (1998). Attacks on sea otters by killer whales. *Marine Mammal Science* 14(4): 888–894.

Heithus, M.R., & Dill, L.M. (2002). Feeding strategies and tactics. In: *Encyclopedia of Marine Mammals*, 2nd ed. (Ed. W.F. Perrin, B. Würsig, & J.G.M. Thewissen), pp. 414–423. Academic Press, San Diego, CA.

Kenyon, K.W. (1969). The sea otter in the eastern Pacific Ocean. *North American Fauna* 68: 1–352.

Krützen, M., Mann, J., Heithaus, M.R., Connor, R.C., Bejder, L., & Sherwin, W.B. (2005). Cultural transmission of tool use in bottlenose dolphins. *Proceedings of the National Academy of Sciences* 102: 8939–8943.

Kruuk, H. (2006). *Otters: Ecology, Behaviour and Conservation.* Oxford University Press, Oxford, UK.

Lariviere, S. (1998). *Lontra felina. Mammalian Species* 575: 1–5.

Love, J.A. (1992). *Sea Otters.* Fulcrum Publishing, Golden, CO.

Monson, D.H., Doak, D.F., Ballachey, B.E., Johnson, A., & Bodkin, J.L. (2000). Long-term impacts of the *Exxon Valdez* oil spill on sea otters, assessed through age-dependent mortality patterns. *Proceedings of the National Academy of Sciences* 97: 6562–6567.

Monson, D.H., Estes, J.A., Bodkin, J.L., & Siniff, D.B. (2000). Life history plasticity and population regulation in sea otters. *Oikos* 90: 457–468.

Riedman, M.L., & Estes, J.A. (1990). The sea otter *Enhydra lutris*: behavior, ecology and natural history. *US Fish and Wildlife Service Biological Report* 90: 1–126.

St. Amant, R., & Horton, T. E. (2008). Revisiting the definition of tool use. *Animal Behaviour* 75: 1199–1208.

VanBlaricom, G. R. (2001). *Sea Otters.* Voyageur Press, Stillwater, MN.

Van Lawick-Goodall, J. (1970). Tool-using in primates and other vertebrates. In: *Advances in the Study of Behavior.* Vol. 3 (Ed. D. Lehrman, R. Hinde, & E. Shaw), pp. 195–249. Academic Press, New York.

Watt, J., Siniff, D.B., & Estes, J.A. (2000). Inter-decadal patterns of population and dietary change in sea otters at Amchitka Island, Alaska. *Oecologia* 124: 289–298.

Williams, T.M., Estes, J.A., Doak, D.F., & Springer, A.M. (2004). Killer appetites: assessing the role of predators in ecological communities. *Ecology* 85: 3373–3384.

Yoxton, G.M. (2006). Caring for wild otters. In: *The Return of the Otter in Europe—Where and How?* (Ed. J.W.H. Conroy, G. Yoxon, A.C. Gutleb, & J. Ruiz-Olmo). International Otter Survival Fund, Broadford, Isle of Skye.

Sirenians

CHAPTER 8

CHAPTER OUTLINE

Distribution
Feeding Behavior and Ecology
Reproduction
Abundance and Status

Exploring the Depths: Conservation of the Florida Manatee
Exploring the Depths: Conservation of the Dugong
Exploring the Depths: Extinction of Steller's Sea Cow

Selected References and Further Reading

Sirenians get their name from the sirens of Greek mythology, aquatic monsters that lured sailors to their doom upon the rocks. The sirenians are split into two extant (living) groups: manatees (Family Trichechidae) and dugongs (Family Dugongidae) (see Chapter 2). Sirenians are the only marine mammals that are obligate herbivores (exclusively feed on plants) and as such have unique life histories and adaptations among the group.

There are three living species of manatee: the West Indian manatee (*Trichechus manatus*), the Amazonian manatee (*T. inunguis*), and the West African manatee (*T. senegalensis*) (**Figure 8.1**). All manatees have rounded tails, and the largest of these is the West Indian manatee, reaching up to 4 m in length and weighing up to 1,500 kg. The West African manatee looks similar to the West Indian manatee but has a blunter snout, more protruding eyes, and more slender body. The Amazonian manatee is the smallest of the manatees at less than 3 m in length and weighing less than 500 kg. They possess a rubbery skin (smoother than other manatees), have no "nails" at the tips of their flippers, and often have white or pinkish belly patches. All manatees move between freshwater and marine habitats.

There is only one living species of dugong (*Dugong dugong*); a second species, Steller's sea cow (*Hydrodamalis gigas*), became extinct in recent history (see Exploring the Depths: Extinction of Steller's Sea Cow). The dugong superficially resembles the manatee but is less than 3 m long and weighs 150 to 300 kg. Dugongs do not have nails on their flippers, possess short tusks, and have a fluked (split) crescent-shaped, dolphin-like tail. Their habitat is exclusively marine.

Like other marine mammals, sirenians possess many adaptations for an aquatic existence (see Chapter 3). They have a fusiform (streamlined/spindle-shaped) body with no pelvic limbs, a dorsoventrally flattened tail that increases surface area for swimming, and forelimbs modified as flippers. They are large mammals, which reduces their heat loss in the water relative to their body size (although cold temperatures are still a problem for sirenians, as discussed below). Their thick impermeable skin reduces water loss (or gain) to their surroundings. Sirenians have large lobular kidneys which improve water and salt extraction. Their nostrils are on top of their muzzles, and muscular flaps can close these off completely. Manatees may be very difficult to see in the wild as sometimes only the tip of their muzzle breaks the water surface. They have thick and heavy bones (pachyosteosclerotic), similar to their distant elephant relatives, and these dense bones provide ballast. Their extensive digestive system often contains copious amounts of gas, and therefore their high bone density counteracts the buoyancy this provides. Lungs are situated dorsally (i.e., just below their "backs"), with two horizontal hemidiaphragms that allow them to empty and fill their lungs independently. This appears to provide additional buoyancy control,

FIGURE 8.1 Three species of manatee: (A) West African, (B) West Indian, and (C) Amazonian with (D) a dugong for comparison.

similar to a submarine that can independently fill and "blow" its ballast tanks.

Distribution

Sirenians live in warm freshwater, brackish, and marine habitats and generally prefer shallow grass beds as feeding areas. They frequently select secluded waterways and lagoons in which to rest and breed, but there is some variation between the different species.

The West Indian manatee is found in coastal and riverine habitats and is euryhaline, being able to inhabit both fresh and saltwater, although it requires a source of fresh water to drink. The species is currently split into two subspecies, the Florida manatee (*T. manatus latirostris*) and the Antillean manatee (*T. manatus manatus*), although some recent DNA work has suggested three geographically distinct lineages may be present (northeastern south America, the northern part of South America and Central America, and Florida and the Greater Antilles). Despite its name, the Florida manatee inhabits coastal water and rivers across the southeastern United States (**Figure 8.2**). During the summer its range actually extends from the coast of Texas (in the Gulf of Mexico) all along the Atlantic Coast, occasionally as far north as Rhode Island. The manatees over-winter toward the southern extent of their range, aggregating in warm-water refuges. Some Florida manatees are found in the Everglades all year, where the water remains above 20°C. The Florida manatee population is believed to have increased in recent years (see Exploring the Depths: Conservation of the Florida Manatee).

The West Indian manatee ranges from the western coast of the Gulf of Mexico, from Mexico through Belize, Guatemala, Honduras, Nicaragua, Costa Rica, Panama, through to the coasts of Colombia, Venezuela, Guyana, Suriname, French Guiana, and Brazil. It is also found around the coasts

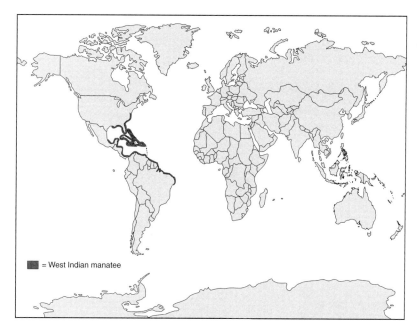

FIGURE 8.2 Map showing the current, known distribution of the West Indian manatee. Data from: IUCN Red List.

of the Caribbean islands, including the Bahamas, Cuba, Jamaica, Hispaniola (Haiti and the Dominican Republic), Puerto Rico, and Trinidad. Christopher Columbus, when traveling to the "New World," noted in his logbook that he saw mermaids (assumed to be manatees) off the coast of Haiti. There were historically manatee populations in Antigua, Barbados, Grenada, Martinique, Tobago, and the Virgin Islands, but these have become extirpated. Of the above manatee populations only the Bahamas, Belize, Cuba, and Puerto Rico populations are believed to be increasing.

As its name suggests, the Amazonian manatee is only found in the fresh water of the Amazon River, from the mouth of the river system in Brazil to Peru, with this range extending through Guyana, Ecuador, Venezuela, and Colombia (**Figure 8.3**).

The West African manatee is found in coastal and riverine areas in 21 West African countries from Senegal in the north to Angola in the south, including coastal and riverine areas of Gambia, Guinea-Bissau, Guinea, Sierra Leone, Liberia, Côte d'Ivoire, Ghana, Togo, Benin, Nigeria, Cameroon,

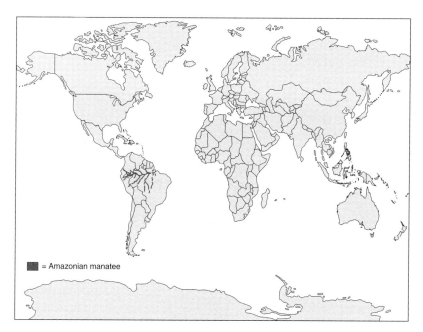

FIGURE 8.3 Map showing the current, known distribution of the Amazonian manatee. Data From: IUCN Red List.

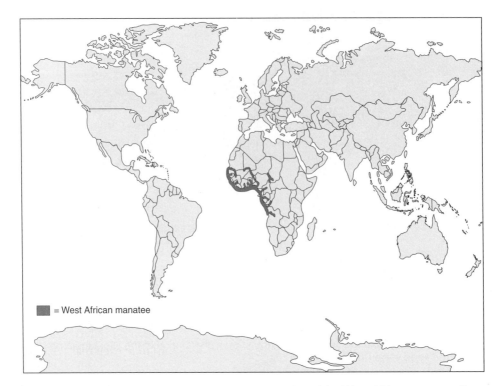

FIGURE 8.4 Map showing the current, known distribution of the West African manatee. Data from: IUCN Red List.

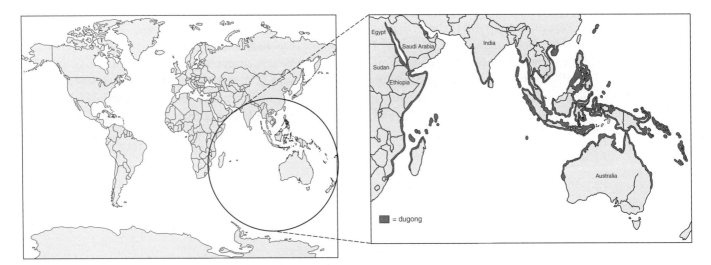

FIGURE 8.5 Map showing the current, known distribution of the dugong. Data from: IUCN Red List.

Gabon, Republic of the Congo, and the Democratic Republic of the Congo (**Figure 8.4**). They are even found in some inshore riverine areas that are cut off from the sea by rapids and dams. Manatees in the Logone and Chari rivers are substantially inland, but in the wet season these rivers connect with the River Niger, which in turn connects to the sea.

Dugongs inhabit coastal waters in the Indo-Pacific. They are found in waters up to 33 m deep, and their distribution is closely associated with the distribution of seagrass beds. In total, dugongs are found in 37 countries from northeastern South Africa and Mozambique, to Pakistan, around the coast of India, the Malaysian Peninsula, Thailand, to Vietnam, then eastward to the Philippines, Indonesia, New Guinea, and south to Australia (**Figure 8.5**).

Feeding Behavior and Ecology

Manatees have a primarily plant-based diet, including benthic seagrass and floating aquatic vegetation, but they very occasionally eat tunicates or sea squirts and even fish captured in nets. Their ridged molar teeth grind up this plant matter. They do not possess incisors or canines but have a fleshy pad in the front of their upper jaw. Manatee teeth, like those of elephants, are continuously replaced as the molars grind down, which happens relatively rapidly if they take in mouthfuls of coarse sand and sediment along with vegetation. The whiskers on their fleshy lips are prehensile and very sensitive, and aid the manatee in both finding and pulling up vegetation. This diet is difficult to digest and so manatees have large intestinal systems that allow longer periods for digestion, aided by symbiotic gut microbes (e.g., bacteria) to break down materials that are relatively indigestible, and have a relatively low energy and nutrient content.

As an adaptation to this nutritionally poor diet, manatees have a metabolic rate 20 to 30% lower than expected for an animal of their size and thus have reduced energy needs compared with other equivalent sized animals. Associated with their lower metabolism, they are relatively slow moving which may make them susceptible to collisions with fast-moving boats. In addition, lower metabolic heat production means they are not well adapted to cold water temperatures. Therefore, they are either found in waters that are constantly warm or must migrate toward the equator or to warm water refuges, such as warm water springs or even to warm water outflows from power stations.

Because of the down-turned muzzle of the dugong, they can only bottom feed on seabed vegetation. Dugongs primarily feed on beds of seagrass and are usually observed singly or in pairs. Occasionally, however, herds of several hundred animals can occur. Like manatees, dugongs can travel great distances (100 to 600 km) and in the winter routinely move tens of kilometers to find warm bays and sheltered habitats.

Reproduction

Female manatees begin to reproduce at 3 to 5 years old, and males become sexually mature at 2 years. Manatees may form large mating herds (typically between February and July) with up to 22 males in a herd. These herds can assemble for up to a month. However, females may only be fertile for a day or two during that period. Female manatees mate many times with several different males (polyandry).

The gestation period for a manatee is approximately 11 to 13 months. When born, the single calf stays close to its mother and is fed on milk via the mother's axillary mammary glands (the teats of the manatee's mammary glands are situated in the "armpit" area). Mothers sometimes nurse their offspring while hovering vertically in the water, holding the calf to the teat with her flipper. It is this behavior that may be responsible for the link between manatees and mermaids, as from a distance this might resemble a human mother suckling a baby.

Calves are weaned at 1 to 2 years but may suckle until 4 years old or more, with an average dependency period of 1.2 years. Calf survival rates are approximately 60 to 70%, and on average there is a period of 2.5 years between births. This means there is a relatively slow growth rate in manatee populations; if an animal dies of unnatural causes (such as a boat strike) it will be several years before that animal is replaced in the population. Adult manatees may live up to 60 years.

Dugongs can either mate like manatees (many males mating with one female), or males occupy defended territories and display to females (lekking). Dugongs generally have their first calf between 10 and 17 years of age. They have a gestation period that is longer than manatees, 18 months, with an intercalf interval of 3 to 7 years. This much lower rate of reproduction in dugongs causes a lower recruitment rate (new individuals being added to the population) than even manatees, and as such populations are even more vulnerable to anthropogenic mortality than manatees. The oldest wild dugong recorded to date was estimated at 73 years.

Abundance and Status

The best-studied manatee populations are undoubtedly those over-wintering in the state of Florida. Aerial surveys have been used to estimate manatee numbers since the 1970s, and counts from 2010 estimated over 5,000 animals. This is one of the highest estimates for manatee abundance in recent years, and it has been suggested that numbers have been increasing in the past decade. However, part of the supposed increase may be an artifact of different survey methods being used over different years, with recent surveys being more efficient. Abundance estimates for the Caribbean region suggest about 2,600 manatees in the entire region, although the number could be higher, as manatees are hard to observe in some of the turbid and murky waters in their range, and because survey effort is low.

Predators such as sharks and crocodiles or alligators occasionally prey upon manatees, but this is rare. Most "natural" mortalities in Florida manatees result from their vulnerability to the cold or through ingesting toxins produced by red tides (dinoflagellate algal blooms). However, these red tides may be more frequent due to increased levels of nutrients entering the manatee's habitats as a result of human activities (e.g., fertilizing lawns, agriculture, sewage, and other sources). Direct anthropogenic causes of mortality in this population include by-catch or entanglement in fishing gear, such as crab pot lines and shrimp nets, and being crushed and/or drowned in floodgates or canal locks.

However, it is collisions with boats that are the largest cause of concern (see Exploring the Depths: Conservation of the Florida Manatee).

The Antillean manatee has been hunted since pre-Columbian times, with manatee dominating the diet in coastal Mayan communities. Likewise, manatees were extensively hunted by Taino and Carib Indians. Upon the arrival of Europeans, hunting continued as manatees were an easily caught source of fresh meat for sailing crews, and during this period of extensive hunting several populations became locally extinct. Hunting still occurs in populations of Antillean manatees in impoverished coastal locations in countries such as Haiti. Many of the problems facing Florida manatees (bycatch, red tides, and boat strikes) also affect Antillean manatees, but habitat loss and degradation, mainly as the result of unchecked coastal development, are the major problems.

Hunting of Amazonian manatees has also been extensive. From 1935 to 1954, between 80,000 and 140,000 Amazonian manatees were hunted primarily for their hides. In Peru and Ecuador manatees were hunted by poachers and the military. Pollution is also a problem in the Amazon, where gold mining in particular has led to elevated levels of mercury in some river systems (mercury is used to extract gold from its ores). Increasingly, oil extraction is degrading remaining pristine areas of the Amazon, especially in Peru and Ecuador, bringing oil pollution, roads, deforestation (leading to increased surface runoff and changes in the biota of the local ecosystems), and more humans into the forest.

Although manatees used to be revered and held sacred in some parts of West Africa, because of a belief that manatees were once humans (killing a manatee was a punishable crime), in the modern age this reverence has waned. Manatees have instead been hunted for food, hides, and body parts for medicines. They have also been culled to protect fisheries and agriculture as manatees sometimes damage fishing nets and browse in rice paddies.

The International Union for Conservation of Nature (IUCN) Red List classifies all manatees as "vulnerable." Amazonian, Florida, and Antillean manatees are listed on the U.S. Endangered Species Act as "endangered," whereas the West African manatee is listed as "threatened." All manatees are protected under the U.S. Marine Mammal Protection Act (although this obviously only has limited effects beyond U.S. waters). Amazonian, Florida, and Antillean manatees are also listed on Appendix I of the Convention on International Trade in Endangered Species of Wild Fauna and Flora (CITES), meaning that international trade in individual manatees (alive or dead) or their products is strictly controlled (see Chapter 17). The West African manatee is listed on CITES Appendix II, which means trade is allowed but it is controlled, with permits being required (see Chapter 17).

Dugongs are subject to many of the same threats as manatees (see Exploring the Depths: Conservation of the Dugong). However, dugongs depend on seagrass beds, and thus loss of this seagrass is a significant issue in their conservation. Because the loss of seagrass is so critical to dugongs, there is greater concern over factors such as coastal construction, siltation, reduced nutrients, and dredging, which all reduce these beds. In 1992 and 1993 it was estimated that 900 km^2 of seagrass beds, and hence dugong habitat, were lost in Australia, primarily due to bad weather and siltation from runoff. Between 1962 and 1999 some 800 dugongs were killed as a result of entanglement in shark nets (which were placed around bathing beaches). However, in terms of dugong conservation, the most effective efforts have probably taken place in Australia. Dugongs are found in the Great Barrier Reef World Heritage Area, and the Great Barrier Reef Marine Park Authority has been active in trying to manage and conserve dugongs in this protected area. This includes bans or restrictions on gillnet use in areas of the park or modifications of fishing gear to reduce entanglement.

Exploring the Depths: Conservation of the Florida Manatee

The primary anthropogenic cause of Florida manatee mortality is through collisions with watercraft, which represented between 20 and 31% of all known manatee deaths annually in Florida between 1999 and 2008 (**Figure B8.1**). In response to the increasing number of vessel-related deaths, state and federal managers began an initiative in the 1980s to increase boater awareness and develop waterway legislation intended to reduce collisions with manatees. Regulations included designating slow-speed (or no-wake) zones and no-entry areas, as well as restraints on the construction of boat-related facilities (such as boat ramps). Initially, boat-related manatee deaths did go down (1992–1993) but rose again (1996–2000) before stabilizing at mostly between 80 and 90 animals a year (roughly 20 to 25% of all deaths).

Exploring the Depths: Conservation of the Florida Manatee (*continued*)

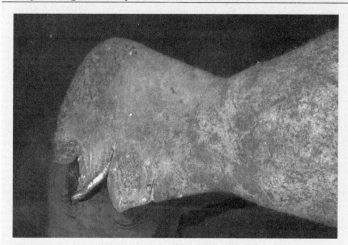

FIGURE B8.1 The tail of a manatee injured by a boat.

Propeller guards have been fitted to some slow commercial vessels and tour boats to try to reduce manatee injuries, but it is often the blunt trauma of impact from collisions with high-speed vessels that cause the most outright deaths. Alarms designed to alert manatees to approaching boats have also been proposed and, to some extent, successfully tested. But there are concerns that these will add sound to an already noisy environment. Furthermore, concerns about boat traffic and manatees are not just limited to fatal collisions. Harassment from boats and swimmers may also drive animals away from preferred sites or induce chronic stress responses, especially in warm water refuges where escape options are limited.

Second only to watercraft-related fatalities are the number of manatees that become crushed or drowned by floodgates and navigation locks. In an attempt to prevent these deaths, the state government has been working to introduce pressure-sensitive devices that detect an object (such as a manatee) blocking their closure, which then causes the gates to reopen. Because of their proximity to developed and developing areas, manatees are subjected to high levels of pollutants, but it is not yet fully known how these pollutants affect the species.

In the United States manatees have been the beneficiaries of many federal and state laws. As early as 1893 Florida imposed a fine of $500 or up to 3-month imprisonment for anyone who harmed or killed a manatee. Both subspecies of the West Indian manatee are protected in the United States under the Marine Mammal Protection Act (MMPA) and as an endangered species under the U.S. Endangered Species Act (ESA).

Because the Florida manatee was protected under the ESA, in 1976 "critical habitats" were designated for the Florida manatee; these areas legally cannot be damaged by any action of the government, businesses, or members of the public. Moreover, the Florida Department of Natural Resources (now the Florida Department of Environmental Protection), among other agencies and nongovernmental organizations, drew up a West Indian manatee recovery plan that came into force in 1980. Funded partly by the U.S. Fish and Wildlife Service (the U.S. government department with responsibility for manatee conservation) and also by the Save the Manatee Club and other nongovernmental sources, there have been three revisions to the plan, the last of which was in June 2001, to focus further on the Florida subspecies.

In 1978 the Florida Manatee Sanctuary Act attempted to reduce the threat that watercraft posed to manatees by introducing speed restrictions and other boating regulations (**Figure B8.2**). It introduced the concept that people could be prosecuted for harming a manatee through negligent actions. Under this legislation lawbreakers initially faced a fine of $1,000 or up to 1 year imprisonment; however, subsequent adjustments reduced this to $600 or 60 days imprisonment for a first offense. However, those injuring or killing a manatee can also face prosecution under the ESA (with fines of up to $50,000 and/or up to a year in a federal prison and up to $25,000 in civil penalties).

By 2001 seven manatee sanctuaries (areas where waterborne activities are prohibited; [**Figure B8.3**]) were established, including part of the larger Crystal River National Wildlife Refuge (within a refuge waterborne activity is restricted). In 2004 an additional protection area was added in Lee County. By 2001 the State of Florida had, through various programs, spent over $500 million acquiring 250,000 acres of land that is important habitat for manatees (as well as other species), particularly along Crystal River and near Blue Spring.

Both state and federal agencies have recently come under fire from those who believe they are not doing enough to meet their responsibility to protect the manatee under the ESA and MMPA. For instance, boaters who break manatee zone speed regulations are subject to little more than a $50 citation. This culminated in January 2000 when a coalition of nongovernmental organizations, including the Save the Manatee Club, filed a legal suit against the government for failing to enforce laws protecting manatees. A settlement was reached in April 2001 requiring the government to establish new boat speed zones and establish safe havens for manatees around Florida. However, there has been more pressure recently to delist the Florida manatee (at least in Florida), again mostly from boaters

FIGURE B8.2 A manatee speed restriction sign.

Exploring the Depths: Conservation of the Florida Manatee (*continued*)

who seek to relax restrictions on their activities. Despite the long fossil record for manatees in the United States (see Chapter 2), one of the main supporting arguments is that the West Indian manatee is "not native." Likewise, their increasing numbers are often cited, regardless of the fact that the population may still be well below a fully sustainable level.

A major education program has been aimed directly at boaters using manatee posters, brochures, and safety tips that are distributed through marinas and environmental kiosks installed at various waterfront locations. Boater safety classes are also offered and fishing line collection sites have been established. However, despite these efforts, collisions continue to occur and boater awareness of the problems faced by manatees, and the laws that protect them, is poor.

FIGURE B8.3 Critical habitat of the Florida manatee currently protected or being considered for protection by the U.S. government. Data from: U.S. Fish and Wildlife Service.

Exploring the Depths: Conservation of the Dugong

The range of the dugong is huge: in coastal areas throughout the Eastern Hemisphere from East Africa to the Solomon Islands and Vanuatu and along the coastlines of 48 countries. Australia has the largest estimated dugong population, at approximately 100,000 animals. Outside Australia dugongs survive in fragmented groups. Little is known about many of these groups outside of incidental sightings and reports from local fishers.

Dugongs are difficult to research. They eat off the seafloor, and only the tip of their muzzle breaks the water surface to breathe (**Figure B8.4**). They are best estimated using aerial surveys, which can be expensive and even dangerous in remote areas. It is difficult to assess a small population of dugongs, especially if water is not clear. Some research can be done from land but not efficiently from boats. Interviews with local community members are a valuable tool. In recent years scientists have attempted to document dugong population numbers, distribution, and conservation issues in various developing countries and have found significant numbers of dugongs along the Andaman coast of Thailand, in the Arabian Gulf, the Red Sea, and New Caledonia.

Dugongs are categorized as "vulnerable" on the IUCN Red List because of population declines, habitat loss and degradation, and human exploitation. As a species with specialized foraging needs for coastal seagrass, dugongs are continually threatened by the proximity of their habitat to coastal human settlements. Even though data on dugongs are lacking, we do know sirenians do not breed quickly and have trouble recovering their numbers if populations become too small. Globally, the most serious threat to dugongs is incidental drowning in fishing nets; however, other serious sources of human-caused mortality to dugongs include habitat degradation, hunting, and boat strikes.

The specific causes of mortality vary by area. In Australia, although dugongs outside of Aboriginal areas are protected by the government, there is a tradition of dugong hunting by indig-

Exploring the Depths: Conservation of the Dugong (*continued*)

FIGURE B8.4 A dugong.

dugong is possibly extirpated in China and certainly in Taiwan. In Okinawa, the Japanese dugong is a symbol for local conservationists protesting U.S. military bases that are already near and anticipated to be moved into areas with dugongs and seagrass. This dugong group is already extremely small, actually too small to even estimate.

Because of the impact of hunting and trade on dugongs, they are listed on CITES Appendix I. Moreover, the Convention on the Conservation on Migrating Species of Wild Animals (see Chapter 17), under the United Nations Environment Programme, has created an intergovernmental Memorandum of Understanding signed by 11 countries to date. The Memorandum of Understanding will attempt to assist national and regional conservation of dugongs and their habitat by raising awareness of dugong conservation needs, encouraging research and habitat monitoring by enhancing international capacity building in scientific and local communities, and working to improve legal protection to reduce dugong mortality.

Ultimately, the global survival of dugongs will depend on the consequences of the depletion and degradation of marine resources, economic insecurity, complicated government jurisdictions, uncertain funding for research, a rapidly increasing human population, and changing social roles and values within human communities.

enous people that is controversial. Scientists, while respecting the right of tradition, believe the number of animals captured is not sustainable in some areas. In Thailand, for example, there is a history of dugong hunting in the past; however, dugong meat is considered delicious, and dugong body parts have a tradition of being used for medicine and amulets. In Vietnam and Cambodia, despite recent legislation, dugong hunting is hard to regulate, as selling meat and body parts is profitable along overpopulated, impoverished, overfished coasts. The

–Contributing Author, Ellen Hines,
San Francisco State University.

Exploring the Depths: Extinction of Steller's Sea Cow

It was in 1741 that Georg Wilhelm Steller first saw a sea cow. He recognized these animals from a description of the Caribbean animals and called them manati, or *morskaia korova* in Russian. Today, the animals are named after him: Steller's sea cows (*Hydrodamalis gigas*) (**Figure B8.5**). On July 12, 1742, after Steller and the crew of the Second Kamchatka Expedition became stranded on what was to be called Bering Island, Steller made an accurate description of a female sea cow. He had been observing the animals during the winter of 1741–1742 but was unable to describe one because of the harsh conditions of the North Pacific winter.

Steller measured the sea cow at 296 inches (7.5 m) in length and 200 Russian pud (3,650 kg). Unlike its extant relatives, sea cows had no teeth at all. They masticated (chewed) seaweed using a horny plate in the mouth. Steller stated that these animals fed almost constantly on kelp in sandy bays or near the mouths of streams and that he never saw any of them submerge completely. Like the modern manatees, the sea cow seemed to have a great fondness for fresh water. The sea cows were the only modern sirenians that inhabited cold water (0–10°C). Unlike extant sirenians, the sea cows had no phalanges (finger bones). The sea cows were more closely related to

Exploring the Depths: Extinction of Steller's Sea Cow (*continued*)

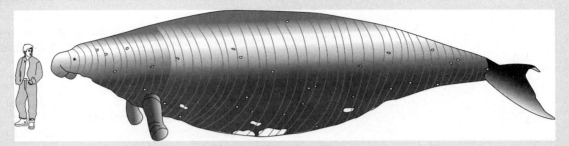

FIGURE B8.5 A Steller's sea cow with a human for scale.

dugongs than to manatees and had a tail much more like the dugong than the broad, flat paddle tail of the manatee.

The Bering Island animals that Steller observed were a remnant population of a much larger group that inhabited the entire North Pacific coastline from Baja California in Mexico to Honshu in Japan during the Pliocene and Pleistocene (**Figure B8.6**). However, in Steller's time the population had become extinct in all of its former range except on Bering and Copper Islands. These islands are located in the Commander (Komandorski) Island group, which is a part of Russia, and they are quite small at only 80 and 56 km long, respectively.

During the winter of 1741–1742, Steller spent many hours observing the sea cows that appeared in great herds near the dwelling he and some of his shipmates had constructed on the beach. He stated that the animals were monogamous. He also endowed the animals with many human characteristics. He called them caring and he noted that when a female was caught and brought onto shore, her mate would remain near her for many days. At times, the males also tried to free females caught by the hunters by thrashing about with their tails and trying to break the ropes attached to the captured animal.

Steller observed that many calves were born in the autumn, and since he saw the sea cows mating in the spring, he surmised that the gestation period of the sea cows was more than a year. He stated that the females were gentle with their calves and spent a lot of time caring for them. Steller also noted that when the animals fed, they would place the young calves in the center of the herd to protect them; he did not state from what the animals needed protection.

During the summer months of 1742 the crew of Bering's wrecked ship *St. Peter* rebuilt another version of the ship to

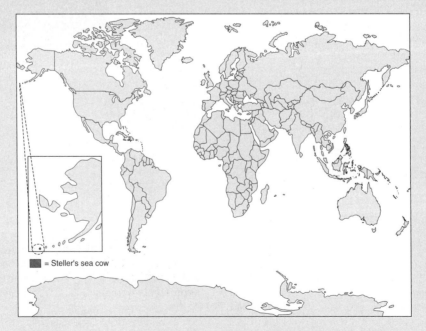

FIGURE B8.6 Map showing the known distribution of the Steller's sea cow in the 1700s.

Exploring the Depths: Extinction of Steller's Sea Cow (*continued*)

return them all to Russia. It did not take them much time to return, but the vessel was quite small, and although Steller prepared a young sea cow to bring back for the Russian Academy of Science Museum collection, it was judged to be too large and had to be left behind. The crew did manage to return with a great many sea otter pelts and tales of the enormous number of otters that remained around Bering Island. It was the subsequent extermination of sea cows by hunters that caused their extinction.

Steller and the crew of the *St. Peter* returned with knowledge of how to hunt the sea cows, and by 1754 no sea cows remained on Copper Island. During the 1760s crews of fur trappers spent up to 9 months on Bering Island hunting sea otters. They used the sea cow almost exclusively for provisioning because the meat was highly palatable and the fat was not only edible but also burned cleanly in lanterns. Leonhard Stejneger estimated that a great many sea cows were also struck and did not die right away. Steller stated there were so many sea cows around Bering Island that they could support the entire population of Kamchatka. However, with the excessive and wasteful hunting by fur trappers, these animals were completely extinct by 1768, just 27 years after their discovery by Europeans. Interestingly enough, Steller also predicted this, stating in a report to the Russian Science Academy that if not protected, the sea cows would surely go extinct very rapidly.

—Contributiong author, Lorelei Crerar, George Mason University.

SELECTED REFERENCES AND FURTHER READING

Aipanjiguly, S., Jacobson, S.K., & Flamm, R. (2002). Conserving manatees: knowledge, attitudes, and intentions of boaters in Tampa Bay, Florida. *Conservation Biology* 17: 1098–1105.

Ames, A.L. & Van Vleet, E.S. (1996). Organochlorine residues in the Florida manatee, *Trichechus manatus latirostris*. *Marine Pollution Bulletin* 32: 374–377.

Belitsky, D.W. & Belitsky, C.L. (1980). Distribution and abundance of manatees *Trichechus manatus* in the Dominican Republic. *Biological Conservation* 17: 313–319.

Bengtson, J.L. & Magor, D. (1979). A survey of manatees in Belize. *Journal of Mammalogy* 60: 230–232.

Blair, I.A. (1983). Manatee metabolism and its influence on distribution in Florida. *Biological Conservation* 25: 315–324.

Buckingham, C.A., Lefebvre, L.W., Schaefer, J.M., & Kockman, H.I. (1999). Manatees response to boating activity in a thermal refuge. *Wildlife Society Bulletin* 27: 514–522.

Carlton, J.T., Geller, J.B., Reaka-Kudla, M.L., & Norse, E.A. (1999). Historical extinction in the sea. *Annual Review of Ecology and Systematics* 30: 515–538.

Colmenero-Rolón, L.C. & Zárate, B.E. (1990). Distribution, status and conservation of the West Indian manatee in Quintana Roo, México. *Biological Conservation* 52: 27–35.

Craig, B.A. & A Reynolds, J.E. (2004). Determination of manatee population trends along the Atlantic coast of Florida using a bayesian approach with temperature-adjusted aerial survey data. *Marine Mammal Science* 20: 386–400.

de Thoisy, B., Spielberger, T., Rousseau, S., Talvy, G., Vogel I., & Vié, J.-C. (2003). Distribution, habitat, and conservation status of the West Indian manatee *Trichechus manatus* in French Guiana. *Oryx* 37: 431–436.

Deutsch, C.J., Self-Sullivan, C., & Mignucci-Giannoni, A. (2007). *Trichechus manatus manatus*. In: *IUCN 2007. 2007 IUCN Red List of Threatened Species*. IUCN, Gland, Switzerland.

Domning, D.P. (1982). Commercial exploitation of manatees *Trichechus* in Brazil c. 1785–1973. *Biological Conservation* 22: 101–126.

Garcia-Rodriguez, A.I., Bowen, B.W., Domning, D., Mignucci-Giannoni, A.A., Marmontel, M., Montoya-Ospina, R.A., Morales-Vela, B., Rudin, M., Bonde, R.K., & McGuiress, P.M. (1998). Phylogeography of the West Indian manatee (*Trichechus manatus*): how many populations and how many taxa? *Molecular Ecology* 7: 1137–1149.

Garrott, R.A., Ackerman, B.B., Cary, J.R., Heisey, D.M., Reynolds, J.E., & Wilcox, J.R. (1995). Assessment of trends in sizes of manatee populations at several Florida aggregation sites. In: *Population Biology of the Florida Manatee* (Ed. T. J. O'Shea, B. Ackerman, & H.F. Percival), pp. 34–55. National Biological Service, U.S. Department of the Interior, Washington, DC.

Hartman, D.S. (1979). Ecology and behavior of the manatee (*Trichechus manatus*) in Florida. *American Society of Mammalogists, Special Publication* 5: 1–153.

Hines, E., Adulyanukosol, K., Duffus, D.A., & Dearden, P. (2005). Community perspectives and conservation needs for dugongs along the Andaman coast of Thailand. *Environmental Management* 36: 654–664.

Hines, E., Adulyanukosol, K., Somany, P., Sam Ath, L., & Cox N. (2008). Community interviews to assess conservation needs of the dugong (*Dugong dugon*) in Cambodia & Phu Quoc, Vietnam. *Oryx* 41: 113–121.

Hines, E., Reynolds, J., Mignucci-Giannoni, A., Aragones, L., & Marmontel, M. (2011). *Sirenian Conservation: Issues and Strategies in Developing Countries.* University Press of Florida, Gainesville, FL.

Homewood, B. (1991). Pollution and hunting threaten Brazil's manatees. *New Scientist* 1760: 15.

Irvine A.B. (1983). Manatee metabolism and its influence on distribution in Florida. *Biological Conservation* 25: 315–324.

Irvine, A.B., Caffin, J.E., & Kochman, H.I. (1981). *Aerial Surveys for Manatees and Dolphins in Western Peninsular Florida.* Fish and Wildlife Service, U.S. Department of the Interior, Gainesville, FL.

Jimenez, I. (2005). Development of predictive models to explain the distribution of the West Indian manatee *Trichechus manatus* in tropical watercourses. *Biological Conservation* 125: 491–503.

Koelsch, J.K. (2001). Reproduction in female manatees observed in Sarasota Bay, Florida. *Marine Mammal Science* 17: 331–342.

Langtimm, C.A., O'Shea, T.J., Pradel, R., & Beck, C.A. (1998). Estimates of annual survival probabilities for adult Florida manatees (*Trichechus manatus latirostris*). *Ecology* 79: 981–997.

Lefebvre, L.W., Ackerman, B.B., Portier, K.M., & Pollock, K.H. (1995). Aerial survey as a technique for estimating trends in manatee population size—problems and prospects. In: *Population Biology of the Florida Manatee* (Ed. T. J. O'Shea, B. Ackerman, & H.F. Percival), pp. 63–74. National Biological Service, U.S. Department of the Interior, Washington, DC.

Lefebvre, L.W., Marmontel, M., Reid, J.P., Rathbun, G.B., & Domning, D.P. (2001). Status and biogeography of the West Indian manatee. In: *Biogeography of the West Indies Patterns and Perspectives*, 2nd ed. (Ed. C.A. Woods & F.E. Sergile), pp. 425–474. CRC Press, Boca Raton, FL.

Marmontel, M., Humphrey, S.R., & O'Shea, T.J. (1996). Population viability analysis of the Florida manatee (*Trichechus manatus latirostris*), 1976–1991. *Conservation Biology* 11: 467–481.

Mignucci-Giannoni, A.A. & Beck, C.A. (1998). The diet of the manatee (*Trichechus manatus*) in Puerto Rico. *Marine Mammal Science* 14: 394–397.

Mignucci-Giannoni, A.A., Montoya-Ospina, R.A., Jiménez-Marrero, N,M., Rodríguez-López, M.A., Williams, E.H., & Bonde, R.K. (2000). Manatee mortality in Puerto Rico. *Environmental Management* 25: 189–198.

Montoya-Ospina, R.A., Caicedo-Herrera, D., Millán-Sánchez, S,L., Mignucci-Giannoni, A.A., & Lefebvre, L.W. (2001). Status and distribution of the West Indian manatee, *Trichechus manatus manatus*, in Colombia. *Biological Conservation* 102: 117–129.

Morales-Vela, B., Olivera-Gómez, D., Reynolds, J.E., & Rathbun, G.B. (2000). Distribution and habitat use by manatees (*Trichechus manatus manatus*) in Belize and Chetumal Bay, Mexico. *Biological Conservation* 95: 167–175.

Mou Sue, L.L., Chen, D.H., Bonde R.K., & O'Shea, T.J. (1990). Distribution and status of manatees (*Trichechus manatus*) in Panama. *Marine Mammal Science* 6: 234–241.

Olivera-Gómez, L.D. & Mellink, E. (2005). Distribution of the Antillean manatee (*Trichechus manatus manatus*) as a function of habitat characteristics, in Bahia de Chetumal, Mexico. *Biological Conservation* 121: 127–133.

O'Shea, T.J., Correa-Viana, M., Ludlow, M.E., & Robinson, J.G. (1988). Distribution, status, and traditional significance of the West Indian manatee *Trichechus manatus* in Venezuela. *Biological Conservation* 46: 281–301.

O'Shea, T.J., Moore, J.F., & Kochman, J.I. (1984). Contaminant concentrations in manatees in Florida. *Journal of Wildlife Management* 48: 741–748.

Powell, J.A., Belitsky, D.W., & Rathbun, G.B. (1981). Status of the West Indian manatee (*Trichechus manatus*) in Puerto Rico. *Journal of Mammalogy* 62: 642–646.

Preen, A. (2004). Distribution, abundance and conservation status of dugongs and dolphins in the southern and western Arabian Gulf. *Biological Conservation* 118: 205–218.

Reep, R.L. & Bonde, R.K. (2006). *The Florida Manatee: Biology and Conservation.* University of Florida Press, Gainesville, FL.

Reynolds, J.E., Szelistowski, W.A., & León, M.A. (1995). Status and conservation of manatees *Trichechus manatus manatus* in Costa Rica. *Biological Conservation* 71: 193–196.

Rizzardi, K. (1997). Toothless? The endangered Florida manatee and the Florida Manatee Sanctuary Act. *Florida State University Law Review* 24: 377–405.

Self-Sullivan, C., Smith, G.W., Packard, J.M., & LaCommare, K.S. (2003). Seasonal occurrence of male Antillean manatees (*Trichechus manatus manatus*) on the Belize barrier reef. *Aquatic Mammals* 29: 342–354.

Smethurst, D. & Nietschmann, B. (1999). The distribution of manatees (*Trichechus manatus*) in the coastal waterways of Tortuguero, Costa Rica. *Biological Conservation* 89: 267–274.

Stavros, H.-C.W., Bonde, R.K., & Fair, P.A. (2008). Concentrations of trace elements in blood and skin of Florida manatees (*Trichechus manatus latirostris*). *Marine Pollution Bulletin* 56: 1221–1225.

Steller, G.W. (1751). *De Bestis Marinis.* (W. Miller & J. E. Miller, translators; P. Royster, transcriber and editor [2005]). University of Nebraska Press, Lincoln, NE.

U.S. Fish and Wildlife Service. (1976). Determination of critical habitat for American crocodile, California condor, Indiana bat, and Florida manatee. *US Federal Register* 4(187): 41914–41916.

U.S. Fish and Wildlife Service (2001). *Technical/Agency Draft, Florida Manatee Recovery Plan* (Trichechus manatus latirostris), third revision. U.S. Fish and Wildlife Service, Atlanta, GA.

Vianna, J.A., Bonde, R.K., Caballero, S., Giraldo, J.P., Lima, R.P., Clark, A., Marmontel, M., Morales-Vela, B., De Sousa, M.J., Parr, L., Rodríguez-Lopez, M.A., Mignucci-Giannoni, A.A., Powell, J.A., & Santos, F.R. (2006). Phylogeography, phylogeny and hybridization in trichechid sirenians: implications for manatee conservation. *Molecular Ecology* 15: 433–447.

Wright, I.E., Reynolds, J.E., Ackerman, B.B., Ward, L.I., Weigle, B.L., & Szelistowski, W.A. (2002). Reproduction in female manatees observed in Sarasota Bay, Florida. *Marine Mammal Science* 18: 259–274.

Wright, S.D., Ackerman, B.B., Bonde, R.K., Beck, C.A., & Banowetz, D.J. (1995). Analysis of watercraft-related mortality of manatees in Florida, 1979–1991. In: *Population Biology of the Florida Manatee* (Ed. T. J. O'Shea, B. Ackerman, & H.F. Percival), pp. 259–268. National Biological Service, U.S. Department of the Interior, Washington, DC.

ﬂ
Pinnipeds

CHAPTER

CHAPTER OUTLINE

Distribution

Fur Seals and Sea Lions
Walruses
True Seals

Exploring the Depths: Guadalupe Fur Seal, Arctocephalus townsendi
Exploring the Depths: Ross seal, Ommatophoca rossii
Exploring the Depths: Lake Seals
Exploring the Depths: Tagging Pinnipeds

Reproduction

Exploring the Depths: Hooded Seal Mating Behavior

Exploring the Depths: Nursing Behavior in Galápagos Fur Seals
Exploring the Depths: Adult Molt

Abundance and Status

Exploring the Depths: Caribbean Monk Seal, Monachus tropicalis
Exploring the Depths: Japanese Sea Lion, Zalophus japonicus

Selected References and Further Reading

Distribution

Pinnipeds are found throughout the world, although the greatest numbers are found in high latitudes and polar regions. Pinnipeds are difficult to observe at sea, and therefore most information on pinniped distribution is based on the haul-out sites where pinnipeds come ashore to breed or molt. However, we are now gaining more information about the movements of pinnipeds at sea through tagging studies (see Exploring the Depths: Tagging Pinnipeds).

Fur Seals and Sea Lions

Breeding sites used by fur seals and sea lions are often remote beaches or islands that have had few predators in the past (**Figure 9.1**). These locations are typically near an area of major marine productivity, such as an upwelling current where nutrients are brought from the ocean depths to the sea surface, resulting in a high abundance of plankton and fish. For several species, such as the Guadalupe (see Exploring the Depths: Guadalupe Fur Seal, *Arctocephalus townsendi*), Juan Fernández, Galápagos, and northern fur seals (see Chapter 16), their isolated but highly concentrated distribution made them vulnerable to human exploitation, as these species were not accustomed to land-based predators. Several species were so depleted they came close to extinction; for example, the Juan Fernández fur seal was thought to be extinct in the early part of the 20th century.

FIGURE 9.1 (A) Sea lions on a crowded beach and (B) a harp seal and pup on ice.

Walruses

All three subspecies of the single member of this family (the Odobenidae) have an Arctic distribution and are typically found in shallow waters, where they can reach benthic prey. Generally restricted to pack ice in summer, they venture further south to over-winter in huge aggregations on rocky beaches or outcrops. For example, the Pacific walrus (*Odobenus rosmarus divergens*) travels between summer ice-bound locations in the Chukchi Sea to winter at various sites in the Bering Sea.

True Seals

Many species of seals haul out to breed on ice, either in the Arctic or Antarctic. This can be either on pack ice or on fast ice where they gain access through a hole in the ice. Seals in warmer habitats may haul out onto rocky or sandy shores. Local prey abundance is not so important for phocids in their choice of a breeding site because these species have large reserves of blubber that last throughout lactation.

Exploring the Depths: Guadalupe Fur Seal, *Arctocephalus townsendi*

This fur seal (**Figure B9.1**) is only found along the coast of California and Mexico and on the Guadalupe islands, where the species breeds (**Figure B9.2**). In fact, the name "Guadalupe" means wolf valley, referring to the fur seals, or "sea wolves." The Guadalupe fur seal was so heavily hunted by humans that it was thought to be extinct at the beginning of 20th century. Despite this, animals were observed in 1926, and two individuals were captured by the San Diego Zoo in 1928. This pair was believed to represent the last of the species, especially as no more were seen until 1949 when a solitary animal was sighted. In 1954 a survey of Guadalupe Island found 14 individuals.

Official protection of Guadalupe Island by the Mexican government did not occur until 1975, but this does seem to have benefited the species. By 1987 there were an estimated 3,259 animals, still a low number, but the situation was not as serious as it had been. In 1997 a colony of more than 1,500 animals was found on the San Benitos Islands, an area that had

FIGURE B9.1 A Guadalupe fur seal.

Exploring the Depths: Guadalupe Fur Seal, *Arctocephalus townsendi* (continued)

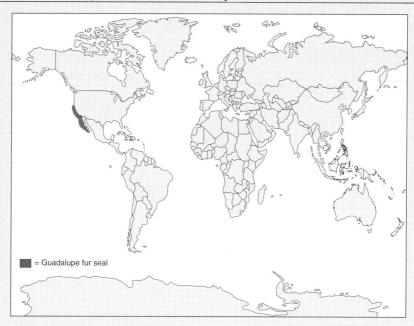

FIGURE B9.2 Map showing the current, known distribution of the Guadalupe fur seal. Data from: IUCN Red List.

historically been a fur seal breeding site. Its recolonization was a good indication that the species was recovering.

Despite the good news, it is very likely that all these animals are descended from just a small number of individuals. The species is said to have gone through a genetic bottleneck. As a result, the species has a very low genetic diversity. This lack of diversity causes concern because it provides the species with fewer resources to overcome changes in environmental conditions or to survive an outbreak of disease. For example, if one animal in the population is susceptible to a particular illness, then it is likely that all the animals in the population will be too. None would possess the necessary genes to fight the disease effectively. The IUCN currently considers the Guadalupe fur seal to be "near threatened."

Exploring the Depths: Ross Seal, *Ommatophoca rossii*

FIGURE B9.3 A Ross seal.

Ross seals (**Figure B9.3**) are one of the most rarely seen pinnipeds because they are solitary and live deep within the consolidated Antarctic pack ice and almost never stray from the Southern Ocean (**Figure B9.4**). Easily identified by their huge eyes, which are thought to aid underwater vision, they are named after the British polar explorer Sir James Ross, who first discovered them in 1840. Ross seals are thought to feed mainly on squid but are also known to eat fish and krill. They are the smallest of the Antarctic seals and have the shortest hair of any seal. Males are smaller than females on average, suggesting (along with their solitary nature) that they are a polygynous species.

One interesting feature of the Ross seal is its throat. When a Ross seal is approached it will throw its head back, open its small mouth (displaying its teeth), and inflate its throat in much the same way as a frog might. At this time the distinctive stripes or streaks that run from the chin to the chest become clearly visible. Ross seals also have distinctive vocalizations,

Exploring the Depths: Ross Seal, *Ommatophoca rossii* (continued)

which are most likely used to attract a mate, maintain a territory, or as part of a threat display. Many of these vocalizations can apparently be produced in both air and water and can travel some distance, in part because of the unique propagation characteristics of the ice-covered underwater environment.

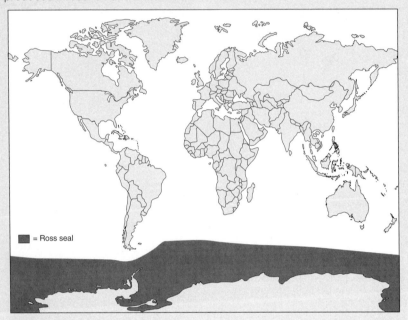

FIGURE B9.4 Map showing the current, known distribution of the Ross seal. Data from: IUCN Red List.

Exploring the Depths: Lake Seals

A number of seal populations and species have moved into isolated lakes and other freshwater environments. The Baikal seal, or nerpa (*Pusa sibirica*), is found only in Lake Baikal in Russia. It is the only exclusively freshwater pinniped. It is also the smallest pinniped, with a maximum recorded length of around 1.5 m. Other "lake seals" include the Caspian seal (*Pusa caspica*) and two subspecies of ringed seal (*Pusa hispida*), including the Saimaa seal (*P. h. saimensis*) and the Ladoga seal (*P. h. ladogensis*) (**Figure B9.5**). The ringed seal also has another isolated subspecies in the northern Baltic Sea (*P. h. botnica*).

A   B
FIGURE B9.5 (A) A Caspian seal and (B) Baikal seals.

Exploring the Depths: Lake Seals (*continued*)

FIGURE B9.5 (*continued*) (C) A Ladoga seal.

All these populations have limited distributions (**Figure B9.6**) and are often highly isolated from other suitable habitats. They are consequently vulnerable to changes in local environmental conditions, including pollution, disease, and climate change. For example, the Baltic ringed seal population depends on ice for breeding and due to its geographical isolation may be unable to reach other regions of sea ice if predicted loss of sea ice occurs in the Baltic due to climate change.

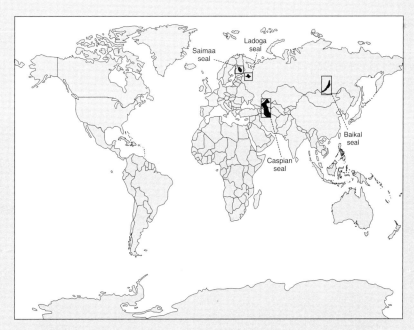

FIGURE B9.6 Map showing the current, known distributions of Caspian, Baikal, Ladoga, and Saimaa seal species. Data from: IUCN Red List.

Studies using satellite telemetry tags are providing scientists with more information about the routes pinnipeds take as they move from their breeding sites to marine feeding grounds as well as their behavior at these feeding sites. For example, tags attached to northern elephant seals (*Mirounga angustirostris*) revealed these animals can travel a total of 18,000 to 21,000 km a year. After breeding, males travel nearly 12,000 km (for up to 123 days) and females travel over 6,000 km (for 73 days) following the California current northward to feed in the northern Pacific, with males and females feeding in different locations. They return to land to molt and after the molt is completed they travel again, with males swimming over 9,000 km and females traveling over 12,000 km. This time, females forage for twice as long as males. Males and females feeding in differing locations may reduce the competition for prey between the sexes. It is possible that males and females may have different nutritional needs and feed in different areas on different prey as a result.

While at the feeding sites, tags have allowed us to determine that northern elephant seals routinely dive to 800 m and can stay underwater for 80 minutes, with only short rest intervals of a few minutes on the surface. Species such as elephant seals spend 80 to 90% of their time underwater after leaving the breeding grounds. Studying anything about the behavior of these animals while in open water was extremely difficult until the pioneering development and deployment of telemetry tags.

Exploring the Depths: Tagging Pinnipeds

In recent years, the movements, diving behavior, and physiology of pinnipeds have been studied in great detail using a variety of tags and instruments attached to free-ranging animals (**Figure B9.7**). The simplest of these are plastic cattle tags that are inserted onto the foreflippers of otariids and hind flippers of phocids. These tags have a unique identification number that will identify the animal, which allows researchers to follow an individual over a number of years. Similar to bands (rings) on birds, tagging provides highly useful and cost effective information on movements and survivorship of pinnipeds. Brands (either hot brands, similar to those used to brand cattle, or freeze brands using liquid nitrogen) are sometimes applied to pinnipeds, marking them with a unique pattern (usually numbers or letters) and making them much easier to identify. (This technique is also occasionally used to mark cetaceans as well.) Such branding is, however, somewhat controversial, because the brand site takes time to heal. Accidents may also occur that result in serious injury to the animal being branded.

Research is underway on the use of passive integrated transponder (PIT) tags that can be implanted subcutaneously. This is currently a widely used technique for the long-term identification of pets. Each microchip is programmed with a unique identification code that becomes activated by a handheld scanner. The disadvantage here is that animals need to be caught or at least scanned at close distance to read the PIT tag. Animals may also need to be marked externally in some way so researchers can identify which animals have been implanted.

Some of the first electronic tags to be attached to pinnipeds were radio transmitters that allowed the location of animals to be determined and thus provided scientists with a way to monitor their movements. These devices emit a VHF radio signal of known frequency that can be detected using a handheld receiver and aerial (from land or boat) or recorded by a remotely operated receiving station. However, the radio signal from these tags cannot be detected over large distances or where the signal is blocked, for example, by coastal features such as headlands. Therefore, satellite telemetry tags are now preferred, because these devices transmit a powerful microwave signal to a satellite, allowing animals to be remotely tracked over large distances. Satellite tags can transmit information

FIGURE B9.7 A harbor seal with a satellite tag.

Exploring the Depths: Tagging Pinnipeds (*continued*)

on the position of a tagged pinniped, via polar orbiting satellites using the ARGOS satellite system and, ultimately, to a scientist's computer. Global Positioning System (GPS) tags are also now being used to track seals because they provide greater accuracy of locations. The problem with all these transmission systems is that none of them can transmit a signal while underwater. This is a particular issue for GPS tags that may take minutes to locate the satellites and determine a location. To deal with this issue, a new generation of GPS tags have been developed that essentially take a rapid snapshot of all satellites located when they are at the surface for even a brief moment. The tag is then able to process this information and store an accurate location while the animal is underwater. The most recent tracking devices use GPS to record and store accurate locations, and then mobile phone technology allows locations and other information gathered to be relayed whenever the animal comes within a mobile phone system.

Tags can be used for much more than simply relaying position. They can also include components to measure behavioral information including orientation, swim speed, and bioacoustics. Likewise, sensors to measure oceanographic features, such as temperature, depth, and conductivity (salinity), can also be included. Deployments on pinnipeds are now providing oceanographic data, particularly in remote polar regions where ship or buoy measurement systems are lacking.

Physiological variables such as heart rate and body temperature can also be measured using loggers. These provide data on the metabolism and energy expenditure of pinnipeds while at sea. Furthermore, some pinnipeds have been fitted with video cameras to record their underwater behavior. One of the first devices, known as a "crittercam," was designed to visually record the prey species seals fed on and how different prey items were caught. Due to their size, these devices are as yet, only suitable for relatively short deployments to minimize any effects of increased hydrodynamic drag on foraging behavior. Most systems have to be recovered from the animal for the video to be downloaded. However, as technology develops, these camera systems will ultimately become smaller and it is hoped that video footage will be transmitted by satellite telemetry (**Figure B9.8**). These advances will lead to further discoveries about the natural behavior of pinnipeds in the open ocean.

It should be noted that tagging of pinnipeds, and other marine mammals, has raised ethical issues because tagging is an invasive process. For example, the process of attaching tags involves handling these wild animals, which may cause stress, and in some cases tag attachment may require surgery. Devices are attached to animals via harnesses, glued onto the pelt, attached via suction cups, or in case of walruses attached to tusks. This may impact on swim speed and dive behavior due to increased drag. Researchers in the United States and most other countries are therefore required to have permits to allow them to tag seals, and part of the licensing evaluation requires consideration of the tagging process and a careful evaluation of this against the importance of the information likely to be gained from the study concerned.

FIGURE B9.8 The isle of North Rona showing a simulated satellite track of a seal (this track is simulated and is not based on actual data for copyright reasons).

Reproduction

We know much more about the reproductive behavior of pinnipeds than their behavior at sea, because all pinnipeds must haul out onto land or ice to give birth. Pinniped breeding systems vary from near monogamy (one male mates with one female), for example, as seen in the crabeater seal (*Lobodon carcinophaga*), ringed seal (*Pusa hispida*), and spotted seal (*Phoca largha*), to polygyny (one male mates with multiple females), as seen in northern elephant seals (*Mirounga angustirostris*). Male seals tend to be larger than females, although this sexual dimorphism (differences in shape and size between the sexes) is slight or even reversed in monogamous species. Most males in the subfamily Phocinae are larger, whereas males in the subfamily Monachinae are generally smaller. The most notable exception to this is the highly polygynous elephant seals, where the males are considerably larger than the females (**Figure 9.2**). For example, male northern elephant seals may be 60 to 70% longer than females and over three times the weight, whereas male ringed seals may only be a few centimeters larger and fractionally heavier than females on average.

In many species that haul out on land (instead of ice) females tend to be clustered in a small area (e.g., one beach). Therefore, one male (sometimes referred to as a "beachmaster") can control a territory containing a "harem" of several females. As a result a male northern elephant seal may impregnate more than 50 females in a season. Males defend their beach territory and harem by displaying to

FIGURE 9.2 A male elephant seal beachmaster next to a female to show the difference in their sizes.

(such as roaring and posturing) and fighting with other males (**Figure 9.3**).

During their efforts to ward off possible competitors, these beachmasters may "trample" pups, which are often injured or even killed as a result. It may seem a little strange for males to show so little concern about the pups in the harems while they try to control many females for mating privileges. However, these pups are from last year's mating and may be less likely to be related to the current beachmaster.

Young male elephant seals more closely resemble females than the highly dimorphic beachmasters. This may

FIGURE 9.3 Fighting male elephant seals.

aid them as they attempt to "sneak" onto the beach and surreptitiously mate with a female toward the edge of a harem. This can attract the attention of the beachmaster, who will then attempt to fight and chase the intruder away. So, although the young male may be able to mate with a female, this opportunistic behavior carries some risk.

Space is less limited for species that breed on ice and as a consequence females are more widely dispersed. Males cannot control a large number of females, unless females are clustered, for example, in the case of the Weddell seal (*Leptonychotes weddellii*) around a breathing hole in the ice. Therefore, ice seals tend to be monogamous, at least sequentially (males may stay with and defend one female until mating has occurred, but then may move on to try to find a second female to breed with). Other variations of mating strategies in pinnipeds include scramble polygyny (in which males mate with females when they find them and then simply move onto the next female) and lekking (where males are found in a group and display to females, who then choose their mate based on these displays, e.g., in walruses).

Because all pinnipeds give birth on land or ice, females and their young offspring are vulnerable to predation, especially the true seals, which are less maneuverable than the fur seals and sea lions. Females may give birth less than 24 hours after hauling out, minimizing the duration of time ashore. Regardless, the young pups are initially unable to swim, are small, and do not have a substantial blubber layer at birth. Some species breed in warmer regions, such as Galápagos fur seals (*Arctocephalus galapagoensis*). In lieu of a thick blubber layer, pinniped pups are usually born with a thick natal coat of fur, called the lanugo. Many seals that breed on the ice are born with a white lanugo, which provides camouflage as well as warmth.

Harbor seal pups are an interesting exception. During gestation the pup develops a lanugo, however at the later stages of gestation the lanugo is molted and the juvenile coat grows. Therefore, at birth they have an intact juvenile coat, which is better adapted for aquatic life, as the pups are able to swim only a few hours after birth. In some areas the juvenile pelage also provides better camouflage on the seaweed-covered rocky shores they inhabit.

Thick lanugo insulates the pups in air until they can build up a substantial blubber layer. However, fur is a poor insulator if it becomes wet, meaning pups must build up their blubber layer and shed the natal coat before they can enter the ocean and learn to swim. It can be fatal for a pup to enter the ocean prematurely, before it has lost its coat and fully developed. In parts of Canada, such as Prince Edward Island, harp seal pups (*Pagophilus groenlandicus*) have been forced into the sea by hunters and have ultimately died as

Exploring the Depths: Hooded Seal Mating Behavior

Male hooded seals (*Cystophora cristata*) patrol the ice edge to find a female who has hauled out to give birth; the male can defend the female against other males until she has given birth and is ready to mate again. When challenged, males inflate a membranous nasal sac, which they "wobble" as a display (**Figure B9.9**). Presumably, the way the hooded seal displays its nasal sacs gives a challenging male an indication of its evolutionary fitness (see Chapter 5). The male attends the female until the pup is weaned, which for hooded seals is rapid, within a matter of days. When the pup is weaned, the female leaves the ice edge and enters the water, where the male finally mates with her.

FIGURE B9.9 A male hooded seal displaying its red nasal sack.

a result. This could be an escalating problem for all ice seal pups, because melting and prematurely breaking ice may lead to pups being cast into the water earlier in the season.

Pinnipeds may be described either as "income" breeders, where females feed during lactation to produce sufficient milk for their pup or, alternatively, as "capital" breeders where prior to pupping, females accumulate all their fat stores to rear their pup. Most otariids are income breeders with a foraging and attendance cycle. The female suckles the pup until it has built up a blubber layer to provide sufficient energy stores for several days. At this time the female comes into estrus and will be mated prior to her departure to sea. The female goes off to sea to feed (typically for 2 to 3 days), after which the mother returns and continues suckling. When the mother returns she has to find her offspring on the normally crowded breeding site. She does this by recognizing the pup's unique call and/or by smell. The alternating cycle of feeding and lactation continues and depending on species, mothers nurse their offspring for 4 months to up to 3 years. This lactation strategy requires a breeding site located near an abundant source of prey and also in an area where there are few predators. Therefore, fur seal and sea lion breeding colonies (also called rookeries) are typically found on islands or remote beaches adjacent to an upwelling or a similar area of high productivity.

Lactation is similar in the walrus, however in this case the female suckles the pup on land for the first few days, before both mother and pup enter the water. Here, the mother can feed and the pup continues to suckle. The pup is generally not weaned for over a year, it can suckle for as much as 2 to 3 years and can spend as long as 5 years with the mother, during which time it learns how to forage. Ovulation in walruses is suppressed until the calf is weaned, and gestation is 15 to 16 months. This means females can only give birth every 2 years at best but are more likely to have 3 or 4 years between offspring. This cycle causes the walrus to have the lowest reproductive rate of any pinniped, which has obvious conservation implications. Most phocids are capital breeders, where the female will remain with the pup after birth until it is weaned, during which time the mother does not go off to feed. The pup is fed on milk with an extremely high fat content, allowing it to grow and gain weight quickly. Weaning is rapid (e.g., 4 days after birth in hooded seals), and the mother can lose up to 50% of its body weight during this time. She is mated and returns to feed at sea, while the pup is abandoned and left to fend for itself. This rapid period of weaning and independence reduces the length of time spent on land/ice and minimizes the risk of predation (e.g., polar bears) and exposure to adverse environmental conditions.

Pinniped milk is typically high in energy and nutrients, much more so than many other mammalian milks (**Table 9.1**). Fat content of pinniped milk is especially high, with walrus milk containing approximately 24% fat (by mass) and other otariid milk up to 40% fat. Phocid milk is the richest, however, containing from 47 to 61% fat. The high fat content of the milk is responsible for the high rate of weight gain and blubber deposition in phocid pups (e.g., a hooded seal pup can gain 7 kg/day).

Exploring the Depths: Nursing Behavior in Galápagos Fur Seals

Galápagos fur seals (*Arctocephalus galapagoensis*) can nurse their pups for over 2 years (**Figure B9.10**). This is possibly an adaptation to cope with the effects of El Niño events, a periodic shift in Pacific weather systems and water bodies that reduces the nutrient-rich cold-water upwelling off western South America and around the Galápagos Islands. The reduction in nutrients causes a reduction in food for marine mammals (and other species) during an El Niño year. However, from August 1982 to July 1983 the El Niño phenomenon was so pronounced only 11% of the normal number of pups were produced and no pups were thought to have survived more than 5 months. The Galápagos sea lions (*Zalophus wollebaeki*) were similarly affected with near total mortality of pups during this period. There are some concerns that climate change may substantially affect El Niño cycles, increasing the severity of events. The consequences of this for species like the Galápagos fur seal could be severe.

FIGURE B9.10 A nursing Galápagos fur seal.

TABLE 9.1 Constituents of Milk (% by Mass)

	Fat	Protein	Carbohydrate	Water
Cow	4%	4%	4%	85%
Human	8%	1%	7%	80%
Seal	30–60%	5–15%	0–1%	30%

Exploring the Depths: Adult Molt

In addition to leaving the water to breed, pinnipeds also haul out to molt. Molting usually occurs at the end of lactation, usually after a period of feeding to recover body condition. Many species do not molt in the water because skin and hair follicles do not grow at low temperature and therefore seals would lose too much heat to the water if they circulated warm blood to the skin surface. An incomplete coat may also increase drag and reduce swimming efficiency. Therefore, animals haul out onto land to minimize heat loss and allow hair to grow quickly. In most pinnipeds the molt largely involves the shedding of hair rather than skin, and the period of molt can last for 1 to 2 months while hair is gradually shed and regrown. Both northern and southern elephant seals have a catastrophic molt where the outer skin layers are molted in addition to the hairs (**Figure B9.11**). Elephant seals must therefore come ashore for several weeks and often seek out mud-filled wallows during the molt, resulting in many animals grouping together. This huddling behavior helps to conserve body heat and energy reserves during the molting fast.

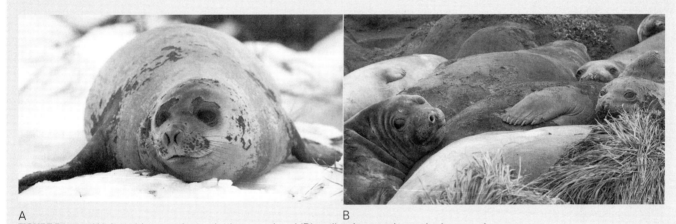

FIGURE B9.11 (A) A molting southern elephant seal and (B) wallowing southern elephant seals.

Abundance and Status

The most abundant pinniped is the crabeater seal (*Lobodon carcinophaga*), with an estimated population size of between 15 and 30 million animals. In fact, the crabeater seal may be the most abundant large wild mammal species on the planet, excluding humans. In contrast, the least abundant pinniped is the Mediterranean monk seal (*Monachus monachus*), with only 350 to 450 individuals. Of 34 species of pinnipeds 11 species (33%) have increasing populations, 4 (12%) are stable, 10 (29%) are decreasing, and the population status of 9 species (26%) is unknown (International Union for Conservation of Nature Red List Version 3.1).

There has been a long history of hunting and exploitation of pinnipeds and also of culling of animals because of perceived conflicts with fisheries. This has already made two species extinct in recent history (the Japanese sea lion and the Caribbean monk seal; see text boxes later in this chapter), and several more species are on the brink of extinction (**Table 9.2**). Chapter 16 describes some of the threats and problems faced by seal populations around the world.

TABLE 9.2 IUCN Category

Species	Category
Japanese sea lion	Extinct
Carribbean monk seal	
Mediterranean monk seal	Critically Endangered
Hawaiian monk seal	
Galápagos fur seal	Endangered
Steller sea lion	
Australian sea lion	
Caspian seal	
Hooded seal	Vulnerable
New Zealand sea lion	
Northern fur seal	

Exploring the Depths: Caribbean Monk Seal, *Monachus tropicalis*

Christopher Columbus reported sighting Caribbean monk seals, or "sea wolves," in his log book during his 1494 voyage to the West Indies. In fact, his crew killed eight of these seals to eat. Historically, Caribbean monk seals were hunted by indigenous natives in the Caribbean, but after Europeans arrived they became a heavily hunted commodity for oil, pelts, and sometimes food. This long history of hunting, possibly in addition to habitat loss and depletion of prey species, led to a decline in the seal population. The last Caribbean monk seal was seen in U.S. waters in 1932. The last record in the Caribbean was in 1952 when there were a small number of animals on Serranilla Bank (an atoll to the southwest of Jamaica). They were officially declared extinct by the U.S. government in 2008.

Exploring the Depths: Japanese Sea Lion, *Zalophus japonicus*

The Japanese sea lion was found on sandy beaches and caves along the coasts of Japan and Korea. Japanese sea lions were thought by some to be a subspecies of the Californian sea lion; however, it is now generally considered to be a species in its own right.

Due to hunting (for oil and products for medicines), entanglement in fishing gear, and habitat degradation the sea lion population was reduced to just a few thousand animals by the early 20th century and to apparently a few tens of animals by the 1930s. The last colony of Japanese sea lions disappeared in the 1950s. However, the sea lion apparently persisted for a little longer, as a juvenile animal was captured in 1974. This was the last official record of the species. There have been suggestions that historical Japanese sea lion sites could be repopulated by reintroduced Californian sea lions. Although this species may be able to repopulate Japanese sea lion habitat, they are genetically and perhaps behaviorally different from the now extinct species.

SELECTED REFERENCES AND FURTHER READING

Berta, A. (2009) Pinnipedia, overview. In: *Encyclopedia of Marine Mammals*. 2nd ed. (Ed. W. F. Perrin, B. Würsig, & J. G. M. Thewissen), pp. 878–885. Academic Press, New York.

Bonadonna, F., Lea, M. A., & Guinet, C. (2000). Foraging routes of Antarctic fur seals (*Arctocephalus gazella*) investigated by the concurrent use of satellite tracking and time-depth recorders. *Polar Biology* 23: 149–159.

Bowen, W. D., Oftedal, O. T., & Boness, D. J. (1985). Birth to weaning in 4 days: remarkable growth in the hooded seal, *Cystophora cristata*. *Canadian Journal of Zoology* 63: 2841–2842.

Boyd. I. L. (2009) Pinniped life history. In: *Encyclopedia of Marine Mammals*. 2nd ed. (Ed. W. F. Perrin, B. Würsig, & J. G. M. Thewissen), pp. 868–873. Academic Press, New York.

Crocker, D. E., & Costa, D. P. (2009). Pinniped physiology. In: *Encyclopedia of Marine Mammals.* 2nd ed. (Ed. W. F. Perrin, B. Würsig, & J. G. M. Thewissen), pp. 873–878. Academic Press, New York.

Daniel, R. G., Jemison, L. A., Pendleton, G. W., & Crowley, S. M. (2003). Molting phenology of harbor seals on Tugidak Island, Alaska. *Marine Mammal Science* 19: 128–140.

De Oliveira, L. R., Meyer, D., Hoffman, J., Majluf, P., & Morgante, J. S. (2009). Evidence of a genetic bottleneck in an El Nino affected population of South American fur seals, *Arctocephalus australis. Journal of the Marine Biological Association of the United Kingdom* 89: 1717–1725.

Erickson, A. W., Bester, M. N., & Laws, R. M. (1993). Marking techniques. In: *Antarctic Seals: Research Methods and Techniques* (Ed. R. M. Laws), pp. 89–118. Cambridge University Press, Cambridge, UK.

Gentry, R. L. (1998). *Behavior and Ecology of the Northern Fur Seal.* Princeton University Press, Princeton, NJ.

Hoelzel, A. R., Fleischer, R. C., Campagna, C., Le Boeuf, B. J., & Alvord, G. (2002). Impact of a population bottleneck on symmetry and genetic diversity in the northern elephant seal. *Journal of Evolutionary Biology* 15: 567–575.

Hooker, S. K., Biuw, M., McConnell, B. J., Miller, P. J., & Sparling, C. E. (2007). Bio-logging science: logging and relaying physical and biological data using animal-attached tags. *Deep Sea Research Part II: Topical Studies in Oceanography* 54: 177–182.

Hooker, S. K., Boyd, I. L., Jessopp, M., Cox, O., Blackwell, J., Boveng, P. L., & Bengtson, J. L. (2002). Monitoring the prey-field of marine predators: combining digital imaging with datalogging tags. *Marine Mammal Science* 18: 680–697.

Kooyman, G. L. (1981). *Weddell Seal, Consummate Diver.* Cambridge University Press, Cambridge, UK.

Lang, S. L. C., Iverson, S. J., & Bowen, W. D. (2005). Individual variation in milk composition over lactation in harbour seals (*Phoca vitulina*) and the potential consequences of intermittent attendance. *Canadian Journal of Zoology* 83: 1525–1531.

Le Boeuf, B. J. (1974). Male-male competition and reproductive success in elephant seals. *American Zoologist* 14: 163–176.

Le Boeuf, B. J., Kenyon, K. W., & Villaramirez, B. (1986). The Caribbean monk seal is extinct. *Marine Mammal Science* 2: 70–72.

Le Boeuf, B. J., & Laws, R. M. (1994). *Elephant Seals: Population Ecology, Behavior and Physiology.* University of California Press, Berkeley, CA.

Lindenfors, P., Tullberg, B. S., & Biuw, M. (2002). Phylogenetic analyses of sexual selection and sexual size dimorphism in pinnipeds. *Behavioral Ecology and Sociobiology* 52: 188–193.

Ling, J. K. (1970). Pelage and molting in wild animals with special reference to aquatic forms. *Quarterly Review of Biology* 45: 16–54.

Boyd, I. L., Bowen, W. D., & Iverson, S. J. (2010). *Marine Mammal Ecology and Conservation. A Handbook of Techniques.* Oxford University Press, Oxford, UK.

McClenachan, L., & Cooper, A. B. (2008). Extinction rate, historical population structure and ecological role of the Caribbean monk seal. *Proceedings of the Royal Society B: Biological Sciences* 275: 1351–1358.

Noren, D. P. (2002). Thermoregulation of weaned northern elephant seal (*Mirounga angustirostris*) pups in air and water. *Physiological and Biochemical Zoology* 75: 513–523.

Riedman, M. (1990). *The Pinnipeds: Seals, Sea Lions & Walruses.* University of California Press, Berkeley, CA.

Sakahira, F., & Niimi, M. (2007). Ancient DNA analysis of the Japanese sea lion (*Zalophus californianus japonicus* Peters, 1866): preliminary results using mitochondrial control-region sequences. *Zoological Science* 24: 81–85.

Shirihai, H., & Jarrett, B. (2006). *Whales, Dolphins and Seals: A Field Guide to the Marine Mammals of the World.* A & C Black, London.

Simmons, S. E., Crocker, D. E., Kudela, R. M., & Costa, D. P. (2007). Linking foraging behaviour of the northern elephant seal with oceanography and bathymetry at mesoscales. *Marine Ecology Progress Series* 346: 265–275.

Timm, R. M., Salazar, R. M., & Peterson, A. T. (1997). Historical distribution of the extinct tropical seal, *Monachus tropicalis* (Carnivora: Phocidae). *Conservation Biology* 11: 549–551.

Trillmich, F. (1981). Mutual mother-pup recognition in Galápagos fur seals and sea lions: cues used and functional significance. *Behaviour* 78: 21–42.

Trillmich, F. & Limberger, D. (1985). Drastic effects of El Niño on Galapagos pinnipeds. *Oecologia* 67: 19–22.

Trites, A. W., Atkinson, S. K., DeMaster, D. P., Fritz, L. W., Gelatt, T. S., Rea, L. D., & Wynne. K. M. (2006). *Sea Lions of the World.* Alaska Sea Grant College Program, University of Alaska Fairbanks, Anchorage, AK.

Weber, D. S., Stewart, B. S., & Lehman, N. (2004). Genetic consequences of a severe population bottleneck in the Guadalupe fur seal (*Arctocephalus townsendi*). *Journal of Heredity* 95: 144–153.

CHAPTER 10
Mysticeti: The Baleen Whales

CHAPTER OUTLINE

Distribution
  Exploring the Depths: Migration
Feeding Behavior and Ecology
Reproduction

Abundance and Status
  Exploring the Depths: Antarctic Minke Whale Abundance
  Exploring the Depths: Public Perceptions of the Status of Baleen Whales
Selected References and Further Reading

The mysticeti are the largest of the marine mammals (**Figures 10.1** and **10.2**). In fact, the blue whale (*Balaenoptera musculus*) is the largest and heaviest animal that has ever lived, as far as we know. In 1909 a 33.6 m female was caught by Antarctic whalers, and in 1947 a 190 metric ton female was captured. These sizes were based on animals that had been butchered and so the measurements may not be completely accurate, but these weights are equivalent to a fully laden Boeing 747. The smallest rorqual is the minke whale, with adults typically 7 to 10 m in length (although the largest "dwarf" minke whale, which may possibly be a subspecies, was 7.8 m). The pygmy right whale (*Caperea marginata*) only grows to 4 to 6.5 m and is the smallest baleen whale.

The longest living mammals are also baleen whales; a bowhead whale aged at 211 years was killed in an aboriginal hunt in Alaska (and thus could theoretically have lived longer). Blue whales and fin whales killed during whaling operations have been aged at 110 and 114 years, respectively. It is possible that there are bowhead whales still alive today that were born while George Washington and Napoleon Bonaparte were still alive.

A   B
FIGURE 10.1   (A) A blue whale and (B) a 747.

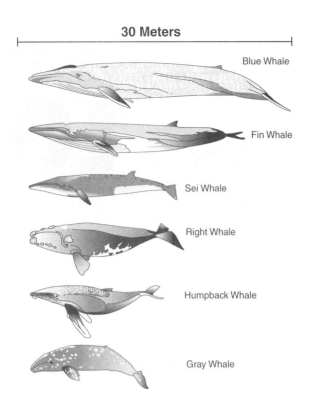

FIGURE 10.2 Drawings of various baleen whales with scale.

Distribution

Many of the baleen whales migrate between polar, subpolar, or temperate regions where they feed during the summer months and tropical or subtropical breeding grounds, following coastlines or submarine mountain chains along their route (see Exploring the Depths: Migration). Although polar waters may not appear to be a particularly hospitable or productive habitat, the long day length in midsummer, coupled with upwelling currents that are high in nutrients, make these regions very productive with dense blooms of phytoplankton and abundant schooling fish and invertebrates. The waters of the Southern Ocean are especially productive, and southern right whales, Antarctic minke whales, humpback whales, sei whales, fin whales, and blue whales can all be found feeding there. Some species migrate little, such as the bowhead whales, the majority of which remain in the Arctic/sub-Arctic, and the Bryde's whale, which has a tropical/subtropical distribution all year round. Several populations of other baleen whales also do not migrate; for example, humpback whales off the coast of Oman (see Exploring the Depths: Migration).

Many baleen whales are found in all the ocean basins (e.g., humpback whales, fin whales, and sei whales; **Figure 10.3**), but some species are confined to just one ocean basin. Right whales in the North Pacific and North Atlantic were recently found to be separate and genetically distinct species (see Chapter 3). The gray whale is only found in the Pacific Ocean as the Atlantic gray whale became extinct in the 17th century.

However, much is still not known about the distribution of baleen whales. For example, although it is known where Antarctic and northern minke whales feed, their breeding grounds are still undiscovered.

Exploring the Depths: Migration

Many large whale species migrate from polar regions (feeding grounds) to the tropics or subtropics (to breed). These journeys can be energetically costly; blue and fin whales have been recorded swimming at speeds of approximately 17 km/h, covering distances of 3,700 km on these migrations. As a result animals may lose 50% of their body mass. So why do they do this? Why use up so much of the energy gained at the feeding grounds?

The most accepted theory is that baleen whale calves, being relatively small in size and having only a thin blubber layer when born, are particularly susceptible to heat loss (see Chapter 4). Therefore, females give birth in warm waters to prevent their calves from becoming hypothermic. The weather in the sheltered tropical bays and shallow water banks where they breed is also considerably calmer than the rough winter waters of the polar regions from which they migrate. However, some baleen whales do not migrate. For example, humpback whales off the coast of Oman occupy an area with an upwelling current that brings nutrients to surface waters. These high levels of productivity allow this population to remain in the same region throughout the year.

It is also possible that baleen whales may migrate to escape predators (specifically killer whales), although in the tropics other predators, such as tiger sharks (*Galeocerdo cuvier*), are common. Another suggestion is that the long migrations are a behavioral relic from when breeding and mating grounds were located close together, but due to the movement of tectonic plates, the two locations have become more and more separated.

However, it seems unlikely that the various migration patterns of current baleen whales could have been produced by the accepted size and direction of plate movements (**Figure B10.1**). How do whales manage to navigate during these long distance migrations? For many species a large proportion of their migration is along the coast, and they probably use visual cues to migrate. It has been suggested that whales at sea may follow submarine mountain chains and other features of the seabed. They may also use sound to help them navigate (see Chapter 5). Some species swim across large stretches of open ocean, however, and we as yet do not know exactly how they navigate over these large distances. Various suggestions have been made, such as whales detecting and navigating by magnetic fields, but nothing has yet been proven.

The record for the longest migration by a marine mammal is held by the humpback whale. Several animals have been recorded traveling from Antarctica to the coast of Costa Rica, a distance of approximately 8,300 km.

Exploring the Depths: Migration (continued)

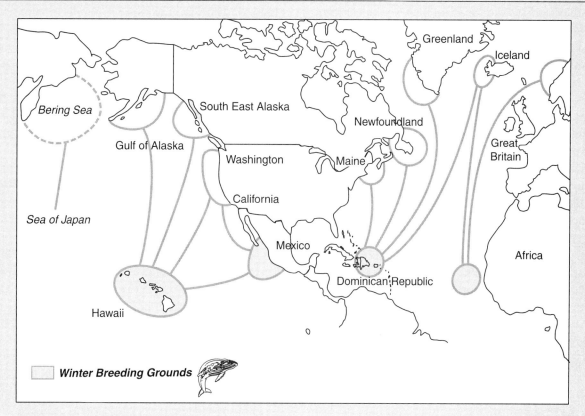

FIGURE B10.1 Humpback whale migration routes.

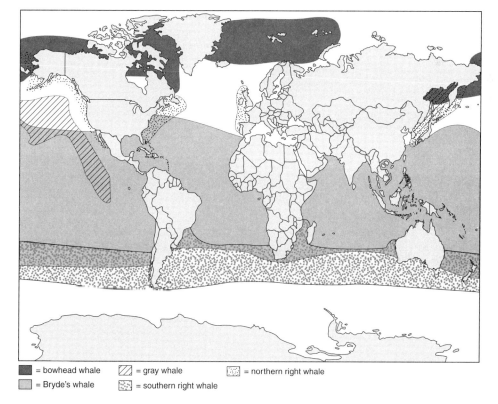

FIGURE 10.3 Map showing the current, known distributions of the bowhead whale, northern and southern right whales, Bryde's whale, and gray whale. Data from: IUCN Red List.

Feeding Behavior and Ecology

Baleen whales possess hundreds of plates of baleen composed of keratin that hang down from their upper jaw. Baleen whales use two main feeding methods to catch their prey: gulping and skimming (and gray whales have their own specific feeding strategy).

Most baleen whales use a gulping strategy when feeding. They engulf schooling fish or invertebrates, which causes the ventral pleats on the underside of the jaws to balloon out, akin to the way the pouch on the underside of a pelican's bill expands when it catches fish. The whale brings its jaws together, leaving a small gap nearly as wide as the length of the baleen plates. The whale then pushes its tongue upward and forward, pushing the water in its mouth cavity out through the gaps in the baleen plates, leaving the engulfed prey trapped behind the plates (**Figure 10.4**).

Most rorqual whales (except the sei whale) are "gulpers." Antarctic minke whales primarily feed on krill, whereas the Northern minke whale feeds on small schooling fish and shrimp, as do Bryde's and humpback whales. Fin whales may also take copepods and schooling squid, whereas the blue whale primarily consumes krill. In the Antarctic summer schools of krill can be several kilometers in diameter, with as many as 30,000 animals per cubic meter, thus providing a major source of food for Antarctic cetaceans, even for the large blue whale. A mouth of a gulping blue whale can hold 90 metric tons of water and prey, allowing the whale to consume an estimated 3.6 metric tons of krill or 40 million individuals every day. Thus, an adult blue whale may consume as much as 1.5 to 3 million calories a day.

Although multiple species of whales feed on krill in the waters of Antarctica or sharing feeding grounds elsewhere, there is no evidence of food competition between species. Indeed, there seems to be ecological niche separation even though species may be sympatric (occurring in the same place at the same time)—that is, species may feed in slightly different locations, on different size classes of prey, or at different depths. An ecological rule, referred to as Hutchinson's rule, suggests that sympatric species that differ by a size of 1.3 (i.e., one species is one-third again larger than the other) seem to be able to coexist without competing. The various baleen whale species that coexist in Antarctic waters seem to fit this rule.

Historically, krill in the waters around Antarctica sustained much greater populations of whales and other marine mammals than can be found today (**Figure 10.5**). However, the amount of krill seems to be decreasing. Researchers analyzing krill catches from the mid-1920s found a 80% decrease in krill abundance over the last 30 years, which has been linked to a decline in winter sea ice (an important habitat for krill) and climate change. Such a large decrease in krill availability could have a major impact on the baleen whale species that depend on Antarctic krill as their main source of food, as well as other species, such as crabeater seals and penguins.

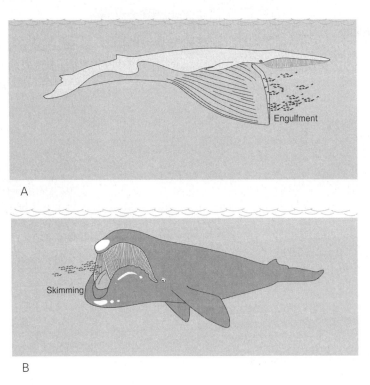

FIGURE 10.4 How baleen plates are used to (A) gulp and (B) skim.

FIGURE 10.5 (A) An individual krill and (B) a krill swarm.

FIGURE 10.6 Humpback whales in a feeding aggregation.

Humpback whales use a special method to increase their feeding success, called bubble netting. They often feed in coordinated groups (of 5–15 animals) in which one individual swims around a school of fish or krill in a spiral, emitting a stream of bubbles from its blowhole. These bubbles cause the prey to bunch closer together, allowing the entire group of humpback whales to swim up from underneath the clumped school and gulp concentrated mouthfuls of prey (**Figure 10.6** and **10.7**). Special calls are produced by the whales when they are just about

FIGURE 10.7 Humpback whale bubble netting.

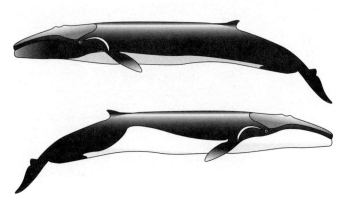

FIGURE 10.8 A drawing of a fin whale showing asymmetrical coloration.

FIGURE 10.9 A skim feeding right whale.

to start bubble netting, which may help to coordinate the group. Bubble netting has been observed primarily in Alaskan and Antarctic humpback whales, but it also occurs in other whale populations.

Other baleen whale species may also encircle and "corral" their prey; fin whales have been observed circling schools of fish. The coloration of fin whales is unusual in that it is asymmetrical, with a white patch on their right lower jaw (the left lower jaw is gray colored) (**Figure 10.8**). Fin whales seem to use this patch of white to help school fish, as they often circle around schools of prey to the right; the white patch may cause the prey to react and bunch together, making it easier for fin whales to catch them.

The Balaenidae, or right whales and bowhead whales, are "skim feeders," as are the pygmy right whales. The only skim-feeding rorqual is the sei whale. Skim-feeding whales swim slowly with their mouths open, usually near the water's surface, trapping zooplankton or small fish and invertebrates in their baleen plates as the seawater flows through them (**Figure 10.9**). The method is similar in many respects to the way many fish extract plankton from seawater, although fish use "gill rakers" instead of baleen plates to sieve the seawater. The right whales primarily feed on zooplankton, especially copepod crustaceans, but also feed on krill. Sei whales and bowhead whales also feed on copepods, krill, and amphipods (another group of planktonic crustaceans) (**Figure 10.10**).

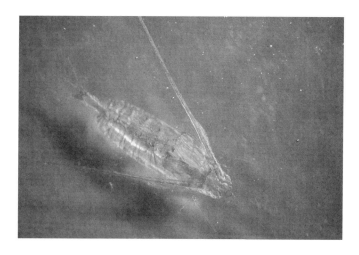

FIGURE 10.10 (A) A copepod and (B) an amphipod.

FIGURE 10.11 Underwater photo of a gray whale feeding.

Gray whales are slightly different in their feeding habits from other mysticetes, because they swim down to the seabed and take in mouthfuls of sediment, filtering out infauna (animals living in the sediment), such as amphipods and tubeworm species, from the seawater and sediment with their baleen plates (**Figure 10.11**). Because they take in seabed sediments, often from shallow areas where there can be high amounts of human activity, the gray whale may be at risk of ingesting large amounts of pollutants and marine debris.

Reproduction

Reproduction for all cetaceans is problematic because seawater is a spermicide. Therefore, female cetaceans have watertight valves in the vagina, reducing the likelihood of water entry. Male cetaceans have retractable, large, and prehensile penises, which allow them to copulate with the female even while swimming. There have also been reported instances of females being mated by two males simultaneously. The breeding season for baleen whales is typically in the winter, when many species migrate to the tropics (see Exploring the Depths: Migration, earlier in the chapter).

Baleen whales are typically polygynous, meaning males will attempt to mate with several females during the breeding season. As a result there can be competition and even aggression between males for access to females. For example, male humpback whales may slap and jostle other males, trying to maneuver their bodies between competing males and a female, sometimes hitting other males with their long flipper. Flippers have hard calcareous callosities on the leading edge, which may act like "knuckledusters"/"brass knuckles." Humpback whales may also blow bubbles as aggressive displays and have been reported blowing bubbles between a female and other males, effectively making a "screen" of bubbles. Humpback whales may also compete via their songs (perhaps to reduce the abovementioned physical competition), although singing may be a way for females to choose the fittest, most appropriate mate, similar to the use of song in birds (see Chapter 5).

Right whales are also quite aggressive when trying to gain access to females and will roll and even leap on other males. The hard callosities on their foreheads may even act akin to antlers when they "head butt" other males.

Another aspect of male reproductive competition is sperm competition. This is when males produce copious amounts of sperm, some of which is specialized to compete with or "wash out" the sperm deposited by a male that had previously mated with a female. Species that engage in sperm competition usually have larger testes than would be expected for their size. For instance, male right whales may have testes weighing a massive 1 metric ton that can produce nearly 20 liters of ejaculate.

Most baleen whales become sexually mature at 5 to 10 years of age. The gestation period is approximately 11 to 12 months for most species, and there is typically a 2- to 3-year interval between calves. Only one calf is born. The calf generally emerges tail first and is pushed to the surface by the mother to breath. Calves are approximately one-fourth of the length of the mother when born but can be less than one-twentieth of the weight. Blue whales, for example, are born at lengths of over 7 m while weighing 3 metric tons.

The calf is fed on milk with a high fat content (e.g., 40 to 50% fat by volume), and lactation is via two retractable nipples that reside in slits on either side of the mother's genital slit. The mother's mammary glands are contained within the body and are not externally visible, which aids in giving the female a more hydrodynamic shape. Weight gain can be very rapid in baleen whale calves. For example, a blue whale calf can drink 200 liters of milk per day, gaining 90 kg/day (4 kg/hour), and can reach 16 m in length and weigh more than 20 metric tons by the time it is weaned at 7 months. Conversely, during lactation the mother can lose up to 50 metric tons in weight.

Baleen whales are weaned at 4 to 11 months after birth, depending on species, and calves typically stay close to the mother during this period. Many species may predate baleen whale calves, including killer whales, and females can be aggressive toward potential predators. Some whalers used to take advantage of the mother–calf bond in baleen whales and would harpoon the slower, more easily tired calf to catch the mother as she stayed with the calf's carcass.

Abundance and Status

Extensive work has been done by the Scientific Committee of the International Whaling Commission to estimate the abundance of the various stocks of baleen whales. As a result much better data are available on many baleen whale abundances than exist for most odontocetes (**Table 10.1**). However, there are still controversies associated with some of the estimates, for example, those on Antarctic minke whales (see Exploring the Depths: Antarctic Minke Whale Abundance).

Almost all populations, known by whalers and consequently also some management systems as "stocks," of baleen whales, were depleted by commercial whaling (see Chapter 13 and 14). As a result many species and stocks have been listed as endangered (**Table 10.2**). Some species have been recovering since whaling bans (e.g., the whaling moratorium) were put in place (see Chapter 14). For example, the humpback whale was downlisted in 2008 from "vulnerable" to "least concern" by the IUCN.

One of the most depleted populations of baleen whales was the blue whale; an estimated 378,200 whales (representing more than 95% of the global population) were killed during whaling operations. The largest population, in Antarctica, was estimated to have been reduced to just over one-thousandth of its original population size. In 1964 the International Whaling Commission banned whaling of blue whales, but illegal whaling by Russian vessels continued until 1973, and the last known hunted blue whale was caught off the coast of Spain in 1978. Although blue whales have been recovering globally, the numbers of several populations are nowhere close to full recovery (e.g., at present Antarctic blue whale numbers are only at approximately 1% of their prewhaling abundance). The blue whale is considered by the IUCN to be "endangered."

Right whales in the northern hemisphere were also heavily depleted by a long history of whaling (see Chapters 13 and 14). Although commercial whaling of these animals has been banned since 1935, with the last hunted individual taken in 1967, the species has not recovered. There are currently only an estimated 300 to 350 right whales in the North Atlantic, and they face a variety of problems, including lack

TABLE 10.1 The Estimated Abundance of Various Whale Stocks

Species and Stock	Year of Abundance Estimate	Estimated Abundance	Range of estimate*
MINKE WHALES			
Southern Hemisphere	1982/83 and 1988/89 2003–2004	761,000 See Textbox	510,000–1,140,00
North Atlantic (Central & Northeastern)	1996–2001	174,000	125,000–245,000
West Greenland	2005	10,800	3,600–32,400
North West Pacific	1989–90	25,000	12,800–48,600
BLUE WHALES			
Southern Hemisphere	1997/98	2,300	1,150–4,500

TABLE 10.1 The Estimated Abundance of Various Whale Stocks (*continued*)

FIN WHALES			
North Atlantic (Central & Northeastern)	1996–2001	30,000	23,000–39,000
West Greenland	2005	3,200	1,400–7,200
GRAY WHALES			
Eastern North Pacific	1997/98	26,300	21,900–32,400
Western North Pacific	2007	121	112–130
BOWHEAD WHALES			
Bering-Chukchi-Beaufort Seas	2001	10,500	8,200–13,500
West Greenland	2006	1,230	490–2,940
HUMPBACK WHALES			
Western North Atlantic	1992/93	11,600	10,100–13,200
Southern Hemisphere (partial estimate)	1997/98	42,000	34,000–52,000
North Pacific	2007	at least 10,000	not yet available
RIGHT WHALES			
Western North Atlantic	2001	about 300	—
Southern Hemisphere	1997	about 7,500	—
BRYDE'S WHALES			
Western North Pacific	1998–2002	20,501	—

* 95% confidence limits
Data from: International Whaling Commission.

Exploring the Depths: Antarctic Minke Whale Abundance

In the austral summer of 1982–1983 and 1988–1989 circumpolar surveys estimated that more than 750,000 Antarctic minke whales existed in the southern hemisphere. However, in the most recent survey (2003–2004) the estimate was less than 40% of that number (approximately 338,000, although the specific number estimates vary according to the analysis method used).

Why was there such a big decrease? Perhaps there were problems with the methodology of previous surveys that overestimated numbers? A major flaw in the methodology would also cast doubt on other whale survey estimates using similar methods. Japanese scientists at the International Whaling Commission claim that minke whales have moved into the ice where survey vessels cannot travel, possibly as a result of competition with the increasing population of Antarctic humpback whales. There is little evidence, however, to suggest a major change in distribution or competition between the species as yet.

Has there been a real decrease in the number of whales in the Antarctic? If so, why might this be? Could it be an effect of climate change, depleted food sources, or increased predation by killer whales? Perhaps the removal of minke whales by "scientific whaling" (see Chapter 14) is reducing numbers? We simply do not know, although it is quite likely that some combination of these factors, and others, are at work.

Under International Union for Conservation of Nature (IUCN) criteria an actual, observed, or implied 50%, or 20% decrease of whales within a 10-year period would make the species fall in the "endangered" or "vulnerable" category respectively, and technically Antarctic minke whales should be eligible for such a designation. However, scientific and political arguments at the International Whaling Commission and other bodies make any precautionary conservation action for the Antarctic minke whale unlikely.

---

of food and entanglement in fishing gear. Mortalities due to ship collisions are a particular concern (see Chapter 15). As such, this species appears to be declining further, making them and the similarly afflicted North Pacific right whale two of the most threatened whale species.

Although the gray whale is currently considered to be relatively stable and is not threatened across most of its range, the western Pacific population is not faring so well. There are only approximately 100 animals in the population, of which only 20 to 30 are mature females, making

TABLE 10.2 Status of Baleen Whales Under the IUCN and U.S. Endangered Species Act

Species	IUCN Listing	U.S. Endangered Species Act
Balaena mysticetus – Bowhead whale	Least Concern	Endangered
Eubalaena glacialis – North Atlantic right whale	Endangered	Endangered
Eubalaena japonica – North Pacific right whale	Endangered	Endangered
Eubalaena australis – Southern right whale	Least Concern	Endangered
Caperea marginata – Pygmy right whale	Data Deficient	—
Eschrichtius robustus – Gray whale	Least Concern	—
Megaptera novaeangliae – Humpback whale	Least Concern	Endangered
Balaenoptera acutorostrata – Northern minke whale	Least Concern	—
Balaenoptera bonaerensis – Antarctic minke whale	Data Deficient	—
Balaenoptera borealis – Sei whale	Endangered	Endangered
Balaenoptera brydei – Bryde's whale	Data Deficient	—
Balaenoptera edeni – Dwarf Bryde's or Eden's whale	Not Listed	—
Balaenoptera omurai – Omura's whale	Data Deficient	—
Balaenoptera musculus – Blue whale	Endangered	Endangered
Balaenoptera physalis – Fin whale	Endangered	Endangered

their ability to increase in number severely limited. Little is known about this population. For example, its breeding area has yet to be identified. However, there are considerable concerns about degradation of this animal's feeding habitat as the result of oil and gas exploration. The western Pacific gray whale population is considered to be "critically endangered" by the IUCN and "endangered" under the U.S. Endangered Species Act.

Because of the enormous impact of commercial whaling on baleen whales, all species are listed on Appendix I of the Convention on International Trade in Endangered Species of Wild Fauna and Flora and thus the international commercial trade of these animals is prohibited (except by Norway, Iceland, and Japan, who have cited reservations against Appendix I listing of minke whales. Iceland also has a similar reservation against the listing of fin whales; see Chapter 14). More details about the conservation and status of cetaceans, including baleen whales, are summarized in Chapters 14 and 15.

Exploring the Depths: Public Perceptions of the Status of Baleen Whales

Thanks to the "save the whales" movement in the 1970s and campaigns by various environmental groups, most members of the public realize there are conservation concerns for baleen whales. However, recent studies indicate that the general public may not be very knowledgeable about which species are most threatened. In 2008 a survey of students at a U.S. university found that most students thought the humpback whale was the most threatened, despite this species being downlisted by the IUCN. Nearly one-fourth thought the blue whale was most threatened, which is perhaps more understandable as the decimation of the blue whale population is reasonably well known. One-fifth thought that northern minke whales were the most threatened, but less than 5% recognized a right whale as the most threatened species.

SELECTED REFERENCES AND FURTHER READING

Atkinson, A., Siegel, V., Pakhomov, E., & Rothery, P. (2004). Long-term decline in krill stock and increase in salps within the Southern Ocean. *Nature* 432: 100–103.

Branch, T.A. (2006a). Abundance estimates for Antarctic minke whales from three completed circumpolar sets of surveys, 1978/79 to 2003/04. Paper presented to the Scientific Committee at the 58th Meeting of the International Whaling Commission, 26 May to 6 June 2006, St. Kitts.

Branch, T.A. (2006b). Possible reasons for the appreciable decrease in abundance estimates for Antarctic minke whales from the IDCR/SOWER surveys between the second and third circumpolar sets of cruises. Paper presented to the Scientific Committee at the 58th Meeting of the International Whaling Commission, 26 May to 6 June 2006, St. Kitts.

Branch, T.A., & Butterworth, D.S. (2001). Southern Hemisphere minke whales: standardised abundance estimates from the 1978/79 to 1997/98 IDCR/SOWER surveys. *Journal of Cetacean Research and Management* 3: 143–174.

Branch, T.A., Matsuoka, K., & Miyashita, T. (2004). Evidence for increases in Antarctic blue whales based on Bayesian modelling. *Marine Mammal Science* 20: 726–754.

Bryant, P.J. (1995). Dating remains of gray whales from the eastern North Atlantic. *Journal of Mammalogy* 76: 857–861.

Clapham, P.J., & Brownell, R.L. (1996). The potential for interspecific competition in baleen whales. *Report of the International Whaling Commission* 46: 361–367.

Corkeron, P.J., & Connor, R.C. (2006). Why do baleen whales migrate? *Marine Mammal Science* 15: 1228–1245.

Cummings, W.C., & Thompson, P.O. (1971). Underwater sounds from the blue whale, *Balaenoptera musculus*. *Journal of the Acoustical Society of America* 50: 1193–1198.

Friedlaender, A., Lawson, G.L., & Halpin, P.N. (2006). Evidence of resource partitioning and niche separation between humpback and minke whales in Antarctica: implications for interspecific competition. Paper presented to the Scientific Committee at the 58th Meeting of the International Whaling Commission, 26 May to 6 June 2006, St. Kitts.

George, J.C., Bada, J., Zeh, J., Scott, L., Brown, S.E., O'Hara, T., & Suydam, R. (1999). Age and growth estimates of bowhead whales (*Balaena mysticetus*) via aspartic acid racemization. *Canadian Journal of Zoology* 77: 571–580.

Hutchinson, G.E., & MacArthur, R.H. (1959). A theoretical ecological model of size distributions among species of animals. *American Naturalist* 93: 117–125.

Ingebrigtsen, A. (1929). Whales caught in the North Atlantic and other seas. *Rapports et Proces-verbaux des Réunions. Conseil International pour l'Éxploration de la Mer* 56: 1–26.

International Whaling Commission. (2001). Report of the workshop status and trends of western North Atlantic right whales. *Journal of Cetacean Research and Management* 2: 61–87.

Mate, B., Duley, P., Lagerquist, B., Wenzel, F., Stimpert, A., & Clapham, P. (2005). Observations of a female North Atlantic right whale (*Eubalaena glacialis*) in simultaneous copulation with two males: supporting evidence for sperm competition. *Aquatic Mammals* 31: 157–160.

Parsons, E.C.M., Rice, J.P., & Sadeghi, L. (2010). Awareness of whale conservation status and whaling policy in the US—a preliminary study on American youth. *Anthrozoös* 23: 119–127.

Piper, R. (2007). *Extraordinary Animals: An Encyclopedia of Curious and Unusual Animals*. Greenwood Press, Westport, CT.

Rasmussen, K., Palacios, D.M., Calambokidis, J., Saborío, M.T., Dalla Rosa, L., Secchi, E.R., Steiger, G.H., Allen, J.M., & Stone, G.S. (2007). Southern Hemisphere humpback whales wintering off Central America: insights from water temperature into the longest mammalian migration. *Biology Letters* 3: 302–305.

Reeves, R.R. (2005). *Report of the Independent Scientific Review Panel on the Impacts of Sakhalin II Phase 2 on Western North Pacific Gray Whales and Related Biodiversity*. IUCN, Gland, Switzerland.

Waring, G.T., Josephson, E., Fairfield-Walsh, C.P., & Maze-Foley, K. (2008). *U.S. Atlantic and Gulf of Mexico Marine Mammal Stock Assessments—2007*. NOAA Technical Memorandum NMFS NE 205. National Marine Fisheries Service, Woods Hole, MA.

Odontoceti: The Toothed Whales

CHAPTER 11

CHAPTER OUTLINE

Sperm Whales
- Distribution
- Reproduction
- Feeding Behavior and Ecology
- Abundance Status
- *Exploring the Depths: Pygmy and Dwarf Sperm Whales*

Beaked Whales
- *Exploring the Depths: Battling Beaked Whales*
- Distribution
- Deep Diving and Sonar
- Status

River Dolphins
- Indian River Dolphins
- Amazon River Dolphins
- Yangtze River Dolphin
- *Exploring the Depths: Franciscana or La Plata Dolphin,* Pontoporia blainvillei

Beluga Whales
- Distribution
- Abundance and Status
- *Exploring the Depths: Cook Inlet Beluga Whale*

The Narwhal
- *Exploring the Depths: The Narwhal's Tusk*

Porpoises
- *Exploring the Depths: Dall's Porpoise and Japanese Catches*
- *Exploring the Depths: Harbor Porpoise,* Phocoena phocoena
- *Exploring the Depths: A Particularly Perplexing Porpoise*

Selected References and Further Reading

The Odontoceti are a very large taxonomic group and make up the majority of the cetacean species. This chapter reviews the various characteristics of the main groups of toothed whales. Chapter 12 will then look in more detail at the largest taxonomic group within the Odontoceti, the Delphinidae or dolphins. The largest toothed whale species is the sperm whale, with its distinctive "submarine" shape and massive head. The smallest is the vaquita, or Gulf of California harbor porpoise, detailed at the end of this chapter.

Sperm Whales

"The sperm whale is nothing more than a nose with an outboard motor."

Bertel Mohl

Sperm whales (*Physeter macrocephalus*, but sometimes referred to as *Physeter catadon* because of a dispute in the scientific priority of the two names) get their common name because whalers believed the waxy substance in the whale's head was sperm and the spermaceti organ was effectively a large gonad (**Figure 11.1**). However, the organ, which in males can be almost one-third of the animal's length, is instead a modified nasal structure that is used in echolocation (see Chapter 5) and possibly also buoyancy control (see Chapter 4). The sperm whale was also referred to as the cachalot, which is possibly derived from a Portuguese word meaning "big head." In China the sperm whale is called the "perfume whale" because of the ambergris, a waxy substance found in their stomachs, possibly a byproduct of digesting squid beaks. Historically, this was literally worth more than its weight in gold as it was used as a fixative for perfumes. Although synthetic substitutes are now available, it is still highly prized by the industry today when it washes up on beaches.

The species is sexually dimorphic in that males grow to be considerably larger (18 m) than females (12 m). In the

FIGURE 11.1 A sperm whale.

past, males may have been larger still, as there is a jaw from a sperm whale in the Nantucket Whaling Museum that is allegedly from a male that was 24 m in length.

Distribution

Sperm whales tend to be found in waters greater than 1 km, and often more than 3 km, deep. They are most commonly seen along the edges of continental shelves, submarine canyons, or islands with steep slopes down to the abyssal planes. The sperm whale has a single blowhole located at the tip of the head, slightly on the left side. Male sperm whales have been recorded diving to over 2 km, staying underwater for up to 1.5 hours or possibly longer. However, it has been suggested that they can dive to 3 km or more, because carcasses have been found entangled in submarine cables.

Reproduction

The distribution of the species also varies by gender, with adult males located primarily in cooler temperate and subpolar regions and with females more limited to warmer temperate and tropical waters. Accordingly, males move to the tropics to breed. Finally, mature males are generally found alone, whereas adult females usually travel together in "nursery pods" (groups), with other females and young whales of both sexes. Sperm whales are thought to have a polygynous breeding system, with dominant males mating with multiple females in a "harem."

Occasionally, predators, including killer whales, have been observed attacking the nursery pods. In response, females have been known to adopt the "Marguerite formation." In this formation, females arrange themselves in a circle around a group of young whales, with the heads of the females pointing toward the cluster of young and their tails pointing outward. The formation almost resembles a daisy, with the young as the center and the females arranged like the petals of the flower. The females use their powerful tails to swipe at and try to dissuade the attacking predators from seizing any of their calves from below. Somewhere between 4 and 20 years of age males leave the nursery pod to join all-male "bachelor pods."

Feeding Behavior and Ecology

In their deep-water habitat, male sperm whales hunt deepwater squid, including giant squid (*Architeuthis* spp.; up to 13 m long) and colossal squid (*Mesonychoteuthis hamiltoni*; up to 14 m long). Evidence of encounters with these great squid species is sometimes found on the flanks of whales in the form of large, dinner plate-sized scars caused by the squids' suckers. Females usually dive to comparatively shallower depths, around 500 to 1,000 m and prey on smaller squid. Sperm whales of both sexes do, however, also take fish, and in parts of Alaska and the Southern Ocean there have been conflicts with fishermen, as some whales have learned to take fish right off the longlines. These whales have learned to approach the boats when they hear the sound of the longline winches being operated. One video recorded a whale plucking the line with its teeth until it dislodged the fish, and then swimming after the prey. However, other removal techniques may also be used. In more natural settings, sperm whales are thought to catch their prey using suction by expanding throat muscles and creating a vacuum to draw in their prey.

The lower jaw of the sperm whale is thin and has 20 to 26 thick conical teeth, which may help to gain purchase on their struggling, slippery prey. Their upper jaw does not have any erupted teeth. Instead, it has sockets into which the teeth of the lower jaw are inserted. However, fully fed sperm whales have been observed with old injuries, including broken or partially missing lower jaws, so these are apparently not essential for successful foraging.

Abundance and Status

Heavily depleted by a long history of whaling (see Chapter 14), with an estimated 1 million sperm whales having been taken over the history of whaling activities, sperm whales are currently at a fraction of their historical numbers. Although commercial hunts of these whales have been banned since 1984, some sperm whales have been taken since in the North Pacific in "scientific" catches (see Chapter 14). As a top predator, accumulation of toxic contaminants is a cause of concern for sperm whales (see Chapter 15). Many animals have also been found to have ingested marine litter and debris (see Chapter 15), possibly taken incidentally with their prey through their suction method of feeding.

Collisions with ships are another problem. For example, in the Canary Islands many sperm whales are struck by high-speed ferries. Sperm whales seem to be particularly susceptible because they are one of the few cetacean species

Exploring the Depths: Pygmy and Dwarf Sperm Whales

A

B

FIGURE B11.1 (A) Dwarf and (B) pygmy sperm whales.

The pygmy (*Kogia breviceps*) and dwarf sperm whales (*Kogia sima*, changed from *Kogia simus* to make the Latin grammar correct) look extremely similar to each other (**Figure B11.1**) and are often recorded in the field as *Kogia* sp., even by experts. It has also been suggested, based on genetic differences, that *K. sima* may in fact be two separate species: an Atlantic and an Indo-Pacific species.

As their name suggests, they are much like miniature sperm whales (approximately 2.7 m when adult). Like their larger relatives they have a small narrow lower jaw with no teeth in their upper jaw. However, their heads are not quite as square and protruding as that of their larger cousins. Instead they are a little more "wedge" shaped. Both species feed on deep-water squid, like the sperm whale, although the size of the squid they tackle is considerably smaller. The dwarf sperm whale has a slightly larger dorsal fin in the middle of the back than the pygmy sperm whale. Both have a tropical to temperate distribution, although the pygmy sperm whale generally occurs slightly farther north and slightly farther south (**Figure B11.2**). Another characteristic shared by

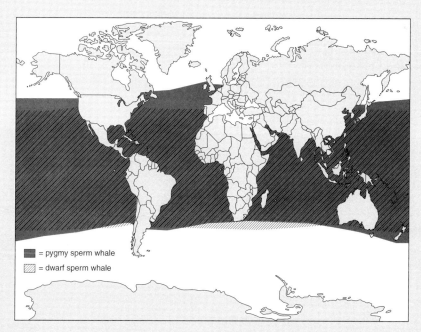

FIGURE B11.2 Map showing the current, known distribution of the pygmy and dwarf sperm whale. Data from: IUCN Red List.

Exploring the Depths: Pygmy and Dwarf Sperm Whales (*continued*)

both dwarf and pygmy sperm whales is the distinctive white "c-shaped" mark on each side of their heads just behind their eyes. These are known as "false gills," because they are thought to mimic the gill slit of a shark, or some other large predatory fish. This is presumably so they are less likely to be preyed on by other marine predators, particularly when they are resting on the surface after a long dive. This mimicry is definitely enhanced by their tendency to bend from side-to-side instead of up-and-down during periods of relative inactivity at or near the surface, giving them a distinctly shark-like look from above (and presumably also from below). Indeed, pygmy and dwarf sperm whales have been killed on several occasions by humans who have mistaken the whales for sharks, with their wedge-shaped heads, distinct dorsal fins, and false gill markings.

Other possible problems for these whales may include marine trash and debris (several animals have stranded or died after ingesting such items) and, similar to beaked whales, underwater noise (several pygmy sperm whales have stranded during mass stranding events associated with military activities: see Chapter 5). Both *Kogia* species are listed in Appendix II of CITES, meaning that international trade in the species is controlled, and both are considered to be "data deficient" by the IUCN.

---

that are almost totally inactive when they sleep. Virtually all cetaceans studied to date shut down only half of their brain at a time when they sleep, partly because they are voluntary breathers. This also allows them to continue swimming, albeit in a much less active way. However, sperm whales sleep underwater, floating motionless near the surface. This makes the whales invisible to oncoming ships. Unfortunately, there are also reports that the sleeping whales appear to be unable to detect and/or respond to the approaching vessels in time to avoid them. Finally, there is also concern about the possible impacts of underwater noise on sperm whales. For example, whales have been reported to change their behavior as a result of exposure to sound produced during oil and gas exploration (see Chapter 5).

Sperm whales are listed under Appendix I of the Convention on International Trade in Endangered Species of Wild Fauna and Flora (CITES), meaning that international trade in sperm whales or their products is prohibited. The International Union for Conservation of Nature (IUCN) considers sperm whales to be "vulnerable," and they are listed as "endangered" under the U.S. Endangered Species Act (see Chapter 17).

Beaked Whales

The beaked whales (Family Ziphiidae) are an extremely diverse but generally understudied group of toothed whales. This is primarily because they spend the majority of their time at depth. Even when at the surface they are very difficult to observe in all but the best weather conditions and are easy to misidentify. The family is split into two subfamilies. The first, Ziphiinae, includes only four species: the Cuvier's beaked whale, Arnoux's beaked whale, Baird's beaked whale (or North Pacific bottlenose whale), and Shepherd's (or Tasman) beaked whale (see Chapter 3). The second subfamily, Hyperoodontinae, contains the remaining species of the beaked whales and the bottlenose whales (**Figure 11.2**).

Most beaked whale species superficially resemble very large dolphins (the largest is the Baird's beaked whale at 12–13 m), possessing both a dorsal fin and an extended rostrum (as do most dolphins), as well as being long and slender. Identification of beaked whales at sea is extremely difficult. Many species can only be identified by examining the position of teeth in the skulls of males (females of many beaked whale species do not have teeth). Males of most beaked whale species only have one or two pairs of teeth, which can be situated at the tips or midsections of their beaks. The shape of these teeth can vary greatly from species to species, ranging from narrow and pointed (e.g., Cuvier's beaked whale), to wide and leaf-shaped (e.g., ginkgo-toothed beaked whale), or even shaped like large, flattened bananas (e.g., strap-toothed or Layard's beaked whale: **Figure 11.3**). Shepherd's beaked whale, however, has a mouth full of teeth, similar to a dolphin.

FIGURE 11.2 Bottlenose whales in Broadford Bay, Scotland.

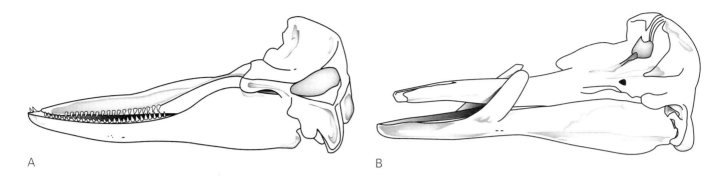
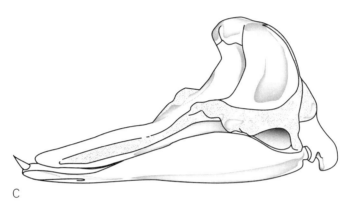

FIGURE 11.3 Skulls of (A) Shepherd's, (B) strap-toothed, and (C) Cuvier's beaked whales.

Exploring the Depths: Battling Beaked Whales

The beaked whales are a poorly known family of whales that spend much of their time making long dives to depths of more than 1,000 m to feed. As a result it is very difficult to study their behaviors, especially rare behaviors such as fights between males over females. Luckily (at least for the scientists), there are plenty of other ways to learn about these behaviors other than seeing them firsthand. Beaked whales are unusual for cetaceans in that they have one or two pairs of tusks. In most species these tusks only erupt to become functional in adult males. By examining the morphology of these tusks and associated structures, we can get a good idea of how they are used because their form is related to their function. These tusks are set in the lower jaw and are supported by reinforced bone. Similarly, the bone of the upper jaw is also reinforced with a bony tissue that is the densest bone found anywhere in the animal kingdom. This makes it very resistant to injuries incurred from directly above but not head on. From this, we can work out that beaked whales probably fight by making a series of passes at each other (like humans jousting on horseback) during which they try to rake each other with their tusks.

This explanation of how these fights occur is supported by examining the scars that are observed almost exclusively in mature males. The backs of adult male beaked whales are typically covered by a tangled mass of long linear scars from where the tusks of other males have apparently been dragged along their backs. These scars indicate intense and dangerous battles. As a result, it would be of benefit to males to have some way to advertise their fighting ability. They may do this by showing off their scars from past fights. Unlike many other cetaceans, the scars of beaked whales heal but do not regain pigment. This means the scars remain clearly visible as white lines along the dark backs of the animals for the rest of their lives. Only particularly strong or vicious males can survive many intense fights, so the amount of scarring could be used to assess the fighting prowess of potential rivals. This means that males could avoid getting into unevenly

Exploring the Depths: Battling Beaked Whales (*continued*)

matched fights. It is also possible that males (or females) can glean information about size, and thus strength, through eavesdropping on the echolocation sounds made by other males.

One aspect of tusk morphology does not seem to be linked to fighting. Different species have different shapes of tusks that are in different positions along the lower jaw. All species seem equally adept at fighting, so having tusks of different shapes and positions does not seem to relate to fighting ability. Rather, beaked whales all look very similar (especially those of the genus *Mesoplodon*). As a result, females need some way of ensuring they only mate with males of their own species. To do this, they are thought to use the shape and position of the highly visible tusks for species recognition. We believe this because species that occur in the same ocean differ in the shape and position of their tusks, whereas some species that live in different oceans do not.

Although the above explanations concerning tusks apply to most beaked whale species, there is one that clearly differs: the northern bottlenose whale. Males of this species have a unique structure in the form of massive bony crests on their upper jaws not found in other species. In addition, their tusks never erupt beyond the gums, and they lack the heavy scarring found in other beaked whale species. This suggests they fight in a very different manner. The structure of the bony crests is remarkably similar to terrestrial mammals that fight by head-butting, and so we infer this may be the way male northern bottlenose whales fight. Although we may not be able to directly observe beaked whale behaviors, such as fights between males, we can still learn a lot about them by looking at the structures that have evolved complementary to fighting behaviors to facilitate intraspecific (within the same species) fighting and assessment of the prowess of potential rivals.

–Contributing author, Colin MacLeod, University of Aberdeen.

This difference in dentition is linked to the cetacean's ecology. Shepherd's beaked whales are fish eaters, whereas the other beaked whales predominantly feed on squid (and use suction to catch their prey, rather than grasping prey with their teeth).

Male beaked whales may have such diverse teeth because the teeth are a secondary sexual characteristic, similar to the elaborate tail of a male peacock or antlers of deer, with females choosing males based on their teeth size and location. They may also play a role in species recognition.

▪ Distribution

The distribution of beaked whales depends on the species, from the very widespread, such as Cuvier's beaked whale (found in most tropical, temperate, and sub-polar waters), to those restricted in distribution, such as Shepherd's beaked whale (found around the coast of Tasmania and the southern tip of South America: **Figure 11.4**). Their preferred habitats appear to be similar to those of the sperm whales (see above), because their distribution is often linked to submarine features such as continental shelf edges, submarine canyons, or precipitous drop-offs around islands. They are usually associated with waters 250 m or deeper but have, on very rare occasions, been observed in shallower waters. It may, however, be that these instances are anomalous (or possibly the result of exposure to predators or noise; see Deep Diving and Sonar section for details). However, for some beaked whales, little, if anything, is known about their behavior, ecology, or distribution (e.g., Perrin's beaked whale).

▪ Deep Diving and Sonar

Beaked whales, like sperm whales, are deep divers, diving to depths of 1.5 km or more for up to 80 minutes. Studies have found that when some beaked whales surface after a deep dive (more than 0.5 km), they take a series of shallower dives (less than 200 m). This superficially appears to be similar to the "decompression stops" that deeper-diving human scuba divers must make to dump nitrogen from their systems and avoid decompression sickness (the "bends"). Indeed, this purpose has been suggested for these dives. However, it would be difficult for nitrogen to pass from the blood into the lungs for "offloading" at 200 m. This is because the lungs of whales collapse when they reach a depth of around

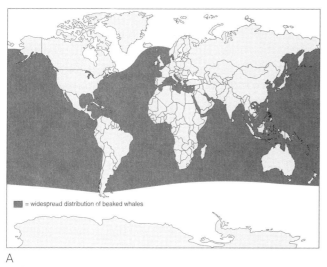
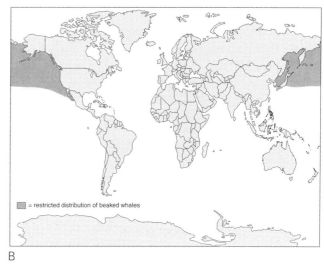

FIGURE 11.4 Maps showing (A) widespread and (B) restricted beaked whale distribution. Data from: IUCN Red List.

70 m, leaving very little, if any, air in their lungs and thus preventing their bloodstream from contacting nitrogen at high partial pressures (see Chapter 4).

Another explanation for the shallower dives may be that the whales' physiology is intolerant of lower pressures (i.e., they may be more physically comfortable at depth). A further alternative is that they are avoiding potential predators that frequent shallower depths (e.g., killer whales) because they are generally solitary and still too small to survive a pack attack on their own (i.e., they are psychologically more comfortable at depth). However, disruption of this shallow-diving behavior does appear to be linked to the appearance of decompression sickness–like conditions in beaked whales. Thus, it has also been suggested that they are reducing the time spent at or near the surface to reduce the time over which pressures are low enough for nitrogen to bubble out of the blood, while remaining close enough to the surface to breath frequently after their long, likely anaerobic deep dives.

Beaked whales, especially Cuvier's beaked whales, appear to be especially susceptible to impacts from exposure to military sonar because there have been many mass strandings coincident with military exercise activity (see Chapter 5). As evidence came in and ideas were developed, it was initially thought that frightened beaked whales might be coming to the surface too quickly and getting "the bends" in much the same way as human divers. However, current thinking is that beaked whales are affected by military sonar and thus strand because the whales stay too near to the surface between deep dives. In any effect, this release of pressure seems to result in bubble-like lesions similar to those seen in bends victims. Another idea that is not mutually exclusive (i.e., both mechanisms could play a part) is that the sonar signals vibrate the nitrogen within the blood with enough energy that it bubbles out of solution, which could also lead to bubbles in tissues and "bends"-like lesions.

It has also been suggested that the pressure at the deep depths reached by beaked whales may have neurological effects on the whales, making them "hyperexcitable" or especially "skittish." It would follow that they could be more likely to react strongly to an unusual stimulus, such as military sonar, which could exacerbate the situation. Beaked whales have also been documented reacting to shipping noise and in general may be particularly sensitive to underwater noise.

Status

Besides sound, other problems facing beaked whales include mortality as a result of bycatch in fishing gear. Beaked whales were also caught during commercial whaling operations (for example, northern bottlenose whales in the North Atlantic), although not as extensively as baleen whales. The largest exploiter of beaked whales was probably Japan, taking nearly 4,000 Baird's beaked whales before the whaling moratorium. Indeed, Japanese whaling operations continue to harvest this species (approximately 60 a year), but Japan does not consider these catches to be governed by the International Whaling Commission (IWC) because beaked whales are "small cetaceans" (see Chapter 14).

All but four of the beaked whale species are listed by the IUCN as "data deficient." Arnoux's and Baird's beaked whales, like the northern and southern bottlenose whales, are considered to be "lower risk, conservation dependent" and are also listed under CITES Appendix I (see Chapter 17). All other beaked whale species are effectively listed under Appendix II of CITES.

River Dolphins

The river "dolphins" (they are not true dolphins in as much as they do not belong to the family Delphinidae) were once classified together in the Family Platanistidae, but now only the Indian River dolphins (genus *Platanista*) remain in that family (for a current classification see Chapter 3). The river dolphins consist of several species that, although taxonomically and genetically very different, look similar in shape and have similar ecologies by virtue of the riverine environment in which they each live. All river dolphins tend to grow to around 2.6 m in length (Amazon river dolphins are slightly larger, at up to approximately 2.8 m), have large, square melons; long, thin beaks; reduced dorsal fins (their fins may be little more than ridges or bumps); and large pectoral flippers. They often also have limited eyesight. Moreover, their coloration tends to be similar as well, being pale gray to off-white, with no discernable markings of color patterns. Many of these features are simply adaptations to restricted, turbid (sediment-filled) waters of rivers and estuaries. Their large flippers make them more maneuverable while the stabilizing effect of a dorsal fin is not really required as they do not achieve sustained high speeds (and the fin might in fact be a hindrance in their shallow, riverine environment. Their prey is often found in benthic mud, making a long beak helpful to catch prey. Finally, the turbid waters in which they live make vision less useful, but echolocation even more important as a sense for navigation in addition to finding prey.

Indian River Dolphins

The so-called Indian river dolphins are usually split into two species (although some authorities only recognize one species, *Platanista gangetica*, with two subspecies). These are the Indus River dolphin (*P. g. minor*), or Susu, found in Pakistan, and the Ganges River dolphin (*P. g. gangetica*) found in Nepal, India, and Bangladesh (**Figure 11.5**).

They have a dorsal ridge instead of a true dorsal fin and small eyes that lack lenses, limiting their vision. Their beak is thin and narrow but, unlike the other river dolphins, their teeth are not all the same size and get larger toward the tip of the beak, the end of which is slightly bulbous (**Figures 11.6 and 11.7**). Both species feed on river fish, prawns, invertebrates, and possibly even turtles and birds.

Threats to both species include accidental entanglement in fishing gear and deliberate capture. Indian River dolphins are sometimes used as bait for catfish fisheries and are occasionally consumed by humans for food. They are also hunted for body parts (for medicines or aphrodisiacs).

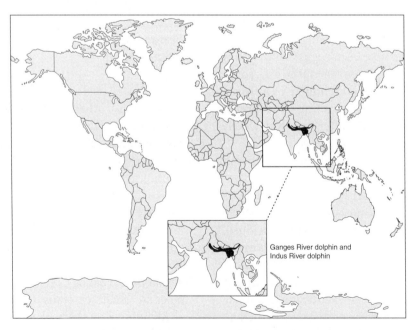

FIGURE 11.5 Map showing the current, known distribution of the Ganges and Indus River dolphin. Data from: IUCN Red List.

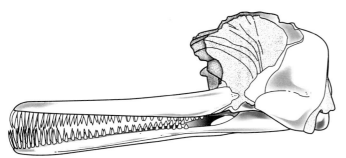
FIGURE 11.6 An Indian River dolphin skull.

FIGURE 11.7 A Ganges River dolphin.

Pollution is another problem (both chemical as well as noise pollution), as one might expect in river systems that drain large areas of often dense human occupation and industrial activities. Organic and trace element pollutants (see Chapter 15) are definitely causes for concern, and high levels have been found in both species. Another major problem is the large numbers of dams and barrages on both the Indus and Ganges River systems. These dams have altered, and are altering, river dolphin habitat substantially, causing high levels of disturbance, as well as physically separating and fragmenting populations. Finally, climate change may be a problem for these river dolphins. The flow of the Ganges and Indus rivers depends on regular rates of ice water melt in the Himalayas. There are concerns that melting and retreating glaciers in the mountain chain may ultimately lead to irregular, or decreased, volumes of water flow in both river systems.

There are estimated to be just under 1,000 Indus River dolphins and between approximately 1,000 and 2,000 Ganges River dolphins. Because of their low numbers, fragmented habitat, and current or projected declines in populations they are considered to be "endangered" by the IUCN. They are also listed as "endangered" under the U.S. Endangered Species Act and are on CITES Appendix I, which bans international trade in these species (see Chapter 17).

Amazon River Dolphins

Amazon River dolphins (*Inia geoffrensis*), or boto, are found, as their name suggests, in the Amazon River basin in Ecuador, Peru, and Brazil, but they are also found in the Bolivian tributaries, as well as in the Orinoco River basin in Colombia and Venezuela (**Figures 11.8** and **11.9**). They are now considered to consist of three subspecies: *I. geoffrensis humboldtiana* in the Orinoco, *I. geoffrensis geoffrensis* in the majority of the Amazon and *I. geoffrensis boliviensis* in Bolivia. The three subspecies (and most notably the latter two) are effectively separated by sets of rapids and waterfalls in the river systems.

Amazon River dolphins are similar in appearance to other river dolphins. They are grayish in color, but are known to turn pink as blood vessels dilate in their skin, which is a way of cooling down. They feed on benthic fish and will, during the flood season, actually swim into the flooded forest and hunt for food in and among the roots of trees.

Only part of the Amazon River dolphins' habitat has been surveyed. Consequently, the total number of individuals is unknown, but it is probably not more than the low tens of thousands combined across the subspecies. In some areas of the Amazon, the river dolphins have been historically revered or ascribed with magic powers (see Chapter 13). This reverence is still seen in some areas today, although not so much in others. Nearly 100 Amazon River dolphins were removed in the 1970s and 1980s for the public display industry in Europe and Japan. Animals have also been deliberately killed to provide body parts for medicines and charms and also to provide meat for catfish fisheries. The bait fishery is still believed to be removing substantial numbers in some parts of their habitat. Incidental bycatch in fishing gear also kills some animals. Pollution is likely to be another problem, as is the case for many other riverine species. In particular, gold mining in the Amazon region uses mercury. Substantive amounts of this toxic heavy metal have been found in the river environment, which could be a noteworthy threat to this species. The Amazon River dolphin is listed under CITES Appendix II (see Chapter 17), and although considered to be vulnerable by the IUCN from 1988 to 2008, it is now listed as "data deficient."

Yangtze River Dolphin

The occurrence of the Yangtze River dolphin (*Lipotes vexillifer*; **Figure 11.10**), or baiji, was restricted to the Yangtze River in China (**Figure 11.11**). However, the animal is now

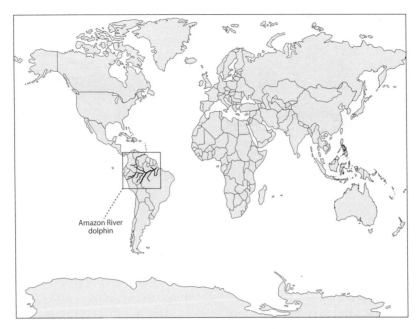

FIGURE 11.8 Map showing the current, known distribution of the Amazon River dolphin. Data from: IUCN Red List.

FIGURE 11.9 An Amazon river dolphin.

FIGURE 11.10 A Yangtze River dolphin.

considered to be "functionally extinct" in that any possible remaining animals are unlikely to be able to repopulate a sustainable population. The baiji had more of a dorsal fin than the other river dolphins and, similar to the Indian River dolphins, had rudimentary eyes. The Yangtze River dolphin was once venerated and was called "the goddess of the river," but during the Chinese Cultural Revolution such traditions were frowned upon and the animals were captured so their skins could be worked into leather.

Entanglement in fishing gear was one major cause of baiji mortality, in particular entanglement in so-called rolling hooks: lines of hooks placed on the riverbed intended to impale and entangle bottom-living fish. Electric fishing was also used in the Yangtze, which consisted of connecting two conducting poles to the battery of a boat or to a generator, shocking the fish (and any baiji) in an area of the river, and then scooping them up. Explosives were also used in the river to catch fish and also during engineering projects to widen the river, which led to further baiji mortalities. Several animals were struck by boats and, as the Yangtze is one of the busiest rivers in China (**Figure 11.12**), the noise from shipping was probably also a substantial problem for this species (see Chapter 5).

One-third of the population of China lives in the Yangtze River valley (approximately 400 million people). With such a high density of humans, pollution (whether sewage,

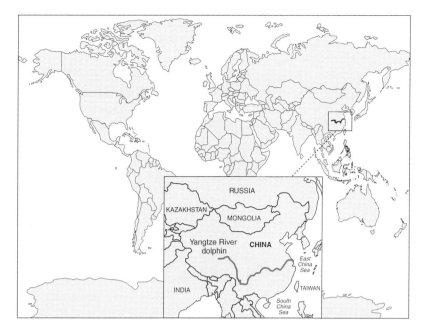

FIGURE 11.11 Map of Yangtze river dolphin distribution. Data from: IUCN Red List.

chemical, or debris) entering the Yangtze is substantial. In addition to all of this, the final nail in the coffin of the Yangtze River dolphin was almost certainly the Three Gorges Dam project (**Figure 11.13**). The largest civil engineering project of its time, this huge hydroelectric power project not only fragmented habitat for a large number of species but also substantially affected the flow and characteristics of the Yangtze both upriver and downriver of the project.

Several projects attempted to rescue the Yangtze River dolphin from extinction. One of the first involved trying to establish a captive breeding facility in Wuhan at the Institute of Hydrobiology. A total of six animals were captured from the wild, but most died within a few weeks or months of capture. One female (Zhen Zhen) survived for 2.5 years but did not breed. The longest surviving animal, a male baiji named Qi Qi, was captured in 1980 and lived until 2002.

A second rescue attempt occurred near Shishou, where an oxbow lake (a body of water that was once part of the river but had since become a separate body of water) was considered to be a viable area for a baiji reserve. The lake was 21 km long and 2 km wide; initially, finless porpoises

FIGURE 11.12 Ships on the Yangtze River.

FIGURE 11.13 The Three Gorges Dam Project on the Yangtze River.

were placed in the reserve to ascertain its suitability as a cetacean habitat. From 1990, 34 finless porpoises were captured for the reserve, but most died during or shortly after capture. Two died because of entanglement in rolling hooks that were still present in the reserve, and seven were killed during a radio tagging experiment. In 1 year, 14 escaped when the reserve flooded. The finless porpoise experiment was deemed to be effective, and so in 1995 a female baiji was placed in the reserve. It was initially thought that the male baiji in Wuhan would be transferred to the reserve to provide a breeding pair; however, this did not happen. In any event the female died in 1996. She was found entangled in a net after only 7 months in the reserve.

In situ efforts to protect the baiji included laws prohibiting the killing of the species. Also, several areas of the river were designated as baiji reserves (a total of 350 km of river) where certain fishing activities were supposed to be limited; five small patrol boats monitored the river for animals and to keep a look out for illegal fishing activity. However, these were little more than public relations exercises and paper

Exploring the Depths: Franciscana or La Plata Dolphin, *Pontoporia blainvillei*

Although the Franciscana has a coastal or estuarine distribution, it is typically grouped with the river dolphins and is not a member of the Delphinidae. They are found in shallow, coastal waters off Argentina, Brazil, and Uruguay, including the La Plata estuary (**Figure B11.3**). They feed on shallow water fish, crustaceans, and cephalopods.

The abundance of this species is not known throughout most of its range, but at least 3,000 Franciscana dolphins are killed every year as a result of entanglement in coastal fishing gear. Several animals have also been reported to have swallowed marine debris and trash. Considering their coastal distribution, pollution (including noise) and general habitat degradation are likely to be problems for this species. The IUCN considers this species to be "vulnerable" based on the level of bycatch it experiences, and it is listed on CITES Appendix II.

FIGURE B11.3 Map showing the current, known distribution of the Franciscana. Data from: IUCN Red List.

parks, as unregulated and uncontrolled development in the Yangtze region and lack of funding for enforcement efforts for these reserve areas left them almost totally ineffective.

The Yangtze River dolphin was considered to be "critically endangered" by the IUCN, listed on Appendix I on CITES thereby prohibiting international trade, and listed as "endangered" under the U.S. Endangered Species Act (see Chapter 17). However, in 2006 a comprehensive survey by both Chinese and international scientists failed to observe any Yangtze River dolphins throughout the Yangtze, and the species was declared "functionally extinct." Chinese officials claim that sightings of baiji in the Yangtze have occurred, but conservation efforts have effectively ceased as most of these sightings are unconfirmed and any remaining population will be unsustainable in any case.

The loss of the Yangtze River dolphin represents not only the loss of a species, or even the loss of a genus, but the loss of an entire family and (depending upon taxonomic details) an entire superfamily. It is also the first marine mammal to go extinct since the Japanese sea lion and the Caribbean monk seal disappeared in the 1950s, the first marine mammal to go extinct since the establishment of the majority of the worlds' endangered species legislation, and the first cetacean extinction attributed to human causes. In short, the world, and in particular the Chinese government, knowingly let the species go extinct. It has become the tragic icon of management inaction, which was arguably illegal under domestic Chinese legislation. Hopefully, lessons will be learned and other similar events can be avoided.

Beluga Whales

The family Monodontidae contains the beluga whale (*Delphinapterus leucas*) and the narwhal (*Monodon monoceros*). Both are Arctic species, but fossil remains of monodontids from Mexico show that members of this family were once more widely distributed, although they probably did not extend into the southern hemisphere. It has been suggested that the two living species might be able to breed together to produce hybrids, based on an unusual skull containing characteristics of both species found in Greenland.

The beluga whale, or white whale, has a dorsal ridge instead of a dorsal fin and a rounded head with prominent "lips" but no beak (**Figure 11.14**). The animals grow to approximately 5.5 m and are gray when born, get paler with age, and become white at about 5 years old. The white coloration possibly acts as camouflage in their Arctic habitat. Their blubber layer is substantial, up to 15 cm thick, as one would expect for an Arctic species. The species is sexually dimorphic, with males being approximately 25% longer than females. They are unusual cetaceans in that their neck

FIGURE 11.14 Beluga whales.

vertebrae are not fused, which gives them much more flexible necks than the majority of other odontocetes (river dolphins also have flexible necks, especially the Amazon river dolphins). They are also very vocal and have been nicknamed the "sea canary" because of their wide range of calls.

Distribution

The species is found in Arctic and sub-Arctic waters, with resident populations in the colder temperate waters of the St. Lawrence Estuary in Canada and the Cook Inlet, near Anchorage, Alaska. However, they do move beyond this range, at least on occasion. For example, one beluga spent several weeks in the Delaware River in the U.S. Mid-Atlantic region in 2005, apparently with no ill effects. Belugas frequently travel in groups of 2 to 10 whales but may, at times, gather in groups of hundreds or even thousands. They can be found in deep waters, among pack ice, near coastal areas, and substantial distances up river systems. They feed on bottom fish, cephalopods, and crustaceans. They may come into extremely shallow rivers, sometimes apparently to molt by attempting to remove skin by rubbing and rolling on riverbeds. The freshwater content of the rivers may also aid in the sloughing of the old skin. Animals in these rivers have even been stranded by the tide and have been aground on sandbanks or in shallow waters until the flood tide returned. This affinity to shallow waters may explain why belugas seem to adapt to captivity better than many other cetacean species. As a consequence they have been a target species for many aquariums (see next section).

Abundance and Status

Beluga whales have been relatively well researched throughout most of their range, compared with many other cetacean species. Approximately 150,000 beluga whales exist globally. Beluga whales have been captured for public display since 1861, when P.T. Barnum exhibited belugas briefly in New York, until a fire resulted in the death of the animals. Subsequently, animals have been removed from the wild for aquariums from several locations; the last known removals were 36 animals in Russia (eight were caught during the last capture in 2008). These eight animals were exported to Marineland in Ontario, Canada. Beluga whales in captivity have been known to live up to 30 years of age, and this was thought to be equivalent to their longevity in the wild. However, recent research has found that the belugas can live to 60 years of age in the wild.

The species is also hunted by native communities in several locations in the Arctic (see Chapter 15), with approximately 2,000 beluga whales being taken in Russia, Canada, and West Greenland and roughly 200 more taken in Alaska annually. Some populations of belugas are believed to be declining as a result of overhunting. The most significantly depleted population is in Cook Inlet, Alaska (see Exploring the Depths: Cook Inlet Beluga Whale).

Pollution has been a particular concern for beluga whales, with animals in the St. Lawrence Estuary found to be highly contaminated with organic contaminants. These include, of particular note, polychlorinated biphenyls, which could be impairing the reproductive and immune systems of the animals, and polyaromatic hydrocarbons, a carcinogenic class of contaminants linked to high rates of tumors found in the St. Lawrence population (see Chapter 15). The species also seems to be particularly sensitive to underwater noise, with animals observed avoiding the sounds from ice-breaking ships at considerable distances.

The species as a whole was listed by the IUCN as "vulnerable" in 1996 but this was changed to "near threatened" in 2008. However, the Cook Inlet population was heavily depleted by hunting in the 1980s and 1990s, from which they have not recovered. Consequently, the IUCN listed the Cook Inlet population of beluga whales as "critically endangered" in 2006. The same population had previously been listed as "depleted" under the U.S. Marine Mammal Protection Act in 2000, and was subsequently listed as "endangered" under the U.S. Endangered Species Act in 2008 (see Exploring the Depths: Cook Inlet Beluga Whale and Chapter 17).

The Narwhal

The narwhal was once referred to as the "corpse whale" by Norwegian sailors because of its white and gray dappled coloration, which they thought resembled a drowned, decaying corpse. Narwhal calves are gray when born and get paler as they get older, with gray spots dappling their dorsal side. As adults get older still the gray dappling recedes, so that the oldest animals have just a thin streak of dappling on their backs. They lack both a dorsal fin and a pronounced rostrum and do not share the lips of the beluga. Their tail flukes are "B" shaped rather than the "C" shape of most cetaceans (**Figure 11.15**).

Exploring the Depths: Cook Inlet Beluga Whale

The Cook Inlet beluga whale is a geographically isolated and genetically distinct population and was listed as "endangered" under the U.S. Endangered Species Act in October 2008. The population is currently estimated at 321 individuals, a decline of over 50% since 1993. The initial decline is largely attributed to unregulated subsistence hunting; however, the population has failed to recover as expected after a voluntary moratorium on hunting was initiated in 1999. Historically, Cook Inlet beluga whales ranged widely throughout the entire inlet. Their current summer range has contracted substantially, and they are now highly concentrated in the Upper Inlet in Knik and Turnagain Arms, where anadromous fish (which migrate from the sea to fresh water to spawn) runs supply the majority of their summer food source. Their winter range is less well known, but they are believed to move into the lower inlet as ice formation impedes movement in the Arms. Little is known about their feeding habits in the winter, although evidence suggests they may feed more on benthic prey after the summer fish runs have subsided. Current threats to the population may include coastal zone development, including oil and gas exploration; pollution; disease; and increased predation pressure from killer whales (*Orcinus orca*).

—Contributing author, Leslie Cornick, Alaska Pacific University.

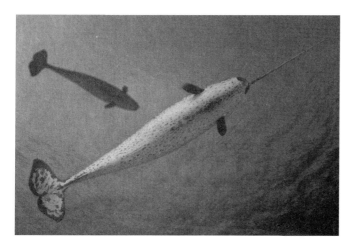

FIGURE 11.15 Narwhals showing their tail shape.

although mixed groups or solitary males can also be found. Narwhals feed on fish, squid, and shrimp. They are accomplished divers, swimming down to depths of over 1.8 km, similar to the depths achieved by beaked whales. Unlike most other cetacean species that visit the Arctic, narwhals remain in the region over winter and spend much time diving under the ice, using cracks and holes in the ice to breath. Occasionally, they can be trapped within the ice as it freezes, which has led to several mass mortality events. Polar bears and killer whales are their main natural predators. Polar bears may also hover at the edge of ice holes hoping to catch a surfacing narwhal (or beluga).

Similar to beluga whales, narwhals are hunted by subsistence whalers, with approximately 1,000 being taken in Western Greenland and Canada annually. Two areas in Greenland have been designated at "sanctuaries" where narwhals are not hunted. Although there may be local management regimes covering narwhal takes, hunting of this species is not governed by the IWC as narwhals are considered to be "small cetaceans" (see Chapters 14 and 15). In addition to hunting, climate change is another threat that may be of particular concern because of their exclusively Arctic distribution. Narwhals are currently listed under

Much less is known about these inaccessible animals in comparison with their white cousins. The narwhal has an Arctic distribution, and they move seasonally with the ice; as the ice retreats narwhals move to higher latitudes. Narwhals are usually found in groups of 5 to 10 individuals. Females, calves, and immature narwhals often travel together, whereas adult males travel in their own groups,

Exploring the Depths: The Narwhal's Tusk

Probably the most fascinating characteristic of the narwhal is its tusk (**Figure B11.4**). The tusk is spiraled and can be up to 3 m long. It is usually only found in males, but very occasionally females can also be found with a tusk. Furthermore, there are rare examples of males with two tusks. The tusk is in fact one of the male's two upper teeth, the left tooth to be specific. Males use these tusks when competing; they "fence" with the tusks, pushing at each other with their tusks crossed, although the pushing tends to be against the widest part of the tusk, at the base. The whales do not attack other males by "spearing" them with the end of the tusk, which would be far more dangerous. Instead, the "fencing" is more akin to wrestling or a test of strength.

It has been suggested that narwhal tusks could be used to break through thin ice to clear spaces to come up to breathe. It may also be used to deter predators, perhaps to fend off waiting polar bears around the edge of an ice hole. Harvard dental researcher Martin Nweeia suggested that because of the nervous enervation of the tusk and its structure, the tusk may function as a sensory organ. Being able to detect changes in temperature or pressure would be particularly important for an animal that swims under the ice and searches for areas of clear water or thin ice cover. However, this is unlikely to be the prime function of the tusk, because echolocation signals could provide much of the same information and females cope very well without a tusk.

The presence of the tusk also seems to force males to feed upside down, so that their sonar can be directed more toward potential prey items in the bottom sediment without

FIGURE B11.4 A narwhal's tusk.

Exploring the Depths: The Narwhal's Tusk (*continued*)

incident. However, it is quite possible they can also make use of it as a guide to bring escaping prey toward their mouths.

A type of dentistry may be practiced among narwhals in that the broken tusk of one individual may be plugged with the tip of another's tusk. Although some have described this as being the result of "head-on collisions," this is unlikely considering the accuracy that would have to be involved (or, more precisely, the extremely low probability of such collisions occurring coincidentally). Because this "capping" of the tusk is unlikely to have happened by chance, it is possible that this is an altruistic act. There are accounts by Arctic indigenous people of seeing young narwhals aligning the tips of their tusks with the broken tusks of older males and, apparently, deliberately snapping the very tips of their tusks off to block open tooth cavities. A narwhal's tusk is hollow and filled with pulp and contains a sizeable nerve, which if exposed would presumably be extremely painful. One could see why this unusual behavior might develop in a social cetacean species. Broken tusks have also been found with the tips crammed with sand and gravel, perhaps from an attempt to cap the broken tusk by jamming the broken tooth into the seabed.

CITES Appendix II and classified as "near threatened" by the IUCN.

Porpoises

The term "porpoise" is sometimes used to describe generic small cetaceans, especially in parts of the United States. It was also once applied to some delphinids by biologists as well in the earlier days of marine mammal science (e.g., the "bottlenose porpoise"). However, porpoises are now more clearly defined as a separate taxonomic group within the odontocetes. Porpoises all have flat, spade-shaped, or incisor-like teeth, as opposed to the pointed, conical teeth of the delphinids and most of the other odontocetes. They also tend to be relatively small (1.5–2.5 m) with rounded heads. Their metabolic rate is higher than the dolphins, which is not surprising because porpoises are smaller animals and often live in cool waters (see Chapter 4). This high metabolic rate would help them to generate more heat, but much like sea otters (see Chapter 7), porpoises may also need to consume relatively more prey and could be susceptible to starvation if prey stocks decrease. Indeed, a spate of stranded harbor porpoises along the North Sea coast showing signs of starvation may be the result of this high metabolic need for prey combined with a climate-induced decrease in prey species in the area.

The fastest and perhaps strangest looking porpoise is the Dall's porpoise (*Phocoenoides dalli*). This porpoise is found in the cooler temperate and sub-polar waters of the North Pacific. This species grows up to 2.4 m in length and has a body shape that slightly resembles an arrowhead, with a distinctive black and white coloration. The front edge of the porpoise's dorsal fin is straight, which is another unusual characteristic. The porpoise's heavily muscled tail stalk allows it to produce fast, short bursts of speed (up to 55 km/hour), presumably to escape its major predator, the killer whale. The black and white coloration may also be a defense against the killer whale. From a distance, hunting orcas (who use vision rather than sound when hunting acoustically sensitive cetaceans) might mistake the Dall's porpoise for another killer whale, giving the porpoise a chance to escape in the confusion.

Exploring the Depths: Dall's Porpoise and Japanese Catches

Few doubt that the 1986 international moratorium on commercial whaling (see Chapter 14) saved many great whale species from extinction. It did not, however, protect smaller cetaceans: dolphins, porpoises, and whales of the suborder Odontoceti (sperm whales excepted). These species are globally at risk from bycatch, chemical pollution, noise pollution, and habitat destruction. In addition, some populations, particularly in Japan, are still subject to large-scale directed hunts (**Figure B11.5**). Since the 1986 moratorium more than 450,000 small cetaceans have been hunted in Japan's coastal waters. During this time Dall's porpoises have constituted almost 90% of the catch.

The Dall's porpoise is predominantly black with a white patch on the belly and lower flanks. Two distinct sub-species are distinguished by the size of the white flank patch: a larger patch denoting the *"truei"* form and a smaller patch indicating the *"dalli"* type.

In Japan, a hand harpoon hunt involving up to 200 boats currently catches around 15,000 Dall's porpoises each year (**Figure B11.6**). The meat and various blubber products are

Exploring the Depths: Dall's Porpoise and Japanese Catches

FIGURE B11.5 Dall's porpoises in a Japanese market.

FIGURE B11.6 A Japanese porpoise hunter.

then sold for human consumption. Two main populations are thought to be targeted in the hunt: a *truei*-type population, hunted off the northeast Sanriku coast of Japan from November to April; and a population of *dalli*-type porpoises, hunted off the Hokkaido coast in May and June and August to October.

Historically, catches were below 10,000 animals per year until the moratorium on commercial whaling led whale meat companies in Japan to seek new "whale" products to trade. The Dall's porpoise catch swiftly increased, to over 40,000 in 1988. An outcry at the IWC resulted in the catches being reduced and an annual catch quota of 17,700 was imposed on hunters for the first time in 1993.

Over the past decade and a half the Scientific Committee of the IWC has repeatedly called on Japan to reduce the catch to sustainable levels. The most recent abundance estimates for the targeted stocks, from surveys in 2003, are 173,638 *dalli*-type porpoises (coefficient of variation = 0.212) and 178,157 *truei* porpoises (coefficient of variation = 0.232). These estimates are both lower than the previous estimates of 226,087 and 216,611, respectively, which were used to set the 17,700 catch quota. A slightly lower quota of 16,875 porpoises was set for 2006–2007 in response to the new abundance estimates and further adjusted to 15,748 (8,084 *dalli*-type and 7,664 *truei*-type porpoises) for the 2009–2010 hunt season. However, these quotas represent more than 4% of the estimated populations and far exceed internationally accepted safe mortality limits for the survival of small cetacean populations.

Discussions in the Scientific Committee of the IWC have also drawn attention to problems with the abundance estimates, including extrapolation into unsurveyed areas based on 1991 survey data and other old data, as well as a lack of data

Exploring the Depths: Dall's Porpoise and Japanese Catches (continued)

on animals struck-and-lost and levels of bycatch. The Scientific Committee has expressed serious concern about the sustainability of the hunt on countless occasions, leading to several Resolutions from the IWC's political body, first in 1990 and subsequently in 1999 and 2001. The 2001 Resolution urged Japan to "halt the directed takes of Dall's porpoises until a full assessment by the Scientific Committee has been carried out." Sadly, the Japanese government has refused to provide the necessary data to carry out a status review or revoke the catch quotas, however the hunting areas were devastated by the March 2011 tsunami and it appears the hunt has stopped for now.

—Contributing author, Clare Perry, Environmental Investigation Agency.

Exploring the Depths: Harbor Porpoise, *Phocoena phocoena*

The harbor, or common, porpoise is one of the smallest of all the cetaceans (**Figure B11.7**), second only to another porpoise, the vaquita (*Phocoena sinus*; see below). At a maximum length of around 1.9 m, the harbor porpoise is certainly small relative to the great majority of the other cetaceans. Its small size gives it a high surface to volume ratio, which means that it lives on an energetic "knife-edge" in the colder waters where it is found. Consequently, it must spend a lot of its time feeding to maintain its body temperature and other energy needs. The harbor porpoise is also relatively short-lived, with few surviving past 12 years of age.

These porpoises are typically (but not always) low profile in the wild. Usually, the viewer just spots a curving back and small triangular fin. Often, encounters are with what appear to be solitary animals, but sometimes a smaller dorsal fin marks a calf following close behind its mother. Small groups are also not unusual. Porpoises rapidly become difficult to see in increasingly rough seas due to this surfacing behavior, which makes them challenging to study. A lack of natural markings (e.g., minor dorsal fin damage commonly seen in bottlenose dolphins or scar patterns in beaked whales) means that they are rarely recognizable as individuals, further hindering scientific study. Like other porpoises, the harbor porpoise has no beak. Its back, lips, and chin are black, and it has a white or pale gray belly. It also has small, dark, slightly rounded flippers. Porpoise calves are generally grayer than their parents.

The word "porpoise" is partially derived from the Latin word for pig (*porcus*), and the harbor porpoise was sometimes called the "puffing pig" because of the exhalation sound it may make when surfacing to breath. People sometimes use the terms "dolphin" and "porpoise" interchangeably, and there is certainly confusion about the animals to which these words apply. However, the evolution of porpoises and dolphins (and hence their taxonomy) is quite distinct. There are various anatomical differences (as noted above) and, in general, dolphins are larger and more socially complex animals, living in various types of groups. Dolphins also produce a range of whistles and tweets that are used in communication, as well as clicks for echolocation. Like other porpoises, harbor porpoises only produce clicks and appear to be far less social.

FIGURE B11.7 A harbor porpoise.

Harbor porpoises are widespread in many sea areas in the North Atlantic and North Pacific. They are virtually absent from the Mediterranean except for a small little-studied population off the Greek coast. Surprisingly, they also occur in the Black Sea where they are regarded as critically endangered (**Figure B11.8**).

These small cetaceans appear to be especially vulnerable to nets set on the seabed, probably because they forage there. There is also growing evidence, for example in the United Kingdom, that their health is being impacted by bioaccumulating organic pollutants (see Chapter 15). Harbor porpoises are currently listed under CITES Appendix II, and although listed as "vulnerable" by the IUCN in 1996, were down-listed in 2008 to "least concern." However, the European Union has listed the harbor porpoise under Annex II of the EU "Habitats Directive" (see Chapter 17) which obligates EU member nations to designate important areas of harbor porpoise habitat as "Special Areas of Conservation."

Exploring the Depths: Harbor Porpoise, *Phocoena phocoena* (continued)

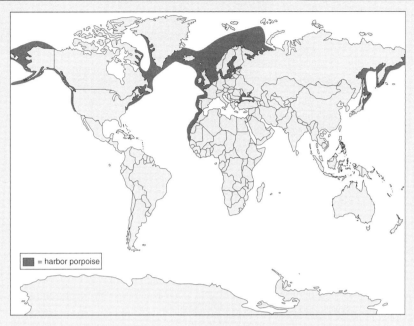

FIGURE B11.8 Map showing the current, known distribution of the harbor porpoise. Data from: IUCN Red List.

The spectacled porpoise (*P. dioptrica*) is another black and white porpoise, with white rings around the eyes that gives the animal its name. This 2.3 m porpoise is found off the southeastern coast of South America but has also been sighted elsewhere in the Southern Ocean and might be more widely distributed (**Figure 11.16**). Burmeister's porpoise (*P. spinipinnis*) is slightly smaller (2 m), is found in the coastal waters of South America (Figure 11.16), and has a distinctive slanting dorsal fin that resembles a jackknife. Both species have been little researched and are considered to be "data deficient" by the IUCN.

The finless porpoise (*Neophocaena phocaenoides*) is found in the coastal waters of the Indo-Pacific. It has a coastal and riverine distribution and, as its name suggests,

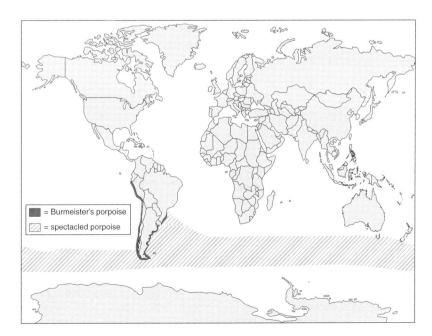

FIGURE 11.16 Map showing the current, known distribution of spectacled and Burmeister's porpoises. Data from: IUCN Red List.

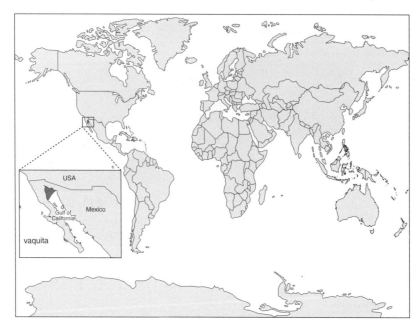

FIGURE 11.17 Map showing the current, known distribution of the vaquita. Data from: IUCN Red List.

lacks a dorsal fin, having instead a ridge or groove, depending on the population. In Japan calves are born dark gray but then get paler with age, becoming almost white in adulthood, and they have a long ridge along their back and tail. Conversely, populations in China are born pale gray/blue and turn darker with age, becoming gray/black, and may have a wide, shallow grooves instead of a long ridge on their backs. In fact, three species of finless porpoises have been suggested to occur in China (see Exploring the Depths: A Particularly Perplexing Porpoise).

The smallest porpoise is the 1.5-m-long vaquita or Gulf of California porpoise (*P. sinus*). Another potential candidate for the hardest cetacean to study it, like the finless porpoise, is shy and cryptic. The vaquita has relatively large flippers for its size and a distinctive black patch around its eye. It is also the most endangered porpoise, likely numbering less than 150 animals, making it one of the most endangered marine mammal species at the present time. The vaquita has extremely low genetic diversity and six digits on each flipper. This may suggest the species may have already survived the problems of "inbreeding depression" that plague small populations that repeatedly breed together, such as congenital disorders. However, the lack of genetic diversity probably makes them vulnerable to disease or parasites, because something that affects one individual will likely affect them all. Furthermore, the vaquita reproduces only once every 2 years and has an extremely limited distribution (**Figure 11.17**). This highly unusual and enigmatic species faces extinction in the near future due to bycatch in the gill nets of various fisheries in the Northern Gulf of Mexico unless verbal conservation commitments and paper planning can be turned into more solid action quickly. As a result, the vaquita is considered to be "critically endangered" by the IUCN.

Exploring the Depths: A Particularly Perplexing Porpoise

Finless porpoises are widely distributed in most coastal shallow waters of Asia from Japan to the Persian Gulf and south to Indonesian waters (including a freshwater population in the Yangtze River of China, where the now-extinct baiji once lived). Few people, however, have seen the finless porpoises, dead or alive, and usually the animals are only known from images in books (**Figure B11.9**). There are two main reasons for this: most marine mammal researchers are in the Western world; and the finless porpoise is intrinsically cryptic in nature (surfaces briefly, lacks a dorsal fin, usually exposes only a tiny part of its

Exploring the Depths: A Particularly Perplexing Porpoise (*continued*)

small body at the surface, and avoids boats). These are, arguably, one of the most challenging cetacean species to study at sea. Even the most basic piece of information of how many species exist is lacking. A decent understanding of taxonomy is important for assessing the conservation status of finless porpoises throughout their range, especially because human activities (particularly fisheries) have already decimated some populations (e.g., in the Yangtze River and the Inland Sea of Japan).

Since the first description of the finless porpoise in 1829, the taxonomy of these animals has continued to perplex researchers mainly because of the great number of morphological variants that exist. Superficially, finless porpoises resemble miniature beluga whales. However, adult coloration can vary greatly from being almost white (Japanese porpoises) to nearly black (the Indo-Pacific species). They can be long and thin or short and chunky, but all have relatively large flukes and flippers (usually ending in a bluntly rounded tip) and a bulbous forehead that can overhang the snout above a small mouth. The most striking characteristic (and unique among true porpoises) is the lack of a dorsal fin, which is instead usually replaced by a dorsal groove or a long low ridge, which can begin anywhere from above the flippers to about the center of the back and runs onto the tailstock.

There is also a patch of numerous small prickles (or tubercles) that run down the length of the middle of the back. The widest point of the patch can span from 0.2 to 14 cm and have anywhere from 1 to 25 tubercles across. Based on unique combinations of the features of the ridge and tubercle patch, three forms of finless porpoises have been identified. Maximum adult size also varies tremendously by region from about 1.5 to 2.3 m in length and weighing between about 60 and at least 110 kg (the smallest and largest animals are found in the Indian Ocean and northern China, respectively).

Much of the scientific debate about the taxonomy of finless porpoises occurred in the 1970s when Professor Giorgio Pilleri and colleagues proposed the existence of two and then later three species. Authors of later studies critiqued the studies of Pilleri and colleagues and an uneasy agreement of one species with three subspecies was reached among marine mammal taxonomists. Unfortunately, some of the later studies also suffered from serious conceptual flaws in study design and analyses, as well as inadequate sampling, which led to erroneous conclusions.

A new approach to analyzing and interpreting DNA and morphological patterns provided strong evidence that the two forms of porpoises occurring in the same region (i.e., in sympatry) were not interbreeding with each other. Under the definition of "biological species," these two forms should thus be considered distinct species. Furthermore, reproductive isolation of these two forms appeared to have started about 18,000 years ago, coinciding with the last major glaciation period.

Two species are now recognized: the Indo-Pacific finless porpoise (*Neophocaena phocaenoides*) and the narrow-ridged finless porpoise (*N. asiaeorientalis*). *Neophocaena asiaeorientalis* consists of two subspecies (as currently recognized): the Yangtze finless porpoise (*N. a. asiaeorientalis*) and the East Asian finless porpoise, or sunameri (*N. a. sunameri*). However, our understanding of finless porpoise taxonomy is far from complete. A full determination of the number of existing species, subspecies and populations requires further research.

FIGURE B11.9 A group of finless porpoises.

–Contributing author, John Wang, Formosa Cetus Research and Conservation Group.

SELECTED REFERENCES AND FURTHER READING

Aguilar De Soto, N.A., Johnson, M., Madsen, P.T., Tyack, P.L., Bocconcelli, A., & Borsani, J.F. (2006). Does intense ship noise disrupt foraging in deep diving Cuvier's beaked whales (*Ziphius cavirostris*)? *Marine Mammal Science* 22: 690–699.

Banguera-Hinestroza, E., Cardenas, H., Ruiz-Garcia, M., Marmontel, M., Gaitan, E., Vazquez, R., & Garcia-Vallejo, F. (2002). Molecular identification of evolutionary significant units in the Amazon River dolphin *Inia* sp. (Cetacea: Iniidae). *Journal of Heredity* 93: 312–322.

Barlow, J., & Gisiner, R. (2006). Mitigating, monitoring and assessing the effects of anthropogenic sound on beaked whales. *Journal of Cetacean Research and Management* 7: 239–249.

Beland, P., & Martineau, D. (1988). About carcinogens and tumors. *Canadian Journal of Fisheries & Aquatic Science* 45: 1855–1856.

Best, R.C., & da Silva, V.M.F. (1989). Amazon River dolphin, boto *Inia geoffrensis* (de Blainville, 1817). In: *Handbook of Marine Mammals, Vol. 4: River Dolphins and the Larger Toothed Whales* (Ed. S.H. Ridgway & R. Harrison), pp. 1–24. Academic Press, New York.

Braulik, G.T. (2006). Status assessment of the Indus River dolphin, *Platanista gangetica minor*, March–April 2001. *Biological Conservation* 129: 579–590.

Brownell, R.L. (1984). Live-capture fisheries for small cetaceans in South American waters. *Reports of the International Whaling Commission* 34: 747.

Bruemmer, F. (1993). *The Narwhal, Unicorn of the Sea*. Key Porter Books Limited, Hong Kong.

Caretta, J.V., Barlow, J., & Enriqez, L. (2008). Acoustic pingers eliminate beaked whale bycatch in a gill net fishery. *Marine Mammal Science* 24: 956–961.

Chen, P., Zhang, X., Wei, Z., Zhao, Q., Wang, X., Zhang, G., & Yang, J. (1993). Appraisal of the influence upon baiji *Lipotes vexillifer* by the Three-Gorge Project and conservation strategy. *Acta Hydrobiologica Sinica* 17: 101–111.

Chivers, S.J., LeDuc, R.G., Robinson, K.M., Barros, N.B., & Dizon, A.E. (2005). Genetic variation of *Kogia* spp. with preliminary evidence for two species of *Kogia sima*. *Marine Mammal Science* 21: 619–634.

Cox, T.M., Ragen, T.J., Read, A.J., Vos, E., Baird, R.W., Balcomb, K., Barlow, J., Caldwell, J., Cranford, T., Crum, L., D'Amico, A., D'Spain, G., Fernandez, A., Finneran, J., Gentry, R., Gerth, W., Gulland, F., Hildebrand, J., Houser, D., Hullar, T., Jepson, P.D., Ketten, D., MacLeod, C.D., Miller, P., Moore, S., Mountain, D.C., Palka, D., Ponganis, P., Rommel, S., Rowles, T., Taylor, B., Tyack, P., Wartzok, D., Gisiner, R., Mead, J., & Benner, L. (2006). Understanding the impacts of acoustic sound on beaked whales. *Journal of Cetacean Research and Management* 7: 177–187.

Da Silva, V.M.F. (2009). Amazon River dolphin *Inia geoffrensis*. In: *Encyclopedia of Marine Mammals*, 2nd ed. (Ed. W.F. Perrin, B. Würsig, & J.G.M. Thewissen), pp. 26–28. Academic Press, New York.

Dalebout, M.L., Steel, D., & Baker, C.S. (2008). Phylogeny of the beaked whale genus *Mesoplodon* (Ziphiidae: Cetacea) revealed by nuclear introns: implications for the evolution of male tusks. *Systematic Biology* 57: 857–875.

Danilewicz, D., Rosas, F., Bastida, R., Marigo, J., Muelbert, M., Rodriguez, D., Lailson-Brito, J., Ruoppolo, V., Ramos, R., Bassoi, M., Ott, P.H., Caon, G., Rocha, A.M., Catão-Dias, J.L., & Secchi, E.R. (2002). Report of the working group on biology and ecology. *Latin American Journal of Aquatic Mammals (Special Issue on the Biology and Conservation of Franciscana)* 1: 25–42.

Dolman, S., Parsons, E.C.M., & Wright, A.J. (2011). Cetaceans and military sonar: a need for better management. *Marine Pollution Bulletin* 63: 1–4.

Donoghue, M., Reeves, R.R., & Stone, G.S. (2003). *Report of the Workshop on Interactions Between Cetaceans and Longline Fisheries*. New England Aquarium Press, Boston.

Dudgeon, D. (2005). Last chance to see: *ex situ* conservation and the fate of the baiji. *Aquatic Conservation* 15: 105–108.

Finley, K.J., Miller, G.W., Davis, R.A., & Greene, C.R. (1990). Reactions of belugas *Delphinapterus leucas* and narwhals *Monodon monoceros* to ice-breaking ships in the Canadian High Arctic. *Canadian Bulletin of Fisheries and Aquatic Sciences* 224: 97–117.

Gero, S., & Whitehead, H. (2007). Suckling behavior in sperm whale: observations and hypotheses. *Marine Mammal Science* 23: 398–413.

Heide-Jørgensen, M.P. (2009). Narwhal *Monodon monoceros*. In: *Encyclopedia of Marine Mammals*, 2nd ed. (Ed. W.F. Perrin, B. Würsig & J.G.M. Thewissen), pp. 754–758. Academic Press, New York.

Heide-Jørgensen, M.P., & Reeves, R.R. (1993). Description of an anomalous monodontid skull from West Greenland: a possible hybrid? *Marine Mammal Science* 9: 258–268.

Heyning, J.E. (1984). Functional morphology involved in intraspecific fighting of the beaked whale, *Mesoplodon carlhubbsi*. *Canadian Journal of Zoology* 62: 1645–1654.

Hobbs, R.C., Lairdre, K.L., Vos, D.J., Mahoney, B.A., & Eagleton, M. (2005). Movements and area use of belugas, *Delphinpaterus leucas*, in a subarctic Alaskan estuary. *Arctic* 58: 331–340.

Hohn, A.A., Rotstein, D.S., Harms, C.A., & Southall, B.L. (2006). Multispecies Mass Stranding of Pilot Whales (*Globicephala macrorhynchus*), Minke Whale (*Balaenoptera acutorostrata*), and Dwarf Sperm Whales (*Kogia sima*) in North Carolina on 15–16 January 2005. *NOAA Technical Memorandum NMFS SEFSC 537*. National Marine Fisheries Service, Washington DC.

Hooker, S.K., & Baird, R.W. (1999). Deep-diving behaviour of the northern bottlenose whale, *Hyperoodon ampullatus* (Cetacea: Ziphiidae). *Proceedings of the Royal Society of London, Series B* 266: 671–676.

Hucke-Gaete, R., Moreno, C.A., & Arata, J. (2004). Operational interactions of sperm whales and killer whales with the Patagonian toothfish industrial fishery off southern Chile. *CCAMLR Science* 11: 127–140.

Iwasaki, T. (2008) Japan. Progress report on small cetacean research April 2007 to March 2008, with statistical data for the calendar year 2007. Paper presented to the Scientific Committee at the 60th Meeting of the International Whaling Commission, June 1-19, 2008, Santiago, Chile. Japan Prog. Rep. SM/2008. Retrieved from http://www.jfa.maff.go.jp/j/whale/w_document/pdf/h19_progress_report.pdf.

Japan Fisheries Agency. (2009). Dall's porpoise, *Phocoenoides dalli*. Retrieved from http://kokushi.job.affrc.go.jp/H20/H20_46.pdf.

Jefferson, T.A., & Hung, S.K. (2004). *Neophocaena phocaenoides*. *Mammalian Species* 746: 1–12.

Jefferson, T.A., & Wang, J.Y. (2011). Revision of the taxonomy of finless porpoises (genus *Neophocaena*): the existence of two species. *Journal of Marine Animals and Their Ecology* 4: 3–16.

Jepson, P.D., Arbelo, M., Deaville, R., Patterson, I.A.P., Castro, P., Baker, J.R., Degollada, E., Ross, H.M., Herráez, P., Pocknell, A.M., Rodríguez, F., Howiell, F.E., Espinosa, A., Reid, R.J., Jaber, J.R., Martin, V., Cunningham, A.A., & Fernández, A. (2003). Gas bubble lesions in stranded cetaceans: was sonar responsible for a spate of whale deaths after an Atlantic military exercise? *Nature* 425: 575–576.

Kannan, K., Sinha, R.K., Tanabe, S., Ichihashi, H., & Tatsukawa, R. (1993). Heavy metals and organochlorine residues in Ganges River dolphins from India. *Marine Pollution Bulletin* 26: 159–162.

Kasuya, T. (1999). Finless porpoise *Neophocaena phocaenoides* (G. Cuvier, 1829). In: *Handbook of Marine Mammals: the Second Book of Dolphins and the Porpoises*. (Ed. S.H. Ridgway & R. Harrison), pp. 411–412. Academic Press, London.

Laidre, K.L., & Heide-Jørgensen, M. (2005a). Arctic sea ice trends and narwhal vulnerability. *Biological Conservation* 121: 509–517.

Laidre, K.L., & Heide-Jørgensen, M.P. (2005b). Winter feeding intensity of narwhals (*Monodon monoceros*). *Marine Mammal Science* 21: 45–57.

Laidre, K.L., Heide-Jørgensen, M.P., & Dietz, R. (2002). Diving behaviour of narwhals (*Monodon monoceros*) at two coastal localities in the Canadian High Arctic. *Canadian Journal of Zoology* 80: 624–635.

Laist, D.W., Knowlton, A.R., Mead, J.G., Collet, A.S., & Podesta, M. (2001). Collisions between ships and whales. *Marine Mammal Science* 17: 35–75.

Lal Mohan, R.S., & Kunhi, K.V.M. (1996). Fish oils as alternative to river dolphin, *Platanista gangetica* (Lebeck) oil for fishing catfish *Clupisoma garua* in the River Ganges, India. *Journal of the Bombay Natural History Society* 93: 86–88.

Liu, R., Wang, D., Yang, J., & Zhang, X. (1997). Some new considerations for the conservation of *Lipotes vexillifer* and *Neophocaenoides* in China. *IBI Reports* 7: 39–44.

Liu, R., Yang, J., Wang, D., Zhao, Q., Wei, Z., & Wang, X. (1998). Analysis on the capture, behavior, monitoring and death of the baiji (*Lipotes vexillifer*) in the Shishou Semi-Natural Reserve at the Yangtze River, China. *IBI Reports* 8: 11–22.

Lockyer, C., Hohn, A.A., Doidge, W.D., Heide-Jørgensen, M.P., Mads, P., & Suydam, R. (2007). Age determination in belugas (*Delphinapterus leucas*): a quest for validation of dentinal layering. *Aquatic Mammals* 33: 293–304.

MacLeod, C.D. (1998). Intraspecific scarring in odontocete cetaceans: an indicator of male "quality" in aggressive social interactions? *Journal of Zoology* 244: 71–77.

MacLeod, C.D. (2000). Species recognition as a possible function for variations in position and shape of the sexually dimorphic tusks of *Mesoplodon* whales. *Evolution* 54: 2171–2173.

MacLeod, C.D. (2002). Possible functions of the ultra-dense bone in the rostrum of Blainville's beaked whale (*Mesoplodon densirostris*). *Canadian Journal of Zoology* 80: 178–184.

MacLeod, C.D., Santos, M.B., Reid, R.J., Scott, B.E., & Pierce, G.J. (2007). Linking sandeel consumption and the likelihood of starvation in harbour porpoises in the Scottish North Sea: could climate change mean more starving porpoises? *Biology Letters* 3: 185–188.

Madsen, P.T., Johnson, M., Miller, P.J.O., Aguilar Soto, N., Lynch, J., & Tyack, P. (2006). Quantative measures of air gun pulses recorded on sperm whales (*Physeter macrocephalus*) using acoustic tags during controlled exposure experiments. *Journal of the Acoustical Society of America* 120: 2366–2379.

Manire, C.A., Rhinehart, H.L., Barros, N.B., Byrd, L., & Cunningham-Smith P. (2004). An approach to the rehabilitation of *Kogia* sp. *Aquatic Mammals* 30: 257–270.

Martineau, D., Beland, P., Desjardins, C., & Lagace, A. (1987). Levels of organochlorine chemicals in tissues of beluga whales (*Delphinapterus leucas*) from the St. Lawrence Estuary, Quebec, Canada. *Archives of Environmental Contamination & Toxicology* 16: 137–147.

Martineau, D., De Guise, S., Fournier, M., Shugart, L., Girard, C., Lagace, A., & Beland, P. (1994). Pathology and toxicology of beluga whales from the St. Lawrence Estuary, Quebec, Canada: past, present and future. *Science of the Total Environment* 154: 201–215.

Martineau, D., Legace, A., Beland, P., Higgins, R., Armstrong, D., & Shugart, L. R. (1988). Pathology of stranded beluga whales (*Delphinapterus leucas*) from the St. Lawrence

Estuary, Quebec, Canada. *Journal of Comparative Pathology* 98: 287–311.

Miller, P.J.O., Aoki, K., Rendell, L.E., & Amano, M. (2008). Stereotypical resting behavior of the sperm whale. *Current Biology* 18: R21–R23.

Muir, D.C.G., Ford, C.A., Rosenburg, B., Norstrom, R.J., Simon, M., & Beland, P. (1996). Persistent organochlorines in beluga whales (*Delphinaperus leucas*) from the St. Lawrence River Estuary I. Concentrations and patterns of specific PCBs, chlorinated pesticides and polychlorinated dibenzo-p-dioxins and dibenzofurans. *Environmental Pollution* 93: 219–234.

Muir, D.C.G., Ford, C.A., Stewart, R E.A., Smith, T.G., Addison, R.F., Zinck, M.E., & Beland, P. (1990). Organochlorine contaminants in belugas *Delphinapterus leucas*, from Canadian waters. *Canadian Bulletin of Fisheries & Aquatic Sciences* 224: 165–190.

Nielsen, J. B., Nielsen, F., Joergensen, P.-J., & Grandjean, P. (2000). Toxic metals and selenium in blood from pilot whales (*Globecephala melas*) and sperm whales (*Physeter catodon*). *Marine Pollution Bulletin* 40: 348–351.

Nolan, C.P., Liddle G.M., & Elliot J. (2000). Interactions between killer whales (*Orcinus orca*) and sperm whales (*Physeter macrocephalus*)with a longline fishing vessel. *Marine Mammal Science* 16: 658–664.

O'Corry-Crowe, G.M. (2009). Beluga whale *Delphinapterus leucas*. In: *Encyclopedia of Marine Mammals*, 2nd ed. (Ed. W.F. Perrin, B. Würsig, & J.G.M. Thewissen), pp. 108–112. Academic Press, New York.

Ott, P.H., Secchi, E.R., Moreno, I.B., Danilewicz, D., Crespo, E.A., Bordino, P., Ramos, R., Di Beneditto, A.P., Bertozzi, C., Bastida, R., Zanelatto, R., Perez, J.E., & Kinas, P.G. (2002). Report of the working group on fishery interactions. *Latin American Journal of Aquatic Mammals (Special Issue on the Biology and Conservation of Franciscana)* 1: 55–64.

Parsons, E.C.M., & Wang J.Y. (1998). A review of finless porpoises (*Neophocaena phocaenoides*) from the South China Sea. In: *The Marine Biology of the South China Sea* (Ed. B. Morton), pp. 287–306. Hong Kong University Press, Hong Kong.

Parsons, E.C.M., Dolman, S., Wright, A.J., Rose, N.A., & Burns, W.C.G. (2008). Navy sonar and cetaceans: just how much does the gun need to smoke before we act? *Marine Pollution Bulletin* 56: 1248–1257.

Pilleri, G., & Gihr, M. (1975). On the taxonomy and ecology of the finless black porpoise, *Neophocaena* (Cetacea, Delphinidae). *Mammalia* 39: 657–673.

Pitman, R.L., Ballance, L.T., Mesnick, S.I., & Chivers, S.J. (2001). Killer whale predation on sperm whales: observations and implications. *Marine Mammal Science* 17: 494–507.

Reeves, R.R., & Chaudhry, A.A. (1998). Status of the Indus river dolphin *Platanista minor*. *Oryx* 32: 35–44.

Reeves, R.R. & Gales, N.J. (2006). Realities of baiji conservation. *Conservation Biology* 20: 626–628.

Reeves, R.R., Smith, B.D., & Kasuya, T. (2000). Biology and Conservation of Freshwater Cetaceans in Asia, IUCN/SSC Occasional Paper No. 23. IUCN, Gland, Switzerland.

Reeves, R.R., Wang, J.Y., & Leatherwood, J.S. (1997). The finless porpoise, *Neophocaena phocaenoides* (G. Cuvier, 1829): a summary of current knowledge and recommendations for conservation action. *Asian Marine Biology* 14: 111–143.

Rice, D.W. 1998. *Marine Mammals of the World: Systematics and Distribution*. Society for Marine Mammalogy, Lawrence, KS.

Rommel, S.A., Costidid, A.M., Fernandez, A., Jepson, P.D., Pabst, D.A., McLellan, W.W., Houser, D.S., Cranford, T.W., Van Helden, A.L., Allen, D.M., & Barros, N.B. (2006). Elements of beaked whale anatomy and diving physiology and some hypothetical causes of sonar-related stranding. *Journal of Cetacean Research and Management* 7: 189–209.

Secchi, E.R., Kinas, P.G., & Muelbert, M. (2004). Incidental catches of franciscana in coast gillnet fisheries in the Franciscana Management Area III: period 1999–2000. *Latin American Journal of Aquatic Mammals* 3: 61–68.

Secchi, E.R., Ott, P.H., Crespo, E.A., Kinas, P.G., Pedraza, S.N., & Bordino, P. (2001). A first estimate of franciscana (*Pontoporia blainvillei*) abundance off southern Brazil. *Journal of Cetacean Research and Management* 3: 95–100.

Secchi, E.R., Ott, P.H., & Danilewicz, D. (2002). Report of the fourth workshop for the coordinated research and conservation of the Franciscana dolphin (*Pontoporia blainvillei*) in the western South Atlantic. *Latin American Journal of Aquatic Mammals (Special Issue on the Biology and Conservation of Franciscana)* 1: 11–20.

Secchi, E.R., & Wang, J.Y. (2002). Assessment of the conservation status of a franciscana (*Pontoporia blainvillei*) stock in the Franciscana Management Area III following the IUCN Red List process. *Latin American Journal of Aquatic Mammals (Special Issue on the Biology and Conservation of Franciscana)* 1: 183–190.

Senthilkumar, K., Kannan, K., Sinha, R.K., Tanabe, S., & Giesy, J.P. (1999). Bioaccumulation profiles of polychlorinated biphenyl congeners and organochlorine pesticides in Ganges river dolphins. *Environmental Toxicology and Chemistry* 18: 1511–1520.

Simmonds, M.P., & Issac, S. (2007). The impacts of climate change on marine mammals: early signs of significant problems. *Oryx* 41: 19–26.

Smith, B.D., Braulik, G., Strindberg, S., Mansur, R., Diyan, M.A.A., & Ahmed, B. (2009). Habitat selection of freshwater cetaceans and the potential effects of declining freshwater flows and sea-level rise in waterways of the Sundarbans mangrove forest, Bangladesh. *Aquatic Conservation* 19: 209–225.

Stewart. R.E.A., Campana, S.E., Jones, C.M., & Stewart, B.E. (2006). Bomb radiocarbon dating calibrates beluga (*Delphinapterus leucas*) age estimates. *Canadian Journal of Zoology* 84: 1840–1852.

Tarpley, R.J., & Marwitz, S. (1993). Plastic debris ingestion by cetaceans along the Texas coast: two case reports. *Aquatic Mammals* 19: 93–98.

Turvey, S.T. (2008). *Witness to Extinction: How We Failed to Save the Yangtze River Dolphin.* Oxford University Press, Oxford.

Turvey, S.T., Barrett, L.A., Braulik, G.T., & Wang D. (2006). Implementing the recovery programme for the Yangtze River dolphin. *Oryx* 40: 257–258.

Turvey, S.T., Pitman, R.L., Taylor, B.L., Barlow, J., Akamatsu, T., Barrett, L.A., Zhao, X., Reeves, R.R., Stewart, B.S., Pusser, L.T., Wang, K., Wei, Z., Zhang, X., Richlen, M., Brandon, J.R., & Wang, D. (2007). First human-caused extinction of a cetacean species? *Biology Letters* 3: 537–540.

Tyack, P.L., Johnson, M., Aguilar Soto, N., Sturlese, A., & Madsen, P.T. (2006). Extreme diving of beaked whales. *Journal of Experimental Biology* 209: 4238–4253.

Viale, D., Verneau, N., & Tison, Y. (1992). Stomach obstruction in a sperm whale beached on the Lavezzi islands: macropollution in the Mediterranean. *Journal de Recherche Oceanographique* 16: 100–102.

Wang, J.Y., Frasier, T.R., Yang, S.C., & White, B.N. (2008). Detecting recent speciation events: the case of the finless porpoise (genus *Neophocaena*). *Heredity* 101: 145–155.

Wang, J.Y., & Yang, S.C. (2006). Unusual cetacean stranding events of Taiwan in 2004 and 2005. *Journal of Cetacean Research and Management* 8: 283–292.

Watkins, W.A., Daher, M.A., Fristrup, K.M., Howald, T.J., & Notarbartolo-di-Sciara, G. (1993). Sperm whales tagged with transponders and tracked underwater by sonar. *Marine Mammal Science* 9: 55–67.

Watwood, S.L., Miller, P.J.O., Johnson, M., Madsen, P.T., & Tyack, P.L. (2006). Deep-diving foraging behaviour of sperm whales (*Physeter macrocephalus*). *Journal of Animal Ecology* 75: 814–825.

Whitehead, H. (2003). *Sperm Whales Social Evolution in the Ocean.* University of Chicago Press, Chicago.

Whitehead, H. (2009). Sperm whale *Physeter macrocephalus*. In: *Encyclopedia of Marine Mammals*, 2nd ed. (Ed. W.F. Perrin, B. Würsig & J.G.M. Thewissen), pp. 1091–1097. Academic Press, New York.

Wright, A.J., Deak, T., & Parsons, E.C.M. (2010). Size matters: Management of stress responses and chronic stress in beaked whales and other marine mammals may require larger exclusion zones. *Marine Pollution Bulletin* 63: 5–9.

Yang, G., Bruford, M.W., Wei, F., & Zhou, K. (2006). Conservation options for the baiji: time for realism? *Conservation Biology* 20: 620–622.

Zhang, X., Wang, D., Liu, R., Wei, Z., Hua, Y., Wang, Y., Chen, Z., & Wang, L. (2003). The Yangtze River dolphin or baiji (*Lipotes vexillifer*): population status and conservation issues in the Yangtze River, China. *Aquatic Conservation* 13: 51–64.

Zhao, X., Barlow, J., Taylor, B.L., Pitman, R.L., Wang K., Wei Z., Stewart, B.S., Turvey, S.T., Akamatsu, T., Reeves, R.R., & Wang, D. (2008). Abundance and conservation status of the Yangtze finless porpoise in the Yangtze River, China. *Biological Conservation* 141: 3006–3018.

Zhou, K. (2009). Baiji *Lipotes vexillifer*. In: *Encyclopedia of Marine Mammals*, 2nd ed. (Ed. W.F. Perrin, B. Würsig, & J.G.M. Thewissen), pp. 71–76. Academic Press, New York.

Zhou, K., & Wang, X. (1994). Brief review of passive fishing gear and incidental catches of small cetaceans in Chinese waters. *Reports of the International Whaling Commission, Special Issue* 15: 347–354.

Zhou, K., & Zhang, X. (1991). *Baiji, the Yangtze River Dolphin and Other Endangered Animals of China.* Yilin Press, Nanjing, China. Zhou, K., Sun, J., Gao, A., & Würsig, B. (1998). Baiji (*Lipotes vexillifer*) in the lower Yangtze River: movements, numbers, threats and conservation needs. *Aquatic Mammals* 24: 123–132.

Zirbel, K., Balint, P., & Parsons, E.C.M. (2011). Navy sonar, cetaceans and the US Supreme Court: a review of cetacean mitigation and litigation in the US. *Marine Pollution Bulletin* 63: 40–48.

Delphinidae: The Oceanic Dolphins

CHAPTER 12

CHAPTER OUTLINE

Killer Whales

Killer Whale Ecotypes
Exploring the Depths: Killer Whale Dorsal Fins
Exploring the Depths: Killer Whale Behavior
Abundance and Status
Exploring the Depths: Killer Whale Longevity

The "Blackfish"

Pilot Whales
Exploring the Depths: Strandings

"Blunt-Headed" Dolphins

Irrawaddy and Snubfin Dolphins
Risso's Dolphin

Right Whale Dolphins

"Beakless" and "Short-Beaked" Dolphins

Genus *Cephalorhynchus*
Exploring the Depths: Hector's Dolphin, Cephalorhynchus hectori
Genus *Lagenorhynchus*
Exploring the Depths: Climate Change and Conservation of Lagenorhynchus *Species*

Humpback Dolphins

Exploring the Depths: Critically Endangered Eastern Taiwan Strait Sousa chinensis *Population*

"Long-Beaked" Dolphins

Genera *Stenella* and *Steno*
Exploring the Depths: Rough-Toothed Dolphin, Steno bredanensis
Genus *Delphinus*
Exploring the Depths: Common Dolphin Bycatch in Europe
Exploring the Depths: Tucuxi and Costero, Genus Sotalia

Bottlenose Dolphins

Exploring the Depths: Sperm Competition
Exploring the Depths: Dolphin Intelligence
Exploring the Depths: Culture in Cetaceans

Selected References and Further Reading

The dolphins are the largest group of odontocetes. For the purposes of describing this large taxonomic group of animals, we split these various species into groups based on their external resemblance:

- Killer whales (*Orcinus* spp.) and similar species
- Pilot whales (*Globicephala* spp.)
- "Blunt-headed dolphins" (*Orcaella* spp. and *Grampus griseus*)
- Right whale dolphins (*Lissodelphis* spp.)
- "Beakless" dolphins (*Cephalorhynchus* spp.)
- "Short-beaked" dolphins (*Lagenorhynchus* spp.)
- Humpback dolphins (*Sousa* spp.)
- "Long-beaked" dolphins (*Stenella*, *Steno*, and *Delphinus* spp.)
- Bottlenose dolphins (*Tursiops* spp.)

Killer Whales

The killer whale, or orca, gets its name because it is the "killer of whales," a reference to their habit of attacking baleen whales, which was first described by the Roman naturalist Pliny the Elder (70 AD) in his work, *Naturalis Historia*. There are no known records of killer whales deliberately killing humans in the wild, although several humans have been injured or killed by killer whales in captivity.

FIGURE 12.1 A killer whale group.

Killer whales have a distinctive black and white coloration, with a white patch near their eyes, white ventral patches, and gray "saddles" (**Figure 12.1**). Their flippers are large and rounded, and they are sexually dimorphic, which means males grow larger than females (9.8 m vs. 8.5 m). Females and juveniles have curved, sickle-shaped dorsal fins, whereas adult males have large straight dorsal fins that are between 1 and 1.8 m in height (see Exploring the Depths: Killer Whale Dorsal Fins).

Killer Whale Ecotypes

Killer whales in the northeast Pacific seem to have three different ecological strategies. These whale "ecotypes" consume different prey and although they are found in the same region, they have different habitats, patterns of movement, group sizes, and genotypes. These ecotypes are colloquially known as "resident," "transient," and "offshore." The generally near-shore resident ecotype specializes in

---

Exploring the Depths: Killer Whale Dorsal Fins

The large dimorphic fin of the male killer whale has been suggested to be a secondary sexual characteristic (**Figure B12.1**), that is, a way for females to assess the fitness of a male killer whale. The large size and straight leading edge of the male dorsal fin (as opposed to smaller curved fins in females and juveniles) would produce substantial drag for an adult male, rather than actually assisting in stabilization. If a male is in poor condition, injured, or diseased, this might cause a reduction in nutrient intake and blubber thickness and could lead to the bending and collapse of the dorsal fin. Bent or collapsed dorsal fins in male killer whales are relatively rare in the wild. Less than 5% of killer whales in British Columbia have abnormal fins, with less than 1% having collapsed fins in Norway. Two of three animals reported in Alaska with collapsed fins occurred shortly after the exposure of these animals to the *Exxon Valdez* oil spill. One population in New Zealand, however, was reported to have 7 of 30 adult male killer whales with bent dorsal fins. However, only one of these, who had suffered an injury as the result of entanglement, had a completely collapsed fin.

Conversely, all adult male killer whales in captivity have collapsed dorsal fins. It has been variously claimed that this is the result of genetics, the result of gravity acting on the fins, or because the fins of the animals are submerged less frequently and thus exposed to warm air more often, causing a breakdown of cartilage and tissue in the fin. Killer whales in captivity also spend a lot of time dormant in the water, and this lack of movement may affect tissue tone in the dorsal fin. Even females in captivity have bent and fully collapsed dorsal fins, which is rarely observed in the wild.

A B
FIGURE B12.1 (A) Killer whales in the wild with erect dorsal fins and (B) a killer whale in captivity showing a "flopped over" dorsal fin.

herding and feeding on fish. They hunt cooperatively using echolocation and are found in larger social groups. The transient killer whales tend to be physically larger, move longer distances, and are found in smaller groups. They specialize in feeding on marine mammals such as sea lions, fur seals, Dall's porpoises, or migrating baleen whale calves, generally without using echolocation. The offshore ecotype is smaller than transients and residents, forms larger social groups than residents (typically 20–100 animals and up to 200 individuals), and has a wide ranging distribution in deeper offshore waters from Alaska to California. The offshore ecotype is predominantly piscivorous. They are genetically distinct from the other two ecotypes but because of their offshore distribution have been more difficult to study.

Research suggests two genetically, morphologically, and behaviorally distinct forms of killer whale in the North Atlantic: a large (up to 8.5 m) marine mammal specialist and a smaller (6.6 m) generalist feeder. Also, several forms of killer whales are found in the Antarctic that are slightly different morphologically and ecologically. Most, like the transient forms, are marine mammal specialists (feeding on Antarctic minke whales or seals) or penguin consumers, but there is also a fish specialist. It has even been suggested that this latter ecotype may actually be a separate species (*Orcinus glacialis*). These Antarctic killer whales can, however, migrate substantial distances, with tagged seal/penguin specialist whales having been recorded travelling from Antarctica to subtropical waters off Uruguay and Brazil.

Exploring the Depths: Killer Whale Behavior

Killer whale (*Orcinus orca*) behavior is diverse. This largest of the dolphin species is cosmopolitan in its distribution and is long-lived, with a low reproductive rate, long dependency period, and cultural transmission of behavior that varies depending on local ecology. Some killer whale populations (e.g., the transients noted above) specialize in hunting other marine mammals, whereas others (e.g., residents) specialize in foraging for fish; even specific species of fish, such as herring or salmon, require different hunting techniques. Hunting is usually cooperative, with prey species generally determining group size. For example, transients tend to travel in smaller groups to maximize efficiency in hunting and energy transfer per group member, whereas residents can energetically afford to travel in larger groups.

Prey type even determines vocal behavior. Transients are often silent (to avoid startling alert seals or porpoises) and usually restrict vocal exuberance to the period immediately after a successful kill. In contrast, residents are highly vocal; they are rarely silent and generate a dizzying array of calls, whistles, and clicks. For both resident and transient ecotypes these vocalizations are not random. Each maternal lineage has its own dialect, producing stereotyped calls that can be clearly distinguished by a trained human listener. Although much remains to be learned about killer whale vocalizations, their calls appear to identify individuals to each other and help coordinate group behavior, as well as signal behavioral state and location.

Regardless of a population's local culture, all killer whales have a long dependency period, during which dialects, hunting techniques, and other important knowledge are taught to juveniles by adult pod members. Among residents, the bond between mother and offspring extends well beyond weaning. In some populations offspring remain with their mothers for life. In these populations sons never disperse, mating with females outside the immediate family but always returning to their mothers, whereas daughters eventually gain independence as they build their own families but associate often with their mothers. Socializing groups of such killer whales can contain up to four generations of a family.

Within ecotypes, hunting and foraging techniques are as widely varied as vocal behavior. In at least two separate populations, killer whales targeting pinnipeds have developed a method for snatching unwary pups off a rookery (pinniped breeding ground). This involves intentionally beaching themselves on an incoming tide and surging from the waves to grab an unsuspecting victim. It is a hazardous undertaking, because of the risk of stranding, but clearly the benefits outweigh the costs. In other habitats killer whales work together to harry and herd fish such as herring (genus *Clupea*) into tight balls and then take turns swinging their tails to stun the fish on the outside. All members of the group can then, at their leisure, snatch and swallow fish as they float helplessly in the water column.

When a killer whale calf is first born, it remains close to its mother almost continuously. As the calf grows older but is still nursing, the mother begins to allow her infant more independence. Calves are quite curious, approaching boats, birds, and adult whales, both related and unrelated. As the calves age the mothers increasingly allow these explorations. Typical distances between them grow, from almost continuous echelon swimming to hours of separation, at distances of up to hundreds of meters.

Until the mother has a new calf (on average after 5 years), she provides almost full-time parental care. However, she will occasionally leave a calf, even one as young as a few weeks of age, with a sibling while she forages or rests. Calves younger than 10 years of age are rarely seen swimming more than a few

Exploring the Depths: Killer Whale Behavior (*continued*)

body lengths from an older brother or sister, although alloparental care (parental care given by a nonparent) can also be provided by unrelated young nulliparous females (those who have not yet given birth). In the former, it is likely kin selection but in the latter case such behavior is probably maintained through reciprocal altruism (mutually positive behavior/assistance); the mother gains time to rest or forage and the young female gains experience in caring for a calf. Sons providing alloparental care to younger siblings may provide a key advantage for a mother by allowing sons to remain with her into adulthood, whereas the sons gain access to reproductive females with whom the mother socializes.

– Contributing author, Naomi Rose,
Humane Society International.

Abundance and Status

The killer whale is a widespread species, and there is no estimate of the total numbers in the wild. However, good information is available on the abundance and status of several coastal, resident populations. For example, the southern resident population, found primarily in U.S. waters in the northeastern Pacific, numbers 80 to 90 individuals, whereas there are approximately 210 to 220 animals in the northern resident population of Canada.

Threats to killer whales have included deliberate culls due to perceived conflicts with fisheries, including one infamous event in Iceland in the 1950s when killer whales were machine-gunned. The southern resident killer whale population in the northeastern Pacific is depleted because of the removal of 45 animals between 1962 and 1976 for aquariums (11 of which died during the capture process). Although the last removal of killer whales in the United States was in 1972 (captures in Canada continued for a few more years), killer whales from another population in Icelandic waters were captured up until the 1980s. Captures have also occurred in Japan as recently as 1992, and permits for the capture of 6 to 10 animals have been issued in Russia since 2001. There, the only successful capture was a female, in 2003, which died just 23 days later. It should be noted that a juvenile killer whale was killed during the same capture attempt.

As a top predator, particularly a species that may consume other marine mammals, killer whales have been found to accumulate high body burdens of pollutants, especially polychlorinated biphenyls (see Chapter 15). These high levels may affect the ability of whales to reproduce and could also increase their susceptibility to disease.

Some killer whales have reportedly been killed by boat strikes. The disturbance and noise that boats cause is another potential problem, especially with the high level of whale watching activity that is focused on killer whales in some parts of the world (**Figure 12.2**). This disturbance may reduce the ability of killer whales to communicate, feed, and rest and has led to attempts to manage and control whale watching activity in areas such as British Columbia (Canada). Killer whales are listed on Appendix II of the Convention on International Trade in Endangered Species of Wild Fauna and Flora (CITES) and are considered to be of "least concern" by the International Union for Conservation of Nature (IUCN). However, the northern resident

Exploring the Depths: Killer Whale Longevity

The life span of killer whales has become a somewhat controversial subject due to the substantial differences between their longevity in captivity versus in the wild. The average life expectancy for a male killer whale in the wild is approximately 30 years, but they can reach 60 to 70 years of age. Female longevity averages approximately 50 years in the wild but can reach 80 to 90 years. In fact, killer whales (as well as pilot whales) are one of the few mammal species, beside humans, that are known to become postmenopausal. In comparison, longevity of killer whales in captivity is substantively shorter. To date, only two female killer whales have exceeded 35 years of age. No captive male has exceeded this age, to date. Most killer whales that have died in marine parks and aquariums have been under 30. Moreover, several studies published in scientific journals show that the annual rates of mortality for captive killer whales are some three times higher than those of animals in the wild (i.e., 6.2–7% in captivity vs. 2.3% in the wild).

FIGURE 12.2 Disturbance from high levels of whale watching activity is a cause of concern for some killer whale populations.

population of killer whales on the west coast of Canada is considered to be "vulnerable" under the Canadian Species at Risk Act, and the southern resident population of killer whales that moves between California and Alaska/British Columbia is listed as "endangered" under the U.S. Endangered Species Act and under Canadian law.

The "Blackfish"

Several species resemble killer whales, at least superficially, and together with the pilot whales are often called, rather nonscientifically, "blackfish." The false killer whale (*Pseudorca crassidens*) is smaller (up to 6 m) and more slender than the killer whale, with no white patches and slight "humps" on the leading edges of the pectoral fins. It tends to be found in deeper, pelagic waters and feeds on fish, squid, and, occasionally, marine mammals. The pygmy killer whale (*Feresa attenuata*) is much smaller, closer to the size of other dolphins (2.6 m), and has distinctive white "lips" around the mouth and a white patch on the ventral side.

Both species have a tropical and subtropical distribution, with the false killer whale occurring at slightly higher latitudes into warmer temperate waters. The melon-headed whale (*Peponocephala electra*) superficially resembles the pygmy killer whale but grows just a little larger (to 2.75 m). It has a dark patch on its back and on the side of its head, with a pale stripe down the front of its head. It feeds on small fish and squid. Like the false and pygmy killer whales, it too has a deep pelagic distribution in the tropical and subtropical waters. In 2004 melon-headed whales drew the attention of conservationists when 200 were found in unusually shallow water in Hanalei Bay, Hawaii. At the same time, the United States and other navies were conducting a sonar tracking exercise, with sonar being activated just before the so-called milling event occurred. A report by the U.S. government concluded that the sonar use was a "plausible, if not likely" cause of the unusual event, thereby expanding concerns for the impacts of military sonar on not just beaked whales (see Chapter 5 and 11) but on other species too.

Pilot Whales

Two living species of pilot whale are the long-finned pilot whale (*Globicephala melas*) and the short-finned pilot whale (*G. macrorhynchus*). Both pilot whales have distinct, long, boomerang-shaped flippers, with the long-finned species, as their name suggests, having slightly longer fins. The long-finned pilot whale is slightly bigger than the short-finned whale (6.7 vs. 6.1 m). The dorsal fin of both species is curved, with a long base, shaped a little like the crest of a wave, which distinguishes pilot whales from other species. The head has a bulbous melon, which resembles a black upturned cauldron, and has earned them the nickname "pot heads."

The short-finned pilot whale has a tropical to subtropical distribution, whereas the long-finned pilot whale is found in temperate to subpolar waters in both hemispheres. Both species tend to be found in deeper, open waters but can also be found close to land in areas where the continental shelf edge is close to the coast. Both feed on squid, and they are found in large social groups that tend to be matriarchal (i.e., led by dominant females).

Pilot whales are perhaps infamous for two things. First, they have an unfortunate tendency to strand in large numbers (see Exploring the Depths: Strandings) in many parts of the world. Second, they are the target of controversial hunts in the Faroe Islands (see Chapter 15) and in Japan. Like almost all cetacean species, pilot whales have been killed as bycatch in fishing gear. Pilot whales have stranded at the same time that military activities were being conducted in several parts of the world, raising concerns that this species might also be susceptible to underwater noise pollution. Other forms of pollution are an issue as well: high levels of heavy metals (especially mercury and cadmium) and organochlorine contaminants (see Chapter 15) have been recorded in long-finned pilot whales. Some concern also exists about disturbance and chronic stress impacts of high levels of whale watching activity concentrated on this species in some locations, such as the Canary Islands. Both species are listed on CITES Appendix II and are considered to be "data deficient" by the IUCN.

"Blunt-Headed" Dolphins

Irrawaddy and Snubfin Dolphins

The Irrawaddy dolphin (*Orcaella brevirostris*) is light gray in color, grows up to 2.8 m in length, has a rounded head, and sports a small dorsal fin. In many ways it more resembles a large gray porpoise than a dolphin species. Inhabiting

---

Exploring the Depths: Strandings

Given the media treatment of the recent strandings of beaked whales in association with exposure to certain navy sonar (see Chapter 5 and 11), one may believe strandings are an unusual occurrence. Although it is true that the mass strandings of beaked whales, and the circumstances surrounding them, are indeed unusual (see Chapter 11), many cetacean species have been recorded as having stranded, alive or dead, upon beaches around the world. Various explanations have been offered for why cetaceans might find their way to land. Animals that die at sea, through natural causes or due to human activities, such as being caught in fishing nets, may come ashore in due course simply as a result of ocean currents. However, the reasons for live strandings of cetaceans are often less clear and have been widely debated.

One obvious possibility is that they are seriously ill. Indeed, these air-breathing animals may seek shallow waters if they become sick. This may be the result either of natural conditions related to pathology or as a consequence of exposure to the various pollutants present in the marine environment. Other suggestions include disruption by, or avoidance of, meteorological and oceanographic disturbances such as hurricanes and disorientation or pressure-related trauma resulting from undersea earthquakes. Astronomical factors have also been suggested, with patterns of strandings linked to lunar cycles as well as to periods of solar activity and sunspots. Lunar phases alter the light available on any particular night and influence the depth and thus availability of food for cetaceans. Solar activity could disrupt the Earth's geomagnetic field. If cetaceans use this field to aid their navigation, as has been suggested by some, this could lead to potentially lethal wrong turns. The most likely explanation is that multiple causes are involved.

Pilot whales in particular are known for their regular mass strandings, where over a hundred seemingly healthy whales may beach themselves at one time. It is thought that these highly social animals try to remain with a sick or distressed animal when an individual moves into shallow waters. Inevitably, these supportive efforts occasionally result in the entire group becoming beached.

If stranding response teams can arrive quickly enough, many stranded pilot whales can often be refloated and saved. However, once stranded, cetacean bodies lose the natural support provided by water and instead are subject to unnatural pressures on their internal organs. The larger the whale, the greater the problem this pressure presents, and for all stranded cetaceans it is usually a race to get them back into their natural environment as soon as possible before organ and tissue damage become irreversible. (Stranded whales and dolphins should only be moved by experts; it is important when encountering a live stranded cetacean to get experts quickly to the scene.)

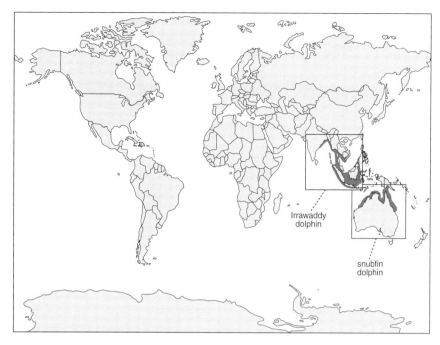

FIGURE 12.3 Map showing the current, known distribution of the Irrawaddy and snubfin dolphin. Data from: IUCN Red List.

coastal, brackish, and freshwater areas of southeast Asia (including rivers), the Irrawaddy dolphin feeds on fish, cephalopods (squid), and crustaceans. A study split the species into two genetically different but nearly identical species morphologically (i.e., external physical characteristics): the Irrawaddy dolphin (*O. brevirostris*) and the Australian snubfin dolphin (*O. heinsohni*).

O. brevirostris is distributed from western Indonesia through Malaysia, westward to eastern India and eastward to Thailand, Cambodia, and the Philippines, and even far up river systems, into land-locked Laos. *O. heinsohni* is distributed from eastern Indonesia and Papua New Guinea to northern Australia (**Figure 12.3**).

Many problems faced by river dolphins (see Chapter 11) are also faced by Irrawaddy dolphins, such as fragmentation of habitat by dams, pollution, decreasing water flow, and bycatch in fishing gear. Concerns over removal of animals for aquariums and marine parks led to *Orcaella* species being uplisted to CITES Appendix I. The IUCN classifies the Irrawaddy dolphin as "vulnerable." Less is known about the newly recognized snubfin dolphin, but bycatch is an issue and their coastal distribution makes them vulnerable to human activities; thus, their IUCN status is "near threatened."

Risso's Dolphin

Risso's dolphin (*Grampus griseus*) has a distinct rounded head, but this species is larger than the Irrawaddy dolphin (3.8 m). It is dark to light gray or even white in coloration. Its color tends to fade with age. It has a tall triangular dorsal fin, which, from a distance, looks similar to that of a killer whale. It feeds on squid, other cephalopods, and crustaceans. Probably the most distinctive feature of this species, however, is their scarring (**Figure 12.4**). Individuals are covered in white scratches, which are caused by the teeth of other Risso's dolphins. Unlike superficial scars and scrapes in other dolphin species that typically fade away after they heal, Risso's dolphins appear to be permanently marked when scarred, with a loss of skin pigmentation. Why does this happen? It has been suggested that this is a way animals

FIGURE 12.4 A Risso's dolphin showing the extensive scarring that this species accumulates.

can assess the fitness of others. An individual covered in white scars has obviously been in many fights and survived and so is likely to be dominant and possibly aggressive, whereas an unmarked animal is likely to be younger, more inexperienced, or less aggressive (see also beaked whales in Chapter 11).

Right Whale Dolphins

The two species of right whale dolphins are unusual looking in that they appear longer and more serpentine than other dolphins, with a small sloping melon and no dorsal fin. They get their name because their dark, finless backs were sometimes mistaken for surfacing right whales by whalers. Both species have similar color patterns, sporting a white underside and a black dorsal side (**Figure 12.5**). However, southern right whale dolphins (*Lissodelphis peronii*) have a white beak and a white patch on their sides that the northern species (*L. borealis*) lacks. The northern right whale dolphin is slightly larger than the southern right whale dolphin (3.1 vs. 3.0 m).

Their distributions are also different. The northern right whale dolphins inhabit the temperate North Pacific, whereas the southern right whale dolphin is found in the waters off southern Australia and South Africa as well as off South America, south into the Southern Ocean (**Figure 12.6**).

Both species are on CITES Appendix II. *L. borealis* is listed by the IUCN as "least concern," whereas *L. peronii* is considered "data deficient."

FIGURE 12.5 (A) Southern right whale dolphins and (B) a northern right whale dolphin.

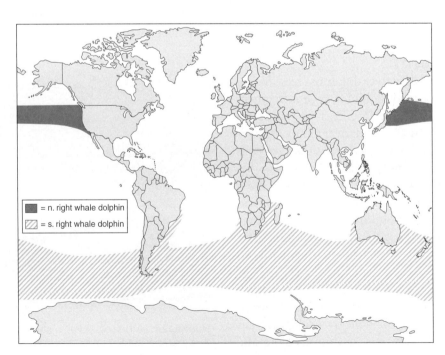

FIGURE 12.6 Map showing the current, known distribution of the northern and southern right whale dolphin. Data from: IUCN Red List.

"Beakless" and "Short-Beaked" Dolphins

Genus *Cephalorhynchus*

There are four members of the genus *Cephalorhynchus* (which literally means "head beak"): Commerson's dolphin (*C. commersonii*), which is found mainly in the coastal waters of southern Argentina; Hector's dolphin (*C. hectori*), which is found in New Zealand; Heaviside's dolphins (*C. heavisidii*), found in the coastal waters off the southwestern tip of Africa (*N.B.*, this species should really be called Haviside's dolphin after the person who collected the first official specimen); and the black, or Chilean, dolphin (*C. eutropia*), which as the name suggests is found in the coastal waters of Chile (**Figures 12.7** and **12.8**).

These round-headed, porpoise-like dolphins all have small discrete ranges and tend to have black and white coloration. There has not been a great deal of research on these species, but Commerson's dolphins and Heaviside's dolphins are considered to be "data deficient" by the IUCN and the Chilean dolphin is "near threatened." However, because of their restricted ranges there are conservation concerns about these species, in particular the "endangered" Hector's dolphin (see text box below).

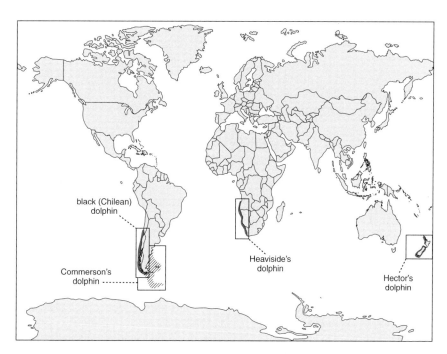

FIGURE 12.7 Map showing the current, known distributions of the Heaviside's dolphin, Commerson's dolphin, black (Chilean) dolphin, and Hector's dolphin. Data from: IUCN Red List.

A   B

FIGURE 12.8 (A) A Heaviside's dolphin and (B) Commerson's dolphins.

FIGURE 12.8 (*continued*) (C) A Hector's dolphin.

Exploring the Depths: Hector's Dolphin, *Cephalorhynchus hectori*

Hector's dolphins are endemic to New Zealand. Their distribution includes North Island (111 individuals; coefficient of variation, 0.44) and South Island waters (7,270 individuals; coefficient of variation, 0.16) and is fragmented into small subpopulations of tens or hundreds of individuals. There are at least four genetically distinct regional populations, and North Island Hector's dolphins are a separate subspecies known as Maui's dolphin (*Cephalorhynchus hectori maui*).

Like the other species in the *Cephalorhynchus* genus, Hector's dolphins are small (adults grow to about 1.5 m) and live in relatively shallow, coastal waters (**Figure B12.2**). The newborns are very large (about 0.6 m) compared with the size of the mother, suggesting this is the smallest possible size for a dolphin calf (perhaps due to thermal and energetic constraints). Hector's dolphins are usually found in small groups of 2 to 10 individuals, and it is common to see several small groups in the same area with groups merging and splitting again. The smallest subgroups (2 to 5 individuals) are often either male-only or female-only groups. Larger groups usually contain both sexes. This helps explain why the frequency of social and sexual behaviors (per minute, per individual) increases significantly after two previously separate groups meet.

Seasonal changes in distribution have been observed in several areas, with dolphins relatively spread out in winter and strongly concentrated close to shore during summer. This is likely to be due to greater fish abundance close to shore in summer, when several Hector's dolphin prey species are spawning. Home ranges are small at just under 50 km of coastline. This makes the species more vulnerable to human impacts, as individuals removed from one population are unlikely to be replaced by immigrants from neighboring populations. Since at least the early 1970s, Hector's dolphins have been caught in commercial gillnet fisheries. Bycatch in trawl fisheries and recreational gillnets goes much further back in time.

Hector's dolphins overlap with gillnet and trawl fisheries throughout their geographic range. The species is listed as "endangered" by the IUCN. In addition, the northern subspecies Maui's dolphin has been listed as "critically endangered." The current population is clearly depleted, at an estimated 27% of the 1970 population.

New protection measures introduced in 2008 (specifically the banning of gillnet and trawl fisheries in some areas of Hector's dolphin habitat) are a major improvement, with at least some protection in most of its range. However, a lack of protection in some areas (e.g., southern part of North Island) and ineffective protection measures in others (e.g., South Island west coast) increase the risk of continued population decline. Without fishing mortality all populations

FIGURE B12.2 Hector's dolphins.

Exploring the Depths: Hector's Dolphin, *Cephalorhynchus hectori* (continued)

would be expected to slowly recover, with the total population almost doubling by 2050 to around 15,000 individuals (half of 1970 population size). Under current management several populations are expected to continue to decline and others could take more than 1,000 years to recover from the impact fisheries bycatch has had over the past 40 years.

—Contributing author, Liz Slooten, Otago University.

Genus *Lagenorhynchus*

The genus *Lagenorhynchus* contains several species of distinctively colored dolphins with short beaks; in fact, the Latin name of their genus literally means bottle-beaked (*lagenos* = bottle; *rhynchus* = bill or beak). Atlantic white-sided dolphins (*L. acutus*), as their name suggests, have a thick pale creamy or yellowish-colored stripe or patch on their sides behind their dorsal fin. They are found in temperate and cooler waters of the North Atlantic, from Northern Europe across to Canada and Greenland (**Figures 12.9** and **12.10**). Pacific white-sided dolphins (*L. obliquidens*) have a similar temperate and cool water distribution in the northern Pacific and have a pale white patch on their sides covering the sides of the head and front part of the body. The white-beaked dolphin (*L. albirostris*) is often confused with the Atlantic white-sided dolphin (its closest genetic relative in the genus) because it has a similar distribution (although its range extends a little further northwards into subpolar waters) and looks similar, but, as its name suggests, it has a stubby white beak and lacks the yellowish patch on its side.

Dusky dolphins (*L. obscurus*) look similar to Pacific white-sided dolphins but are found in cooler waters in the southern hemisphere such as the continental shelf off Argentina, Chile, and Peru as well as in the waters of southern Africa, New Zealand, and several more remote offshore islands. The range of Peale's dolphin (*L. australis*) is also in the southern hemisphere but is more restricted, only being found in the cool waters of South America (specifically Chile and Argentina). The species superficially resembles other Pacific species but with a slightly different pale patch and stripe distribution. Likewise, the hourglass dolphin (*L. cruciger*) can be found in the waters of Chile and Argentina, but its range extends into the even colder subpolar and polar waters of the Southern Ocean. The species gets its name from a distinct white pattern on its side that resembles a very flattened and wavy figure eight.

Although they are currently placed together in the same genus, this is largely due to their morphological appearance. Genetically, they are different, with white-sided dolphins being separate from the others (possibly a separate genus, i.e., *Leucopleurus acutus*), the four Pacific-dwelling species being more closely related, potentially in a different genus again (genus *Sagmatias* has been suggested), and one that is genetically closer to the right whale dolphins and *Cephalorhynchus*. Molecular studies have suggested a closer relationship and postulated that Peale's dolphins or the hourglass dolphins might actually belong in the genus *Cephalorhynchus*.

In terms of their conservation status, the *Lagenorhynchus* dolphins are mostly considered to be of "least concern" by the IUCN, although Peale's dolphin and dusky dolphins are considered to be "data deficient." Several of these species have been hunted either in the Atlantic (e.g., Atlantic white-sided dolphins in Norway, Newfoundland, and the Faroes) or Pacific (Pacific white-sided dolphins in Japan; see Chapter 15). Dolphins in Chile, such as Peale's dolphin, have been hunted to supply bait for fishing. Moreover, there are some concerns about the possible disturbance impacts of dolphin watching (see Chapter 18) on some populations, such as dusky dolphins in New Zealand. Climate change is perhaps a particular conservation concern for this group of dolphins (see Exploring the Depths: Climate Change and Conservation of *Lagenorhynchus* Species).

Humpback Dolphins

The two currently recognized species of humpback dolphins are the Atlantic humpback dolphin (*Sousa teuszii*) and the Indo-Pacific humpback dolphin (*S. chinensis*).

178 PART II: Ecology and Status

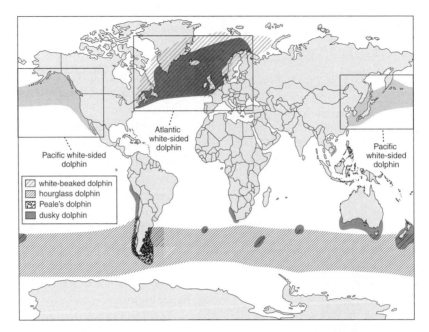

FIGURE 12.9 Map showing the current, known distribution of the Atlantic white-sided, Pacific white-sided, white-beaked, dusky, Peale's, and hourglass dolphins. Data from: IUCN Red List.

FIGURE 12.10 (A) Atlantic white-sided, (B) Pacific white-sided, (C) white-beaked, (D) dusky, (E) Peale's, and (F) hourglass dolphins.

Exploring the Depths: Climate Change and Conservation of *Lagenorhynchus* Species

The genus *Lagenorhynchus* consists of six species found in temperate to polar waters in various parts of the world. Three of these species are only found in the relatively shallow shelf waters that surround the continents. This fact, combined with their preference for cooler waters, makes *Lagenorhynchus* species extremely vulnerable to the effects of climate change. This can be seen by looking at what is currently the best-studied case of climate change–induced impacts on delphinids: the population of Atlantic white-beaked dolphins found in the shallow waters of the North Sea and neighboring shelf areas of northwest Europe. Unlike other small cetaceans, where bycatch is a major concern, this population is currently most affected by rising water temperatures.

Studied since the 1970s, these white-beaked dolphins were the most common dolphin species in these shelf areas. The region is thought to hold around 20% of all the world's white-beaked dolphins, making this area of global importance for the conservation of the species.

White-beaked dolphins in this region are only common in waters cooler than 14°C and are completely absent in waters warmer than 18°C. If the water temperature in any particular area increases, the dolphins become much less common there and can even disappear from that location altogether. Over the last 20 years temperatures in the shelf waters of northwest Europe have increased by 1 to 2°C. This increase is enough to make waters too warm for the local white-beaked dolphins. As a result their distribution is gradually shifting northward. Most recently, the white-beaked dolphins have disappeared from much of western Scotland, which once held some of the highest densities of this species found anywhere in the world.

Although such shifts in species distribution are not always a cause for concern, the ability of white-beaked dolphins to shift their distribution northward is limited by the fact that almost all waters to the north are too deep for them to feed in. Therefore, this population of white-beaked dolphins is effectively trapped in the shelf waters of northwest Europe. Thus, as water temperatures increase, the size of the area suitable for them is gradually shrinking. Some models predict that because of global climate change, all the shelf waters in northwest Europe will be too warm for white-beaked dolphins in as little as 30 years and this population will disappear completely. Although other populations of this species will remain in other parts of the North Atlantic, such as around Iceland and the Barents Sea, the loss of this population will represent a substantial decline in the global population size and a loss of important genetic diversity. Other less well-studied populations of *Lagenorhynchus* species that also only occur in cool shelf waters, such as the dusky dolphin and Peale's dolphin, are likely to face a similar threat from rising water temperatures associated with global climate change.

—Contributing author, Colin MacLeod,
University of Aberdeen.

The former is found in coastal waters of western Africa, whereas the latter is found from South Africa, along the coast of the Indian Ocean, through southeast Asia to Shanghai, China in the northeast, and to northern Australia in the southeast (they have not been reported, however, in the Philippines). Both species have a coastal and estuarine distribution.

S. chinensis can be divided into two morphological types (**Figure 12.11**). From South Africa to the west coast of India, dolphins are dark gray or almost black in color, with a definite "hump" (referred to as the *plumbea* type). From the east coast of India through the rest of their range, animals do not have a "hump" and are white or pale in color when adult (and gray as calves), sometimes with blue-gray spots or flecks (referred to as the *chinensis* type). When active, the *chinensis*-type dolphins may turn a bright "bubble-gum" pink color as vessels carry blood to the body surface, presumably to help cool down the animals. *Plumbea*-type humpback dolphins can be found in coastal waters, bays, and estuaries in waters that are typically less than 15 m deep (**Figure 12.12**). They are often found feeding in areas with rocky reefs and rocky shores. *Chinensis*-type humpback dolphins are also typically found in estuaries, especially around the coast of China and eastern Asia, although animals in Australia can be more coastal. It has been suggested that this more estuarine distribution may be a way to avoid predatory sharks (especially tiger sharks, *Galeocerdo cuvier*) that are less tolerant of fresh water, whereas a major predator of the *plumbea*-type dolphin is the freshwater-tolerant bull shark (*Carcharhinus leucas*). The *chinensis* type of dolphin is usually found in shallow waters too (less than 10 m deep).

As coastal and estuarine species inhabit waters next to many heavily populated developing countries, both species of humpback dolphins face problems from human activities, including entanglement in fishing gear, injuries from boat strikes, underwater noise, and habitat degradation (including the reduction of freshwater input into estuarine habitats). Several populations, particularly the population inhabiting the waters of Hong Kong and the Pearl River Delta, have been found with high levels of

FIGURE 12.11 (A) An Atlantic humpback dolphin, (B) a *plumbea*-type humpback dolphin, (C) *chinensis*-type humpback dolphins, and (D) a pink *chinensis*-type humpback dolphin.

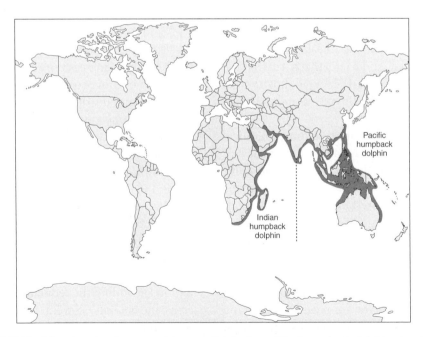

FIGURE 12.12 Map showing the current, known distribution of the humpback dolphin showing the divide between Indian and Pacific (*plumbea* and *chinensis*, respectively) types. Data from: IUCN Red List.

pesticides (notably DDT) and heavy metals (especially mercury) in their tissues. The Atlantic humpback dolphin is listed as "vulnerable" by the IUCN, and the Indo-Pacific humpback dolphin is considered to be "near threatened," although the population that inhabits the coastal waters of western Taiwan, which number possibly less than 100 animals, is considered to be "critically endangered" (see Exploring the Depths: Critically Endangered Eastern Taiwan Strait *Sousa chinensis* Population). Both species are listed under CITES Appendix I (see Chapter 17).

"Long-Beaked" Dolphins

Genera *Stenella* and *Steno*

The various members of the genus *Stenella* look like most people's idea of a dolphin, with long beaks and blue, white, and gray coloration, often in striking patterns, swirls, stripes, or spots. They are sometimes found in large groups, especially the pelagic or offshore species.

Although each species in the genus is now relatively well established in the literature, the taxonomy of *Stenella* was still uncertain until the 1980s, mostly due to similarities in their physical appearance. The genus is still considered to be paraphyletic (i.e., it is an artificial assemblage of species that come from different genetic lineages and do not share a single common ancestor). Because of earlier confusion over the species in this genus, there are likely many errors in the historical records of sightings and occurrences of these dolphins.

The pantropical spotted dolphins (*S. attenuata*), as their name suggests, are found throughout the tropical and temperate waters of the world and are usually spotted

---

Exploring the Depths: Critically Endangered Eastern Taiwan Strait *Sousa chinensis* Population

In the waters of China and Taiwan, in the eastern part of their range, about eight populations of Indo-Pacific humpback dolphins are estimated. Among these, the best understood is the Pearl River Estuary (PRE) population, which has an estimated range of more than 4,000 km² in a broad stretch of water less than 40 m deep. Studies have also provided valuable data on the Eastern Taiwan Strait (ETS) population, which is resident along the west coast of Taiwan in waters less than 25 m deep. This population has a narrow, linear range of less than 600 km² within only about 3 km of the shore.

Humpback dolphins in this region are born dark gray and turn white or pink as they mature (see above), with blue-gray spots that gradually disappear in the course of this transition. However, differences have been observed between the spotting patterns of ETS humpback dolphins and two other nearby populations, including the PRE dolphins, suggesting the ETS population is distinct and isolated. Other common features include a robust, medium-sized body up to about 2.8 m in length; large, round-tipped flippers; and a long, thin beak. The dorsal fin is short, slightly curved, and sits on a base that is broad but not a prominent hump like that seen on dolphins in the western part of the global range of this species.

ETS and PRE humpback dolphins feed on estuarine, littoral (intertidal), and demersal (bottom dwelling) fish and, rarely, on cephalopods (e.g., squid). Dive duration is generally less than about 4 to 5 minutes, and they swim slowly in small groups that usually comprise fewer than 10 individuals.

Living in coastal and estuarine waters humpback dolphins, in many parts of their global range, occur in close proximity to dense human populations, and their habitats are therefore often affected by a variety of anthropogenic activities. The PRE and ETS humpback dolphins are no exception and indeed share many common threats.

Western Taiwan and the coastal areas in and around the PRE (including Hong Kong) have extremely dense human populations and intense levels of industrial and infrastructure development and fishing. The main threats to the PRE and ETS populations include interactions with boats and fishing nets, prey depletion, loss of habitat due to ongoing land reclamation, and air, water, and noise pollution. The damming and diversion of rivers feeding into the estuaries along western Taiwan are also altering the habitat of the ETS population and their prey.

Despite similarities in the types of threat faced by PRE and ETS humpback dolphins, the outlook for these two populations is different. Although the PRE population is estimated at more than 2,500 individuals and believed to be viable, the ETS population is estimated to be in the high tens and was listed in 2008 as "critically endangered" under the IUCN Red List of Threatened Species.

Although there is now increasing public awareness of the plight of the ETS humpback dolphins and the likely impacts of the above-mentioned activities on the health of many other species, including humans, there are currently no effective protection measures in place. Meanwhile, consumption by people around the world of high-tech goods and other products made in Taiwan continues to fuel further land reclamation, water diversion, and pollution-intensive development. Decisions made by Taiwan's government within the next few years will likely determine whether this population survives in the long term.

—Contributing author, Christine MacFarquhar, Wild at Heart.

as adults. However, they look similar to the Atlantic spotted dolphin (*S. frontalis*) and their distribution overlaps in the tropical Atlantic, although the latter species is found slightly further northward and southward into warm temperate waters (**Figures 12.13** and **12.14**). The clymene dolphin (*S. clymene*) and spinner dolphin (*S. longirostris*) look very similar, and both have pale gray horizontal stripes on their sides. However, the clymene dolphin has a slightly more curved dorsal fin and is not quite as long (approximately 2 m as an adult vs. 2.3 m). The clymene dolphin has a distribution similar to the Atlantic spotted dolphin, whereas the spinner dolphin's distribution broadly resembles that of the pantropical spotted dolphin.

Spinner dolphins are represented by a number of subspecies (see Chapter 3) and get their common name from their tendency to spin longitudinally (around the axis of their length) as they jump. Spinning in this way has, so far, not been seen in any other dolphin species, except the clymene dolphin. Spinner dolphins spin during several jumps in quick succession, in some of the most impressive acrobatic displays of any cetacean (**Figure 12.15**). Possible explanations for this behavior include attempts to attract a mate, efforts to remove parasites, general communication and coordination with other spinner dolphins (perhaps also through the generation of sound and bubbles on landing), and perhaps just for "fun."

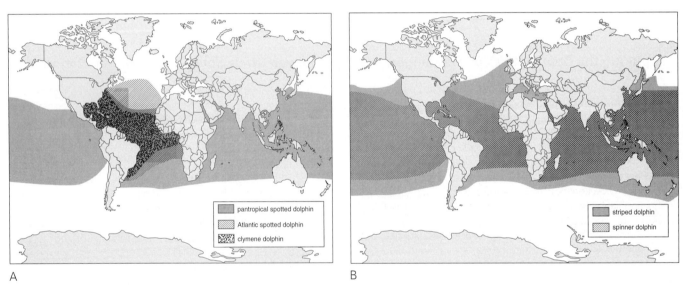

FIGURE 12.13 (A) Map showing the current, known distribution of the pantropical spotted, Atlantic spotted, and clymene dolphin. (B) Map showing the current, known distribution of the striped and spinner dolphin. Data from: IUCN Red List.

FIGURE 12.14 (A) Pantropical spotted dolphins and (B) Atlantic spotted dolphins.

FIGURE 12.14 (*continued*) (C) clymene dolphins, (D) striped dolphins, and (E) spinner dolphins.

The striped dolphin (*S. coeruleoalba*) is a little more distinctive (**Figure 12.16**), with a dramatic pattern of clear horizontal stripes on its sides. It also has the most cosmopolitan distribution, being found throughout the world's tropical and temperate waters. The range of this species seems to be broadening as the result of climate change; for example, it has moved further north in northern Europe, into the waters of northern Scotland, as waters get warmer.

The IUCN considers the conservation status of the pantropical spotted dolphin and striped dolphin to be of "least concern," whereas the other three species of *Stenella*

FIGURE 12.15 A spinner dolphin leaping and spinning.

FIGURE 12.16 Striped dolphins showing their distinctive color pattern.

are considered to be "data deficient." Both spinner and pantropical spotted dolphins were particularly affected by purse seine fisheries for tuna in the eastern tropical Pacific, where until 1992, an estimated half a million dolphins were killed every year (see Chapter 15).

The rough-toothed dolphin (*Steno bredanensis*) is often, somewhat unjustly, referred to as one of the ugliest species of dolphin. Although they do have a distinctly reptilian look about their faces, the two-toned mottled appearance they gain as adults (they look very much like bottlenose dolphins when calves) and their long snouts give them features just as "becoming" as any other dolphins. Relatively little is known about this species because they had not been studied systematically until quite recently.

Exploring the Depths: Rough-Toothed Dolphin, *Steno bredanensis*

The rough-toothed dolphin is the only species in the genus *Steno*. It is easy to recognize by its smoothly sloping melon and the lack of a crease in the upper beak (**Figure B12.3**). Its body is countershaded with a black to dark gray back, medium gray sides, and a white to pinkish belly. Adults often have yellowish white spots and scarring on their sides. The tip and sides of the beak of the adults is white-pinkish, which makes them very distinctive. The dolphin gets its name from its unique tooth morphology, which consists of fine ridges that give the teeth a rough feeling.

The rough-toothed dolphin can be found in tropical to warm-temperate waters, although knowledge about its exact distribution is inadequate (**Figure B12.4**). Rough-toothed dolphins are offshore, deepwater dolphins and are mostly found in low abundances.

In spite of their worldwide occurrence, little is known about their ecology, social organization, behavior, vocal repertoire, and taxonomy. Like most dolphin species, rough-toothed dolphins are social and live in groups. They are known for their synchronized swimming in small (sub)groups in a tight-ranking formation, but the biological significance of this behavior is not understood. Rough-toothed dolphins have been characterized as the most intelligent cetacean species. In captivity the species has been described as "bold and inventive," with one captive individual surprising its trainers by inventing novel behaviors. Their bold behavior has, however, led to problems;

FIGURE B12.3 Rough-toothed dolphins.

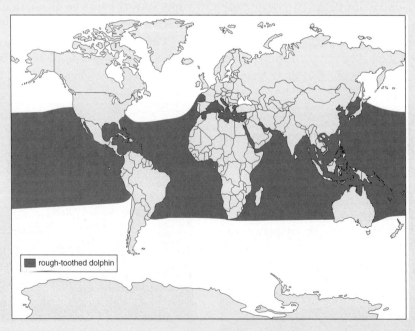

FIGURE B12.4 Map showing the current, known distribution of the rough-toothed dolphin. Data from: IUCN Red List.

Exploring the Depths: Rough-Toothed Dolphin, *Steno bredanensis* (continued)

they have been known to steal bait and fish from fishing lines, which has made them unpopular with fishermen.

Rough-toothed dolphins are incidentally caught by tuna purse seines in the eastern tropical Pacific Ocean, but they are also taken as bycatch in gillnet and driftnet fisheries and direct dolphin fisheries. Mass strandings of rough-toothed dolphins have occurred in several areas, but the causes are generally unknown.

This species has been observed in shallow and coastal waters of Hawaii, the Canary Islands, Honduras, Aruba, Brazil, and French Polynesia and has been shown to exhibit high-site fidelity in most of these areas. The rough-toothed dolphin is classified in its own (monotypic) genus, and it is still uncertain to which other species it is most closely related. It is also unclear if there are genetic differences among rough-toothed dolphins in different ocean systems and different ecozones (e.g., offshore vs. inshore). The locations where rough-toothed dolphins have been observed close to shore now provide a unique opportunity to start investigating the biology of rough-toothed dolphins in detail.

–Contributing author, Jolanda Luksenburg, George Mason University.

Genus *Delphinus*

The so-called common dolphins (genus *Delphinus*) are currently split into the short-beaked common dolphin (*D. delphis*) and the long-beaked common dolphin (*D. capensis*) (**Figure 12.17**). Both have a distinct "hourglass" pattern on their sides with a yellow to tan patch, followed by a lighter gray patch. They are virtually identical and almost impossible to separate in the field, but, as their name suggests, the long-beaked dolphin has a slightly larger snout and higher tooth count. Their distribution is also slightly different (**Figure 12.18**). The short-beaked common dolphin is found from northwest Africa to Europe, New Zealand, Australia, Japan, Hawaii, and the eastern tropical Pacific. The long-beaked common dolphin is also found in Japan, eastern Asia, and northwest Africa but does not extend to Europe. It is also found in South Africa, Madagascar, Southeast Asia, and eastern South America.

The two species were only recognized as being separate in 1994, based on morphological differences. More recent genetic analyses, however, suggest the taxonomy may be more complicated and that long-beaked common dolphins may perhaps just be a specialized form of short-beaked common dolphins. A third species of common dolphin (*D. tropicalis*) that occurs in the Indo-Pacific has been

A
B

FIGURE 12.17 (A) A short-beaked common dolphin and (B) a long-beaked common dolphin.

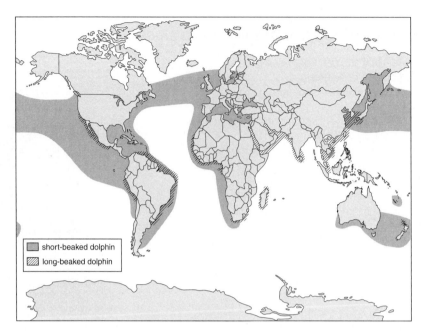

FIGURE 12.18 Map showing the current, known distribution of short-beaked and long-beaked dolphins. Data from: IUCN Red List.

Exploring the Depths: Common Dolphin Bycatch in Europe

At the end of February 1989 more than 600 dead dolphins stranded over 2 days on the coasts of the Landes and Vendée regions in France. Some were freshly dead, whereas others had been decomposing out at sea for many weeks. In the years immediately following, large numbers of dead dolphins continued to come ashore on some parts of the French Atlantic coast. The sight and smell of them caused disgust and outrage, and this situation was mirrored on the coasts across the English Channel on the shores of Cornwall and Devon, although the numbers were only a fraction of those seen in France (probably the result of the prevailing waves and currents). The dead bodies raised questions about what was killing the dolphins and what could be done about it.

Fairly quickly the finger of blame was pointed towards fishing operations. Some dolphin bodies washed ashore were tied together with rope, some were weighted down, and others had been mutilated in attempts to make the bodies disappear below the waves (or ease their removal from nets). That the carcasses, which were mainly common dolphins (*D. delphis*), mostly appeared in mid-winter indicated something happening offshore in that season, when the dolphins enter the English Channel primarily to follow the spawning sea bass (*Dicentrarchus labrax*).

In fact, bycatch (the incidental capture of nontarget species) in fisheries had long been acknowledged as a major threat to the conservation of cetaceans in the northeast Atlantic region by bodies ranging from national governments to the European Commission. In general, the harbor porpoise, *Phocoena phocoena*, which tends to be found in the shallower waters on the continental shelf, had been seen as the most vulnerable species because of capture in bottom-set gillnet and tangle net fisheries. The more oceanic species like the common dolphins were thought to be less impacted although vulnerable to pelagic (i.e., mid-water) trawls and pelagic driftnets. However, they were also known to be killed in a number of different types of fishing gear within the same sea area.

The situation in the 1990s seemed to be linked to the pelagic trawl fisheries that operated in the northeast Atlantic. In these fisheries modern technology allows two vessels working in tandem to pull large nets (some the size of aircraft hangers) swiftly through the water. However, gaining hard evidence about which fishery was to blame for the dolphin deaths, and then what might be done, proved difficult. Various studies were conducted on the fishing fleets, and postmortem examinations (or necropsies) were carried out on some of the fresh bodies coming ashore.

The first notable peak in common dolphin strandings in the United Kingdom happened in 1992 when 118 dolphin carcasses came ashore in Cornwall and Devon; nearly half of these were common dolphins. Examination of the bodies determined that most of them had died as a result of incidental capture in fisheries. From 1990 to 2002 over 95% of the stranded bycaught common dolphins found in the United Kingdom were in the southwest of England (i.e., an area close

Exploring the Depths: Common Dolphin Bycatch in Europe (*continued*)

to France), and the majority stranded in the first 3 months of the year. Most had net marks and other injuries consistent with entanglement in small-meshed mobile gear (i.e., trawl netting) (**Figure B12.5**).

A number of international bodies have considered what level of cetacean bycatch could be considered "sustainable." For example, the Scientific Committee of the International Whaling Commission considered an annual bycatch level of 2% of estimated abundance to be unsustainable and of concern for harbor porpoise populations. Similarly, the Parties to the Agreement on the Conservation of Small Cetaceans of the Baltic, North East Atlantic, Irish, and North Seas adopted 1.7% of abundance as a general definition of the threshold of "unacceptable interactions." However, to apply such thresholds good knowledge is required about the size of the cetacean population and the bycatch level, and such information is difficult to gain. Cetacean bycatch is not only a conservation issue but also a significant animal welfare problem. Injuries sustained by bycaught cetaceans typically include broken beaks, torn and severed fins and flukes, and cuts and abrasions on the skin.

Although some debate remains about the causes of the dolphin mortalities, the United Kingdom responded in December 2004 by banning its own pair-trawling fishery from the 12-mile inshore zone around Devon and Cornwall, a move that probably pushed the fishery further offshore and had no affect on the fleets of other nations. Nonetheless, in recent years the number of bodies coming ashore has declined, although bycatch still occurs in pair trawls, and stranded common dolphins have been found on U.K. coasts outside of the pair-trawling season, suggesting other types of bycatch are also a source of dolphin mortality.

FIGURE B12.5 An example of a bycaught dolphin. This photo shows a striped dolphin.

Exploring the Depths: Tucuxi and Costero, Genus *Sotalia*

The small gray South American dolphin, called the tucuxi (pronounced too-koo-shi), was split into two species (**Figure B12.6**). The tucuxi (*S. fluviatilis*) is distributed in the Amazon River system, whereas the newly recognized "costero," estuarine or Guiana/Guyana dolphin (*S. guianensis*), is distributed in the coastal waters of eastern Central America and northeast South America (**Figure B12.7**). Many problems facing the Amazon River dolphin are also faced by the tucuxi, such as mercury pollution from mining, dams, bycatch in fishing gear, and being hunted for food, medicines, or to produce folkloric charms. The coastal costero likewise faces some direct hunting to provide bait for fishing operations, and animals are lost through entanglement in fishing gear. Coastal development and human coastal overpopulation are also of concern, as they are for most coastal species inhabiting waters next to developing countries. The IUCN considers both species of *Sotalia* to be "data deficient" because of a lack of information on population sizes or trends. However, CITES lists *Sotalia* species on Appendix I, thus prohibiting trade of live animals or their body parts.

FIGURE B12.6 A tucuxi.

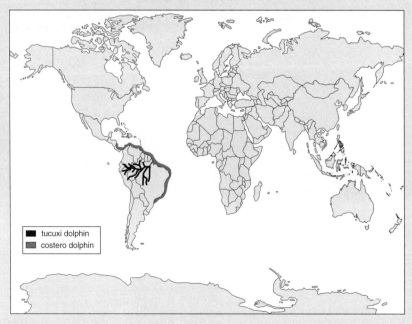

FIGURE B12.7 Map showing the current, known distribution of the tucuxi and costero. Data from: IUCN Red List.

suggested, based on morphology. However, some authorities consider it to be a slightly different form of *D. capensis*.

Bottlenose Dolphins

The animal that most people think of when one says "dolphin" is the common bottlenose dolphin (*Tursiops truncatus*). This is partly due to their cosmopolitan distribution, being found throughout most of the world (except cold polar and subpolar regions), and their often coastal habitats that bring them into close contact with humans. However, a bottlenose dolphin was also the star of the American television series "Flipper" (see Chapter 13) and, as such, has become the iconic dolphin in the minds of the general public (**Figure 12.19**). Moreover, most captive dolphins are bottlenose dolphins. Due in no small part to their coastal distributions and wide geographic range, bottlenose dolphins are also probably the most studied of the dolphins.

Bottlenose dolphins range considerably in size from 1.9 to 3.8 m. Generally, bottlenose dolphins in warmer water populations tend to be smaller in physical size, whereas cooler water animals are larger. They can weigh up to 650 kg and are sexually dimorphic, with males being larger. They have a medium-sized, relatively wide beak, which gives them their name: the beak was thought to resemble the stubby milk bottles that were once delivered to doorsteps in several parts of the world. Bottlenose dolphins are found worldwide in temperate and tropical waters from 10 to 32°C. There is considerable variation in their patterns of habitat use, which may differ even between individuals within a population and may change depending on time of day, tidal state (they may use tides to travel or feed during certain tidal states), reproductive status (e.g., mothers with young calves may be found in more inshore and sheltered areas), food supply, weather conditions (some may shelter in enclosed bays or shallow waters in rough conditions), or season (there may be distinct differences in distribution, behavior, and habitat use during summer and winter). Overall, they seem to be quite adaptable, and the behavior of specific populations may be linked to local conditions.

Bottlenose dolphins were previously thought to be one species; however, studies on the morphology, skeleton, and genetics of bottlenose dolphins have shown at least two, or possibly more, species (see Chapter 3): the common bottlenose dolphin (*T. truncatus*) and the Indo-Pacific bottlenose dolphin (*T. aduncus*).

Although the two recognized species of tursiops look almost identical, they are different genetically. It has even been suggested that the Indo-Pacific bottlenose dolphin should be in a different genus from the common bottlenose dolphin (see Chapter 3).

The Indo-Pacific bottlenose dolphin is, in general, slightly more slender and smaller than the common bottlenose, has a slightly longer beak in proportion to its body length, and it also tends to be spotted on its underside. It inhabits coastal waters from South Africa and northern Australia in the south, through the Indo-Pacific to eastern China and Taiwan in the north. The animals that are currently classified as *T. aduncus* might in fact be more than one species. For example, a study published in 2011 suggested that a small population of bottlenose dolphins in southeastern Australia are genetically distinct from *T. aduncus* to such a degree that perhaps could be a new species (i.e., *T. australis* or the "Burrunan dolphin").

The common bottlenose dolphin too has been suggested at various times to contain a number of distinct subspecies, if not different species. Although they have not been divided into separate (sub)species yet, there are also two distinct forms, or "morphotypes," of common bottlenose dolphin in the western North Atlantic: coastal and offshore. These two forms are actually more genetically divergent than *Delphinus delphis* and *D. capensis* noted previously, so this could mean at least one further species of bottlenose dolphin being recognized in the future.

The two different types of common bottlenose dolphin are also different behaviorally and ecologically. The coastal form is frequently seen in harbors, bays, lagoons, estuaries, and river mouths. In general, their distribution often varies according to age, sex, or reproductive status. Nursing mothers and calves often live closest to the shore, with adolescent animals inhabiting waters slightly further offshore and adult males and females without calves being found further out still. Group size is usually fewer than 20 individuals for coastal animals.

FIGURE 12.19 A common bottlenose dolphin.

Offshore groups can be much larger than coastal groups, up to 1,000 individuals. They often move substantial distances, presumably searching for ephemeral patches of prey in the open ocean; for example, specific groups have been recorded moving between locations that are 4,000 km apart. They are often found on the edge of the continental shelf and over the shelf break.

The social structure of bottlenose dolphin groups can be fluid, with groups coalescing or breaking apart. Some animals associate with each other for long periods of time; others only have fleeting associations. One study found that certain animals in a bottlenose dolphin pod may act like the social "glue" that keeps pods together, and if these animals are removed, groups can break up. Groups are frequently formed according to the sex of the animals (e.g., all males or all females), age status (juvenile groups, adults only, adults and calves), reproductive condition (immature, mature, pregnant), and family relationships (mother and calves, siblings, aunts; bottlenose dolphins tend to be matrilineal). Past associations (animals that have foraged together before or "friends" that just tend to associate) also play a role. Bonds between mothers and calves are often strong, and there are many matrilineal associations. Interestingly, bonds between males may also be strong, with some males consistently associating with each other, often in pairs. In short, individuals may associate with certain animals more than others. Also, during any particular day a bottlenose dolphin may interact with a variety of other animals depending on the context (e.g., feeding, resting, mating).

Bottlenose dolphins have also been observed associating with other species of cetaceans in both peaceable and aggressive encounters. For example, Indo-Pacific humpback dolphins and bottlenose dolphins have formed mixed groups when feeding in Australia. Mixed groups of Atlantic spotted dolphins and common bottlenose dolphins occasionally occur in the Dominican Republic, but in the nearby Bahamas bottlenose dolphins are more likely to attack spotted dolphins and consequently are often chased away by aggressively vocalizing mobs of the smaller species when bottlenoses approach. Probably the most famous instances of aggressive bottlenose dolphin behavior were reported in the Moray Firth in Scotland, where dolphins were observed attacking and killing harbor porpoises. Similar behavior has now been reported in other parts of the world. Bottlenose dolphins have even been reported killing members of their own species, as infanticide of dolphin calves has been observed in several locations, including the coastal waters of Virginia. Thus, despite their "friendly" reputation, bottlenose dolphins can be antagonistic, especially larger males that tend to be more aggressive and dominant (**Table 12.1**).

As suggested by their varied habitat use and foraging strategies, bottlenose dolphins tend to be opportunistic feeders. They eat a wide range of fish, crustaceans, and cephalopods, with prey species tending to be the most locally abundant or nutritious species available. They can feed individually or cooperate to herd prey in groups. Several specialized techniques also have been observed in bottlenose dolphins. They may even feed in association with other species, as noted above. Some of the most specialized foraging behavior may be found in the creeks of South Carolina. Here, certain dolphins chase fish onto shallow mudflats and then rush forward, beaching both themselves (always on their right-hand side) and the fish. The dolphins then capture these stranded fish in their beaks, before wriggling and sliding themselves back down the muddy banks into the water. A similar behavior has been reported in Shark Bay, Australia, where dolphins chase fish into shallow water and then swim at rapid speeds, adjusting the pectoral fins so they "hydroplane" or skim along in water just a few centimeters deep, occasionally stranding themselves in the process. Also in Shark Bay, certain dolphins have been observed carrying pieces of sponge on their beaks. These sponges are used as a protective shield when the dolphins dive down to catch and feed on spiny benthic fish. This sponge-carrying behavior has been put forward as an example of "tool use" (see Chapter 7) in cetaceans and has been highlighted as evidence that bottlenose dolphins are especially intelligent (see Exploring the Depths: Dolphin Intelligence).

As with many cetacean species, bottlenose dolphins have a mating system in which both males and females may mate with multiple partners, and it is believed that sperm competition (see Exploring the Depths: Sperm Competition) is one way for males to ensure their reproductive success. Dominance hierarchies may also be important, with large dominant males trying to gain as much access to females as possible and defending females against other males, but again there is much variability within the species. Sexual behavior is common in bottlenose dolphins, and male calves have been reported to attempt to mate with their mothers even before they are weaned. Same-sex sexual behavior is also common, with males frequently attempting

TABLE 12.1 Examples of "Aggressive" Behavior in a Bottlenose Dolphin.

Biting (another animal or in air)
Swimming with mouth open
Hitting (with tail or flipper)
Slapping water (with tail or flipper)
Jerking head backwards/forwards or sideways
Swimming quickly straight at an opponent
Jumping on top of an opponent
Trying to keep an opponent underwater

Exploring the Depths: Sperm Competition

Males may reproductively compete with each other via intrasexual selection (see Chapter 5) through aggression, possibly through vocalizations, but also through sperm competition. That is to say, the male that produces the most sperm increases the likelihood that one of his sperm will impregnate the female. Sperm competition frequently occurs in insect species. It is also believed to be important in some primates, such as chimpanzees (*Pan troglodytes*) and especially bonobos (*Pan paniscus*). In primates, relative size of testes in males seems to be related to sperm competition. Highly dimorphic species, where large males guard a harem of females (and sperm competition is not an important factor), have relatively small testes (e.g., *Gorilla gorilla*). However, primates that are promiscuous, with males and females both mating frequently with multiple partners (polygynandry), tend to have much larger testes, and the males produce large amounts of sperm (e.g., bonobos). Many cetaceans, and in particular delphinids, have extremely large testes for their body mass and can produce large quantities of sperm. For example, dusky dolphins (*Lagenorhynchus obscurus*) have testes that weigh 8% of their body mass, which is 100 times the relative size of human testes. To put this into context, if human males had testes that were the same proportion of their body mass as dusky dolphins, testes would weigh 6.5 kg. Thus, for cetaceans this mating strategy and this form of intrasexual selection is probably important.

---

to mate with other males or even members of either sex of other species of cetaceans (e.g., Atlantic spotted dolphins). The famous dolphin researcher, Dr. Ken Norris, reportedly once said that "sex to dolphins is like a handshake to humans" and likely plays a social role as much as a reproductive one.

Sexual maturity in male bottlenose dolphins occurs between 9 and 14 years of age, and female sexual maturity occurs between 5 and 13 years. Depending on the population, calving can occur year round or can be seasonal. As with other cetaceans only a single calf is born. The gestation period is approximately 12 months, and the calves are suckled for another 12 to 18 months. However, calves may stay with their mothers for 3 years or more, learning important skills, such as foraging. Females usually have calves every 3 to 6 years, which means a relatively low recruitment rate: if a dolphin dies, it can take several years for that animal to be replaced in the population.

Maximum life span for bottlenose dolphins in the wild is from 40 to 45 years for males and up to 50 years for females. Shark predation, disease, and human activities are the major causes of death. The life spans of captive common bottlenose dolphins in U.S. aquariums and marine theme parks are approximately equal to those in the wild.

Globally, bottlenose dolphins are still plentiful and are classified as "least concern" by the IUCN (*T. aduncus* is classified as "data deficient"), but in some areas numbers have been greatly reduced and populations may be nearing local extinction (known as extirpation). For example, the Black Sea subspecies (*T. truncatus ponticus*) is vulnerable after a long history of removals. Between 1946 and 1983, 20,000 to 30,000 animals were killed through hunting and culling due to perceived competition with fisheries. More recently, additional removals of notable numbers of animals from the region have occurred for marine theme parks and aquariums. The resulting decline in numbers to low levels has led to environmentalists and scientists alike calling to have this population listed on Appendix I of CITES (see Chapter 17) and so prohibit trade of this particular population of bottlenose dolphins. (Similar to other cetaceans, the common bottlenose dolphin is listed on CITES Appendix II, which allows limited trade, although a "zero" quota has been established for Black Sea bottlenose dolphins.)

In the United States the bottlenose dolphins are protected by the Marine Mammal Protection Act (MMPA) (see Chapter 17). Bottlenose dolphins in the Mid-Atlantic are considered "depleted" under the MMPA (see Chapter 17) because a disease outbreak (morbillivirus) in 1987 and 1988 killed over 1,000 dolphins on the east coast of the United States. Also in the United States removal of coastal dolphins to supply animals for marine theme parks and aquariums drastically reduced some other populations (e.g., the southeastern coast and the Gulf of Mexico), leading to a moratorium on live captures from U.S. waters.

In Europe bottlenose dolphins are considered a conservation priority species and are listed under Annex II of the EU Habitats Directive (see Chapter 17). European countries with populations of bottlenose dolphins have an obligation to establish Special Areas of Conservation (SACs) for this species. In the United Kingdom two such SACs currently exist in the Moray Firth in northeastern Scotland and in Cardigan Bay, Wales.

Bottlenose dolphins are accidentally caught in a variety of fishing gear, including gillnets, shrimp trawls, and purse seines used to catch tuna (see above and Chapter 15). These dolphins are also occasional victims of harpoon and drive fisheries, for example, in Taiji, Japan. This controversial fishery was recently the subject of the 2010 Oscar-winning documentary *The Cove* (see Chapter 15).

Their coastal distribution means that many populations of bottlenose dolphins are vulnerable to pollution, habitat alteration, and human disturbance (e.g., from boats, including whale watching activities; see Chapter 18). Several "die-offs" of bottlenose dolphins have occurred due to disease outbreaks. Diseased dolphins often have elevated levels of chemical pollutants (especially polychlorinated biphenyls). These contaminants may increase their susceptibility to disease (see Chapter 15). Moreover, high rates of mortality seen in first-born calves in some bottlenose dolphin populations (e.g., in Sarasota Bay, Florida) are believed to be partly due to high (organochlorine) pollutant loads in mothers, who offload these pollutants onto their calves through the placenta and via contaminated milk (for details see Chapter 15). In recent years dolphin mortalities from algal blooms in locations such as Florida have also been a cause for concern. Elevated levels of nutrients (typically due to fertilizer run-off from farms, lawns, and gardens) have caused levels of certain algae to increase. Some algal species (especially dinoflagellates) produce toxins that can be potentially lethal to various marine species and can contaminate prey species that, if ingested in sufficient quantities, can even be deadly to top predators such as turtles, pinnipeds, and cetaceans.

Exploring the Depths: Dolphin Intelligence

How does one measure intelligence? If you were to look at brain size, the sperm whale has the largest brain in the animal kingdom (up to 7.8 kg), but then again the sperm whale is the largest odontoceti and you would expect larger animals to have larger brains. To factor in the size of animals, you have to look at the ratio between the mass of the brain and the mass of the animal. This is referred to as the encephalization quotient, or EQ. Animals with a brain size expected for an animal of their particular mass have an EQ of 1. Animals with EQ values greater than 1 have brains that are larger than expected based on their size alone. Humans have an EQ of 7.0. The smaller dolphins have EQs ranging from 3.24 to 4.56, which is equivalent to several early hominids, ancestors of modern humans (for example, *Homo habilis* had an EQ of 4.4). The sperm whale, despite having the largest brain, has a relatively low EQ (0.58), and baleen whales have EQs that are even lower (e.g., humpback whale, 0.44; blue whale, 0.21).

It should be noted, however, that EQs do not account for the high proportion of a cetacean's mass that is blubber, a tissue that does not require neurological control and thus has no brain mass dedicated to it. Likewise, the spermaceti organ is a large mass of waxy tissue that substantially increases the mass of a sperm whale while requiring little, if any, neural control. To demonstrate the importance of this, we could use EQs to rank the intelligence of different countries. We would find that, on average, the people of the United States, with a high rate of obesity, would have much lower average EQs than many other countries, because of a low body mass to brain mass ratio! In fact, if we were to rank intelligence solely by EQs, ants would be considered to be far more intelligent than humans, having one of the largest brains relative to their size of any animal on Earth.

Another consideration is that marine animals can be disproportionately large in size. The buoyant nature of water supports them and has allowed the baleen whales in particular to achieve body masses that would not be possible in terrestrial mammals, but this increase in mass has occurred in ways that do not necessarily require an increase in brain size.

Moreover, EQs do not factor into account the structure of animal brains. For example, folding in the neocortex, the outer layer of the brain in which much of the so-called higher functions of the brain occur, is complex in humans. This increases the surface area of the neocortex, which is often given as a structural indicator of the higher levels of intelligence in humans. However, cetacean brains, in particular those of the delphinids, have even more substantial folding of the neocortex than humans. Furthermore, special "spindle" neurons possessed by both humans and great apes that are believed to be involved in cognition have now also been found in humpback whales and may be present in other whales. The neurons in several whale brains are also arranged uniquely in "islands" or clusters, which are again believed to promote interconnections and speed of processing.

Thus, the brain structure of cetaceans has several indicators of greater processing ability. However, it must not be forgotten that the structure and function of cetacean brains may not be readily comparable with terrestrial mammals, such as primate and humans, because of several factors: their sensory systems are primarily acoustic; they are able to rest one side of their brains (see Chapter 4), which means there may be some duplication of neurological function across the halves; and the evolution in, and adaptation of cetaceans to, a marine environment.

Tool use by animals has been used to imply high levels of intelligence in those species, including the bottlenose dolphin (see Chapter 7). However, it is the study of linguistic ability and self-awareness that has provided perhaps the most convincing evidence that cetaceans, particularly dolphins, have high levels of intelligence. Work in these areas had limited success at first because of their strong anthropocentric basis. For example, one of the first studies to investigate cetacean linguistic abilities involved trying to teach bottlenose dolphins to mimic human speech. Researcher John Lilly built a facility in which humans lived with dolphins for 2.5 months and attempted to teach them language. The experiment was not successful, and unfortunately Lilly made several claims about language and abilities of dolphins beyond what he actually observed, which gave dolphin language studies a bad reputation for many years. Although this attempt failed, beluga whales have subsequently been found to be able to mimic some human speech sounds, and bottlenose dolphins have mimicked computer-generated sounds.

More recently, Louis Herman and his research group in Hawaii conducted some interesting research into linguistic abilities. For example, common bottlenose dolphins were taught simple sign language and a computer-generated sound language and were thus able to construct simple sentences (structured with a subject, verb, and object). As a result it was found that dolphins could understand artificial symbolic languages and, moreover, could understand simple sentences and novel combinations of words. Most interestingly, it demonstrated that dolphins could comprehend sentence structure or syntax.

Exploring the Depths: Dolphin Intelligence (*continued*)

This showed that dolphins have the potential for advanced linguistic abilities, especially when you consider the symbols used involved something unfamiliar to a cetacean—hands.

One of the most compelling pieces of evidence for cetacean intelligence is their seeming ability to comprehend how other individuals see the world, and experiments have demonstrated they have self-awareness. For example, one study had a dolphin observing two trainers. The first trainer was able to observe an event, but the second trainer was obscured from observing the event by a screen. However, the dolphin appeared to be aware that the second trainer had not seen the event and used this knowledge to its own advantage (i.e., getting extra food). This ability to understand how others perceive the environment is usually not seen in humans until approximately 4 years of age, which would at least put dolphins on a cognitive par with a human child.

Several studies investigated the ability of various species to recognize their own image in a mirror, because this ability is assumed to be a measure of self-awareness and therefore higher cognitive ability. In humans, the ability to recognize oneself in a mirror does not appear until about 24 months of age. The most usual test involves animals recognizing their image in a mirror and using that image to investigate part of their body that has been marked in some way. However, without hands or a trunk or the flexibility to touch themselves with their fins or beaks, this presents a problem for testing cetaceans. Instead, tests with bottlenose dolphins have consisted of marking or, more importantly, *not marking* the animals with zinc oxide cream or nontoxic marker pens on their bodies. The latter action consisted of going through all the motions and touching the animals in exactly the same way but leaving no mark behind. The dolphins could then inspect themselves using a mirror placed in their pools (**Figure B12.8**). At first, the dolphins would stop by the mirror when they noticed the mark. Later, the animals would swim directly to the mirror to investigate the place where they thought they had been marked, often twisting and turning to expose the proper spot. These experiments not only showed that the dolphins recognized they were being reflected in the mirrors but that they knew they could use the mirrors to investigate what had been done to them. These results indicate a significant level of self-awareness.

Despite their successes (or perhaps, more accurately, *because* of them) studies like these raise some major ethical issues about the way humans treat cetaceans. If these animals are self-aware and have cognitive abilities equal to a human child, what are the moral implications of permitting these animals to be held in captivity? Similarly, how ethical is it to allow cetaceans to be injured or killed incidentally by human activities (for example, by allowing takes under the MMPA; see Chapter 17)? Perhaps most importantly, what are the moral implications of allowing inhumane methods of directed slaughter of these animals in drive fisheries in locations such as Japan (see Chapter 15)?

FIGURE B12.8 A dolphin during a mirror self-awareness experiment.

Exploring the Depths: Culture in Cetaceans

There are many definitions of culture but most scientists consider it to be behavioral variations between groups of animals, where these behaviors are passed on and maintained by social learning. Culture effectively involves transmission of a behavior, which involves both teaching and learning by the animals concerned. Culture was previously considered to be the province of humans or, at most, higher primates. However, evidence of culture has now been identified in several cetacean species. Take killer whales in the eastern North Pacific, for instance. Killer whales in this region share many of the same calls, but each group has a number of calls unique to their group members. This is essentially a vocal dialect. Similar dialects have also been reported in sperm whales. The continuously changing, but learned, nature of humpback whale song (see Chapters 5 and 10) is another often-given example of culture in cetaceans.

Cultural transmission can occur in one of three ways. The first is a spread of novel and complex behaviors among members of the same generation, which seems to be what occurs in male humpback whales when they learn songs from other males (see Chapters 5 and 10). The second type of cultural transmission is between a mother and her calf (**Figure B12.9**). Examples of this can be found when a female killer whale teaches her offspring how to strand itself to catch elephant seal or southern sea lion pups, or when a bottlenose dolphin teaches her calf how to use sponges to protect itself when hunting for prey (see above). The third type of cultural transmission can occur more widely between any members of a group, regardless of generation. One example of this may be the vocal dialects in sperm or killer whales, although there is likely a mother–offspring component to this as well.

Exploring the Depths: Culture in Cetaceans (*continued*)

FIGURE B12.9 A humpback whale mother with her calf.

SELECTED REFERENCES AND FURTHER READING

Baird, R.W., Webster, D.L., Mahaffy, S.D., McSweeney, D.J., Schorr, G.S., & Ligon, A.D. (2008). Site fidelity and association patterns in a deep-water dolphin: rough-toothed dolphins (*Steno bredanensis*) in the Hawaiian Archipelago. *Marine Mammal Science* 24: 535–553.

Beasley, I., Robertson, K.M., & Arnold, P. (2005). Description of a new dolphin, the Australian snubfin dolphin *Orcaella heinsohni* sp. n. (Cetacea, Delphinidae). *Marine Mammal Science* 21: 365–400.

Birkhead, T.R., & Hunter, F.M. (1990). Mechanisms of sperm competition. *Trends in Ecology & Evolution* 5: 48–52.

Brakes, P., & Simmonds, M.P. (2011). *Whales & Dolphins: Cognition, Culture, Conservation and Human Perceptions.* Earthscan, London, UK and Washington, DC.

Buckland, S.T., Smith, T., & Cattanach, K.L. (1992). Status of small cetacean populations in the Black Sea: a review of current information and suggestions for future research. *Reports of the International Whaling Commission* 42: 513–516.

Caballero, S., Trujillo, F., Vianna, J.A., Barrios-Garrido, H., Montiel, M.G., Beltrán-Pedreros, S., Marmontel, M., Santos, M.C., Rossi-Santos, M., Santos, F.R., & Baker, C.S. (2007). Taxonomic status of the genus *Sotalia*: species level ranking for "Tucuxi" (*Sotalia fluviatilis*) and "Costero" (*Sotalia guianensis*) dolphins. *Marine Mammal Science* 23: 358–386.

Charlton-Robb, K., Gershwin, L., Thompson, R., Austin, J., Owen, K., & McKechnie, S. (2011). A new dolphin species, the Burrunan dolphin *Tursiops australis* sp. nov., endemic to southern Australian coastal waters. *PLOS One* 6 (9) e24047: 1–17.

Dahlheim, M.E., Schulman-Janiger, A., Black, N., Ternullo, R., Ellifrit, D., & Balcomb, K.C. (2008). Eastern temperate North Pacific offshore killer whales (*Orcinus orca*): occurrence, movements, and insights into feeding ecology. *Marine Mammal Science* 24: 719–729.

Da Silva, V.M.F., & Martin, A.R. (2000). The status of the freshwater form of tucuxi *Sotalia fluvialis* (Gervais,

1853): a review of available information. Paper presented to the Scientific Committee at the 52nd Meeting of the International Whaling Commission, June 11–28 2000, Australia.

Dam, M., & Bloch, D. (2000). Screening of mercury and persistent organochlorine pollutants in long-finned pilot whale (*Globicephala melas*) in the Faroe Islands. *Marine Pollution Bulletin* 40: 1090–1099.

DeMaster D. P., & Drevenak, J. K. (1988). Survivorship patterns in three species of captive cetaceans. *Marine Mammal Science* 4: 297–311.

Dunn, D.G., Barco, S.G., Pabst, D.A., & McLellan, W.A. (2002). Evidence for infanticide in bottlenose dolphins of the western North Atlantic. *Journal of Wildlife Diseases* 38: 505–510.

Durban, J.W., & Pitman, R.L. (2011). Antarctic killer whales make rapid, round-trip movements to subtropical waters: evidence for physiological maintenance migrations? *Biology Letters*. Published online (doi:10.1098/rsbl.2011.0875). Retrieved November 1, 2011.

Fisher, S.J., & Reeves, R.R. (2005). The global trade in live cetaceans: implications for conservation. *International Journal of Wildlife Law & Policy* 8: 315–340.

Foote, A.D., Newton, J., Piertney, S.B., Willerslev, E., & Gilbert, M.T.P. (2009). Ecological, morphological and genetic divergence of sympatric North Atlantic killer whale populations. *Molecular Ecology* 18: 5207–5217.

Ford, J.K.B. (1991). Vocal traditions among resident killer whales (*Orcinus orca*) in coastal waters of British Columbia. *Canadian Journal of Zoology* 69: 1454–1483.

Ford, J.K.B. (2009a). Dialects. In: *Encyclopedia of Marine Mammals* (Ed. W.F. Perrin, B. Würsig, & J.G.M. Thewissen), pp. 310–311. Academic Press, New York.

Ford, J.K.B. (2009b). Killer whale *Orcinus orca*. In: *Encyclopedia of Marine Mammals*, 2nd ed. (Ed. W.F. Perrin, B. Würsig, & J.G.M. Thewissen), pp. 640–657. Academic Press, New York.

Ford, J.K.B., Ellis, G.M., & Balcomb, K.C. (2000). *Killer Whales: The Natural History and Genealogy of* Orcinus orca *in the Waters of British Columbia and Washington*. UBC Press and University of Washington Press, Vancouver and Seattle.

Ford, J.K.B., Ellis, G.M., Barrett-Lennard, L.G., & Morton, A.B. (1998). Dietary specialization in two sympatric populations of killer whales (*Orcinus orca*) in coastal British Columbia and adjacent waters. *Canadian Journal of Zoology* 76: 1456–1471.

Gannier, A., & West, K.L. (2005). Distribution of the rough-toothed dolphin (*Steno bredanensis*) around the Windward Islands (French Polynesia). *Pacific Science* 59: 17–24.

Götz, T., Verfuß, U.K., & Schnitzler, H.U. (2006). "Eavesdropping" in wild rough-toothed dolphins (*Steno bredanensis*)? *Biology Letters* 2: 5–7.

Hale, P.T., Barreto, A.S., & Ross, G.J.B. (2000). Comparative morphology and distribution of the *aduncus* and *truncatus* forms of bottlenose dolphin *Tursiops* in the Indian and Western Pacific Oceans. *Aquatic Mammals* 26: 101–110.

Herman, L.M. (1986). Cognition and language competencies of bottlenosed dolphins. In: *Dolphin Cognition and Behavior: A Comparative Approach* (Ed. R.J. Schusterman, J.A. Thomas & F.G. Woods), pp. 221–252. Lawrence Erlbaum Associates, Hillsdale, NJ.

Heyning, J.E., & Perrin, W.F. (1994). Evidence for two species of common dolphins (genus *Delphinus*) from the eastern North Pacific. *Natural History Museum LA County Contributions in Science* 442: 1–35.

Hoelzel, A.R., Dahlheim, M., & Stern, S.J. (1998). Low genetic variation among killer whales (*Orcinus orca*) in the eastern North Pacific and genetic differentiation between foraging specialists. *Journal of Heredity* 89: 121–128.

Hoelzel, A.R., Potter, C.W., & Best, P. (1998). Genetic differentiation between parapatric "nearshore" and "offshore" populations of the bottlenose dolphin. *Proceedings of the Royal Society of London B* 265: 1–7.

Hof, P.R., & Van der Gucht, E. (2007). Structure of the cerebral cortex of the humpback whale, *Megaptera novaeangliae* (Cetacea, Mysticeti, Balaenopteridae). *Anatomical Record* 290: 1–31.

Jefferson, T.A. (2000). Population biology of the Indo-Pacific hump-backed dolphin in Hong Kong waters. *Wildlife Monographs* 144:1–65.

Jefferson, T.A. (2009). Rough-toothed dolphin: *Steno bredanensis*. In: *Encyclopedia of Marine Mammals*, 2nd ed. (Ed. W.F. Perrin, B. Würsig, & J.G.M. Thewissen), pp. 990–992. Academic Press, New York.

Jefferson, T.A., & Hung, S.K. (2004). A review of the status of the Indo-Pacific humpback dolphin (*Sousa chinensis*) in Chinese waters. *Aquatic Mammals* 30: 149–158.

Jefferson, T.A., Hung, S.K., & Lam, P.K.S. (2006). Strandings, mortality and morbidity of Indo-Pacific humpback dolphins in Hong Kong, with emphasis on the role of environmental contaminants. *Journal of Cetacean Research and Management* 8: 181–193.

Jefferson, T.A., & Karczmarski, L. (2001). *Sousa chinensis*. *Mammalian Species* 655: 1–9.

Jefferson, T.A., & Van Waerebeek, K. (2002). The taxonomic status of the nominal dolphin species *Delphinus tropicalis* van Bree, 1971. *Marine Mammal Science* 18: 787–818.

Jerison, H.J. (1973). *Evolution of the Brain and Intelligence*. New York: Academic Press.

Kasuya, T. (2007). Japanese whaling and other cetacean fisheries. *Environmental Science & Pollution Research* 10: 39–48.

Kingston, S.E., & Rosel, P.E. (2004). Genetic differentiation among recently diverged Delphinid taxa determined using AFLP markers. *Journal of Heredity* 95: 1–10.

Klinowska, M. (1988). Cetacean "navigation" and the geomagnetic field. *Journal of Navigation* 41: 52–71.

Krützen, M., Mann, J., Heithaus, M.R., Connor, R.C., Bejder, L., & Sherwin, W.B. (2005). Cultural transmission of tool use in bottlenose dolphins. *Proceedings of the National Academy of Science USA* 102: 8939–8943.

Lahvis, G.P., Wells, R.S., Kuehl, D.W., Stewart, J.L., Rhinehart, H.L., & Via, C.S. (1995). Decreased lymphocyte responses in free-ranging bottlenose dolphins (*Tursiops truncatus*) are associated with increased concentrations of PCBs and DDT in peripheral blood. *Environmental Health Perspectives* 103: 67–72.

Leatherwood, S.J., & Reeves, R.R. (1990). *The Bottlenose Dolphin*. Academic Press, San Diego, CA.

Leduc, R., Perrin, W., & Dizon, E. (1998). Phylogenetic relationships among the delphinid cetaceans based on full cytochrome b sequences. *Marine Mammal Science* 15: 619–648.

Lilly, J.C. (1961). *Man and Dolphin*. Doubleday Press, New York.

Lilly, J.C. (1962). Vocal behaviour of the bottlenose dolphin. *Proceedings of the American Philosophical Society* 106: 520–529.

Lusseau D., & Newman M.E.J. (2004). Identifying the role that animals play in their social networks. *Proceedings of the Royal Society of London B* 271: 477–481.

MacLeod, C.D. (2009). Global climate change, range changes and potential implications for the conservation of marine cetaceans: a review and synthesis. *Endangered Species Research* 7: 125–136.

MacLeod, C.D., Bannon, S.M., Pierce, G.J., Schweder, C., Learmonth, J.A., Reid, R.J., & Herman, J.S. (2005). Climate change and the cetacean community of northwest Scotland. *Biological Conservation* 124: 477–483.

MacLeod, C.D., Weir, C.R., Santos, M.B., & Dunn, T.E. (2008). Temperature-based summer habitat partitioning between whitebeaked and common dolphins around the United Kingdom and Republic of Ireland. *Journal of the Marine Biological Association UK* 88:1193–1198.

Mann, J., Sargeant, B.L., Watson-Capps, J.J., Gibson, Q.A., Heithaus, M.R., Connor, R.C., & Patterson, E. (2008). Why do dolphins carry sponges? *PLoS ONE* 3: e3868.

Marino, L. (2009). Brain size evolution. In: *Encyclopedia of Marine Mammals*, 2nd ed. (Ed. W.F. Perrin, B. Würsig, & J.G.M. Thewissen), pp. 149–152. Academic Press, New York.

Marten, K., & Psarakos, S. (1995). Evidence of self-awareness in the bottlenose dolphin (*Tursiops truncatus*). In: *Self-Awareness in Animals and Humans: Developmental Perspectives* (Ed. S.T. Parker, R.W. Mitchell, & M.L. Boccia), pp. 361–379. Cambridge University Press, Cambridge, UK.

Matkin, C., Ellis, G., Saulitis, E., Barrett-Lennard, L., & Matkin, D. (1999). *Killer Whales of Southern Alaska*. North Gulf Oceanic Society, Homer, AK.

May-Collado, L., & Agnarsson, I. (2006). Cytochrome *b* and Bayesian inference of whale phylogeny. *Molecular Phylogenetics & Evolution* 38: 344–354.

Mead, J.G., & Potter. C.W. (1995). Recognizing two populations of the bottlenose dolphin (*Tursiops truncatus*) off the Atlantic coast of North America: morphological and ecological considerations. *IBI Reports* 5: 31–44.

Monteiro, N.C., Alves, J.T.T., Avila, F.J.C., Campos, A.A., Costa, A.F., Silva, C.P.N., & Furtado, N.M.A.A. (2000). Impact of fisheries on the tucuxi (*Sotalia fluviatilis*) and rough-toothed dolphin (*Steno bredanensis*) populations off Ceara state, northeastern Brazil. *Aquatic Mammals* 26: 49–56.

Natoli, A., Cañadas, A., Peddemors, V.M., Aguilar, A., Vaquero, C., Fernández-Piqueras, P., & Hoelzel, A.R. (2006). Phylogeography and alpha taxonomy of the common dolphin (*Delphinus* sp.). *Journal of Evolutionary Biology* 19: 943–954.

Natoli, A., Peddemors, V.M., & Hoelzel, A.R. (2004). Population structure and speciation in the genus *Tursiops* based on microsatellite and mitochondrial DNA analyses. *Journal of Evolutionary Biology* 17: 363–375.

Olesiuk, P.F., Bigg, M.A., & Ellis, G.M. (1990). Life history and population dynamics of resident killer whales (*Orcinus orca*) in the coastal waters of British Columbia and Washington State. *Reports of the International Whaling Commission* (Special Issue) 12: 209–243.

Olesiuk, P.F., Ellis, G.M., & Ford, J.K.B. (2005). *Life History and Population Dynamics of Northern Resident Killer Whales* (Orcinus orca) *in British Columbia*. CSAS Research Document 2005/045. Fisheries and Oceans Canada, Nanaimo.

Olson, P.A. (2009). Pilot whales *Globicephala melas* and *G. macrorhynchus*. In: *Encyclopedia of Marine Mammals*, 2nd ed. (Ed. W.F. Perrin, B. Würsig & J.G.M. Thewissen), pp. 847–852. Academic Press, New York.

Parra, G., Azuma, C., Preen, A.R., Corkeron, P. J., & Marsh, H. (2002). Distribution of Irrawaddy dolphins, *Orcaella brevirostris*, in Australian waters. *Raffles Bulletin of Zoology* 10: 141–154.

Parsons, E. (1993). Some fin to think about. *BBC Wildlife* 11(3): 93.

Parsons, E.C.M. (2004a). The potential impacts of pollution on humpback dolphins—with a case study on the Hong Kong population. *Aquatic Mammals* 30: 18–37.

Parsons, E.C.M. (2004b). The behaviour and ecology of the Indo-Pacific humpback dolphin (*Sousa chinensis*). *Aquatic Mammals* 30: 38–55.

Parsons, E.C.M., Dolman, S., Wright, A.J., Rose, N.A., & Burns, W.C.G. (2008). Navy sonar and cetaceans: just how much does the gun need to smoke before we act? *Marine Pollution Bulletin* 56: 1248–1257.

Patterson, I.A., Reid, R.J., Wilson, B., Grellier, K., Ross, H.M., & Thompson, P.M. (1998). Evidence for infanticide in bottlenose dolphins: an explanation for violent interactions with harbour porpoises? *Proceedings of the Royal Society of London B* 265: 1167–1170.

Perrin, W.F., Donovan, G.P., & Barlow, J. (1994). *Gillnets and Cetaceans. Reports of the International Whaling*

Commission. Special Issue 15. International Whaling Commission, Cambridge, UK.

Pitman, R.L., & Ensor, P. (2003). Three forms of killer whales (*Orcinus orca*) in Antarctic waters. *Journal of Cetacean Research & Management* 5: 131–139.

Pitman, R.L., Perryman, W.L., LeRoi, D., & Eilers, E. (2007). A dwarf form of killer whale in Antarctica. *Journal of Mammalogy* 88: 43–48.

Pitman, R.L., & Stinchcomb, C. (2002). Rough-toothed dolphins (*Steno bredanensis*) as predators of mahimahi (*Coryphaena hippurus*). *Pacific Science* 56: 447–450.

Pryor, K.W., Haag, R., & O'Reilly, J. (1969). The creative porpoise: training for novel behavior. *Journal of Experimental and Analytical Behaviour* 12: 653–661.

Rayment, W., Dawson, S.M., & Slooten, E. (2010). Seasonal changes in distribution of Hector's dolphin at Banks Peninsula, New Zealand: implications for protected area design. *Aquatic Conservation* 20: 106–116.

Reiss, D., & Marino, L. (2001). Mirror self-recognition in the bottlenose dolphin: a case for cognitive convergence. *Proceedings of the National Academy of Sciences* 98: 5937–5942.

Rendell, L., & Whitehead, H. (2001). Culture in whales and dolphins. *Behavioral and Brain Sciences* 24: 309–382.

Reynolds, J.E., Wells, R.S., & Eide, S.D. (2000). *The Bottlenose Dolphin: Biology and Conservation*. University Press of Florida, Gainesville.

Richards, D.G., Wolz, J.P., & Herman, L.M. (1984). Vocal mimicry of computer generated sounds and vocal labeling of objects by a bottlenose dolphin, *Tursiops truncatus*. *Journal of Comparative Psychology* 98: 10–28.

Ritter, F. (2002). Behavioural observations of rough-toothed dolphins (*Steno bredanensis*) off La Gomera, Canary Islands (1995–2000), with special reference to their interactions with humans. *Aquatic Mammals* 28: 46–59.

Ross, A., & Isaac, S. (2004). *The Net Effect? A Review of Cetacean Bycatch in Pelagic Trawls and Other Fisheries in the North-East Atlantic*. Whale & Dolphin Conservation Society, Chippenham, UK. Retrieved 1 November 2011 from http://www.wdcs.org/publications.php.

Ross, H.M., & Wilson, B. (1996). Violent interactions between bottlenose dolphins and harbour porpoises. *Proceedings of the Royal Society of London B* 263: 283–286.

Ross, P.S., Ellis, G.M., Ikonomou, M.G., Barrett-Lennard, L.G., & Addison, R. (2000). High PCB concentrations in free-ranging Pacific killer whales, *Orcinus orca*: effects of age, sex and dietary preference. *Marine Pollution Bulletin* 40: 504–515.

Rozanova, E.I., Alekseev, A.Y., Abramov, A.V., Rassadkin Y.N., & Shestopalov, A.M. (2007). Death of the killer whale *Orsinus* [sic] *orca* from bacterial pneumonia in 2003. *Russian Journal of Marine Biology* 33: 321–323.

Schwacke, L.H., Voit, E.O., Hansen, L.J., Wells, R.S., Mitchum, G.B., Hohn, A.A., & Fair, P.A. (2002). Probabilistic risk assessment of reproductive effects of polychlorinated biphenyls on bottlenose dolphins (*Tursiops truncatus*) from the southeast United States coast. *Environmental Toxicology and Chemistry* 21: 2752–2764.

Secchi, E.R., & Vaske, T.J. (1998). Killer whale (*Orcinus orca*) sightings and depredation on tuna and swordfish longline catches in southern Brazil. *Aquatic Mammals* 24:117–122.

Simmonds, M.P. (2006). Into the brains of whales. *Applied Animal Behaviour* 100: 103–106.

Slooten, E., & Dawson, S.M. 2010. Assessing the effectiveness of conservation management decisions: Likely effects of new protection measures for Hector's dolphin. *Aquatic Conservation* 20: 334–347.

Small, J., & DeMaster, D.P. (1995). Survival of five species of captive marine mammals. *Marine Mammal Science* 11: 209–226.

Smith, B.D., Shore, R.G., & Lopez A. (2007). *Status and Conservation of Freshwater Populations of Irrawaddy Dolphins*. Wildlife Conservation Society, New York.

Smolker, R.A., Richards, A., Connor, R., Mann, J., & Berggren, P. (1997). Sponge-carrying by Indian Ocean bottlenose dolphins: possible tool-use by a delphinid. *Ethology* 103: 454–465.

Soulsbury, C.D., Graziella Iossa, G., & Harris, S. (2008). *The Animal Welfare Implications of Cetacean Deaths in Fisheries*. School of Biological Sciences, University of Bristol, UK. Retrieved November 1, 2011 from http://www.wdcs.org/publications.php.

Southall, B.L., Braun, R., Gulland, F.M.D., Heard, A.D., Baird, R.W., Wilkin, S.M., & Rowles, T.K., (2006). Hawaiian melon-headed whale (*Peponocephala electra*) mass stranding event of 3–4 July 2004. NOAA Technical Memorandum NMFS-OPR-31. Office of Protected Resources, NOAA, Silver Spring, MD.

Stacey, P.J., & Leatherwood, S. (1997). The Irrawaddy dolphin, *Orcaella brevirostris*: a summary of current knowledge and recommendations for conservation action. *Asian Marine Biology* 14: 195–214.

Visser, I.N. (1998). Prolific body scars and collapsing dorsal fins on killer whales (*Orcinus orca*) in New Zealand waters. *Aquatic Mammals* 24: 71–81.

Visser, I.N. (1999). Propeller scars on and known home range of two orca (*Orcinus orca*) in New Zealand waters. *New Zealand Journal of Marine and Freshwater Research* 33: 635–642.

Wang, J.Y., Chou, L.S., & White, B.N. (1999). Mitochondrial DNA analysis of sympatric morphotypes of bottlenose dolphins (genus: *Tursiops*) in Chinese waters. *Molecular Ecology* 8: 1603–1612.

Wang, J.Y., Chou, L.S., & White, B.N. (2000a). Osteological differences between two sympatric forms of bottlenose dolphins (genus *Tursiops*) in Chinese waters. *Journal of Zoology* 252: 174–162.

Wang, J.Y., Chou, L.S., & White, B.N. (2000b). Differences in the external morphology of two sympatric species of bottlenose dolphins (Genus *Tursiops*) in the waters of China. *Journal of Mammalogy* 81: 1159–1165.

Wang, J.Y, Yang, S.-C., Hung, S.K., & Jefferson, T.A. (2007). Distribution, abundance and conservation status of the eastern Taiwan Strait population of Indo-Pacific humpback dolphins, *Sousa chinensis*. *Mammalia* 71: 157–165.

Webster, R.A., Dawson, S.M., & Slooten, E. (2009). Evidence of sex segregation in Hector's dolphin (*Cephalorhynchus hectori*). *Aquatic Mammals* 35: 212–219.

Wells, R.S., & Scott, M.D. (2002). Common bottlenose dolphin (*Tursiops truncatus*). In: *Encyclopedia of Marine Mammals*, 2nd ed. (Ed. W.F. Perrin, B. Würsig, & J.G.M. Thewissen), pp. 249–255. Academic Press, New York.

Wells, R.S., Tornero, V., Borrell, A., Aguilar, A., Rowles, T.K., Rhinehart, H.L., Hofmann, S., Jarman, W.M., Hohn, A.A., & Sweeney, J. C. (2005). Integrating life-history and reproductive success data to examine potential relationships with organochlorine compounds for bottlenose dolphins (*Tursiops truncatus*) in Sarasota Bay, Florida. *Science of the Total Environment* 349: 106–119.

Whitehead, H. (2009). Culture in whales and dolphins. In: *Encyclopedia of Marine Mammals*, 2nd ed. (Ed. W.F. Perrin, B. Würsig, & J.G.M. Thewissen), pp. 292–294. Academic Press, New York.

Whitehead, H., Rendell, L., Osborne, R.W., & Würsig, B. (2004). Culture and conservation of non-humans with reference to whales and dolphins: review and new directions. *Biological Conservation* 120: 431–441.

Williams, R., Bain, D.E., Ford, J.K.B., & Trites, A.W. (2002). Behavioural responses of male killer whales to a "leapfrogging" vessel. *Journal of Cetacean Research & Management* 4: 305–310.

Williams, R., Trites, A.W., & Bain. D.E. (2002). Behavioural responses of killer whales (*Orcinus orca*) to whale-watching boats: opportunistic observations and experimental approaches. *Journal of Zoology* 256: 255–270.

Woodley, T. H., Hannah, J. L., & Lavigne, D. M. (1997). A comparison of survival rates for free-ranging bottlenose dolphins (*Tursiops truncatus*), killer whales (*Orcinus orca*), and beluga whales (*Delphinapterus leucas*). Technical Report No. 97-02. International Marine Mammal Association, Guelph, Ontario.

Worthy, G.A.J., & Hickie, J.P. (1986). Relative brain size of marine mammals. *American Naturalist* 128: 445–459.

Wright, A.J. (2005). Lunar cycles and sperm whales (*Physeter macrocephalus*) strandings on the North Atlantic coastlines of the British Isles and Eastern Canada. *Marine Mammal Science* 21: 145–149.

Würsig, B. (2009). Intelligence and cognition. In: *Encyclopedia of Marine Mammals*, 2nd ed. (Ed. W.F. Perrin, B. Würsig, & J.G.M. Thewissen), pp. 616–623. Academic Press, New York.

Yurk, H., Barrett-Lennard, L., Ford, J.K.B., & Matkin, C.O. (2002). Cultural transmission within maternal lineages: vocal clans in resident killer whales in southern Alaska. *Animal Behaviour* 63: 1103–1119.

Conservation

A humpback whale entangled in fishing gear.

PART
II
Conservation

CHAPTER 13

Humans and Marine Mammals

CHAPTER OUTLINE

Marine Mammals in Folklore and History
  Exploring the Depths: Narwhals and Unicorns

Marine Mammal Hunts and Early Whaling
  Exploring the Depths: Whales and Whaling in the United Kingdom
  Commercial Whaling: The Early Years

  Exploring the Depths: Life as a Whaler
  Exploring the Depths: The Essex

Changing Attitudes Toward Marine Mammals
  Exploring the Depths: Public Opinion, Whaling, and Seal Culls

Selected References and Further Reading

FIGURE 13.1 One of the carvings at Bangudae, Korea, showing a whale.

Humans have long had a fascination with marine mammals, particularly cetaceans. They appear in many legends and folk tales and have also been frequent artistic muses. Early rock carvings found in Norway dating back approximately 9,000 years ago depict cetaceans. Likewise, rock etchings in Bangudae near Ulsan, Korea, carved 4,000 years ago, feature a variety of cetaceans, including what appear to be Dall's porpoises and gray whales (**Figure 13.1**).

There has also been a long history of using cetaceans as resources: their meat for food; their blubber for oil; and their baleen, teeth, and bones as materials to construct clothes, jewelry, tools, or even buildings. Many situations in which early humans used whales as a resource were probably opportunistic, from when a carcass stranded on a coast. However, a recent bone carving found in the Russian Arctic appears to depict a canoe full of hunters pursuing a whale, suggesting humans may have been actively hunting whales in the region for 3,000 years. This chapter describes the changing role of marine mammals in human eyes, from creatures of legend, to resources to be hunted, and finally to marine mammals as stars of stage and screen.

Marine Mammals in Folklore and History

In the Western world possibly one of the most famous tales of a whale is the biblical (and the Koran and Torah) story of Jonah and the whale (*Jonah* 1:17) (**Figure 13.2**). The Old Testament states that Jonah was instructed by God to preach to gentiles in Nineveh (modern Assyria). However, Jonah rebelled and boarded a ship in Joppa. During Jonah's voyage there was a terrible storm and Jonah was swallowed by a great fish, sea monster, or, as it is generally described in

FIGURE 13.2 A medieval image of Jonah and the whale.

more recent versions of the Bible, a whale (the word used in early translations of the Bible could refer to any large sea creature). The whale carried Jonah for 3 days and nights in its stomach before eventually regurgitating him on land.

Also in the Bible, the creature called Leviathan (*Job* 41) is often depicted as a monstrous whale.

Another marine mammal is mentioned in some editions of the Bible. The covering of both the Tabernacle (*Exodus* 26:14) and the Ark of the Covenant (*Numbers* 4) are said to be made of sea cow or dugong hide; the thick hides of dugongs were often used to make tent materials and sandals in the Middle East.

Whales and dolphins feature prominently in the historical culture of the northern Mediterranean. The ancient Phoenicians were believed to have hunted sperm whales in the Mediterranean as early as 1,000 BC. Probably one of the most famous images from ancient Mediterranean and European history related to cetaceans is the mural of dolphins (probably short-beaked common dolphins as they have yellow markings on their sides) in the Queen's Room in the Minoan capital of Knossos in Crete (**Figure 13.3**). These dolphins were painted 3,600 years ago and are astonishingly lifelike.

The images of dolphins appear frequently in subsequent ancient Greek art, possibly because of the role that dolphins played in Greek mythology. For example, when

FIGURE 13.3 The Queen's room dolphin mural from the palace in Knossos, Crete.

the god Apollo founded the oracle at Delphi he assumed the form of a dolphin and carried priests to this location on his back. Moreover, the hunter Orion was carried into the stars on the back of a dolphin when he became a constellation. The constellation Delphinus, the dolphin, was placed in the sky by Poseidon (the Greek god of the sea), according to legend, as a reward for Delphinus playing matchmaker between Poseidon and the nerid (sea nymph) Amphitrite who, although reluctant at first, became Poseidon's wife. Another constellation is *cetus*, sometimes called the whale, although in ancient Greek *cetus* referred to any large marine species: fish, whale, or even sea monster.

Another legend tells of how Taras, who was said to be the son of Poseidon, was shipwrecked, but a dolphin sent by his father saved him and brought him to shore, allowing Taras to found the city that bore his name. In fact, many legends tell of dolphins that came to aid those in peril at sea. Phalanthos, the leader of a group of Spartan refugees, was rescued by a dolphin when he also became shipwrecked off the coast of southern Italy. Another tale has a dolphin saving Arion, a one-time royal court musician. Arion was traveling home with the wealth he had amassed at court as a favored entertainer. Unfortunately, the sailors on the ship tried to threaten and murder their wealthy passenger. Arion's final wish was to sing a final song, which the sailors permitted, and on its completion he threw himself into the sea. However, Arion was rescued by a dolphin, who had become enchanted by his final song, and the dolphin bore him away to land and safety.

In 180 AD the Roman poet Oppian wrote that the ancient Greeks believed dolphins had once been humans, but the god Dionysus (the Greek god of wine) had changed them into marine creatures. Because they were once humans, they were believed to retain human thoughts and undertake human deeds. Oppian also wrote that dolphins stranded themselves so that mortal humans could bury them and remember their friendship with dolphins. Possibly because of this and other legends, the ancient Greeks held dolphins in high regard. This is exemplified in the legal code in which killing a dolphin was considered to be equal to killing a human, and both crimes were punishable by death.

In terms of understanding the nature of cetaceans, one of the first to study and write about cetaceans was Aristotle (384–322 BC). Although more famous for his philosophical works, Aristotle was one of the first marine biologists and in his treatise, *Historia Animalium*, recognized several distinct varieties of cetaceans, including dolphins, killer whales, sperm whales, and right whales. He also noted that cetaceans were all air-breathing and were not fish. Moreover, he observed a distinction between toothed and baleen whales, with the latter possessing "hairs that resemble hog bristles" instead of teeth.

Several hundred years later, before his untimely demise during the eruption of Mount Vesuvius, the Roman writer Pliny the Elder (23–79 AD) noted in his work *Naturalis Historia* that whales migrate to different places at different times of the year and that they were sometimes attacked by killer whales. However, unlike Aristotle, Pliny thought cetaceans were fish. Pliny's work effectively remained the main factual text on cetaceans until the 17th century. Although Pliny's work had factual observations, he also peppered his writing with reported anecdotes and tales, including a story of a peasant boy and a solitary dolphin called Simo. The boy used to feed Simo, and in return the dolphin would give the boy rides on his back across the bay. However, the boy became ill and died. The dolphin, however, returned to their meeting place for many days until Simo also died "of a broken heart."

Moving to another part of Europe, marine mammals have been culturally significant in Scotland since at least the Iron Age. Throughout eastern Scotland you can find carvings on standing stones of creatures referred to as "the Pictish beast" (**Figure 13.4**). However, the dolphin-like beak, the triangular shape on their backs like a dorsal fin, and the rounded, flipper-like limbs indicate the Pictish beast highly resembles a bottlenose dolphin (or possibly, as some researchers have suggested, a beaked whale). These stones carved between 300 and 842 AD are found in an area of eastern Scotland adjacent to the Moray Firth and along the coast of northeastern Scotland to Aberdeen that certainly coincides with the present day distribution of bottlenose dolphins in this area.

Seals are also featured in Scottish culture. Selkies, or seal people, were said to be able to take the form of humans but upon donning their magical seal skin would turn into seals. There are many tales of Selkie men and women leaving the seal people for, often tragic, romances with humans on the land. Sometimes beautiful seal maidens were

FIGURE 13.4 An artist's representation of the "Pictish beast."

captured and coerced by human males who found and took their seal skins, threatening to destroy them. The founder of clan Macfie was said to have married a Selkie woman, and thus their descendants are said to carry seal blood to this day. Similar legends are also found in Ireland, England, Iceland, Norway, and the Faroe islands.

Despite these beliefs, Scottish attitudes toward seals were not necessarily reverential in the Middle Ages. For example, the monks of the Isle of Iona Monastery, which was founded by the famous Saint Columba, harvested seals for their meat. Because seals were considered to be "fish," they could be eaten on holy days when eating "meat" was forbidden. Saint Columba had several encounters with cetaceans as well, according to legend. For example, when he and his monks were traveling from the Isle of Iona to the Isle of Tiree they encountered a "monster of the deep," which turned out to be "a whale of extraordinary size, which rose like a mountain above the water, its jaws open to show an array of teeth" (pp. 125–126 in Sharpe [1995]; see Selected References and Further Reading). Probably the most famous encounter of Saint Columba's was with a cetos, a sea monster or whale, in the River Ness near the Moray Firth. The sea creature was swimming toward a man who had fallen in the water and was threatening to eat him, but Saint Columba's words caused the beast to back away. There may be some truth in this as the Moray Firth is inhabited today not only by bottlenose dolphins but also by minke whales. However, in more modern telling, "cetos" was translated as "monster" and instead of being seen in the River Ness the creature was taken to have been encountered in Loch Ness. Thus, a major boost to the Scottish tourism industry was born!

Another Gaelic saint, St. Brendan, the Abbot of Clonfert in the 6th century, had another famous cetacean encounter. He encountered a "fish of enormous size … spouting foam from its nostrils," which followed and threatened St. Brendan's vessel, until it was itself attacked by another sea monster. St. Brendan was also said to have landed on an island, and when he tried to light a fire, the island started to sink, revealing itself to be a monstrous whale (**Figure 13.5**). A similar whale island is encountered by the Middle-Eastern hero Sinbad in the "Thousand and One Nights" (or "Arabian Nights") stories. Such stories might also have a grain of truth; sleeping right whales may look like small black rocks, and people have been known to climb onto the back of a sleeping right whale after approaching quietly in a canoe or kayak, that is, until the whale wakes up!

The Nordic peoples have also had a long association with marine mammals, including hunting (see Marine Mammal Hunts and Early Whaling). By the 9th century

FIGURE 13.5 Image of St. Brendan on the back of a whale.

Norwegians had identified 23 different species of cetacean. Moreover, the Icelandic text *Speculum Regale* (*The Mirror of Royalty*), written in the 13th century, highlights the profusion of whales off the Icelandic coast, noting species such as sperm whales, narwhals, and species called "orcs," which were aggressive and would attack other marine mammals (i.e., killer whales). In fact, *Speculum Regale* went so far as to say that these marine mammals were the only interesting sight Iceland had to offer.

There are many North American legends related to whales as well. The Haida people of northwest America told of an evil ocean people that used killer whales as canoes. To protect themselves from these raiders, the Haida turned one of their chiefs into a killer whale to protect them. Killer whales also feature prominently in the culture of the Tlingit people. Killer whales appear on totems, masks, carvings, and textiles. The Tlingit did not hunt killer whales because they too believed orcas protected their people.

In the subpolar regions of North America the Inuit people have hunted marine mammals for several thousand years. Although the Inuit have recently been associated with cetacean hunts due to their battles for hunting quotas at the International Whaling Commission, seal meat is the most important component (and often largest part) of an Inuit diet (see Marine Mammal Hunts and Early Whaling). In fact, the Inuit diet may include many marine mammals (in addition to fish and some local plants), such as harp seal, harbor seal, ringed seal, bearded seal, walrus, beluga whale, and polar bear. Hunting has thus become a huge part of Inuit culture too. Traditionally, when an Inuit boy killed his first seal or caribou a feast was held. Similarly, a number of ceremonies have historically included the returning of the skull of the hunted whale to the sea to ensure the immortality and reincarnation of the whale and thus future hunting success.

The Inuit legend of Sedna, the goddess of the sea, also features marine mammals. Sedna as a young girl refused to accept and marry a number of possible suitors and instead fell in love with and married a dog. Sedna's rejected suitors in their fury kidnapped her, whisked her away in a boat, and tried to throw her into the sea. However, Sedna grabbed hold of the side of the boat to stop herself from drowning. The wicked suitors, however, grabbed a blade and chopped off her fingers, which fell into the sea and transformed into pinnipeds and cetaceans.

In some regions of South America, Amazon River dolphins, or boto, are said to be able to transform into human men and come onto land to woo and seduce young women during the night and during fiestas. Their gaze is said to be able to enchant young women and make them powerless or even paralyze them. This belief is so strong that many "unexplained" pregnancies and resulting children are attributed to the dolphins. As a result of this belief, unfortunately, some boto are hunted; their genitals are used in love potions and their eyes as a fetish to increase the holder's attractiveness to the opposite sex.

In Australia a dog-faced sea monster called the bunyip has been linked to sightings of dugongs or possibly encounters of transient pinnipeds swimming into unusually northern waters. There has been a long history of hunting dugongs by the indigenous peoples of Australia, which continues today. However, dolphins have largely been considered to be sacred animals, and killing one would evoke the wrath of the gods. The sacred nature of dolphins is illustrated by the activities in Mornington Island, northern Australia, where one tribe of Australian Aborigines has a shaman who "speaks" to bottlenose dolphins in coastal waters, thereby ensuring the happiness and good fortune of the tribe.

Also in the southern hemisphere, cetaceans, especially sperm and right whales, feature prominently in the culture of the indigenous Maori people. The Nagāti Porou people of northeastern New Zealand tell of their ancestor Paikea being rescued and carried across the Pacific to New Zealand on a whale, a legend that is recounted in the 2002 movie *Whale Rider*. The Nagāti Tahu people are said to be descended from a relative of Paikea and consider sperm whales to be *taonga* (treasure); stranded animals were prayed over and the lower jaw removed for ceremonial carving and placement. Some members of the Nagāti Tahu are still involved with whales through whale watching activities in New Zealand, including sperm whale watching in Kaikoura.

A final example of cetaceans being held in high and special regard by humans can be found in Vietnam. A legend tells of members of the Vietnamese royal family (of the Nguyen Dynasty) who were at sea when a storm sank the royal yacht. However, it is said the royal family survived this disaster because dolphins pushed the shipwrecked royals to shore. Since then, cetaceans were decreed to be sacred and stranded whales were interred with reverence in special "whale temples." If a fisherman finds a whale carcass, he is obliged to help inter the whale and mourn it for 3 years, after which he will be rewarded with bumper fish catches.

Although we mentioned dugongs briefly above, any review of marine mammals in human folklore would not be complete without discussing manatees. Sirenians appear in folklore and legend, frequently being credited as the basis of many mermaid legends around the world (**Figure 13.6**). This is why they are named after the sirens of Greek mythology who lured sailors onto rocks with their voices (although sirens in Greek mythology lived on land and were part-woman, part-bird, it was through the Romans and medieval tales that they became associated with the sea and became more mermaid-like). Mermaids appear frequently in Arab and Greek legend, and the Roman writer Pliny the Elder mentions nereids in his work *Naturalis Historia*, describing them as rough skinned, with a tail of a fish. The Norwegian text *Speculum Regale* features mermaids too. A creature called the *margygr*, which was found near Greenland, had a fish-like tail but breasts, joined flipper-like fingers, a sloping forehead, wrinkled

FIGURE 13.6 Alexander the Great descends in an early diving bell and encounters a mermaid. In this medieval image the mermaid looks very similar to a dugong.

Exploring the Depths: Narwhals and Unicorns

Medieval folklore from Scotland tells of the *Biasd na Srogaig*, or sea unicorn. It is possible that during the "Little Ice Age" (not a real ice age but a cool climate period that extended from the 13th to the mid-19th centuries) of the late Middle Ages that narwhals could have extended their range southward so that occasional vagrants came into the waters of Scotland. It has been suggested that the occurrence of narwhals in Scottish waters and a trade in their tusks led to the unicorn being adopted in the royal crest of Scotland. Mary Queen of Scots allegedly had a croquet mallet made from narwhal horn, which it was said she never lost a match while using.

There was trade in "unicorn" horn in the middle ages, because ground horn was said to be a cure against poisons.

Unicorn horn was also seen to be a highly prestigious item. For example, in the 15th century the Duke of Burgundy had a sword with a pommel and scabbard made of "unicorn horn." A 12th century bishop's crosier, likewise said to be made from a narwhal tusk, is currently housed in Salzburg Cathedral. The Danish "unicorn" throne was also constructed with horns (**Figure B13.1**), as was the scepter of the Austrian ruling family of the Habsburgs. So valuable were they that Queen Elizabeth I paid £10,000 (approximately £2 million in modern currency) for a unicorn horn, which was turned into a scepter. These horns could have come from narwhals stranded or hunted in Greenland, Iceland, or possibly northern Norway during this time.

FIGURE B13.1 The Danish unicorn throne.

cheeks, and a large mouth, features that sound very similar to the Steller's sea cow. There are several stories of mermaids being seen off the coast of Scotland, with one tale specifically stating that the mermaid had a rounded tail, which again might relate to a real-life sighting of a vagrant manatee.

One of the most famous sightings of a mermaid was by Christopher Columbus in January 1493 who noted seeing three mermaids off the coast of what is now Haiti and an area inhabited by manatees. He wrote that the mermaids were "not as pretty as they are depicted, for somehow in the face they look like men." A few hundred years later in 1614 the explorer John Smith (of Pocahontas fame) also reported seeing a mermaid off the coast of Massachusetts, an area to the extreme north of the Florida manatee's range.

Finally, on June 15, 1608 explorer Henry Hudson, who gave his name to the Hudson River, reported in his log book that two of his crew had sighted a mermaid off the coast of Russia, saying it had long black hair, breasts like a woman, white skin, and was "speckled like a macrell" but had a porpoise-like tail. From the Arctic location this could possibly have been a Steller's sea cow.

Marine Mammal Hunts and Early Whaling

The Norse (Norwegians) commonly used whale products from stranded whales, and they were probably one of the first groups of people to actively hunt both small cetaceans and larger whales from boats. One of the first accounts of actively hunting marine mammals comes from King Alfred of Wessex's text *Orosius*, which dates to 890 AD. He tells of the merchant and chieftain Ohthere (or Ottar) who lived near what is now Tromsø and actively hunted walruses for their ivory. In *Orosius* it states that Ohthere traveled as far north as the "whale hunters" travel, giving us one of the earliest dates known in which Norwegians were actively hunting whales. By the 10th century Norwegians had introduced laws determining the ownership and distribution products of a whale carcass. Hunting techniques used by the Norse involved lancing whales with spears or shooting whales with crossbow bolts coated in a culture of the bacterium *Clostridium* spp., which led to the whale succumbing to septicemia and beaching.

However, in Europe the Basques, who inhabit northeast Spain and southwest France around the Bay of Biscay, were really the masters of early whaling. The waters of the bay were inhabited by northern right whales, and these became an important resource for the Basques. In the 7th and 8th centuries they began to build watch towers for whales along the coast and, upon sighting one, would launch small boats from the shore to catch the animals. Within a few hundred years the Basques were exporting whale meat to the rest of Europe. Whale meat, like seal meat, was considered to be "fish" and therefore could be eaten on holy and feast days. In the Middle Ages holy and feast days could account for over one-third of the year, and thus there was a ready market. Other whale products were also exported, including bone and baleen, which were used to make the plumes and crests on the helmets of knights. By 1052 the Basques had effectively started commercial whaling in Europe (see Commercial Whaling: The Early Years, below).

In the Arctic the Inuit have hunted pinnipeds for at least 4,000 years and cetaceans for over 2,000 years, with indigenous people hunting whales possibly as early as 3,000 years ago on the Arctic coast of Russia. The eastern coast of the United States has had a relatively long history of whale hunting. In 1605 George Waymouth, an early English colonist sailing the coast of New England, reported seeing Native Americans actively hunting a whale using a harpoon attached to a rope and dispatching the creature with a hail of arrows. When the Mayflower Pilgrims first encountered Native Americans, they were busy cutting up a black "grampus" on the beach, possibly a part of a group of stranded long-finned pilot whales. The indigenous people of that area used larger stranded cetaceans as a resource, which early settlers in New England also began to exploit, as a source of oil in particular. The Pilgrims often used local natives to butcher and process the carcasses in exchange for alcohol and metal items because it was a smelly and unpleasant job that early settlers often thought was beneath them.

In Japan bones of dolphins at an archaeological site on the coast of the Sea of Japan suggest drive fisheries for dolphins existed at least as early as 3200 BC. From the same era harpoon heads and dolphin remains from northeastern Kyushu implied that hand-held harpooning of small cetaceans also occurred. A number of relics from Hokkaido, such as whale skeletons, harpoons, and paintings depicting whaling, show that whales were being hunted in the Okhotsk Sea from the 5th until the 14th century. Altogether, skeletal remains indicate 13 species were used in historic Japan (either after hunting or stranding), including North Pacific right whales, humpback whales, sei whales, northern minke whales, sperm whales, and a variety of small cetaceans. Whaling with harpoons started elsewhere in Japan from the 15th and 16th centuries (**Figure 13.7**). In 1677 whalers in Taiji introduced a new form of "net" whaling in which whales were entangled first and were then harpooned or stabbed with a lance. Calves were often targeted first because mothers would stay with and attend the entangled calf, making it easier to capture the adult. However, whaling activities in Japan were relatively low scale and regional

Exploring the Depths: Whales and Whaling in the United Kingdom

The use of cetaceans as a resource in the United Kingdom dates back to the Stone Age. For example, whale bones were used as a building material at the Neolithic site at Skara Brae, Orkney. On the west coast of Scotland, on the island of South Uist, Bronze Age remains of bottlenose whales and sperm whales and Iron Age remains of northern minke whales and bottlenose dolphin indicate that at least these species were used. However, because there is a lack of harpoons or similar weapons in the archaeological record, these animals were most likely stranded carcasses or perhaps hunted opportunistically rather than hunted through any active whaling. Also, these whales seem to have been mainly used as a source of materials for crafted items, such as tools and pots, or for building rather than as a source of food, adding weight to the hypothesis of using beached rather than hunted animals.

During the early Middle Ages the Norse occupied northern and western Scotland and brought with them a tradition of using cetaceans; a variety of bones from bottlenose dolphins, killer whales, long-finned pilot whales, bottlenose whales, and northern minke whales, as well as even larger species such as sperm and blue whales, have all been found at Nordic archeological sites. The large numbers of bones suggest active hunting of cetaceans during this period, but there is no direct archeological evidence to support anything more than just the use of stranded carcasses. Norwegian laws on whale ownership were introduced into medieval Scotland at least as early as the 11th century and were used up until 1611.

South of the border, there was also a history of marine mammal hunting in England dating back to the Anglo-Saxons. In 731 AD the "venerable" Bede wrote that, "seals as well as dolphins are frequently captured, and even whales." The so-called Franks Casket (**Figure B13.2**), a box constructed from the bones of a whale in 700 AD, is an example of whales being used as a building material, although the bone that was used in the casket was from a whale that was not specifically hunted but instead stranded on the Northumbria coast. *Aelfrics Colloquoy*, a text written in the late 10th century, states that fishermen refused to catch whales because they were very dangerous and were likely to sink the boats sent out to catch them.

By the 11th century there were active fisheries for porpoises. In fact, the name porpoise comes from *porc poisson*, or pig fish, because they were hunted as a source of bacon and meat to be consumed on religious feast days, when only "fish" could be consumed. The so-called scrag (meaning shriveled) whale, probably the now extinct Atlantic gray whale, was also sometimes caught in coastal waters, such as the Severn Estuary.

In the early 14th century cetaceans were declared to be "*Fishes Royale*" in England by Edward III, which meant any captured or stranded cetaceans (or sturgeon) were automatically the property of the crown. Many English kings and queens ate cetaceans at royal banquets, especially porpoises, including Henry VII, Henry VIII, and Elizabeth I. The *Fishes Royale* decree was extended to Scotland by the 17th century, but in the islands off the west coast of Scotland Nordic-style drive fisheries (such as those practiced today in the Faroe Islands; see Chapter 15) and hunts for small cetaceans continued right up until the early 20th century.

FIGURE B13.2 The Franks casket.

until the 20th century (see Commercial Whaling: The Early Years, below).

Commercial Whaling: The Early Years

Commercial whaling in Europe was led by the Basques of France and Spain, who hunted primarily northern right whales in the Bay of Biscay. It is said that the right whale got its name because it was the "right" whale to hunt because it did not sink when it was harpooned, unlike most other whale species. Also, the right whale was slow, could be found close to the coast, provided a substantial yield of oil thanks to its thick blubber layer, and had a mouth full of long plates of baleen. The Basques used most every part of a whale, including the feces, which were used to dye clothes. Copepods, the main prey of right whales, contain caretenoids (orange-red pigments), giving right whale feces orange and red pigment. As a result Basque flags feature red or orange prominently (and this is also why the color orange became associated with the Dutch, another whaling nation albeit slightly later, as noted below). The Basques also hunted Atlantic gray whales and, around 1540, began to hunt bowhead whales off the coasts of Labrador and Newfoundland (bowhead whales are rare in this region

FIGURE 13.7 A painting showing early Japanese whaling.

today but appear to have extended this far south during the cooler temperatures of the so-called Little Ice Age).

British ships also ventured into the waters around Labrador and Newfoundland searching for whales, but in 1607 the explorer Henry Hudson reported an abundance of bowhead whales off the Arctic island of Spitzbergen, in the Svalbard Archipelago. In 1610 the British sent whaling ships to the islands and began commercial whaling in earnest, followed shortly thereafter by the Dutch. Between 1610 and 1669 the Dutch took approximately 15,000 whales (mostly bowheads), although they took nearly four times that number from 1669 to the end of the 18th century. The British took comparable numbers of whales and by 1732 had a fleet of 137 whaling vessels. By the 1800s the British had overtaken the Dutch as the main whaling nation.

Attempts to hunt whales in America began (in the year 1614) with John Smith (of Pocahontas fame) who set off on an expedition to establish a whaling operation in what is now Maine. However, Smith was unsuccessful, and although whales were seen, they were not the right whales the whalers wished for but the faster and much harder to catch rorquals. From the 1630s colonists in New England began to actively search for stranded whales along their beaches, because the oil they provided was a valuable economic resource. By the 1650s the colonists had started to actively hunt whales with boats launched from the shoreline.

The American capital of whaling was arguably the island of Nantucket in New England. The local colonists realized in the late 1660s that whales could be a useful resource when a "scrag" (gray) whale came into their vicinity, which they duly butchered. In 1690 they hired an experienced shore whaler to teach the people of the town how to catch whales, and by 1700 most of the inhabitants became involved in the whaling industry. Most whales taken in American whaling were coastal northern right whales, although Atlantic gray whales and humpback whales were also taken.

According to folk tales, in 1712 sperm whaling started when Nantucket Captain Christopher Hussey, whose ship was searching for right whales, was blown further out to sea and came across a group of sperm whales, which they harpooned and took back to shore. However, as Christopher Hussey was only 6 years old at the time, this seems to be a local myth, although a man called Bachelor Hussey was a whaler in the town, and the two names could have been mixed up somewhere along the line.

Sperm whales soon became highly sought after due to the high (and high-quality) oil yields from the spermaceti organ in their heads. To meet the demand whaling in the United States changed from shore-based whaling to deep-sea whaling, primarily targeting sperm whales. Sperm whaling grounds included the Caribbean and such far-flung locations as Indonesia and the Galapagos Islands in the Pacific, with voyages to these remote locations lasting 4 or 5 years (see Exploring the Depths: Life as a Whaler).

Spermaceti oil could be burned with a cleaner, less smoky flame than other oils, so that it was often used to produce high-quality candles or for lamps. The oil was also used to lubricate clockwork, especially watches and ships' chronometers, as the oil was also less likely to gum up the workings than lower grade oils. It was also used initially as a medicine or for cosmetics. The most valuable product from sperm whale hunts, however, was the ambergris, lumps of waxy material that were sometimes extracted from sperm whale stomachs. The substance was put to a variety of uses (Charles II of England allegedly ate a dish composed of eggs and ambergris), but primarily it was used in the perfume industry. Another product from sperm whales was scrimshaw (**Figure 13.8**), carved sperm whale teeth or bone that the whaling crew would sell for extra money or perhaps give as gifts to loved ones. These started as distractions to relieve the monotony of whaling cruises, but some pieces became quite sought after and valuable.

FIGURE 13.8 A piece of scrimshaw carved from the tooth of a sperm whale.

Exploring the Depths: Life as a Whaler

Life on a whaling boat was a hard one. Injuries were common, especially when crews had harpooned a whale, because the animal might smash one of the catching boats to flinders with its tail or jaws. A standard American whaling boat of this era might have a crew of 30 to 35 men, comprising a captain, four mates (officers), four harpooners, a steward to keep records of supplies and oil yields, a cooper to make barrels to store the oil, a blacksmith to make rings for the barrels, harpoons, lances, and other items, and a crew of 15 to 20 sailors. The harpooners were highly skilled and were often Native Americans or from indigenous groups in the Caribbean or Pacific Islands, who were experts in throwing spears at a moving target. The harpooners often had a cabin of their own and were frequently excused from many of the routine chores on board a whaling vessel, because if they were injured, the ability of the crew to catch whales, and thus their ability to garner profits, would be greatly diminished.

Whaling crews caught their prey by rowing up to resting sperm whales in small rowing boats. When they got close enough, the harpooner would hurl the harpoon at the whale. The harpoon would sometimes have a swiveling head, locked in place by a wooden peg. On hitting the whale the peg would break, and the harpoon head would swivel, embedding and anchoring it in the side of the animal. The end of the harpoon would be attached by rope to the rowing boat or wooden floats called drogues. The sperm whale, not surprisingly, would react to this attack and would swim rapidly, dragging the catching boat behind, in what became called a "Nantucket sleigh ride." Crews would have a hatchet at the front of the boat to cut the harpoon rope if the whale threatened to dive and drag the catching boat under with it. When the whale became exhausted, the boat would get closer and the animal would be dispatched by a long lance stabbed into the animal until vital organs such as the lungs were hit and the animal finally expired, which could take hours. Many boat crews perished as the whale lashed out in its death throes. Of course, having a working blacksmith's forge on board a ship smeared and stoked full of oil was a risky proposition and added to the dangers that were rife on board a whaling ship, even when there were no whales to be seen.

The most famous account of life on a whaling vessel is probably Herman Melville's book, *Moby Dick*, published in 1851 (**Figure B13.3**). Moby Dick was a white sperm whale, although not an albino, as the book describes unusually white pigmentation on part of his body. Several whales have been seen in the wild with similar white coloring due to a rare genetic abnormality. Melville had presumably heard of the whale reported off the coast of Chile that had been named "Mocha Dick." Melville

Exploring the Depths: Life as a Whaler (*continued*)

had worked on several sperm whaling boats and had picked up much information about the behavior of whales on these trips, as well as tales about the life on whaling boats, including the story of the *Essex* (see Exploring the Depths: The *Essex*), a whaling boat that had been sunk by a whale.

As noted above, the voyages of these sperm whaling boats could last several years. The crew's food mainly consisted of salted pork or beef and thick crackers. Interestingly, whale meat was not consumed, because it was considered to be distasteful. This way of life was maintained until the middle of the 19th century when new technology increased the safety and efficiency with which whales could be killed, heralding the age of modern commercial whaling (see Chapter 14).

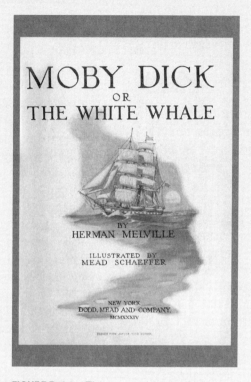

FIGURE B13.3 The cover of *Moby Dick*.

Exploring the Depths: The *Essex*

The *Essex* was a Nantucket sperm whaling boat captained by George Pollard that encountered a group of sperm whales nearly 4,000 km off the west coast of South America. A male whale rammed the *Essex* twice, holing the vessel and causing it to sink. Twenty-one crew members escaped on their small catching boats and sailed to the deserted Henderson Island. Within a week they had exhausted the food supply on the Island, and all but three got back into the catching boats and sailed out to sea. With the food situation becoming dire, the crews of the boats turned to cannibalism and took lots as to who was to die to feed their crewmates and who was to dispatch their imminent meal. It was 3 months until the surviving crew members were all rescued, during which time 7 men had been killed and eaten.

One of the surviving crew, Owen Chase, wrote about the incident. Herman Melville eventually met Chase's son, who gave him a copy of the book and inspired Melville to write his own fictional tale about an obsessed whaling captain and a vengeful whale.

The United States, and New York in particular, was also the setting for a rather unusual debate about the taxonomy of the whale. An 1818 lawsuit challenged the religious-based belief that the whale was a fish in an effort to circumvent the state requirement for inspection and taxation of fish oil on whale oil. This pitted science against religion over four decades before the release of Charles Darwin's *Origin of Species*, but also sparked intense public debate. The verdict, perhaps unsurprisingly, sided with religion, as the jury effectively declared whales to be fish.

Changing Attitudes Toward Marine Mammals

From the middle of the 20th century onward the public's attitudes toward marine mammals in the West began to change. No longer were they considered to be just a resource to be exploited but rather as animals that were somewhat "special," with pictures and articles appearing in popular magazines about their unique appearance and behaviors from the 1940s onward. Some environmental and animal welfare groups (such as Friends of the Earth and the Humane Society of the United States) started protesting against whaling in the 1950s and 1960s and began to promote whale conservation.

A major factor that changed public perceptions of cetaceans was the 1964 American TV series "Flipper," which featured the Ricks family and a friendly Florida bottlenose dolphin. In the TV show Flipper was portrayed as extremely intelligent, loyal, and affectionate, a portrayal that influences many of the public's attitudes to cetaceans even today. About this time public attention became focused on the large number of dolphins being killed in the tuna fisheries of the eastern Tropical Pacific (see Chapter 15), which led to the drafting and enacting of the U.S. Marine Mammal Protection Act in 1972 (see Chapter 17) and ultimately led to the introduction of "dolphin friendly" or "dolphin safe" labeling of tuna in the 1980s (see Chapter 15).

In the early 1970s calls for whale conservation reached their peak with the Animal Welfare Institute and the Fund for Animals initiating the now famous "Save the Whales" campaign in 1971 (**Figure 13.9**), which called for a boycott of Japanese goods to force an end to commercial whaling (see Chapter 14). The discovery of humpback whale song and the subsequent release of a recording of "songs of the humpback whale" recorded by marine biologist Roger Payne increased public interest and concern over whales.

In 1975 a newer environmental group called Greenpeace expanded its environmental protests from nuclear materials and took up the anti-whaling cause with the "Project Ahab" campaign. This campaign gained a large amount of media attention, and other environmental and animal welfare groups joined the "save the whales" bandwagon. Eventually, public pressure and scientific data showing severe depletion of whales led to the international commercial whaling moratorium being passed in 1982 (see Chapter 14).

Today, although whaling continues in some countries (see Chapter 14), by far the largest economic activity involving cetaceans is whale watching (see Chapter 18), which has become a billion dollar industry worldwide. Although whale watching is seen by many as a benign use of cetaceans (especially compared with whaling) and is actively promoted by many environmental groups, it nonetheless can sometimes have significant impacts on the cetaceans being watched, and there are many concerns about the risk that tourism poses to some populations of marine mammals (see Chapter 18).

Attitudes and awareness of cetaceans and their conservation varies considerably between countries and even within regions of countries. For example, several surveys in Scotland have discovered that only a small proportion of the public was aware of the diversity of cetaceans in Scottish waters, with those in urban areas being even less aware. The public had relatively good knowledge as to whether specific species occurred in Scotland, with over one-half knowing that bottlenose dolphins and harbor porpoises could be found in Scottish waters but only one-third knowing about the most common baleen whale in the region, the northern minke whale.

However, surveys investigating threats to cetaceans showed that the public was concerned about whale conservation. Nearly one-half of those surveyed considered that cetaceans were not sufficiently protected and favored the introduction of special protection laws for cetaceans. The public was more concerned about some issues such as oil spills and their impacts than were a panel of experts who were also surveyed. Conversely, the experts were more concerned about issues related to climate change and underwater noise than the general public. A more recent study on American college students found that levels of knowledge about cetacean conservation were low, with most of those surveyed believing the humpback whale was the most endangered whale species and only a few percent recognizing right whales as being extremely endangered. The lack of connection between public awareness and the conservation status of whales and dolphins is somewhat troubling and certainly makes conservation efforts more difficult.

FIGURE 13.9 A "Save the Whales" rally/march.

Exploring the Depths: Public Opinion, Whaling, and Seal Culls

In 1994 Milton Freeman and Stephen Kellert conducted an international study investigating public attitudes to whaling and found that 60% of Australians, 54% of Germans, 48% of Americans, and 43% of British surveyed were opposed to whaling under any circumstance, as compared with only 24% of Japanese and 21% of Norwegians. Five years later Kellert completed another survey of American attitudes and found that 70% of Americans opposed whaling, a sizeable increase in opposition. On the other side of the Atlantic, a survey in 2000 in Scotland found that over 95% of the public surveyed was opposed to whaling, and yet another survey found that many tourists would boycott going to a country where whaling was being conducted. In the Caribbean a more recent study found that a country having a strong conservation stance for whales and dolphins would make four-fifths of European and American tourists more likely to visit that country on holiday, whereas conversely a pro-whaling stance would make tourists less likely to visit the country.

The same U.S. study by Kellert found that three-fifths of Americans disapproved of culls of pinniped populations to protect fish stocks and even disagreed with killing seals if they caused damage to fishing gear. In Scotland a similar survey indicated that three-fourths of the public surveyed disagreed with seal culls, with more than one-half strongly objecting. A study of tourists also found that substantial numbers of those vacationing might boycott the country if seal culls were introduced, which could potentially cost more than £100 million (~U.S. $150–200 million) in lost tourism revenue.

Cetaceans and seals are clearly an emotive issue, and there is strong opposition to whaling and killing seals, at least in North America and many European Countries, such as the United Kingdom. Moreover, several studies appear to suggest that this opposition has the potential to have an economic impact on tourism and trade for countries that engage in these activities.

SELECTED REFERENCES AND FURTHER READING

Brennand, M., Parker Pearson, M., & Smith, H. (1998). *The Norse Settlement and Pictish Cairn at Kilphaeder, South Uist. Excavations in 1998.* Department of Archaeology & Prehistory, University of Sheffield, Sheffield, UK.

Burnett, D. G. (2007). *Trying Leviathan: The Nineteenth Century New York Court Case that Put the Whale on Trial and Challenged the Order of Nature.* Princeton University Press, Princeton, NJ.

Clark G. (1947). Whales as an economic factor in prehistoric Europe. *Antiquity* 21: 214–232.

Chase, O. (1821) (reprinted 1999). *Shipwreck of the Whale-ship Essex.* The Lyons Press, Guilford, CT.

Colgrave, B.C., & Mynors, R.A.B. (1969). *Bede's Ecclesiastical History of the English People.* Clarendon Press, Oxford, UK.

Cravalho, M. A. (1999). Shameless creatures: an ethnozoology of the Amazon River dolphin. *Ethnology* 38: 47–58.

Dolin, E. J. (2007). *Leviathan. The History of Whaling in America.* WW. Norton & Co., New York.

Fraser, F. C. (1977). Fishes Royal: the importance of dolphins. In: *Functional Anatomy of Marine Mammals* (Ed. J. Harrison), pp. 1–41. Academic Press, London.

Freeman, M. M. R, & Kellert, S. R. (1994). International attitudes to whales, whaling and the use of whale products: a six country survey. In: *Elephants and Whales: Resources for Whom?* (Ed. M. M. R. Freeman & U. P. Kreuter). Gordon and Breach, London.

Fullem, G. D. (1995). Norway, the international whaling moratorium, and sustainable use: a modern environmental policy conundrum. *Willamette Bulletin of International Law and Policy* 3: 79–115.

Gardiner M. (1997). The exploitation of sea mammals in medieval England: bones and their social context. *Archaeology Journal* 154: 173–195.

Gravena, W., Hrbek, T., Da Silva, V. M., & Farias, I. P. (2008). Amazon River dolphin love fetishes: from folklore to molecular forensics. *Marine Mammal Science* 24: 969–978.

Howard, C., & Parsons, E. C. M. (2006). Attitudes of Scottish city inhabitants to cetacean conservation. *Biodiversity and Conservation* 15: 4335–4356.

Hutchins, J. (1968). *Discovering Mermaids and Monsters.* Shire Publications, Colchester, England.

Kasuya, T. (2009). Japanese whaling. In: *Encyclopedia of Marine Mammals*, 2nd ed. (Ed. W. F. Perrin, B. Würsig, & J. G. M. Thewissen), pp. 643–649. Academic Press, New York.

Kellert, S. R. (1999). *American Perceptions of Marine Mammals and Their Management.* Humane Society of the United States, Washington, DC.

Lindquist, O. (1995). Whaling by peasant fishermen in Norway, Orkney, Shetland, the Faeroe Islands, Iceland and Norse Greenland: mediaeval and early modern whaling methods and inshore legal régimes. In: *Whaling and History* (Ed. B. Basberg, J. E. Ringstad, & E. Wexelsen), pp. 17–54. Kommador Chr. Christensens, Sandefjord, Norway.

Lund, N. (1984). *Two Voyagers, Ohtherea and Wulfstan at the Court of King Alfred.* William Sessions, York, UK.

McLeod, B. A., Brown, M. W., Moore, M. J., Stevens, W., Barkham, S. H., Barkham, M., & White, B. N. (2008). Bowhead whales, and not right whales, were the primary target of 16th- to 17th-century Basque whalers in the western North Atlantic. *Arctic* 61: 61–75.

Mulville, J. (2002). The role of cetacea in prehistoric and historic Atlantic Scotland. *International Journal of Osteoarchaeology* 12: 34–48.

Parsons, E. C. M. (2003). Seal management in Scotland: tourist perceptions and the possible impacts on the Scottish tourism industry. *Current Issues in Tourism* 6: 540–546.

Parsons, E. C. M. (2004). Sea monsters and mermaids in Scottish folklore: can these tales give us information on the historic occurrence of marine animals in Scotland? *Anthrozoös* 17: 73–80.

Parsons, E. C. M. (2012). From whaling to whale watching: a history of cetaceans in Scotland. *Glasgow Naturalist* (in press).

Parsons, E. C. M., & Draheim, M. (2009). A reason not to support whaling: a case study from the Dominican Republic. *Current Issues in Tourism* 12: 397–403.

Parsons, E. C. M., & Rawles, C. (2003). The resumption of whaling by Iceland and the potential negative impact in the Icelandic whale-watching market. *Current Issues in Tourism* 6: 444–448.

Pluskowski, A. (2004). Narwhals or unicorns? Exotic animals as material culture in medieval Europe. *European Journal of Archaeology* 7: 291–313.

Scott, N. J., & Parsons, E. C. M. (2004). A survey of public awareness of the occurrence and diversity of cetaceans in Southwest Scotland. *Journal of the Marine Biological Association of the United Kingdom* 84: 1101–1104.

Scott, N. J., & Parsons, E. C. M. (2005a). A survey of public opinion on seal management in southwest Scotland. *Aquatic Mammals* 31: 104–109.

Scott, N. J., & Parsons, E. C. M. (2005b). A survey of public opinions in Southwest Scotland on cetacean conservation issues. *Aquatic Conservation* 15: 299–312.

Sharpe, R. (1995). *Adomnán of Iona. Life of St. Columba.* Penguin Books, New York.

Smith, B. D., Jefferson, T. A., Leatherwood, S., Ho, D. T., Thuoc, C. V., & Quang, L. H. (1997). Investigations of marine mammals in Vietnam. *Asian Marine Biology* 14: 145–172.

Szabo, V. E. (2005). Bad to the bone? The unnatural history of monstrous medieval whales. *The Heroic Age: A Journal of Early Medieval Northwestern Europe* 8 (June 2005). Retrieved November 1, 2011 from http://www.heroicage.org/issues/8/szabo.html.

Whaling and the International Whaling Commission

CHAPTER 14

CHAPTER OUTLINE

History of Modern Whaling
   Exploring the Depths: Japanese Commercial Whaling

International Convention on the Regulation of Whaling
   Whaling Quotas and Bans
   Whaling Moratorium
   Norwegian Whaling
   Japanese "Scientific" Whaling
   Exploring the Depths: Structure of the IWC
   Iceland's "Scientific" and Commercial Whaling
   Exploring the Depths: Iceland and Whale Watching

   Revised Management Procedure and Revised Management Scheme
   Exploring the Depths: North Atlantic Marine Mammal Commission
   Exploring the Depths: Cetacean Meat Contamination
   Exploring the Depths: IWC Conservation Committee

Aboriginal Whaling
   Exploring the Depths: Aboriginal or Indigenous Whaling
   Exploring the Depths: Welfare and Whaling

Selected References and Further Reading

History of Modern Whaling

Early whaling was a primitive and dangerous affair. It was also relatively inefficient because of the methods used; whales were generally killed with hand-held harpoons and lances. This all changed in the middle of the 19th century when developments in whaling technology made the activity more efficient and allowed more whales to be harvested, not only in terms of numbers, but also in terms of populations and even whole new species in previously unexploited regions. A main driver behind these changes was the Norwegian Svend Foyn, a seal hunter who noticed the high density of whales encountered on sealing trips that were, at the time, not being hunted. He then considered it to be his life's vocation to develop whale fisheries.

The first major revolution in whaling technology came in the 1840s with the development of the "bomb lance," an explosive harpoon fired from a large shotgun-like weapon. Although increasing the range at which whales could be attacked, the early bomb lances did little to make the profession less perilous and may well have actually made it more dangerous. The lances were inaccurate, and the recoil from the weapons could bowl a man over.

The next major development came in 1859, when the first steam-driven whaling ship was built. With this development, whaling ships were no longer so dependent on the wind for their operation. Despite this, the hunting of whales was still undertaken from small boats. However, in 1863 Svend Foyn built the first steam-powered whaling schooner or catching boat (called *Spes et Fides*). This much larger vessel had deck-mounted guns that fired harpoons and grenades to dispatch the whales. This step allowed the efficient capture of even the fast-swimming rorqual whales for the first time. Foyn also developed a method to insert a tube into a whale carcass and inflate it with compressed air. This meant the carcasses of negatively buoyant whales, such as the rorquals, did not sink upon death, making the carcasses easier to retrieve and handle.

The next major advancement came in 1869, when Svend Foyn introduced the cannon-fired explosive harpoon. This device had flanges to secure the harpoon in the target animal and a grenade at the tip, which would detonate upon hitting the animal. This grenade-tipped harpoon design is similar to the harpoons used today, as the design has changed little in the subsequent 150 years.

FIGURE 14.1 A South Georgia whaling station.

At the same time whaling methods were becoming more efficient, new whale stocks were also being discovered, leading to their exploitation. For example, in the mid-19th century whaling vessels discovered new populations of bowhead whales and North Pacific right whales in the Bering Sea as well as the breeding ground of gray whales in Baja, California. However, it was not until the end of the 19th century that opportunities for a major new source of whales emerged.

In 1882 a speculative expedition was launched from Dundee, Scotland to the Antarctic. The expedition discovered Dundee Island (63°30′S, 55°55′W) and also realized the potential for large-scale commercial whaling in this region. In 1904 a whaling station was established at Grytviken, South Georgia, which opened up an era of Antarctic whaling (**Figure 14.1**). Unsurprisingly, this was soon followed by more shore-based stations on other subpolar islands, such as the South Orkneys.

Of course, none of this would have occurred if there was no market for whale products. Unfortunately, during this period the demand for whale oil actually increased (**Table 14.1**), and the development of a technique that transformed oil by hydrogenation allowed the use of whale oil for soap and margarine production. Whaling vessels in the Antarctic (and other regions) were still limited by how far they could travel to hunt whales and how many could be taken, because they would have to return carcasses to shore stations to process. In 1925, however, the first factory ship was produced (**Figure 14.2**). These vast vessels (at the time some of the largest ships afloat) had ramps to pull up whale carcasses and could cut up and process a blue whale at sea within an hour, alleviating the urgent need to return to shore. These factory ships substantially increased the efficiency with which whales could be killed and allowed whaling in new areas of Antarctica that had previously been unexploited. As a result this development led to the decimation of many whale populations.

Initially, humpback whales were one of the most sought after species because they were often found in more coastal waters, making them easier to capture. However, stocks became quickly depleted, catch rate declined, and efforts were directed instead towards the larger fin and blue whales. Takes of these latter two species escalated with the introduction of factory ships. Blue whale stocks also began to decline in the 1930s, leading to an increase in catches of the slightly smaller fin whale. Whaling effort was substantially reduced during World War II, allowing a few years for some stocks to recover slightly. After the war whaling began again in earnest. Initially, the target species was mostly fin whales because of depleted levels of blue whales. However, fin whales also became depleted, so the main target switched again to sei whales. Finally, when sei

TABLE 14.1 Uses of Whale Products During The History of Commercial Whaling

Uses of Whale Oil	Uses of Baleen	Uses of Whale Tissues	Uses of Whale Meat
cooking fat	brushes	*Blood*	animal feed
crayons	corsets	fertilizer	dog food
detergent	crinolines	sausages	fertilizer
dynamite	fans	*Connective Tissue*	human consumption*
ice cream	fishing rods	jello	
lamp oil	riding crops	photographic film	
linoleum	shoehorns	sweets	
lipstick	shop shutters	*Skin*	
lubricants	umbrella ribs	bootlaces	
margarine	watch springs	bicycle saddles	
paint		handbags	
soap		shoes	
polish		*Tendons*	
shampoo		surgical thread	
		tennis racquets	

*Apart from a brief period just after the Second World War, whale meat has only been extensively eaten in Iceland, Norway, Korea, and Japan.
Data from: International Whaling Commission and other sources.

whale stocks started to decrease in the 1960s and 1970s, catches switched to targeting the smallest of the rorquals, the minke whales.

The decreasing whale stocks in conjunction with the availability of cheaper vegetable and mineral oils led to whaling becoming less and less profitable. U.S. whaling companies went into an early decline, pulling out of the industry in the 1920s and 1930s. By the early 1960s European whaling countries had withdrawn from the industry as well, leaving Japan and Russia as the two main whaling nations. Heavily subsidized government whaling operations allowed activities to continue, even though whaling ceased to be economical. Despite this, the last factory ship sailed in 1978. By the 1980s the industry had become economically nonviable around the globe.

During the history of commercial whaling millions of whales were killed. The approximate numbers per species are as follows:

- 250,000 humpback whales
- 350,000 blue whales

Exploring the Depths: Japanese Commercial Whaling

Although Japan had a long history of small-scale coastal whaling and had undertaken a few offshore whaling expeditions just before World War II, the country was initially a relatively small player in the international whaling scene, with only a small presence on the high seas. It was not until after World War II that the Japanese increased their whaling effort. The war had left Japanese industry and agriculture in a poor state, raising concerns about how the country was going to provide enough food for the population. Whaling offered an alternative source of protein for devastated postwar Japan, and the industry was promoted by then U.S. General, Douglas MacArthur. As a result Japan's whaling industry expanded substantially during the 1950s, and whale meat became a major source of protein for the Japanese population over this period. For example, school lunches for children often included whale meat.

FIGURE 14.2 A whaling factory ship.

- 500,000 fin whales
- 1,000,000 sperm whales

Hundreds of thousands of animals from various other species were also killed. These included bowhead whales, right whales, northern bottlenose whales, Baird's beaked whales, and both species of minke whales.

International Convention for the Regulation of Whaling

As whale stocks declined and ultimately threatened the sustainability of the whaling industry, whalers of several nations joined forces to enact The Convention for the Regulation of Whaling in 1931. This effort was subsequently built upon in 1937 by the International Agreement for the Regulation of Whaling and finally in 1946 by the International Convention for the Regulation of Whaling. This latter convention set up the International Whaling Commission (IWC), the internationally recognized authority on managing whale stocks.

The International Convention for the Regulation of Whaling was one of the first truly global agreements for the management of threatened species, and the Convention's preamble highlights the importance of conserving whale stocks for future generations:

> *Recognizing the interest of the nations of the world in safeguarding for future generations the great natural resources represented by the whale stocks; Considering that the history of whaling has seen over-fishing of one area after another and of one species of whale after another to such a degree that it is essential to protect all species of whales from further over-fishing.*

However, if one reads the above quote again, it can be seen that the Convention was also about maximizing whales as a resource to be exploited.

The sometimes conflicting goals of the Convention to both "provide for the proper conservation of whale stocks" on the one hand and to "make possible the orderly development of the whaling industry" on the other, as well as to maximize the economic efficiency and profitability of the whaling industry at the same time, have led to the polarization of the IWC and a considerable amount of argument and conflict.

The IWC was also one of the first international treaty organizations in which science was supposed to play a major role (the roles of the IWC Scientific Committee are discussed in Exploring the Depths: Structure of the IWC).

TABLE 14.2 The Original Signatories of the International Convention for the Regulation of Whaling and the 15 Major Whaling Nations in 1946

Argentina
Australia
Brazil
Canada
Chile
Denmark
France
Holland (The Netherlands)
New Zealand
Norway
Peru
South Africa
United Kingdom
USSR
USA

Data from: International Whaling Commission.

The 15 major whaling nations at that time were the initial signatories of the International Convention (**Tables 14.2** and **14.3**). However, Japan was not one of these initial signatories (Japan joined in 1951).

Whaling Quotas and Bans

One of the first achievements of the International Convention for the Regulation of Whaling was the introduction of a quota system, the so-called blue whale unit. Quotas were allocated as a total number of units, and 1 blue whale unit was equal to 2 fin whales, 2.5 humpback whales, or 6 sei whales. This quota system continued to be used until 1972.

However, the blue whale unit did not prevent the blue whale and other species from being exploited past the point of economic viability. Blue whale numbers did not recover because whalers targeting fin whales would still take blue whales opportunistically when they were sighted. Similarly, both fin and blue whales were still taken when whalers turned to the smaller sei and minke whales in an effort to maintain their investment in boats and equipment. This ability to shift between species meant that greater catches of the more abundant species subsidized catches of rarer species and ultimately led to greater depletion of stocks than would have happened if single species quotas had been issued.

TABLE 14.3 The Current (as of January 2012) Signatories of the International Convention for the Regulation of Whaling

Antigua & Barbuda	Germany	Nicaragua
Argentina	Ghana	Norway
Australia	Greece	Oman
Austria	Grenada	Palau
Belgium	Guatemala	Panama
Belize	Guinea-Bissau	Peru
Benin	Guinea	Poland
Brazil	Hungary	Portugal
Bulgaria	Iceland	Romania
Cambodia	India	Russian Federation
Cameroon	Ireland	San Marino
Chile	Israel	St. Kitts & Nevis
China	Italy	St. Lucia
Colombia	Japan	St. Vincent & The Grenadines
Congo	Kenya	Senegal
Costa Rica	Kiribati	Slovak Republic
Côte d'Ivoire	South Korea	Slovenia
Croatia	Laos	Solomon Islands
Cyprus	Lithuania	South Africa
Czech Republic	Luxembourg	Spain
Denmark	Mali	Suriname
Dominica	Marshall Islands	Sweden
Dominican Republic	Mauritania	Switzerland
Ecuador	Mexico	Tanzania
Eritrea	Monaco	Togo
Estonia	Mongolia	Tuvalu
Finland	Morocco	UK
France	Nauru	Uruguay
Gabon	The Netherlands	USA
The Gambia	New Zealand	

Data from: International Whaling Commission.

Consequently, the blue whale unit was replaced with a species-specific quota system called the New Management Procedure. This system was similar to fisheries management methods that called for a certain catch quota that would, theoretically, give the maximum yield (or maximum sustainable yield) for a certain population size. However, the New Management Procedure required accurate information on stock boundaries and abundances to work effectively. The problem was that many population estimates for these widely roaming species were (and often still are) plagued by large margins of errors, as well as many uncertainties.

Over the years the IWC introduced several whaling prohibitions for specific species, ultimately culminating in the passing of a total, albeit technically temporary, ban (moratorium) in 1982 (partly in response to declining numbers and partly in response to mounting global anti-whaling sentiment and the original "Save The Whales" campaign). Although passed in 1982, the moratorium actually came into effect in 1986. These specific bans were as follows:

- 1931: Bowhead whales
- 1935: Southern and northern right whales

- 1937: Gray whales
- 1966: Humpback whales and blue whales
- 1979: Sei whales (except in Iceland)
- 1984: Sperm whales

Whaling Moratorium

During the 1970s public and governmental concern for cetaceans increased. For example, in the United States the Nixon Administration passed the Marine Mammal Protection Act (MMPA) that, among other actions, prohibited whaling and the sale of whale products. In 1972 at the U.N. Conference on the Human Environment in Stockholm, the U.S. government proposed a 10-year ban or moratorium on commercial whaling. The resolution was the brainchild of a group called Project Jonah and passed by 53 votes to zero. It took, however, 10 years for the IWC to finally introduce its own moratorium in 1982 (which eventually came into effect in 1986).

The moratorium was put in place ostensibly to allow a comprehensive review of the status of whale stocks and to develop a quota system that would allow a sustainable take of whales by taking into account the nature of the scientific data available on whale populations (see Revised Management Procedure and Revised Management Scheme). However, over 25,000 whales have been killed since the moratorium came into effect, with Norway, Japan, and Iceland still hunting large numbers of whales despite the commercial whaling ban.

Norwegian Whaling

When the whaling moratorium was enacted, Norway put in a reservation. This is a tactic used by many countries that wish to be a member of a treaty but do not wish to be bound by one or more tenets with which they disagree. Norway is therefore not bound by the moratorium and can legally harvest whales. Although Norway initially respected the moratorium, it reinitiated commercial whaling in 1993. In recent years Norwegian whalers have legally taken 600 to 700 northern minke whales a year. However, Norwegian scientists have set annual quotas in excess of 1,000 whales over the last few years, under pressure from the Norwegian government.

Japanese "Scientific" Whaling

Although Japan initially declined, they eventually signed on to the commercial whaling moratorium in 1988 after international political pressure and the threat of sanctions (specifically preventing Japanese fishing access to Alaskan waters). However, in 1988 the Japanese immediately began to catch whales again using a loophole in the moratorium that allows whales to be killed for scientific research. Article 8 of the convention states the following:

> [A]ny Contracting Government may grant to any of its nationals a special permit authorizing that national to kill, take and treat whales for purposes of scientific research.

The number and species of whales that can be taken is "as the Contracting Government thinks fit," and these takes are exempt from other aspects of the Convention, such as the commercial whaling moratorium or any quota system. The Convention calls for these scientific takes to "so far as practicable be processed and the proceeds shall be dealt with in accordance with directions issued by the Government by which the permit was granted." The original intent of this article was to allay waste after scientific samples were obtained. In reality, after blubber and stomach contents and some other tissue samples are taken, meat from the whales is processed and sold in markets, as it would be in commercial whaling. Article 8 of the Convention also requires governments issuing such permits to present the data gathered every year. Accordingly, Japan does so every year at the Scientific Committee of the IWC. However, each year, especially in more recent times, this presentation receives criticism from other IWC scientists (see Exploring the Depths: Structure of the IWC).

The "special permit catches" or so-called scientific whaling programs of Japan are currently supposed to investigate impacts of whales on fisheries, which they do by looking at stomach contents and applying the results to ecosystem models. However, scientific whaling programs have been heavily criticized by scientists, including most

---

Exploring the Depths: Structure of the IWC

The Scientific Committee is made up of approximately 400 scientists who are either invited because of their expertise in an area of interest to the IWC or designated by IWC member countries. The Scientific Committee then gives advice in the form of a report (published every year as a special supplement to the *Journal of Cetacean Research and Management*) that is then read and discussed by the commissioners and their aides. The commissioners are representatives of the IWC member nations and are often politicians or civil servants, although some are also scientists. Many are senior members of the governmental fisheries or environmental/conservation agencies of the member nation, often depending on the stance of the particular country toward whales and whaling. The decisions made by the commissioners are often politically motivated and, although the Commission is ostensibly basing its decision making on science, politics often plays a major role in what is agreed upon.

scientists in the IWC's own Scientific Committee. Criticisms are that methods used are too simplistic and have major biases, sample sizes (i.e., the number of whales taken) have not been scientifically justified, and data could be gathered by nonlethal methods.

Initially, when the Scientific Committee reviewed the results of the scientific whaling program, the proponents of the research (i.e., the scientific whalers) were allowed to comment on the validity of the data. For example, when scientists for several countries criticized the Japanese research, the Japanese scientists opposed the criticisms. It was then noted in the report that some nations criticized and some nations supported the scientific whaling, without clarifying that those supporting the research were also those conducting it. Consequently, the report made the discussion look more balanced than it really was. However, more recently changes have been made to review procedure, with a small group of scientists reviewing the scientific aspects of the special permit research and reporting back to the scientific committee. However, even this approach has been problematic because most panel members are IWC Scientific Committee members and are thus not independent. For example, several panel members have received funds from scientific whaling activities or have worked on projects with the Japanese, creating conflicts of interest.

In any case the Scientific Committee and the IWC can only "suggest" that Japan changes its scientific whaling program because they have no power to prohibit the catches. Moreover, several published scientific papers and reports have already refuted many of the conclusions of the scientific whaling program, such as the claim that whales eat so many fish as to cause declines in human fisheries. Studies by Kristen Kaschner and Daniel Pauly analyzed prey consumption by whales and concluded that there was no substantial competition between cetaceans and human fisheries, that most of the world's whales consume prey that are not commercially sought, and that poor management of human fisheries is largely responsible for the depletion of fish stocks, not consumption by cetaceans.

At present, Japan currently takes northern minke whales, and other species, in the North Pacific (under a research program called JARPN). In 2007, 208 northern minke whales, 100 sei whales, 50 Bryde's whales (although, as noted in Chapter 3, there may be multiple species of Bryde's whale), and 3 sperm whales were caught. In the program in Antarctic waters (called JARPA), Japan took approximately 856 Antarctic minke whales in 2005, but this dropped to 508 and 551 whales in 2006 and 2007, respectively. In 2008 they also took 10 fin whales in the Antarctic under this Antarctic program. Between 1986 and 2007 a total of 11,389 whales were taken for "scientific research" by Japan. The latter takes are particularly controversial because the IWC made the Southern Ocean around Antarctica a whale sanctuary in 1994. Commercial hunting of whales is banned in this area regardless of existence of the moratorium. However, as the JARPA takes are for "scientific research," they are not considered to be commercial whaling, and thus the Japanese government considers that the sanctuary regulations do not apply.

Moreover, the number of Antarctic minke whales is currently in question, with the most recent circumpolar surveys (1991–2004) estimating only 40% of the number of minke whales from the previous surveys (1985–1991). Such a large decline in a short period of time should list the Antarctic minke whale as "endangered" under International Union for Conservation of Nature (IUCN) listing criteria (more than 50% decline within a 10-year period). As yet it is unknown why there was such a large decrease, but this has not stopped the Japanese whalers from proposing an increase in the number of Antarctic minke whales to be taken in the JARPA program.

Much controversy also surrounded the proposal by the Japanese government to take humpback whales in the JARPA program. Since commercial whaling of humpback whales ceased their numbers have been increasing, resulting in them being downlisted from "vulnerable" to "least concern" by the IUCN due to this recovery. This allowed Japan to propose adding humpback takes to their JARPA scientific quota. However, several countries expressed their opposition to this proposal, especially Australia, because the humpbacks Japan proposed to take were the same animals important to the Australian whale-watching industry when the whales migrate to and from Antarctica.

As a result of the Japanese effort to recruit support at the IWC, the influx of new pro-whaling countries into the IWC outnumbered the anti-whaling and neutral nations by the 2006 IWC meeting. At this meeting, hosted by the island of St. Kitts in the Caribbean, the so-called St. Kitts declaration was passed. This declaration called for a resumption of commercial whaling and blamed whales for depletion of fish stocks. However, more anti-whaling countries joined the IWC in 2007, giving the majority to anti-whaling countries again.

Iceland's Scientific and Commercial Whaling

Icelandic whaling activities are a complicated, yet controversial, issue. Iceland was a member of the IWC and agreed to the whaling moratorium. However, from 1986 to 1989 a "scientific" whaling program was undertaken in which 292 fin whales and 70 sei whales were taken (species that were then, and still are, considered to be endangered by the IUCN). In 1992 Iceland left the IWC but in 2002 tried to rejoin. Although Iceland had never taken a reservation against the whaling moratorium (as Norway had done), when it tried to rejoin it did so but this time stating it wanted a reservation. At an intersessional meeting of the IWC there was a vote to decide whether Iceland should be allowed to rejoin, and the vote passed by one. However, that one deciding vote was by Iceland, which voted on its own membership, even though it was not a member. There were many protests over this rejoining, with a reservation to one of the Commissions major tenets (the moratorium) and also the way in which Iceland's membership was determined.

Upon rejoining the IWC Iceland stated it would not start commercial whaling until 2006. However, it almost immediately put a proposal forward for lethal takes of whales for scientific research purposes (initially proposing to take 100 fin whales, 100 northern minke whales, and 50 sei whales). The Icelandic proposal was severely criticized by the IWC Scientific Committee, again because much of the work could be done by nonlethal means, the sampling regime would bias much of the data, and their proposed work on assessing impacts of whales on fisheries had already been deemed inappropriate due to a lack of valid ecosystem models with which to analyze the data. Moreover, Iceland had previously conducted a scientific whaling program in the late 1980s, but none of the data from that program had been published or had even been considered by the IWC.

Nonetheless, in the summer of 2003 Iceland harvested 37 minke whales under a special permit (with a further 25 in 2004 and 39 in 2005). In 2006 Iceland announced it was resuming commercial whaling and took 60 minke whales in that year (and 39 in 2007). In response to this resumption of commercial whaling 25 nations delivered a formal diplomatic protest (called a "demarche") to the Icelandic government on November 1, 2006. The whaling briefly stopped in 2007, primarily due to lack of demand for whale meat that had already been harvested, but whaling resumed in 2008.

Whaling in Iceland has been criticized, particularly from Icelandic tourism businesses (see Exploring the Depths: Iceland and Whale Watching). Moreover, only a limited market exists for whale meat in Iceland; some had to be disposed of as pet food. Because of this Iceland attempted to export whale meat to Norway and Japan. In June 2008 Iceland exported 80 tons of fin whale meat to Japan despite the fact that trade in whale products is restricted under the Convention on International Trade in Endangered Species of Wild Fauna and Flora (CITES). Iceland and Japan have a reservation against the listing of fin whales under CITES Appendix I (see Chapter 17). In 2009 Icelandic whalers caught 125 fin whales and 79 minke whales and stated plans to export up to 1,500 tonnes of fin whale meat to Japan.

Exploring the Depths: Iceland and Whale Watching

Whale watching started in Iceland in 1990 after the cessation of commercial and scientific whaling. By 2002 the new industry was estimated to be worth over U.S.$12 million a year, which was three to four times the economic value of commercial whaling in Iceland at its height. Moreover, it was one of the fastest growing sectors of the tourism industry, with estimates at that time that the income from whale watching would reach over $20 million by 2006, a substantial revenue for a country with only 320,000 inhabitants. Consequently, the Icelandic Tourist Industry Association protested the decision for Iceland to resume scientific whaling in 2003. The Association has also released multiple statements protesting the subsequent resumption of commercial whaling in Iceland in 2009, and expressing their concern about the impact it is having on the country's tourism industry.

Studies in the United Kingdom found that 91% of whale watchers stated they would not go whale watching in a country that hunted whales and that most would boycott visiting such a country altogether. Evidence of this has been seen in Iceland with reports of many tourists canceling their holidays, especially those planning to take whale watching trips, after the announcement of Iceland's proposed whaling activities. A more recent survey based in Iceland since the resumption of whaling suggested that whale watchers there were much less concerned. However, the results of this study are somewhat biased by the fact that only local tourists and those who had already made the decision to come to Iceland were included.

Moreover, the Icelandic Tourist Industry Association has complained that sightings of minke whales, the main target of the whale watching industry as well as the whalers, have been reduced. During the first few years of the scientific whaling program most whales taken were in key whale watching areas (for example, one-third of the animals were taken from Keflavik coastal waters, where several whale watching companies operate). There have also been complaints when tourists on whale watching vessels saw whaling boats returning with dead whales on board, or whales were killed while whale watching vessels were nearby.

Revised Management Procedure and Revised Management Scheme

As noted above, the moratorium was enacted as a precaution to allow whale stocks to recover by prohibiting commercial whaling until a thorough review of whale stocks and a sustainable quota system could be put in place. The new, revised quota calculation method developed was called the Revised Management Procedure (RMP). The RMP was accepted by the IWC Scientific Committee in 1993 and by the Commission itself in 1994, so a quota calculation method is now in place that allows commercial whaling at an allegedly sustainable level.

Exploring the Depths: North Atlantic Marine Mammal Commission

In 1992 the North Atlantic Marine Mammal Commission (NAMMCO) was established by Norway, Iceland, Greenland, and the Faroe Islands. This organization has a similar structure to the IWC, with a scientific committee, and was initially intended as an alternative to the IWC, established by countries that actively hunted cetaceans and/or pinnipeds. However, NAMMCO only contains three fully independent nations (although autonomous to different extents, Greenland and the Faroes both still rely on Denmark for matters of foreign policy) and does not have full regional representation (e.g., although NAMMCO is supposed to cover the North Atlantic, it does not include Canada, the United States, the United Kingdom, Ireland, Finland, Sweden, France, Spain, or Portugal in its membership). It also does not have the international scientific representation of the IWC, meaning it does not have the recognition and standing the IWC possesses on the international political arena as the competent body for the management of whales. However, given the disputed jurisdiction of the IWC over matters of small cetaceans and the lack of jurisdiction whatsoever over pinnipeds, NAMMCO claims authority in these matters in the North Atlantic region.

Many scientists and government agencies indeed believe that the RMP is scientifically accurate, cautious in the numbers generated, and robust to factors that might impact whales (e.g., pollution and climate change, for which trials were done in 1993–1994). However, many other scientists, government agencies, and some environmental groups are more skeptical, despite the acceptance by the IWC Scientific Committee. There are many reasons for this. For example, the RMP requires accurate information on the number of animals in each whale stock, which is frequently unavailable. Additionally, some information about their population demographics and the threats they face is needed as well but is also frequently lacking. For example, the large decrease in the numbers of Antarctic minke whales, noted above, remains unexplained. Finally, there must also be accurate information on how many whales are being or have been harvested. There certainly have been examples of falsification of whale catch data in the past, by Russia and Japan, for example. Also, the potential effects of environmental factors on whale stocks such as climate change, pollution, and disease are hard to predict.

Besides the quota-generating method (RMP), for commercial whaling to resume there must be a management structure in place within which the quotas are implemented. The proposed management structure is referred to as the Revised Management Scheme (RMS). The RMS encompasses management measures such as reporting mechanisms, product tracking, and enforcement. Some countries (e.g., United States) and environmental nongovernmental organizations have tried to negotiate a version of the RMS and reopen commercial whaling, because they hope the RMP will close the loophole allowing scientific whaling and produce total quotas smaller than the number of animals currently being taken by Norway, Iceland, and Japan. Also, scientific whaling is not bound by most of the control mechanisms that the IWC can bring to bear on commercial whaling (such as closed seasons, areas, and the protection of the Southern Ocean and other sanctuaries).

However, other governments and nongovernmental organizations express concerns that if an RMS is implemented, some nations may only abide by parts of the management measures, if any at all. Furthermore, the RMS does not, as it currently stands, prohibit scientific whaling, so it could continue in addition to commercial whaling. Several versions of an RMS have been brought to the IWC, but as yet none has been approved. The pro-whaling countries objected to many of the controls and restrictions embedded within the versions of the RMS proposed by the anti-whaling nations, whereas some anti-whaling nations complained that versions of the RMS from pro-whalers were weak and would do little to rein in some of the historical problems with whaling (such as illegal catches and falsified data). For example, one issue was the lack of an international DNA tracking system to tell if a product was legally taken or not.

As a consequence of this deadlock, and as part of the wider political wrangling surrounding this issue, several governments have expressed concern over the continued viability of the IWC, with Japan threatening on more than one occasion to withdraw completely. Many pro-whalers have argued that the IWC is supposed to facilitate whaling and accuse the anti-whaling nations of impeding progress. However, the anti-whaling countries continue to point out that the IWC is a mechanism for maintaining sustainable whaling, noting that we are unable to achieve this for many species and that most stocks have yet to recover from earlier unregulated activities.

Aboriginal Whaling

Under the International Convention for the Regulation of Whaling and the activities of the IWC, indigenous

Exploring the Depths: Cetacean Meat Contamination

Studies by Japanese researcher Professor Tetsuya Endo have found that dolphin and toothed whale meat for sale in food markets in Japan often contain high levels of heavy metals, such as mercury (**Figure B14.1**). In one of Professor Endo's studies, more than two-thirds of samples analyzed exceeded Japanese health limits for mercury, with one sample having mercury concentrations 36 times higher than allowable levels. In another study kidney damage was reported in laboratory rats fed some of the meat that had been on sale after just one dose of contaminated meat. Since this study several others have also noted high contaminant levels in meat from odontocetes and pinnipeds in various places around the world. Furthermore, other follow-up investigations have found dolphin meat for sale as baleen whale meat, the latter of which generally has a lower contaminant level due to baleen whales being on a lower trophic level than dolphins. This means that contaminant loads do not have the chance to bioaccumulate and biomagnify (see Chapter 15).

Despite this, another study found that one-third of minke whales harvested and analyzed from the scientific whaling program in the North Pacific (JARPN) tested positive for the bacteria *Brucella* spp. These bacteria cause the disease brucellosis in mammals, the symptoms of which include joint and muscle pain as well as the inflammation of the testes in males or the induction of abortion in females. In extreme cases it can even cause heart problems and death. The disease is transferable to humans from animals (it is zoonotic), and there are major health implications for humans who come into contact with contaminated whale meat, especially those who ingest it.

To reduce existing stockpiles the Japanese government has provided large quantities of whale meat to schools and hospitals for lunches, thus potentially exposing children and patients to various pollutants and diseases. This has raised many concerns both in Japan and internationally that has, in combination with the discovery of highly toxic dolphin meat being sold as whale meat, led some local governments in Japan to resist the program.

FIGURE B14.1 Whale meat for human consumption being sold in an Asian market.

Exploring the Depths: IWC Conservation Committee

In 2003 an IWC resolution brought into existence the IWC Conservation Committee, whose aims are as follows:

- Assess progress made in the conservation of whales
- Prepare conservation recommendations for the IWC
- Implement conservation programs
- Help to focus public and private resources on key conservation issues facing cetaceans

Issues discussed by the committee have included whales struck by ships, bycatch in fishing gear, and the management of whale watching. The committee only meets for roughly one-half day each year at the IWC, limiting the amount that can be discussed. Nonetheless, the fact that the committee exists is a very positive step for the IWC toward a more holistic, ecosystem-based approach to whale management and conservation, beyond simply regulating whaling activities.

communities are permitted to hunt a quota of whales for subsistence purposes. These quotas are set if the indigenous peoples can demonstrate a traditional, nutritional, and cultural need for hunting whales (**Table 14.4**). This currently occurs in countries such as the Russian Federation, the United States, Greenland (an autonomous country within the Kingdom of Denmark), and St. Vincent and the Grenadines (Bequia). Aboriginal whale hunts are generally less controversial than scientific whaling, partly because those taking the animals are in need of meat and blubber to survive and partly because they are still governed through the IWC by way of a quota system.

The Inupiat and Yupik people of Alaska have hunted bowhead whales for over 2,000 years (see Exploring the Depths: Aboriginal or Indigenous Whaling) and are allocated a quota of approximately 60 whales a year. Although bowhead whales are listed under the U.S. Endangered Species Act as "endangered" and under the MMPA as "depleted," both Acts contain provisions that exempt subsistence taking from their restrictions (although subsistence takes may still require a formal environmental assessment process under the U.S. National Environmental Policy Act). Most notably, the MMPA not only sets limits on the impacts of human activities on marine mammals, it also restricts the effects of those activities on subsistence hunts as well. Regardless of the legal standing, the Inupiat and Yupik hunt is widely accepted. Traditional methods have been set aside in favor of more humane modern methods, including spotter planes, motorized boats, and exploding harpoons. It has been argued that because of the use of more modern methods, whales are killed faster and thus the hunts are more humane than if traditional methods were used. The meat and blubber from the whales are used to feed the population, and whale bone and baleen are carved to make souvenirs for tourists to provide a source of income for other needed items.

In contrast to the Inupiat and Yupik hunt, the 1996 application for an aboriginal whaling quota of gray whales by the Makah of Washington State (United States) was denied. The Makah argued that they had the right to whale (as granted by treaty between the U.S. Government and Native Americans), that there was a long historical tradition of the Makah hunting whales (see Exploring the Depths: Aboriginal or Indigenous Whaling), that there was overwhelming support for a resumption of whaling in the community, and that there was a nutritional need evidenced by high rates of diabetes and nutritional health problems in indigenous populations (see Renker, 2007) or the whaling "needs statement" by the Makah. However, one technical criticism of the proposed hunt was that the Makah have not hunted whales since 1926 and thus some believed the Makah did not qualify for an aboriginal whaling quota, because there was no nutritional need after a 70-year absence of whale meat in the diet.

The following year the Makah issued a joint proposal with the Chukotka people of Russia, which was viewed more positively and ultimately granted, although some concerns remained (noted below). One gray whale was caught in 1999, but court injunctions prevented further catches.

A number of animal welfare groups protested the Makah hunt on various grounds. One criticism was that this particular hunt is supposed to be traditional and cultural, but the hunters use motorized boats and antitank rifles. However, as noted above, the use of modern equipment may be more humane than traditional methods. There were also concerns that several of the traditional whale hunting rituals had not been followed, for example, abstaining from alcohol and sex before the hunt for those taking part, and for the hunters to be involved with the butchering of the hunted whale. Finally, because of the use of modern

TABLE 14.4 Whales Hunted Under Aboriginal Quotas in Recent Years

Country	Species	2005	2006	2007
Russian Federation	Gray whale	124	134	131
	Bowhead	2	3	0
Greenland	Minke whale	180	184	169
	Fin whale	13	11	12
	Humpback	0	1	0
St. Vincent and The Grenadines	Humpback	1	1	1
United States	Bowhead	68	39	63
	Gray whale			1*

* Illegal catch
Data from: International Whaling Commission.

weapons there were issues related to human safety, with high-powered rifles being used in close proximity to shipping and human habitation.

In December 2002 the U.S. Ninth Circuit Court of Appeals ruled that the Makah whale hunt violated the MMPA. In 2005 the U.S. government received an MMPA waiver request from the Makah. However, in September 2007 five members of the Makah tribe hunted a gray whale without permission and in violation of the MMPA; they were prosecuted in June 2008. Three of the whalers pleaded guilty and were each sentenced to 2 years of probation and over 100 hours of community service. The other two whalers went to trial. One was sentenced to 90 days in prison. However, the judge saw the other as "remorseless" for killing the whale, sentencing him to 5 months in federal prison.

Greenland has an annual hunt of minke whales and fin whales, but this has also been controversial because no assessment of the number of whales found off Greenland had been conducted for well over a decade. However, no sooner had the data been gathered on whale abundances than other controversial issues were raised. Specifically, in 2008 Greenland called for a hunt of 10 humpback whales. These whales were not included in the assessment of Greenland's whales, so it would be impossible to determine whether such a hunt would be sustainable. The request was appropriately turned down in 2008 but in the last moments of the 2009 meeting this request was repeated. The matter was not settled at the main IWC meeting but rather at a subsequent intersessional meeting at which only a few countries attended, partly because funds were not available for the additional travel for many of the poorer countries. At this meeting the request was granted, which caused some disgruntlement with those unable to attend and cast a vote.

There are additional concerns about the Greenlandic hunt because not all whale meat is consumed locally for subsistence needs. A survey by the World Society for the Protection of Animals found that as much as 19% of minke whale meat and 40% of fin whale meat was sold in markets elsewhere in Greenland or even exported to Denmark (the assumption is that this is not a violation of CITES due to the ties that remain between these nations over foreign policy, although there is a legal gray area as to whether this import of whale meat into the European Union (EU) is against European law [Greenland is not a member of the EU, despite its association with Denmark]). Such distribution is contrary to the local subsistence consumption requirements for aboriginal whaling quotas. Moreover, there have been attempts to export meat to Norway. Environmental groups such as the World Society for Protection of Animals have argued for a reduction of the aboriginal whaling quota for Greenland, because there is obviously a surplus supply of meat beyond what was required for local subsistence needs.

Finally, concerns have also been raised about the humane aspects of the hunt. For example, in 2003 a hunted fin whale took 12 hours to die, and in 2002 and 2003 two minke whales each took around 5 hours to die. These data led to several animal welfare groups voicing additional opposition to the Greenland hunts.

In recent years Japan has tried to gain a quota of whales for "small-type coastal whaling." Attempts have been made to draw parallels of this type of whaling with aboriginal whaling, suggesting "village" communities that would conduct this whaling are small and impoverished. However, one community proposed to be involved in this type of whaling is the city of Shimonoseki (**Figure 14.3**), which has a population approximately one-half that of Washington, DC, or Boston, Massachusetts (United States). Moreover, this city is a major port, with many large industrial plants; thus, whaling would only ever be a minute fraction of the city's economic activity. Moreover, as proposed the "small-type coastal whaling" would be conducted with Antarctic-style whale catcher ships. Consequently, this would only be different from current scientific or commercial whaling because of the coastal location where whales would be taken.

FIGURE 14.3 The city of Shimonoseki, which is the host port of many whaling vessels.

Exploring the Depths: Aboriginal or Indigenous Whaling

Appearing in both art and legend, whales have long inspired the indigenous coastal people of this planet. Although the majesty and migratory patterns of certain cetaceans attracted the attention of island and shore-dwelling people, the whaling tradition of aboriginal people varies from a few thousand to a few hundred years. Many of these aboriginal populations took advantage of drift whales and stranded whales for food and material resources. Few developed the daring, expertise, and technology to hunt healthy whales in the coastal maritime environment (**Figure B14.2**). An even smaller segment had the navigational capability to travel out of sight of land and hunt one or more of the great whales: sperm, fin, humpback, and right whales, as well as bowheads and grays.

Prominent among traditional native whalers are the hunters of the north Pacific and Atlantic Oceans. The contemporary Makah Tribe of Washington State, United States, and their Nootkan relatives in Canada, has the most well-documented ethnographic whale-hunting tradition thanks in part to voluminous records of traders, explorers, Indian agents, and early anthropologists. As for archeology, the research generated by the Ozette archeological excavation (1970–1981) provided archeologists with a glimpse of a vigorous whaling village buried by a catastrophic mudslide before contact with non-Native Americans. Although it is impossible to determine the exact age of this whaling tradition, sites along the northern coast of Washington State and Vancouver Island indicate coastal whalers could have been hunting as early as 2,000 to 3,000 years ago.

These intrepid hunters ventured 50 to 100 miles into the Pacific Ocean in cedar canoes. A crew of eight ritually trained men primarily pursued the humpback whale and the migratory gray whale. The Alaskan Eskimo, Canadian Inuit, and Greenlandic Inuit whale hunters also opportunistically pursued great whales that swam in their coastal waters, including the bowhead and fin whales. Smaller minke whales were also a primary prey for Greenlandic natives. Yupik, Eskimo, and Chukchi hunters living along the western Pacific Coast also pursued the gray and bowhead whales miles out to sea. Now known as the Chukotkan whale hunters, these people also have an archeologically demonstrated whaling tradition that dates back at least 2,000 years.

The length of time that aboriginal whaling has occurred in warmer waters is difficult to determine. On the Indonesian islands of Lembata and Solor the sperm whale remains the preferred prey, although large fish, such as whale sharks, are currently hunted with similar techniques. In the Caribbean Islands, where smaller whale species tended to be the primary prey, a significant exception is found among the Bequian whalers of St. Vincent and the Grenadines, where humpbacks are the quarry.

From an ethnographic perspective aboriginal whaling is important and interesting for several reasons. The practice indicates hunters lived in a cohesive, complex economy that could mobilize and coordinate considerable human resources to process, preserve, and distribute the enormous amount of food, byproducts, and raw materials a great whale would bring to the community. In addition, the inherent danger of the pursuit brought the social and supernatural realms of indigenous life together in a complex web of ritual, prestige, and social networks, which was renewed with every hunt.

—Contributing author, Ann Renker, Makah Cultural Resource Center.

FIGURE B14.2 A bowhead whale caught in a managed aboriginal whaling hunt in Alaska.

Exploring the Depths: Welfare and Whaling

The catastrophic mismanagement of whales and whaling in the latter part of the 20th century gave rise to an impassioned anti-whaling movement, a movement concerned with governments' failure to protect species from over-exploitation. "Save the whales!" was predominantly a rallying cry to prevent extinctions. In recent decades, however, there has been a growing awareness and acceptance that our responsibility to protect whales and other animals extends beyond simply maintaining viable populations. Governments, industries, and consumers around the world are slowly but surely embracing the scientific and ethical case for the humane treatment of animals.

Despite decades of research, most notably by the Norwegian government, a reliably humane way to kill whales has not been found. In ongoing commercial/"scientific" hunts in Norway, Iceland, and Japan, whales are killed using exploding penthrite harpoons, a method little changed in over one hundred years. The harpoon is designed to penetrate into the whale's body, before the trigger mechanism fires the grenade. Whalers aim at the upper thorax region, where an accurate strike can result in blast-induced neurotrauma, rendering the whale instantly insensible. It is true to say that a proportion of whales do die in this way, and do not suffer. The fundamental welfare problem with whaling lies with those whales that are not effectively despatched by the initial harpoon strike.

Even the best whaler cannot assure an accurate lethal harpoon shot. Whalers are shooting a moving animal—which may appear on the surface to breathe for only a few seconds—from a moving platform. Add into this equation possibly adverse weather conditions and the fact that whales vary greatly in size, and the result is significant—and unavoidable—margin for error. The last data reported by whaling nations, during the early 2000s, highlighted that around one in five whales in Norwegian hunts, and around three in five whales in Japanese hunts do not die within one minute. Times to death averaged between 2–3 minutes but whales have been recorded as taking over an hour to succumb to their injuries.

Whales struck by an exploding harpoon but not rendered unconscious will inevitably suffer extraordinary pain as they remain anchored to the whaling vessel by the embedded harpoon. Despite catastrophic internal injuries and blood loss, numerous videos of whale hunts show harpooned whales actively attempting to dive away from the threat, aggravating the wound. In some instances, if the harpoon dislodges, whales are said to be "struck and lost," many doomed to die lingering painful deaths from their wounds.

Whalers will of course work to deliver "secondary killing methods"—these may entail further harpoons or rifle fire—but can struggle to bring the stressed animal under control to effectively deploy these. In a Norwegian whale hunt filmed in 2005 it took whalers over 14 minutes and seven rifle shots after the harpoon impact before the animal's last visible sign of life. A 2007 study of Japanese hunt welfare concluded that a significant proportion of whales were in fact killed by asphyxiation, having been winched tightly to the bow of the ship with their blow holes submerged.

In 2009, Japan, Norway, and Iceland collectively killed some 1,681 whales (Norway, 484; Japan, 993; Iceland, 204). Since all three countries have stopped collecting welfare data—they admitted resenting that the data was used against them by anti-whaling organizations and governments—the international community has no way of knowing how many died immediately. However, based on previous statistics it is fair to estimate that many hundreds of these animals will have endured long and painful deaths.

There is a wealth of evidence to demonstrate that there is no reliably humane way to kill whales at sea. This then leaves governments and the International Whaling Commission with the fundamental question of whether whaling is an ethically defensible method of commercial meat production in the 21st century. As countless governments and intergovernmental bodies now decree that animals be treated humanely and spared "unnecessary suffering," the slaughter methods employed by the world's relic whaling industries look increasingly archaic, crude, and unacceptable. In this light, the 1986 ban on commercial whaling is today relevant not only to guard against species loss, but also to protect individual sentient animals from unacceptable and unnecessary suffering.

The true unexaggerated misery of commercial whaling remains best described in an eyewitness account. Dr. Harry Lillie, a ship's physician on an Antarctic whaling trip in the 1940s, wrote: "If we can imagine a horse having two or three explosive spears stuck in its stomach and being made to pull a butcher's truck through the streets of London while it pours blood into the gutter, we shall have an idea of the method of killing. The gunners themselves admit that if whales could scream the industry would stop, for nobody would be able to stand it."

–Contributing author, Claire Bass, World Society for Protection of Animals (WSPA).

SELECTED REFERENCES AND FURTHER READING

Brakes, P., et al, (2004). *Troubled Waters: a review of the welfare implications of modern whaling activities.* World Society for the Protection of Animals.

Clapham, P. J., Berggren, P., Childerhouse, S., Friday, N. A., Kasuya, T., Kell, L., Kock, K., Manzanilla-Naim, S., Sciara, G. N. D., Perrin, W. F., Read, A. J., Reeves, R. R., Rogan, E., Rojas-Brancho, L., Smith, T. D., Stachowitsch, M., Taylor, B. L., Thiele, D., Wade, P. R., & Brownell, R. L. (2003). Whaling as science. *BioScience* 53: 210–212.

Clapham, P. J., Childerhouse, S., Gales, N. J., Rojas-Bracho, L., Tillman, M. F., & Brownell, R. L. (2007). The whaling issue: conservation, confusion, and casuistry. *Marine Policy* 31: 314–319.

Coleman, J. L. (1995). The American whale oil industry: a look back to the future of the American petroleum industry? *Natural Resources Research* 4: 273–288.

Currie, D. (2007). Whales, sustainability and international environmental governance. *RECIEL* 16: 45–57.

Ellis, R. (1991). *Men and Whales.* Knopf, New York.

Endo, A., & Yamao, M. (2007). Policies governing the distribution of by-products from scientific and small-scale coastal whaling in Japan. *Marine Policy* 31: 169–181.

Endo, T., Haraguchi, K., Cipriano, F., Simmonds, M. P., Hotta, Y., & Sakata, M. (2004). Contamination by mercury and cadmium in the cetacean products from Japanese market. *Chemosphere* 54: 1653–1662.

Endo, T., Haraguchi, K., Hotta, Y., Hisamichi, Y., Lavery, S., Dalebout, M. L., & Baker, C. S. (2005). Total mercury, methylmercury, and selenium levels in the red meat of small cetaceans sold for human consumption in Japan. *Environmental Science and Technology* 39: 5703–5708.

Endo, T., Haraguchi, K., & Sakata, M. (2002). Mercury and selenium concentrations in the internal organs of toothed whales and dolphins marketed for human consumption in Japan. *Science of the Total Environment* 300: 15–22.

Endo, T., Haraguchi, K., & Sakata, M. (2003). Renal toxicity in rats after oral administration of mercury-contaminated boiled whale livers marketed for human consumption. *Archives of Environmental Contamination and Toxicology* 44: 412–416.

Gales, N.J., Kasuya, T., Clapham, P. J., & Brownell, R. L. (2005). Japan's whaling plan under scrutiny. *Nature* 435: 883–884.

Gales, N., et al. (2007). Is Japan's whaling humane? *Marine Policy* 32(3): 408–412.

Gambell, R. (1999). The International Whaling Commission and the contemporary whaling debate. In: *Conservation and Management of Marine Mammals* (Ed. J. R. Twiss, Jr., & R. R. Reeves), pp. 179–198. Smithsonian Institution Press, Washington, DC.

Gerber, L., Morisette, L., Kaschner, K., & Pauly, D. (2009). Should whales be culled to increase fishery yield? *Science* 323: 880–881.

Hamazaki, T., & Tanno, D. (2001). Approval of whaling and whaling-related beliefs: public opinion in whaling and nonwhaling countries. *Human Dimensions of Wildlife* 6: 131–144.

Hammond, P. (2006). Whale science and how (not) to use it. *Significance* 3: 54–56.

Harrop, S.R. (2003). From cartel to conservation and on to compassion: animal welfare and the International Whaling Commission. *International Journal of Wildlife Law & Policy* 6: 79–104.

Heazle, M. (2004). Scientific uncertainty and the International Whaling Commission: an alternative perspective on the use of science in policy making. *Marine Policy* 28: 361–374.

Herrera, G.E., & Hoagland, P. (2006). Commercial whaling, tourism, and boycotts: An economic perspective. *Marine Policy* 30: 261–269.

Hirata, K. (2005). Why Japan supports whaling. *Journal of International Wildlife Law & Policy* 8: 129–150.

Holt, S. J. (2006). Propaganda and pretext. *Marine Pollution Bulletin* 52: 363–366.

Holt, S. J. (2007). Whaling: will the Phoenix rise again? *Marine Pollution Bulletin* 54: 1081–1086.

Huelsbeck, D. R. (1994a). Mammals and fish in the subsistence economy of Ozette. In: *Ozette Archaeological Project Reports*, Volume II (Ed. S. Samuels), pp. 17–92. Washington State University, Pullman, WA.

Huelsbeck, D. R. (1994b). The utilization of whales at Ozette, In: *Ozette Archaeological Project Reports*, Volume II (Ed. S. Samuels), pp. 265–304. Washington State University, Pullman, WA.

Iliff, M. (2008a). Normalization of the International Whaling Commission. *Marine Policy* 32: 333–338.

Iliff, M. (2008b). Compromise in the IWC: is it possible or desirable? *Marine Policy* 32: 997–1003.

Kaschner, K., & Pauly, D. (2005). Competition between marine mammals and fisheries—food for thought? In: *The State of the Animals*, Volume III (Ed. D. J. Salem & A. N. Rowan), p. 95–117. Humane Society of the United States Press, Gaithersburg, MD.

Kasuya, T. (2007). Japanese whaling and other cetacean fisheries. *Environmental Science & Pollution Research* 14: 39–48.

Knowles, T.G., & Butterworth, A., (2006). *Immediate immobilisation of a minke whale using a grenade harpoon requires striking a restricted target area.* Animal Welfare 15: 55–57.

Miller, A. R., & Dolšak, N. (2007). Issue linkages in international environmental policy: the International Whaling Commission and Japanese development aid. *Global Environmental Politics* 7: 69–96.

Morishita, J. (2006). Multiple analysis of the whaling issue: understanding the dispute by a matrix. *Marine Policy* 30: 802–808.

Murata, K. (2007). Pro- and anti-whaling discourses in British and Japanese newspaper reports in comparison: a cross-cultural perspective. *Discourse & Society* 18: 741–764.

Ohishi, K., Zenitani, R., Bando, T., Goto, Y., Uchida, K., Maruyama, T., Yamamoto, S., Miyazaki, N., & Fujise, Y. (2003). Pathological and serological evidence of *Brucella*-infection in baleen whales (Mysticeti) in the western North Pacific. *Comparative Immunology, Microbiology and Infectious Diseases* 26: 125–136.

Osherenko, G. (2005). Environmental justice and the International Whaling Commission: Moby-Dick revisited. *Journal of International Wildlife Law & Policy* 8: 221–240.

Parsons, E. C. M., & Draheim, M. (2009). A reason not to support whaling: a case study from the Dominican Republic. *Current Issues in Tourism* 12: 397–403.

Parsons, E. C. M., & Rawles, C. (2003). The resumption of whaling by Iceland and the potential negative impact in the Icelandic whale-watching market. *Current Issues in Tourism* 6: 444–448.

Parsons, E. C. M., Rose, N. A., Bass, C., Perry, C., & Simmonds, M. P. (2006). It's not just poor science—Japan's "scientific" whaling may be a human health risk too. *Marine Pollution Bulletin* 52: 1118–1120.

Powell, E. A. (2009). Origins of whaling, Chukotka Peninsula, Russia. *Archaeology* 62(1). Retrieved November 1, 2011 from http://www.archaeology.org/0901/topten/origins_of_whaling.html.

Renker, A. M. (2002). The Makah tribe: people of the sea and the forest. Digital Collection. University of Washington Libraries, Seattle, WA. Retrieved November 1, 2011 from http://content.lib.washington.edu/aipnw/renker.html

Renker, A. M. (2007). Makah whale hunting: a needs statement. Paper presented to the Commission at the 59th Meeting of the International Whaling Commission, May 2007, Anchorage, Alaska. Retrieved November 1, 2011 from http://iwcoffice.org/_documents/commission/iwc59docs/59-asw%209.pdf.

Renker, A. M., & Gunther, E. (1990). The Makah. In: *The Handbook of North American Indians, Volume 7: The Northwest Coast* (Ed. W. Sturtevant), pp. 422–430. Smithsonian Institution, Washington, DC.

Schweder, T. (2001). Protecting whales by distorting uncertainty: non-precautionary mismanagement? *Fisheries Research* 52: 217–225.

Simmonds, M. P., Haraguchi, K., Endo, T., Cipriano, F., Palumbi, S. R., & Troisi, G. M. (2002). Human health significance of organochlorine and mercury contaminants in Japanese whale meat. *Journal of Toxicology & Environmental Health* 65: 1211–1235.

Stoett, P. J. (2005). Of whales and people: normative theory, symbolism, and the IWC. *Journal of International Wildlife Law & Policy* 8: 151–176.

Waterman, T. T. (1920). Whaling equipment of the Makah Indians. *University of Washington Publications in Anthropology* 1: 1–67.

Wessen, G. (1990). Archaeology of the ocean coast of Washington. In: *The Handbook of North American Indians, Volume 7: The Northwest Coast* (Ed. W. Sturtevant), pp. 412–420. Smithsonian Institution, Washington, DC.

World Society for the Protection of Animals. (2008). *Exploding Myths. An Exposé of the Commercial Elements of Greenlandic Aboriginal Subsistence Whaling.* World Society for Protection of Animals, London.

Threats to Cetaceans

CHAPTER 15

CHAPTER OUTLINE

Direct Takes of Small Cetaceans

Exploring the Depths: Cetacean Culls
Exploring the Depths: Live Takes

Fisheries and Cetaceans

Exploring the Depths: What is Stress?
Exploring the Depths: U.S. Tuna Labeling Controversy

Ship Strikes

Pollution

Heavy Metals
Organohalogens

Exploring the Depths: Butyltin
Exploring the Depths: Radioactive Discharges
Harmful Algal Blooms
Marine Litter and Debris
Oil
Exploring the Depths: Polyaromatic Hydrocarbons
Sewage and Disease

Climate Change

Habitat Degradation

Selected References and Further Reading

Direct Takes of Small Cetaceans

In Chapter 14 directed takes (i.e., deliberate hunting) of large whales were discussed. Several member countries consider the International Whaling Commission (IWC) to be only competent to manage directed takes of the so-called great whales, meaning the sperm whales and all the baleen whales, except for the pygmy right whale. An appendix to the original convention listed these species, but nowhere in the original convention does it emphatically state that the working of the IWC is limited to these species. There is an IWC subcommittee on small cetaceans, in this case meaning all the odontocetes except for the sperm whale plus the pygmy right whale (including beaked whales, dolphins, and porpoises). However, the work of the subcommittee is poorly funded, and Japan and several of its allied nations do not attend these subcommittee meetings, so the amount that can be achieved is limited. There is no international body with wide membership that monitors or manages hunts of small cetaceans. Substantial numbers of small cetaceans are, however, taken around the world, and some of these takes may be threatening some populations with extirpation.

At present, there are aboriginal subsistence hunts for beluga whales and narwhals in Russia, Canada, the United States, and Greenland (see Chapters 11 and 14). Approximately 200 beluga whales are taken in Alaska each year, and a further 2,000 are taken in Canada, Russia, and western Greenland. Roughly 1,000 narwhals are taken in Canada and western Greenland. The blubber and meat from these animals are used for human and sled dog consumption. Native hunters use narwhal tusks for tent poles, sled runners, and lance shafts; narwhal tusks are also sold as curios to tourists. There are concerns about overharvesting of some populations, which have become depleted as a result (e.g., the beluga population of the Cook Inlet, near Anchorage, Alaska).

One particularly controversial hunt is the take of long-finned pilot whales in the Faroe Islands in the "grindadráp" hunt (**Figure 15.1**). The Faroes are an independent protectorate of Denmark where hunts of pilot whales have been conducted since 1584. The pilot whales are driven

FIGURE 15.1 Faroese pilot whale hunt.

into bays by boats. Hook-like gaffs attached to ropes are driven into the animals and then used to drag the whales to the shore. However, concerns about the humaneness of this technique led to the use of a blunt-ended gaff instead, which is inserted into the whale's blowhole, and then the animal is dragged to shore. Despite being considered more "humane," the blowhole is a sensitive part of the whale's anatomy, and the gaffs can damage the skin around it. When secured, a machete-like knife is used to cut into the back of the whale's head to sever the spinal cord. A study monitoring the nature of the whale-killing methods found that nearly one-half of the first attempts to dispatch the whale involve cuts into the wrong part of the body, and the time it took to cut into the whale's spinal cord on average took more than 1 minute.

From 1709 until the present more than 250,000 pilot whales have been hunted using this method, with an average annual hunt during the 20th century of approximately 1,200 pilot whales per year (with numbers ranging from about 1,000 to 3,000 during the first years of the 21st century). The Faroese state that the grindadráp is an important part of Faroese culture and a valuable source of nutrition for the islanders. However, whale meat carries health warnings because of high levels of the toxic heavy metals mercury and cadmium as well as organic pollutants (such as polychlorinated biphenyls [PCBs] and pesticides; see Pollution section) that are found in the meat. Several studies conducted in the Faroe Islands have examined the effects of eating contaminated long-finned pilot whale meat on the population. Reported effects linked with mercury contamination included mental retardation, neurological abnormalities, and brainstem damage in children whose mothers had consumed whale meat. Prenatal exposure to mercury was believed to cause irreversible neurological damage in children in the Faroes. In another study abnormal heart activity was also linked to mercury contamination in whale meat. Moreover, it has been suggested the unusually high rate of Parkinson disease among Faroe Islanders may also be linked to their whale meat consumption. In November 2008 the chief medical officer of the Faroe Islands advised that pilot whale meat should not be considered safe for human consumption because of its contaminant load. This is likely to further reduce the decreasing demand for the meat and may lead to the collapse of the market in the Faroes.

The largest commercial hunts for small cetaceans in recent years have been in Peru and Japan. Before 1996 some 60 ports in Peru were taking approximately 20,000 cetaceans annually, mostly dusky dolphins (*Lagenorhynchus obscurus*) and Burmeister's porpoises (*Phocoena spinipinnis*), for human consumption and as bait for fisheries. Technically, these catches were banned in 1990. However, the ban made little difference due to governmental inaction until 1996, when international pressure and media attention resulted in better enforcement. Despite this, some catches still continue, with perhaps as many as 3,000 animals taken each year.

In Japan at least 16 species of small cetaceans have been hunted in recent years, including approximately 50 Baird's beaked whales (*Berardius bairdii*) per year (hunted using explosive harpoons in the same way minke whales are hunted) and annual catches of between 10,000 and 17,000 Dall's porpoises (*Phocoenoides dalli*). Additionally, approximately half a million porpoises have been hunted since the 1960s. These porpoises are hunted with hand-held harpoons, which are thrown from the bows of fast motor boats. Mothers and calves are often targeted because calves tire and are easier to capture. This results in a disproportionately high catch of reproductively active females, which will have an adverse impact on the ability of the population to recover from these hunts.

In addition 1,000 to 2,000 small cetaceans from various species are killed annually in drive fisheries, where groups are herded by boats and underwater noise into enclosed bays to be killed. In 2008 the Japanese government reported their quotas for small cetacean fisheries for the 2007–2008 season, which included 16,876 Dall's porpoises, 541 Risso's dolphins, 1,018 bottlenose dolphins, 879 spotted dolphins, 685 striped dolphins, 369 short-finned pilot whales, and 100 false killer whales. A few dozen animals caught in this fashion are sold to "dolphinariums" and aquariums (especially young female bottlenose dolphins), whereas the remainder are slaughtered for their meat. The products of some of these animals are sold for human consumption,

Exploring the Depths: Cetacean Culls

Despite a lack of scientific evidence marine mammals, including cetaceans, are often used as a scapegoat for declining fisheries. In fact, this "reason" is often cited in support for the culling of cetaceans. Culls typically do not involve taking cetaceans for their products (such as meat and blubber) but simply involves killing them to reduce numbers. Small cetaceans in the Black Sea were culled in this way in substantial numbers (see Chapter 12). For example, between 20,000 and 30,000 bottlenose dolphins from the Black Sea subspecies *Tursiops truncatus ponticus* were killed, leading to the depletion of this subspecies. The Black Sea culls were most intense between 1931 and 1941 when some 50,000 small cetaceans were killed every year. Culls have occurred in other regions; for example, between 1928 and 1939 beluga whales were actually bombed in Quebec, Canada. In 1956, at the request of the Icelandic government, the U.S. Navy machine-gunned and depth-charged several hundred killer whales in Iceland because of perceived conflicts with fisheries.

Exploring the Depths: Live Takes

Cetaceans were first taken into captivity from the wild for public display over 100 years ago. Currently, at least 25 species are in captivity around the world. Facilities for these animals vary from penned-off areas of the ocean and landscaped lagoon-like pools through state-of-the-art tanks to bare concrete pools and muddy ponds. Of concern is the impact these takes have on wild populations of these species. For example, most dolphins in facilities in the Caribbean were caught from the wild, with many coming from Cuba (e.g., dolphins at the facility in Curaçao) where local authorities have issued capture permits for approximately 15 (and up to 28) bottlenose dolphins per year. Although some scientific studies estimate the abundance of dolphins in Cuban waters, to date, no study has been done to assess the impacts of these takes on the populations in the wild. The small cetacean subcommittee of the IWC expressed its concerns about the sustainability of these captures, and these concerns were echoed by the International Union for Conservation of Nature's (IUCN's) cetacean specialist group. Another incident occurred in the Caribbean in 2002 when eight bottlenose dolphins were illegally captured from the coastal waters of a national park in the Dominican Republic (**Figure B15.1**), leading to local protests and court cases. Across the border in Haiti six dolphins were captured in 2004, although these were released back into the wild after public protests (it should be noted that these protests were based on ethical issues, not specifically conservation concerns).

Live captures of Indo-Pacific bottlenose dolphins in the Solomon Islands have also received much attention in recent years. In 2003 approximately 100 dolphins were captured over the course of several months, and a further 30 to 50 were captured in late 2007. At least three animals are known to have died during the last capture operation, and the media reported that at least nine dolphins died during the 2003 captures. From the initial capture 28 animals were exported to a facility in Cancun, Mexico. At least half of these animals have died since their export. In October 2007 another 28 Solomon Island dolphins were exported to Dubai, in the United Arab Emirates, and 18 more were exported to the Philippines (destined for Singapore) in late 2008. Like the Cuban captures, these operations have been criticized by the small cetacean subcommittee of the IWC and the IUCN cetacean specialist group on the grounds that there is scant scientific data available on the wild population of Indo-Pacific bottlenose dolphins in the Solomon Islands and that these captures are unlikely to be sustainable.

Exploring the Depths: Live Takes (*continued*)

FIGURE B15.1 Dolphins in Manati Park, Dominican Republic.

One of the most contentious captures of bottlenose dolphins (and other species) is by Japanese drive fisheries. Bottlenose dolphins are selected from captured groups for aquariums and the remaining animals are slaughtered. In 1997 five killer whales were also live-captured off Taiji, Japan (**Figure B15.2**), and all were sent to Japanese aquariums. Within 5 months two were dead. Another died in 2004 and a fourth in 2007. Attempts to capture killer whales have also been made in Kamchatka, Russia. Since 2001 quotas from 6 to 10 animals have been issued by the authorities. After several unsuccessful attempts, a single female was caught in 2003 in a capture operation that also resulted in the death of a juvenile orca. Moreover, the captured female only survived for a mere 23 days (see Chapter 12).

The IUCN 2002–2010 Conservation Action Plan for the World's Cetaceans states the following (p. 17, Reeves *et al.*, 2003):

> *Removal of live cetaceans from the wild, for captive display and/or research, is equivalent to incidental or deliberate killing, as the animals brought into captivity (or killed during capture operations) are no longer available to help maintain their populations. When unmanaged and undertaken without a rigorous program of research and monitoring, live-capture can become a serious threat to local cetacean populations. All too often, entrepreneurs take advantage of lax (or non-existent) regulations in small island states or less-developed countries...*

Exploring the Depths: Live Takes (continued)

FIGURE B15.2 Drive hunts for cetaceans in Futo, Japan.

but many have been found to be highly contaminated with mercury. Most of the meat from these small cetaceans is used to produce fertilizer and domestic animal/fish feed.

Fisheries and Cetaceans

Fishing can affect all marine mammals in many ways. First, human fishing activity can deplete the prey species of cetaceans. The United Nations Food and Agriculture Organization estimated that 70% of the world's commercial fish stocks are fully fished, overexploited, severely depleted, or only slowly recovering from depletion. In short, the majority of the world's fish stocks are under strain from human fisheries. Although competition between fisheries and the "great whales" (baleen whales and sperm whales) is minimal (see Chapter 14), there may be some competition between human fisheries and delphinids for schooling fish and benthic species. Additionally, nonselective fishing methods may deplete prey species for cetaceans, even if they are not the target species of the fishery.

However, it is important to realize that prey item sizes may be a major consideration here. Selective pressures on certain life stages (e.g., mature adults) by one predator could, in certain situations, actually boost the availability of other life stages (e.g., juveniles) to another species. This works only if the different life stages of the prey species are actually in competition with each other for food or other resources. As a result, a cetacean consuming a juvenile fish might actually boost the size and number of adults that are the target of a fishery. Similarly, the reverse may be true for some smaller human fisheries, although it is unlikely that the relationship will hold for any large-scale removal of a prey species due to commercial fishery activities.

A more direct way that fisheries impact marine mammal populations is through bycatch of these animals in fishing gear (accidental entanglement) (**Figure 15.2**). Bycatch is a major source of cetacean mortality, killing at least 300,000 whales and dolphins (as well as slightly more pinnipeds) annually. Cetaceans can be entangled in nets that are in active use (i.e., being set by fishing boats) or discarded "ghost" nets. Furthermore, pieces of fishing line and other broken gear cast aside by careless fishermen may also entangle marine mammals, leading to injury or death.

Drift nets are infamous for their ability to indiscriminately entangle cetaceans and other marine species (such as turtles, sharks, and pinnipeds). These nets can be more

FIGURE 15.2 A dolphin caught in fishing gear.

than 50 km (30 miles) long and are largely undetectable to cetaceans. These nets were once widely used, particularly in the 1960s, but as a result of concerns over their impacts on marine life the United Nations instigated a ban in 1992 on nets longer than 2.5 km (1.6 miles). The number of animals caught in these nets could be substantial; for example drift nets for tuna used between the Azores and southwestern waters of the United Kingdom were estimated to have resulted in bycatch of some 1,600 striped and short-beaked common dolphins between 1992 and 1993.

Gillnets are smaller nets typically used most commonly in coastal waters and may float at the surface or sit on the seabed. In Europe and North America gill nets have been particularly problematic for harbor porpoises (*Phocoena phocoena*), causing unsustainable rates of mortality and extirpating some local populations in Europe. Efforts to reduce this mortality have resulted in regulations requiring the attachment of acoustic deterrent devices (or "pingers" [**Figure 15.3**]) to nets, which has reduced rates of bycatch in some areas. Presumably, these devices work by warning porpoises of the presence of nets or making them acoustically more "visible." Although the attachment of pingers has been a successful mitigation measure, they do cause some logistical problems with fishing, and they can sometimes fall off or fail, which means fishermen have to monitor the state of their nets for them to be fully effective. Similarly, some have expressed concern that widespread use of pingers in areas of high fishing activity may be excluding cetaceans from major parts of their habitat and limiting their access to food. Gill nets are almost completely responsible for bringing the vaquita to the brink of extinction (see Chapter 11).

The purse seine nets were introduced in 1959 as a means to catch yellowfin tuna in fisheries in the eastern tropical Pacific. These tuna fisheries in particular have probably killed more dolphins in the last 50 years than any other human activity. In the 1960s and 1970s over half a million dolphins were killed every year, with an estimated 6 to 12 million spinner (*Stenella longirostris*) and spotted dolphins (*S. attenuata*) bycaught throughout that period, in addition to many animals from several other species. The high mortality occurred because the fishery would specifically target schools of dolphins, as schools of tuna associate with groups of dolphins. It is not certain why they do this, but it may be that the dolphins provide the tuna some protection from predators or assist them finding prey.

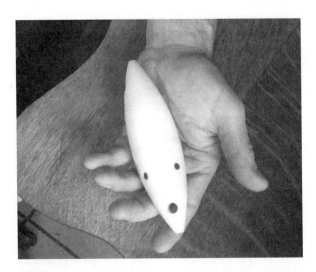

FIGURE 15.3 A "pinger" used to reduce bycatch in fishing gear.

(The dolphins appear to gain no benefits from the tuna.) Because dolphins are far more visible than submarine tuna, especially at distance, the purse seiners chase schools of dolphins and eventually set their nets around them, hoping to encircle not only the dolphins but also a good catch of tuna below. In the early days of the fishery the fishermen would then simply gather in the nets, trapping the target tuna but also entangling the dolphins.

A need to manage this high level of bycatch was one of the reasons behind the introduction of the 1972 U.S. Marine Mammal Protection Act (see Chapter 17). However, despite this legislative act the bycatch was not substantially curtailed. In 1990, as a result of public concern, voluntary "dolphin friendly" or "dolphin safe" tuna labels were introduced into U.S. law. Dolphin safe tuna was initially defined as tuna caught by methods other than the inherently harmful setting of purse seine nets on dolphins. The dolphin safe label quickly became widespread, and as a result of a massive decline in their ability to sell tuna, members of the Eastern Tropical Pacific Fishery began to make efforts to reduce dolphin mortalities.

In 1992 the Inter-American Tropical Tuna Commission, the body governing tuna fishing in the eastern tropical Pacific, adopted the International Dolphin Conservation Program, a voluntary bycatch reduction scheme. This was codified (put into law) as an international agreement in La Jolla, California. The agreement set limits on dolphin mortality for each tuna vessel. The aim was to reduce these limits each year until a zero mortality level was reached.

Measures to monitor and reduce dolphin mortality included an onboard observer program to record dolphin kills and the "backing down" of purse-seine nets (dropping the top of the net several feet below the water surface so encircled dolphins could swim out and over the top of the nets). In 1995 the Panama Declaration made these measures mandatory.

As a result of these actions, bycatch of dolphins in this fishery has reduced substantially, from hundreds of thousands of animals to a few thousand every year. Nonetheless, some animals are still entangled. Furthermore, the dolphin populations have not shown much, if any, sign of recovery. Despite the reduction in bycatch, dolphins are still chased, herded, and encircled by boats. A study into the reasons for the lack of recovery found that calves can easily become separated from their mothers during this chase and the distances involved mean they do not always reunite. Furthermore, the fleeing animals are almost certainly undergoing an extensive, and possibly debilitating, stress response. Given that any single animal may then be targeted several times in a year, chronic stress and the effects of stress on young animals may be also involved in the lack of recovery in these populations.

The regulations for the tuna–dolphin issue were only for the eastern tropical Pacific region. Elsewhere in the world tuna fisheries have largely unregulated bycatches of cetaceans.

The Panama Declaration introduced a new definition of dolphin safe. If observers did not see any dead dolphins

Exploring the Depths: What is Stress?

The term "stress" is colloquially used to refer collectively to the physiological, psychological, and behavioral responses to danger or, more accurately, *perceived* danger. The physiological stress response is initially driven primarily by adrenaline and noradrenaline (also known as epinephrine and norepinephrine), which induces the so-called fight-or-flight mechanism. A longer term recovery phase then sets in, driven by hormones (glucocorticoids) secreted from the adrenal gland (e.g., cortisol, corticosterone). Both phases are designed to maximize the survival potential of the animal as it seeks to eat or to avoid being eaten, generally shifting resources away from activities of long-term interest (such as reproduction and digestion) in favor of a more immediate focus on visual acuity and muscle activity.

Although this is a life-saving mechanism, a perfectly reasonable escape mechanism might become ineffective or even inappropriate in response to anthropogenic disturbance. Consider that a hedgehog curling up into a ball is perfect for avoiding predators but likely to be lethal in response to an approaching car. Similarly, a cetacean hanging slightly below the surface to avoid an unknown predator but maintaining close access to air is at the perfect depth to be hit by an oncoming ship.

Repeated or prolonged stress responses can also become maladaptive. For example, a leading cause of infertility in humans is reported to be stress. Heart disease and a number of other potentially lethal conditions have also been linked to some extent with chronic stress. This all has obvious implications for the conservation of a depleted marine mammal species.

Although behavioral responses associated with physiological stress responses are highly context-based, certain behaviors are generally associated with annoyance in cetaceans, which may indicate the initiation of a stress response. In cetaceans these behavioral responses may include obvious stereotypes, such as "fluke slapping" on the surface or breaching. However, physiological responses may be initiated without any obvious behavioral clues, in much the same way as people living on or near busy roads may suffer from chronic stress without any outward indication.

Exploring the Depths: U.S. Tuna Labeling Controversy

U.S. law states that it must be proven that the purse seine tuna fishery in the eastern tropical Pacific fishery is not causing "significant adverse impacts" on dolphin stocks in order for it to be considered dolphin safe and labeled accordingly (**Figure B15.3**). In 1999 and again in 2002 the U.S. Department of Commerce tried to make such a ruling. Such a determination would allow Mexico and other tuna fishing nations using purse seine nets to label their tuna as dolphin safe, a U.S. requirement for them to sell it in the United States. However, an environmental group, the Earth Island Institute, sued the government in federal court and won, preventing such a determination.

During this controversial case, leaked memos and other documents brought up in the case proved that the U.S. government had conceded (in private) that studies conducted by U.S. government scientists showed that dolphin populations were not recovering and that there was significant stress-related mortality associated with the fishery. Additionally, there was evidence of under-reporting of dolphin mortalities. Political pressure was put on the Commerce Department to ignore the conclusions of its own scientists, and these facts were virtually ignored when the U.S. government made its findings. In total, it was found that attempts had been made to withhold more than 300 important documents from the court record. This was heavily criticized by the courts.

FIGURE B15.3 A tuna can with a dolphin safe label.

in a net, then the tuna from that set would be considered dolphin safe, and their ability to sell that tuna would be greatly enhanced. However, this new definition of dolphin safe used an "observed mortality" standard rather than a "fishing method" standard. The U.S. Dolphin Conservation Act of 1997 stated that the United States would adopt this new definition if the government made a final finding that the chase and encirclement of dolphins were not having significant adverse impacts on depleted stocks. However, given the calf separation and probable chronic stress (revealed initially by U.S. government scientists), reaching this conclusion was not possible. Despite this, the government tried to make such a finding in 1999 and again in 2002. On both occasions the finding was challenged in court, and in both cases the courts ruled against the government because of the scientific information discussed above (see Exploring the Depths: U.S. Tuna Labeling Controversy). As a result, the original definition of dolphin safe has so far been retained within the United States.

Ship Strikes

Reports of ships hitting and killing cetaceans were found as early as the 1800s, but ship strikes were relatively unusual events until the 1950s. Rates of ship collisions with cetaceans have increased since that time, partly due to the volume of shipping but also to the speed at which vessels are traveling. For some species of whales, such as North Atlantic right whales, ship strikes are a major cause of mortality (**Figure 15.4**). In particular, high-speed ferries are a particular cause of concern, with multiple ship strikes reported from the Canary Islands, Japan, and the Mediterranean. Ship strike rates are so high in the Canaries that it has been predicted that the short-finned pilot whales (*Globicephala macrorhynchus*) in coastal waters might be at risk of extirpation if the situation continues unabated.

Ironically, there seems to be a relatively high rate of ship strikes by whale watching vessels, no doubt because they specifically target areas where cetaceans occur and often attempt to get as close to whales and dolphins as possible. The trend for companies in some areas to purchase larger and faster whale watching vessels is thus a serious cause of concern. Faster vessels decrease the time that boat operators (and whales) have to avoid vessel collision. Faster moving vessels also are more likely to cause mortal injuries rather than less severe wounds. It has been suggested that boats traveling at speeds over 13 knots are more likely to kill cetaceans in a collision than boats traveling at slower speeds.

FIGURE 15.4 A ship-struck whale.

Pollution

One of the most insidious and widespread threats to cetacean populations is marine pollution. Pollution comes in many forms:

- Industrial waste
- Agricultural chemicals
- Sewage
- Radioactive discharges
- Litter
- Oil
- Noise

Cetaceans occupy a high trophic level and as such are particularly susceptible to pollutants. Small concentrations of pollutants accumulate and magnify as one gets higher up the food chain. Pollutants such as heavy metals and many pesticides are considered conservative pollutants in that they are not readily broken down by metabolic or environmental processes, making them effectively permanent ecosystem components. Contaminants that cannot be excreted or broken down by an organism are said to bioaccumulate in tissues over that organism's lifetime. Animals feeding on bioaccumulators take in a higher level of contaminants, which in turn accumulate within themselves. It then follows that predators of these animals gain even higher inputs of contaminants and so on, with animals at the highest trophic level obtaining the highest concentrations in a process referred to as biomagnification. Because they are long-living, top predators, marine mammals generally bioaccumulate and biomagnify extremely high levels of contaminants.

Heavy Metals

Heavy metals are a class of contaminants of particular concern for cetaceans. These are high atomic weight metals, such as mercury or lead. Heavy metals are sometimes incorporated into a grouping known as "trace elements" along with other toxic substances that also include both nonmetals and lower atomic weight metals. Many of these trace elements are essential to the body in very low concentrations (for example, cobalt is a component of vitamin B_{12}), but in high concentrations they can be toxic. Some heavy metals, however, have no essential function in the body (e.g., mercury, lead), and any concentrations can be harmful.

Trace elements of particular concern include chromium, cadmium, and mercury. Some forms of chromium are toxic and even carcinogenic. Cadmium is also toxic and can cause depressed growth, cardiac changes, hypertension, kidney damage, fetal deformities, and cancer. Kidney damage linked to high concentrations of cadmium has been reported in cetaceans. Mercury can cause neurological damage, immune system suppression, and fetal abnormalities in mammals. Several industrialized countries including China, Japan, and countries bordering the Mediterranean have reported extremely high levels of mercury contamination in cetaceans. Specific toxic effects of mercury that have been observed in cetaceans include a decreased nutritional state and various lesions and liver abnormalities. High mercury (and other heavy metal) levels have also been associated with disease-induced mortality, as a consequence of immune suppressive action.

▪ Organohalogens

Another class of pollutant of particular concern with cetaceans is the organohalogens, which are organic molecules with halogen atoms (bromine, chlorine, fluorine) attached. Similar to heavy metals, organohalogens are conservative contaminants that bioaccumulate and biomagnify. Moreover, they are lipophilic (fat soluble). Organohalogens include pesticides, such as dichlorodiphenyltrichloroethane (DDT), PCBs, and polychlorinated dibenzodioxin (PCDD).

Introduced in 1939 as a pesticide, DDT is a long-lasting chemical that is relatively inexpensive to make. DDT was widely used in post–World War II Europe to eradicate fleas, the vector of the disease typhus, and in southern and eastern Asia against mosquitoes, the vector of the disease malaria. DDT was also used as a general pesticide in agriculture and forestry. DDT is metabolized by organisms into DDD (dichlorodiphenyldichloroethane) and then the less toxic DDE (dichlorodiphenyldichloroethylene), which is the most common form of this pesticide found in the marine environment. However, scientists, when referring to the chemicals in marine species, often lump DDT and its metabolites together and so when referring to DDT levels are talking about "total" DDT concentrations—often 90% or more of which is in the form DDE.

DDT began to be used internationally in 1940, and in just 20 years it could be found throughout the marine environment, including as a contaminant in the tissues of Antarctic wildlife, such as penguins. High DDT concentrations have been linked to mortalities of marine mammals. For example, pesticides from Salinas Valley, California, eventually became incorporated into the tissues of coastal seals and sea lions, leading to the death of some animals. Concern over the impacts of DDT on wildlife (especially birds) led Rachel Carson to write the book *Silent Spring* in 1963. The subsequent public and scientific attention toward this contaminant led to bans on its use in most countries in the 1970s.

PCBs have been produced since the 1930s and are used as insulating fluid for industrial transformers and capacitors, hydraulic fluids, surface coatings for copy paper, lubricating oils, flame retardants, and as additives in rubber, plastics, paints, and waxes. Until PCBs were banned in the late 1970s and early 1980s, the total world production reached 1.5 million tons, much of which ultimately entered the marine environment.

Polychlorinated dibenzodioxin (PCDD) or simply "dioxins," are some of the most toxic organic compounds and are used in herbicides (such as 2,4-dichlorophenoxyacetic acid and 2,4,5-trichlorophenoxyacetic acid). One of the worst of this class of chemicals is 2,3,7,8-tetrachlorodibenzodioxin (TCDD). The infamous Agent Orange defoliant used in the Vietnam War was a source of dioxins, which led to health problems and disease in the Vietnamese populations that were exposed to it. Dioxins are produced in wood processing and from the combustion of organic materials, such as those that occur in municipal or medical incinerators and heating furnaces.

Many of these organic pollutants are immunosuppressive, damaging the immune system and making mammals more susceptible to disease. Organochlorines, in particular, are hormone mimics and can cause infertility, fetal abnormalities, mental impairment, and growth abnormalities. As with other conservative contaminants, marine mammals are vulnerable to organohalogen contamination because they are long-lived top predators with a thick blubber (fat) layer that can store these lipophilic contaminants. Unfortunately, cetaceans, unlike many other mammals, lack an enzyme that can aid the breakdown of organohalogen contaminants. Mammal-eating killer whales (and also polar bears), which are consuming already contaminated seals, have been seen to biomagnify especially high levels of organochlorines.

Immune system suppression (and disease vulnerability) as a result of organochlorine contamination has been documented in both seals and cetaceans. Organochlorines have also been implicated with skeletal deformities, lipid metabolism abnormalities, testosterone deficiencies, reproductive abnormalities, and reproductive failure in a variety of species of marine mammals.

Organohalogens are lipid soluble and accumulate in the blubber layer of cetaceans. While in the blubber layer the pollutants are effectively inert. However, they can be mobilized during periods of food scarcity, chronic stress, disease, or pregnancy and lactation. Fortunately for the mother, the lipid-soluble pollutants can be passed to the offspring in the womb or to the newborn via lactation. Unfortunately for the calves, this means young cetaceans have been found with extremely high contaminant burdens. These have been linked with high levels of calf mortality, especially in first-born animals that get dosed with the full history of the mothers' contaminant exposure. (Later calves only receive doses equivalent to maybe 2 or 3 years of exposure because the mother could "off-load" the rest into previous calves with each birth.) A study on dolphin milk contamination in Hong Kong estimated that dolphin calves were receiving doses of PCBs and DDT exceeding human health limits if they ingested even tiny quantities of milk (1 ml)—no more than a few drops of milk a day!

Population level effects on marine mammals have also been linked to organohalogens. Probably the most clear-cut and famous example is the decline of beluga whales (*Delphinapterus leucas*) in the St. Lawrence Estuary. Other examples likely include seal mass mortality events in northern Europe (see Chapter 16) as well as cetacean mass mortalities in the North Sea (1988), the Mediterranean (1990–1992), the Gulf of Mexico (1992–1994), and on the east coast of the United States (1987–1988). In these latter cases the mobilization of fat stores, perhaps when animals were stressed, allowed organochlorines to circulate, making it likely that the animals' immune systems were suppressed. Organochlorines therefore likely contributed to these and other epidemic outbreaks and mass mortality events.

Research has found other organohalogens, such as brominated and fluorinated compounds, in marine species, most notably (once again) in top predators, such as marine mammals and birds. Although there are insufficient studies to confirm the particular details of their toxic effects, it is highly likely they have similar impacts to organochlorines and other organohalogens.

Harmful Algal Blooms

Nutrients are essential for life in the oceans, but there is growing concern over the accelerated nutrient enrichment of coastal waters because of human activities. Domestic sewage effluent, including human and household waste; runoff from agriculture; industrial discharges; shipping discharges (both sewage and wastewater); and even fish fecal waste and uneaten feed at aquaculture sites all contribute extra nutrients to the coastal marine environment. This increased nutrient content (especially nitrogen- and phosphorus-containing materials) can lead to an explosion in the growth of marine phytoplankton or algal blooms. Some algal species (for example, dinoflagellates) produce toxins. These toxin-producing algal blooms are referred to as harmful algal blooms.

These toxins are bioaccumulated and biomagnified through the food chain, may also be absorbed directly from the water column, and can even become airborne through spray at the sea surface or along coastlines. Beachgoers often report headaches during harmful algal blooms near to the coast. However, humans have the option to stay off the beaches while marine mammals (that breathe right at the surface) may not realize what is affecting them and/or may not have the option to leave a foraging or breeding area.

The toxins can cause diseases with effects that range from intestinal problems to paralysis or heart attacks. Algal toxins have been suggested as a major factor in several marine mammal mass mortalities (including species such as bottlenose dolphins, humpback whales, seals, and manatees). Algal toxins could not be found in significant quantities in all animals in those mortality events, but they may have added to the problems of already biologically stressed animals (e.g., high organohalogen levels). Moreover, algal toxin testing is not routinely undertaken in necropsies of stranded cetaceans and other dead marine mammals. Consequently, there may be more animals with algal toxin-related diseases than we realize.

Marine Litter and Debris

Marine debris (litter) includes all objects that do not naturally occur in the marine and coastal environment (water surface, water column, seabed, shore) but are nevertheless found there. Plastics are a major source of marine debris and make up the majority of floating debris. Discarded fishing nets are also a particular problem. This debris can entangle marine mammals (**Figure 15.5**), reducing the circulation in limbs if wrapped around them (possibly leading to infected injuries and gangrene) or impairing abilities (for example, whales wrapped in debris may have to expend

---

Exploring the Depths: Butyltin

Since the 1960s butyltin (especially tributyltin) was widely used as an antifouling paint on the hulls of ships and marine structures. The chemical prevents marine organisms, such as barnacles, from settling on painted structures and increasing drag or their weight. In the 1980s it was accepted that butyltins were having a toxic effect on marine organisms, and bans on tributyltin use on boats under 25 m started to be introduced. Despite these regulations, however, tributyltin pollution is still a problem in parts of the world with heavy shipping traffic, especially near ports used by large vessels. For example, high levels of butyltins have been found in Japan and Hong Kong, which are both home to vulnerable marine mammal species. Similar to organochlorines, butyltin is known to disrupt the immune system of mammals, including cetaceans. In this capacity butyltins may have contributed to a mass mortality event of dolphins in Florida. Recently, tributyltin has also been linked to hearing damage in mammals, clearly a problem for highly acoustic cetaceans.

Exploring the Depths: Radioactive Discharges

Although some radionuclides (atoms that have an unstable nucleus and can emit radiation when they decay) come from natural sources in the marine environment, there are far more anthropogenic sources of radioactive materials. These include atmospheric fallout from nuclear weapon detonations, accidental release from nuclear installations, and discharges from nuclear plants. Although the significance for marine mammal health is unknown, radionuclides have been detected in a number of cetaceans around the world, with some of the highest levels found in species around the United Kingdom.

considerable extra energy when swimming because of the drag the debris produces). Entanglement may lead to other injuries, especially amputations, and even death. It is estimated that as many as 100,000 marine mammals and turtles die every year because of entanglement in marine debris.

Debris can also be ingested and cause animals to choke. This is particularly an issue for plastic bags, which can be mistaken for jellyfish. The debris may block the digestive tract (resulting in starvation); fill the stomach, giving the animal a false sense of satiation (a full stomach), thus leading to starvation; and cause ulceration of the stomach and digestive tract. Toxic chemicals can also leach out of the ingested materials. Several species of cetacean, including harbor porpoises and northern minke whales, have been found to have ingested marine debris.

■ Oil

Catastrophic oil tanker spills are one form of pollutant widely covered in the media that really grabs public attention. Although these spills are extremely damaging where they do occur, nearly half of the oil in the oceans is actually from more routine sources. For example, small spills during oil transfer or tanker cleaning account for a greater volume

FIGURE 15.5 A humpback whale entangled in a discarded fishing net.

of oil than major spills every year. Urban sources, such as oil that makes its way into urban storm drainage systems or industrial discharges, account for more still: nearly six times the volume of oil entering the oceans from catastrophic tanker spills. Furthermore, natural seepage of oil through the seabed annually contributes around twice the amount of oil compared with catastrophic spills.

Oil can have several impacts on marine mammals, including respiratory problems if fumes from oil are inhaled (a problem for marine mammals surfacing in spills), accidental ingestion of oil or toxic hydrocarbons especially through eating contaminated prey, or becoming physically coated in oil. The latter may not be such a problem for cetaceans. This is partly because dolphins seem to actively avoid surface spills. Even if they became coated, the oil does not adhere to their smooth body surface for long (although baleen whales may be more affected as their baleen may trap oil). However, fur-bearing marine mammals, especially sea otters (*Enhydra lutris*), may be more vulnerable (see Chapter 7). Although estimates varied, somewhere between 3,500 and 5,500 sea otters were oiled during the *Exxon Valdez* oil spill, as were approximately 200 harbor seals (*Phoca vitulina*). Some populations of sea otters are not recovering from the drastic cut in numbers as was expected. This may be because of long-term chronic health effects due to the initial exposure as well as to contamination from toxic oil byproducts produced as the spill has been broken down in the environment (see Chapter 7).

The infamous *Deepwater Horizon* oil spill in 2010 (which released over 200 million gallons of crude oil into the Gulf of Mexico; **Figure 15.6**) caused concerns about the potential impacts on cetaceans in the Gulf: several groups of dolphins and sperm whales were observed swimming in the oil plume. There were also concerns about the potential effects on wildlife (including cetaceans) of dispersants being used on the oil slick and toxic byproducts of burning oil (a method used to reduce oil coverage). Although dead cetaceans were found during and after the spill, it is hard to say whether there was a direct link without extensive tests of stranded animals (and many were so decomposed that testing of tissues was impossible). A high number of dead dolphin calves (including fetal animals) during the early part of 2011 led to concerns that oil contamination might be linked to this seemingly high stillbirth and calf mortality rate. At the time of this writing, it is too early to tell if there is a link.

Exploring the Depths: Polyaromatic Hydrocarbons

Polyaromatic hydrocarbons (PAHs) are found in fossil fuels and are also emitted by burning organic materials. PAH contamination has therefore been associated mainly with oil spills. Adverse health effects could occur if animals consume PAH-tainted prey in areas impacted by oil spills. They can potentially inhale PAH contaminated air as well. PAHs can cause mental impairment, birth defects, and heart and respiratory abnormalities in humans, and several types of PAH are known to be carcinogenic (e.g., benzo[a]pyrene). Researchers have suggested that the high rate of tumors seen in beluga whales in the St. Lawrence Estuary may be the result of PAH contamination.

Sewage and Disease

Several studies have suggested that marine species may become sick from exposure to pathogens (disease-causing organisms) in domestic or agricultural sewage effluents. For example, a hepatitis outbreak in marine mammals in the United States was attributed to exposure to human sewage. Pet feces may also transfer pathogens into the marine environment. Distemper, toxoplasmosis, and neosporosis are pathogen-induced diseases believed to have been passed to dolphins and other marine mammals from dog or cat feces flushed out through sewage systems. The fact that many marine mammal populations also have pollutant burdens that influence their immune systems may make them more susceptible to sewage-borne diseases.

Climate Change

Climate change is still proving to be a controversial topic. Although the overwhelming majority of experts accept that it is happening and is being driven by human activities, some members of the general public and a small minority

FIGURE 15.6 Oil on a beach in the Gulf of Mexico from the *Deepwater Horizon* oil spill.

of scientists still strongly dispute this. The theory behind climate change is fairly straightforward. The sun provides heat to planet Earth. In the atmosphere certain gases, known as greenhouse gases, help to retain some of this radiation. Industrialized human activities in the last two centuries have added abnormally high levels of greenhouse gases, especially carbon dioxide (which is the most common but not the most efficient greenhouse gas), to the atmosphere. This is trapping more of the sun's energy and causing the planet to warm. The warming effect is not equally distributed over the whole planet. It is most strongly felt at the poles, where sea ice is in decline and glaciers are melting. The melting of the ice, especially over land in Greenland and the Antarctic, and the warming of the seas are causing the sea level to rise. It is also possible that human-caused climate change is increasing the occurrence and severity of extreme weather conditions.

Climate change has been a scientifically recognized international concern for many years. In fact, a group of independent experts, known as the Intergovernmental Panel on Climate Change (IPCC), was established over 20 years ago to advise world governments. The IPCC has issued advice via a series of reports, the most recent (the fourth report) issued in 2007. In that same year the organization was honored with the Nobel Peace Prize for its work. The core science from the IPCC has been brought together and articulated by 152 scientists from more than 30 countries and has been reviewed by more than 600 other experts. The main findings are as follows:

- Global atmospheric concentrations of carbon dioxide, methane, and nitrous oxide (the greenhouse gases) have increased markedly as a result of human activities since 1750 and now far exceed preindustrial values. The global increases in carbon dioxide concentration are due primarily to fossil fuel use and land use change, whereas those of methane and nitrous oxide are primarily due to agriculture.
- The IPCC has "very high confidence" that the global average net effect of human activities since 1750 has been one of warming.
- Warming of the climate system is "unequivocal," and this can be seen from observations of increases in global average air and ocean temperatures, widespread melting of snow and ice, and rising global average sea level.
- At continental, regional, and ocean basin scales numerous long-term changes in climate have been observed. These include changes in arctic temperatures and ice, widespread changes in precipitation amounts, ocean salinity, wind patterns and aspects of extreme weather including droughts, heavy precipitation (rain and snow storms), heat waves, and the intensity of tropical cyclones.
- Continued greenhouse gas emissions at or above current rates will cause further warming and induce many changes in the global climate system during the 21st century that will "very likely" be larger than those observed during the 20th century.

An irreversible change that will result from climate change is species loss, and the IPCC has estimated that 20 to 30% of plant and animal species assessed so far are likely to be at increased risk of extinction if global temperatures rise by more than 1.5 to 2.5°C. This figure increases to 40 to 70% if the global average temperature increase exceeds about 3.5°C. In the marine environment, known or predicted changes not only include an increase in temperature and a rise in sea levels but also a reduction in sea-ice cover and changes in storminess, salinity, acidity, and ocean circulation. For marine mammals in particular, it is likely that the distribution of many species will need to change to keep pace with these changes. It is also very likely that the distribution and abundance of various prey species will be similarly affected. Areas that were previously productive feedings zones may disappear, and species previously common in particular areas may become uncommon or be replaced entirely by other species. Marine mammals using polar habitats, especially sea-ice and ice-edge habitats, such as seals and polar bears or bowhead and Antarctic minke whales, are particularly at risk because these habitats are shrinking and will continue to do so (**Figure 15.7**) (for a discussion on climate change and polar bears see Chapter 6). Others may become trapped in enclosed water bodies, such as the Baltic or Mediterranean, unable to move easily northward toward areas with more appropriate conditions. For these animals the big question will be whether or not they can adapt.

There is also concern about how climate change may influence human behavior in ways that could cause greater conflict with marine mammals. In particular, the retreat of sea ice in the Arctic is opening up the area to vessels, fisheries, and fossil fuel exploration and extraction. All these activities could adversely impact the marine mammal populations there, which have been exposed to relatively little of these activities to date.

The precise implications of climate change are not clear, and various initiatives are now underway to assess likely impacts for particular species and to help identify the most vulnerable ones. However, it is clear that many

FIGURE 15.7 Melting ice and other climate change-induced phenomena are a cause of concern for a variety of marine mammal species.

marine mammals will face uncertain futures as a result of the climate-related changes in the marine environment.

Habitat Degradation

One final issue of concern is the high rate of physical loss and degradation of cetacean habitat. Many cetacean species have precise physical habitat requirements in terms of depth, freshwater access, coastal access, and other aspects. If they lose or are displaced from these habitats, it can have serious consequences for the health and viability of populations, especially if their use is linked with reproduction or foraging. Habitat loss, fragmentation, and degradation are caused by many factors (**Figure 15.8**):

- Land reclamation
- Dams and barrages
- Bridges, harbors, and other coastal and offshore constructions
- Dredging
- Siltation
- Fishing and aquaculture
- Boat traffic
- Other noise pollution (see Chapter 5)

246 PART III: Conservation

FIGURE 15.8 Examples of factors causing habitat loss. (A) Dredging, (B) speed boats, (C) coastal construction, (D) dams, and (E) silty coastal waters.

SELECTED REFERENCES AND FURTHER READING

Addison, R. F. (1989). Organochlorines and marine mammal reproduction. *Canadian Journal of Fisheries and Aquatic Science* 46: 360–368.

Aguilar, A., & Raga, J. A. (1993). The striped dolphin epizootic in the Mediterranean Sea. *Ambio* 22: 524–528.

Archer, F., Gerrodette, T., Chivers, S., & Jackson, A. (2004). Annual estimates of the unobserved incidental kill of pantropical spotted dolphin (*Stenella attenuata attenuata*) calves in the purse-seine fishery of the eastern tropical Pacific. *Fishery Bulletin* 102: 233–244.

Baird, R. W., & Hooker, S. K. (2000). Ingestion of plastic and unusual prey by a juvenile harbour porpoise. *Marine Pollution Bulletin* 40: 719–720.

Baker, J. R. (1989). Pollution-associated uterine lesions in grey seals from the Liverpool Bay area of the Irish Sea. *Veterinary Record* 125: 303.

Barlow, J., & Cameron, G. A. (2003). Field experiments show that acoustic pingers reduce marine mammal bycatch in the Californian drift gill net fishery. *Marine Mammal Science* 19: 265–283.

Bennett, P. M., Jepson, P. D., Law, R. J., Jones, B. R., Kuiken, T., Baker, J. R., Rogan, E., & Kirkwood, J. K. (2001). Exposure to heavy metals and infectious disease mortality in harbour porpoises from England and Wales. *Environmental Pollution* 112: 33–40.

Berrow, S. D., Long, S. C., McGarry, A. T., Pollard, D., Rogan, E., & Lockyer, C. (1998). Radionuclides (137Cs and 40K) in harbour porpoises *Phocoena phocoena* from British and Irish coastal waters. *Marine Pollution Bulletin* 36: 569–576.

Bodkin, J. L., Ballachey, B. E., Dean. T. A., Fukuyama, A. K., Jewett, S. C., McDonald, L., Monson, D., O'Clair, C. E., & VanBlaricom, G. R. (2002). Sea otter population status and the process of recovery from the 1989 *Exxon Valdez* oil spill. *Marine Ecology Progress Series* 241: 237–253.

Borrell, A., & Aguilar, A. (1991). Were PCB levels abnormally high in striped dolphins affected by the western Mediterranean die-off? *European Research on Cetaceans* 5: 88–92.

Bowles, D., & Lonsdale, J. (1994). An analysis of behaviour and killing times recorded during a pilot whale hunt. *Animal Welfare* 3: 285–304.

Buckland, S. T., Smith, T., & Cattanach, K. L. (1992). Status of small cetacean populations in the Black Sea: a review of current information and suggestions for future research. *Reports of the International Whaling Commission* 42: 513–516.

Calmet, D., Woodhead, D., & Andre, J. M. (1992). 210Pb, 137Cs and 10K in three species of porpoises caught in the eastern tropical Pacific Ocean. *Journal of Environmental Radioactivity* 15: 153–169.

Costas, E., & Lopez-Rodas, V. (1998). Paralytic phycotoxins in monk seal mass mortality. *Veterinary Record* 142: 643–644.

Cowan, D. F., & Curry B. E. (2002). *Histopathological Assessment of Dolphins Necropsied Onboard Vessels in the Eastern Tropical Pacific Tuna Fishery*. Southwest Fisheries Science Center, Administrative Report LJ-02–24C. Southwest Fisheries Science Center, La Jolla, CA.

Cowan, D. F., & Curry, B. E. (2008). Histopathology of the alarm reaction in small odontocetes. *Journal of Comparative Pathology* 139: 24–33.

Dam, M., & Bloch, D. (2000). Screening of mercury and persistent organochlorine pollutants in long-finned pilot whale (*Globicephala melas*) in the Faroe Islands. *Marine Pollution Bulletin* 40: 1090–1099.

Das, K., Siebert, U., Fontaine, M., Jauniaux, T., Holsbeek, L., & Bouquegneau, J.-M. (2004). Ecological and pathological factors related to trace metal concentrations in harbour porpoises *Phocoena phocoena* from the North Sea and adjacent areas. *Marine Ecology Progress Series* 281: 283–295.

Dean, T. A., Bodkin, J. L., Fukuyama, A. K., Jewett, S. C., Monson, D. H., O'Clair, C. E., & VanBlaricom, G. R. (2002). Sea otter (*Enhydra lutris*) perspective: mechanisms of impact and potential recovery of nearshore vertebrate predators following the 1989 *Exxon Valdez* oil spill. *Marine Ecology Progress Series* 241: 255–270.

Delong, R. L., Gilmartin, W. G., & Simpson, J. G. (1973). Premature births in California sea lions: associations with high organochlorine pollutant residue levels. *Science* 181: 1168–1170.

Demaster, D. J., Fowler, C. W., Perry, S. L., & Richlen, M. E. (2001). Predation and competition: the impact of fisheries on marine mammal populations over the next one hundred years. *Journal of Mammology* 82: 641–651.

Dubey, J. P., Zarnke, R., Thomas, N. J., Wong, S. K., Van Bonn, W., Briggs, M., Davis, J. W., Ewing, R., Mense, M., Kwok, O. C. H., Romand, S., & Thulliez, P. (2003). *Toxoplasma gondii, Neospora canium, Sarcocystis neurona,* and *Sarcocystis canis*-like infections in marine mammals. *Veterinary Parasitology* 116: 275–296.

Duinker, J. C., Hillebrand, M., Th, M., & Nolting, R. F. (1979). Organochlorides and metals in harbour seals (Dutch Wadden Sea). *Marine Pollution Bulletin* 10: 360–364.

Endo, T., Haraguchi, K., Cipriano, F., Simmonds, M. P., Hotta, Y., & Sakata, M. (2004). Contamination by mercury and cadmium in the cetacean products from Japanese market. *Chemosphere* 54: 1653–1662.

Endo, T., Haraguchi, K., Hotta, Y., Hisamichi, Y., Lavery, S., Dalebout, M. L., & Baker, C. S. (2005). Total mercury, methylmercury, and selenium levels in the red meat of small cetaceans sold for human consumption in Japan. *Environmental Science and Technology* 39: 5703–5708.

Endo, T., Haraguchi, K., & Sakata, M. (2002). Mercury and selenium concentrations in the internal organs of toothed whales and dolphins marketed for human consumption in Japan. *Science of the Total Environment* 300: 15–22.

Fisher, S. J., & Reeves, R. R. (2005). The global trade in live cetaceans: implications for conservation. *International Journal of Wildlife Law & Policy* 8: 315–340.

Flewelling, L. J., Naar, J. P., Abbott, J. P., Baden, D. G., Barros, N. B., Bossart, G. D., Bottein, M.-Y. D., Hammond, D. G., Haubold, D. G., Heil, C. A., Henry, M. S., Jacocks, H. M., Leighfield, T. A., Pierce, R. H., Pitchford, T. D., Rommel, S. A., Scott, P. S., Steidinger, K. A., Truby, E. W., Van Dolah, F. M., & Landsberg, J. H. (2005). Red tides and marine mammal mortalities. *Nature* 435: 755–756.

Fujise, Y., Honda, K., Tatsukawa, R., & Mishima, S. (1988). Tissue distribution of heavy metals in Dall's porpoise in the northwestern Pacific. *Marine Pollution Bulletin* 19: 226–230.

Gearin, P. J., Gosho, M. E., Laake, J. L., Cooke, L., DeLong, R., & Hughes, K. M. (2000). Experimental testing of acoustic alarms (pingers) to reduce bycatch of harbor porpoise, *Phocoena phocoena*, in the state of Washington. *Journal of Cetacean Research and Management* 2: 1–9.

Geraci, J. R. (1990). Physiologic and toxic effects on cetaceans. In: *Sea Mammals and Oil. Confronting the Risks* (Ed. J. R. Geraci, & D. J. St. Aubin), pp. 167–197. Academic Press, San Diego, CA.

Geraci, J. R., St. Aubin, D. J., & Reisman, R. J. (1983). Bottlenose dolphins *Tursiops truncatus* can detect oil. *Canadian Journal of Fisheries and Aquatic Science* 40: 1516–1522.

Gewin, V. (2004). Pacific dolphins make waves for US policy on Mexican tuna. *Nature* 427: 575.

Grandjean, P., Murata, K., Budtz-Jørgensen, E., & Weihe, P. (2004). Cardiac autonomic activity in methylmercury neurotoxicity: 14-year follow-up of a Faroese birth cohort. *Journal of Pediatrics* 144:169–176.

Grillo, V., Parsons, E. C. M., & Shrimpton, J. H. (2004). A review of sewage pollution in Scotland and its potential impacts on harbour porpoise populations. Paper presented to the Scientific Committee at the 53rd Meeting of the International Whaling Commission, July 3–16, 2001, London.

Hall, A. J., Law, R. J., Wells, D. E., Harwood, J., Ross, H. M., Kennedy, S., Allchin, C. R., Campel, C. A., & Pomeroy, P. P. (1992). Organochlorine levels in common seals (*Phoca vitulina*) which were victims and survivors of the 1988 Phocine Distemper Epizootic. *Science of the Total Environment* 115: 145–162.

Harwood, J., & Reijnders, P. J. H. (1988). Seals, sense and sensibility. *New Scientist* 120 (1634): 28–29.

Helle, E., Olsson, M., & Jensen, S. (1976). PCB levels correlated with pathological changes in seal uteri. *Ambio* 5: 261–263.

Hernandez, M., Robinson, I., Aguilar, A., Gonzalez, L. M., Lopez-Jurado, L. F., Reyero, M. I., Cacho, E., Franco, J., Lopez-Rodas, V., & Costas, E. (1998). Did algal toxins cause monk seal mortality. *Nature* 393: 28–29.

Honda, K., Tatsukawa, R., Itano, K., Miyazaki, N., & Fujiyama, T. (1983). Heavy metal concentrations in muscle, liver and kidney tissue of Striped dolphin *Stenella coeruleoalba* and their variations with body length, weight, age and sex. *Agricultural and Biological Chemistry* 47: 1219–1228.

Howard, C., & Parsons, E. C. M. (2006). Attitudes of Scottish city inhabitants to cetacean conservation. *Biodiversity and Conservation* 15: 4335–4356.

Iwasaki, T. (2008). Japan. Progress report on small cetacean research April 2007 to March 2008, with statistical data for the calendar year 2007. Paper presented to the Scientific Committee at the 60th Meeting of the International Whaling Commission, June 1–19, 2008, Santiago, Chile.

Iwata, H., Tanabe, S., Mizuno, T., & Tatsukawa, R. (1995). High accumulation of toxic butyltins in marine mammals from Japanese coastal waters. *Environmental Science and Technology* 29: 2959–2962.

Jefferson, T. A., & Curry, B. E. (1994). A global review of porpoise (Cetacea: Phocoenidae) mortality in gill nets. *Biological Conservation* 67: 167–183.

Jepson, P. D., Bennett, P. M., Allchin, C. R., Law, R. J., Kuiken, T., Baker, J. R., Rogan, E., & Kirkwood, J. K. (1999). Investigating potential associations between chronic exposure to polychlorinated biphenyls and infectious disease mortality in harbour porpoises from England and Wales. *Science of the Total Environment* 243/244: 339–348.

Jepson, P. D., Bennett, P. M., Deaville, R., Allchin, C. R., Baker J. R., & Law, R. J. (2005). Relationships between PCBs and health status in UK-stranded harbour porpoises (*Phocoena phocoena*). *Environmental Toxicology and Chemistry* 24: 238–248.

Kastelein, R. A., Au, W. W. L., & de Haan, D. (1999). Detection distances of bottom-set gillnets by harbour porpoises (*Phocoena phocoena*) and bottlenose dolphins (*Tursiops truncatus*). *Marine Environmental Research* 49: 359–375.

Kastelein, R. A., & Lavaleije, M. S. S. (1992). Foreign bodies in the stomach of a female harbour porpoise (*Phocoena phocoena*) from the North Sea. *Aquatic Mammals* 18: 40–46.

Kasuya, T. (2007). Japanese whaling and other cetacean fisheries. *Environmental Science & Pollution Research* 14: 39–48.

Kawai, S., Fukushima, M., Miyazaki, N., & Tatsukawa, R. (1988). Relationship between lipid composition and organochlorine levels in the tissues of striped dolphin. *Marine Pollution Bulletin* 19: 129–133.

Kenney, R. D. (1993). Right whale mortality: a correction and an update. *Marine Mammal Science* 9: 445–446.

Kraus, S. D. (1990). Rates and potential causes of mortality in North Atlantic right whale (*Eubalaena glacialis*). *Marine Mammal Science* 6: 278–291.

Kraus, S. D., Read, A. J., Solow, A., Baldwin, K., Spradlin, T., Anderson, E., & Williamson, J. (1997). Acoustic alarms reduce porpoise mortality. *Nature* 388: 525.

Kuiken, T., Bennett, P. M., Allchin, C. R., Kirkwood, J. K., Baker, J. R., Lockyer, C. H., Walton, M. J., & Sheldrick, M. C. (1994). PCBs cause of death and body condition in harbour porpoises (*Phocoena phocoena*) from British Waters. *Aquatic Toxicology* 24: 13–28.

Kuiken, T., Simpson, V. R., Allchin C. R., Bennett, M., Codd, G. A., Harris, E. A., Howes, G. J., Kennedy, S., Kirkwood, J. K., Law, R. J., Merrett, N. R., & Phillips, S. (1994). Mass mortality of common dolphins (*Delphinus delphis*) in south-west England due to incidental capture in fishing gear. *Veterinary Record* 134: 81–89.

Lahvis, G. P., Wells, R. S., Kuehl, D. W., Stewart, J. L., Rhinehart, H. L., & Via, C. S. (1995). Decreased lymphocyte responses in free-ranging bottlenose dolphins (*Tursiops truncatus*) are associated with increased concentrations of PCBs and DDT in peripheral blood. *Environmental Health Perspectives* 103: 67–72.

Laist, D. W. (1997). Impacts of marine debris: entanglement of marine life in marine debris including a comprehensive list of species with entanglement and ingestion records. In: *Marine Debris: Sources, Impacts and Solutions* (Ed. J. Coe & D. B. Rogers), pp. 99–139. Springer-Verlag, New York.

Laist, D. W., Knowlton, A. R., Mead, J. G., Collet, A. S., & Podesta, M. (2001). Collisions between ships and whales. *Marine Mammal Science* 17: 35–75.

Law, R. J., Alaee, M., Allchin, C. R., Boon, J. P., Lebeuf, M., Lepom, P., & Stern, G. A. (2003). Levels and trends of PBDEs and other brominated flame retardants in wildlife. *Environment International* 29: 757–770.

Leonzio, C., Focardi, S., & Fossi, C. (1992). Heavy metals and selenium in stranded dolphins of the northern Tyrrhenian (NW Mediterranean). *Science of the Total Environment* 119: 77–84.

MacLeod, C. D. (2009). Global climate change, range changes and potential implications for the conservation of marine cetaceans: a review and synthesis. *Endangered Species Research* 7: 125–136.

Martineau, D., Beland, P., Desjardins, C., & Lagace, A. (1987). Levels of organochlorine chemicals in tissues of beluga whales (*Delphinapterus leucas*) from the St. Lawrence estuary, Quebec, Canada. *Archives of Environmental Contamination and Toxicology* 16: 137–147.

Martineau, D., De Guise, S., Fournier, M., Shugart, L., Girard, C., Lagace, A., & Beland, P. (1994). Pathology and toxicology of beluga whales from the St. Lawrence Estuary, Quebec, Canada. Past, present and future. *Science of the Total Environment* 154: 201–215.

Monson, D. H., Doak, D. F., Ballachey, B. E., Johnson, A., & Bodkin, J. L. (2000). Long-term impacts of the Exxon Valdez oil spill on sea otters, assessed through age-dependent mortality patterns. *Proceedings of the National Academy of Sciences* 97: 6562–6567.

Mooney, T. A., Nachtigall, P., & Au, W. W. L. (2004). Target strength of a nylon monofilament and an acoustically enhanced gillnet: predictions of biosonar detection ranges. *Aquatic Mammals* 30: 220–226.

Murata, K., Weihe, P., Budtz-Jørgensen, E., Jørgensen, P. J., & Grandjean, P. (2004). Delayed brainstem auditory evoked potential latencies in 14-year-old children exposed to methylmercury. *Journal of Pediatrics* 144: 177–183.

Nakata, H., Sakakibara, A., Kanoh, M., Kudo, S., Watanabe, N., Nagai, N., Miyazaki, N., Asano, Y., & Tanabe, S. (2002). Evaluation of mitogen-induced responses in marine mammal and human lymphocytes by in-vitro exposure of butyltins and non-ortho coplanar PCBs. *Environmental Pollution* 120: 245–253.

Parsons, E. C. M. (1999). Trace metal concentrations in the tissues of cetaceans from Hong Kong's territorial waters. *Environmental Conservation* 26: 30–40.

Parsons, E. C. M. (2004). The potential impacts of pollution on humpback dolphins—with a case study on the Hong Kong population. *Aquatic Mammals* 30: 18–37.

Parsons, E. C. M., Bonnelly De Calventi, I., Whaley, A., Rose, N. A., & Sherwin, S. (2010). A note on illegal captures of bottlenose dolphins (*Tursiops truncatus*) in the Dominican Republic. *International Journal of Wildlife Law & Policy* 13(4): 240–244.

Parsons, E. C. M., Clark, J., Warham, J., & Simmonds, M. P. (2010). The conservation of British cetaceans: a review of the threats and protection afforded to whales, dolphins and porpoises in UK Waters, Part 1. *International Journal of Wildlife Law & Policy* 13: 1–62.

Parsons, E. C. M., & Rose, N. A. (2007). Tuna fishing. In: *Encyclopedia of Environment and Society* (Ed. P. Robbins), pp. 1779–1781. Sage, San Antonio, TX.

Parsons, E. C. M., Rose, N. A., & Telecky, T. M. (2010). What, no science? The trade in live Indo-Pacific bottlenose dolphins from Solomon Islands—a CITES decision implementation case study. *Marine Policy* 34: 384–388.

Pesante, G., Zanardelli, M., & Panigada, S. (2000). Evidence of man-made injuries on Mediterranean fin whales. *European Research on Cetaceans* 14: 192–193.

Petersen, M. S., Halling, J., Bech, S., Wermuth, L., Weihe, P., Nielsen, F., Jørgensen, P. J., Budtz-Jørgensen, E., & Grandjean, P. (2008). Impact of dietary exposure to food contaminants on the risk of Parkinson's disease. *Neurotoxicology* 29: 584–590.

Rawson, A. J., Patton, G. W., Hofmann, S., Pietra, G. G., & Johns, L. (1993). Liver abnormalities associated with chronic mercury accumulation in stranded Atlantic bottlenose dolphins. *Ecotoxicology and Environmental Safety* 25: 41–47.

Read, A. J., Drinker, P., & Northridge, S. (2006). Bycatch of marine mammals in the U.S. and global fisheries. *Conservation Biology* 20: 163–169.

Reeves, R. R., Smith, B. D., Crespo, E. A., & Notarbartolo di Sciara, G. (2003). *Dolphins, Porpoises and Whales.*

2002–2010 Conservation Action Plan for the World's Cetaceans. IUCN, Gland, Switzerland.

Reijnders, P. J. H. (1980). Organochlorine and heavy metal residues in harbour seals from the Wadden Sea and their possible effects on reproduction. *Netherlands Journal of Sea Research* 14: 30–65.

Reijnders, P. J. H. (1986). Reproductive failure of common seals feeding on fish from polluted waters. *Nature* 324: 456–457.

Ross, P. S., Ellis, G. M., Ikonomou, M. G., Barrett-Lennard, L. G., & Addison, R. (2000). High PCB concentrations in free-ranging Pacific killer whales, *Orcinus orca*: effects of age, sex and dietary preference. *Marine Pollution Bulletin* 40: 504–515.

Samuels, E. R., Cawthron, M., Lauer, B. H., & Baker, B. E. (1970). Strontium-90 and caesium-137 in tissues of fin whales (*Balaenoptera physalus*) and harp seals (*Pagophilius groenlandicus*). *Canadian Journal of Zoology* 48: 267–269.

Schwacke, L. H., Voit, E. O., Hansen, L. J., Wells, R. S., Mitchum, G. B., Hohn, A. A., & Fair, P. A. (2002). Probabilistic risk assessment of reproductive effects of polychlorinated biphenyls on bottlenose dolphins (*Tursiops truncatus*) from the southeast United States coast. *Environmental Toxicology and Chemistry* 21: 2752–2764.

Scott, N. J., & Parsons, E. C. M. (2005). A survey of public opinions in Southwest Scotland on cetacean conservation issues. *Aquatic Conservation* 15: 299–312.

Siebert, U., Joiris, C., Holsbeek, L., Benkes, H., Failing, K., Frese, K., & Petzinger, E. (1999). Potential relation between mercury concentrations and necropsy findings in cetaceans from German waters of the North and Baltic Seas. *Marine Pollution Bulletin* 38: 285–295.

Silvani, L., Gazo, J. M., & Aguilar, A. (1999). Spanish driftnet fishing and incidental catches in the western Mediterranean. *Biological Conservation* 90: 79–85.

Simmonds, M. P. (1992). Cetacean mass mortalities and their potential relationship with pollution. In: *Proceedings of the Symposium "Whales: Biology-Threats-Conservation". Brussels, June 5–7, 1991.* (Ed. by J. J. Symoens), pp. 217–245. Royal Academy of Overseas Sciences, Brussels.

Song, L., Seeger, A., & Santos-Sacchi, J. (2005). On membrane motor activity and chloride flux in the outer hair cell: lessons learned from the environmental toxin tributyltin. *Biophysical Journal* 88: 2350–2362.

Steuerwald, U., Weihe, P., Jorgensen, P. J., Bjerve, K., Brock, J., Heinzow, B., Budtz-Jorgensen, E., & Grandjean, P. (2000). Maternal seafood diet, methyl mercury exposure, and neonatal neurologic function. *Journal of Pediatrics* 136: 599–605.

Subramanian, A. N., Tanabe, S., & Tatsukawa, R. (1987). Age and size trends and male-female differences of PCBs and DDE in Dalli-type Dall's porpoises, *Phocoenoides dalli* of the north-western North Pacific. *Proceedings of the National Institute of Polar Research Symposium on Polar Biology* 1: 205–216.

Subramanian, A. N., Tanabe, S., Tatsukawa, R., Saito, S., & Miyazaki, N. (1987). Reduction in testosterone levels by PCBs and DDE in Dall's porpoises of the north-western North Pacific. *Marine Pollution Bulletin* 18: 643–646.

Swart, R., Ross, P., Vedder, L., Timmerman, H., Heisterkamp, S., Loveren, H., Vos, J., Reijnders, P. J. H., & Osterhaus, A. (1994). Impairment of immune function in harbour seals (*Phoca vitulina*) feeding on fish from polluted waters. *Ambio* 23: 155–159.

Takahashi, S., Le, L. T. H., Saeki, H., Nakatani, N., Tanabe, S., Miyazaki, N., & Fujise, Y. (2000). Accumulation of butyltin compounds and total tin in marine mammals. *Water Science and Technology* 42: 97–108.

Tanabe, S. (1988). PCB problems in the future: foresight from current knowledge. *Environmental Pollution* 50: 5–28.

Tanabe, S. (1999). Butyltin contamination in marine mammals: a review. *Marine Pollution Bulletin* 39: 62–72.

Tanabe, S., Tatsukawa, R., Maruyama, K., & Miyazaki, N. (1982). Transplacental transfer of PCBs and chlorinated hydrocarbon pesticides from a pregnant striped dolphin, *Stenella coeruleoalba*, to her fetus. *Agricultural and Biological Chemistry* 46: 1249–1254.

Tanabe, S., Wantanabe, S., Kan, H., & Tatsukawa, R. (1988). Capacity and mode of PCB metabolism in small cetaceans. *Marine Mammal Science* 4: 103–124.

Tarpley, R. J., & Marwitz, S. (1993). Plastic debris ingestion by cetaceans along the Texas coast: two case reports. *Aquatic Mammals* 19: 93–98.

Tregenza, N., Aguilar, A., Carrillo, M., Delgado, I., Díaz, F., Brito, A., & Martin, V. (2000). Potential impact of fast ferries on whale populations: a simple model with examples from the Canary Islands. *European Research on Cetaceans* 14: 195–197.

Tregenza, N. J. C., Berrow, S. D., Hammond, P. S., & Leaper, R. (1997). Common dolphin *Delphinus delphis* L., bycatch in bottom set gill nets in the Celtic Sea. *Report of the International Whaling Commission* 47: 835–839.

Tregenza, N. J. C., Berrow, S. D., Leaper, R., & Hammond, P. S. (1997). Harbour porpoise *Phocoena phocoena* L. bycatch in set gill nets in the Celtic Sea. *ICES Journal of Marine Science* 54: 896–904.

Tregenza, N. J. C., & Collet, A. (1998). Common dolphin *Delphinus delphis* bycatch in pelagic trawl and other fisheries in the North East Atlantic. *Reports of the International Whaling Commission* 48: 453–459.

Twiss, J. R., & Reeves, R. R. (1999). *Conservation and Management of Marine Mammals.* Smithsonian Institution Press, Washington, DC.

Van Dolah, F. M. (2000). Marine algal toxins: origins, health effects, and their increased occurrence. *Environmental Health Perspectives* 108(Suppl. 1): 133–141.

Van Dolah, F. M., Doucette, G. J., Gulland, F. M. D., Rawles, T. L., & Bossart, G. D. (2003). Impacts of algal toxins on marine mammals. In: *Toxicology of Marine Mammals* (Ed. J. G. Vos, G. D. Bossart, M. Fournier, & T. J. O'Shea), pp. 247–269. Taylor and Francis, London.

Van de Vijver, K. I., Hoff, P. T., Das, K., van Dongen, W., Esmans, E. L., Jauniaux, T., Bouquegneau, J-M., Blust, R., & de Coen, W. (2003). Perfluorinated chemicals infiltrate ocean waters: link between exposure levels and stable isotope ratios in marine mammals. *Environmental Science and Technology* 37: 5545–5550.

Van Waerebeek, K., Sequeira, M., Williamson, C., Sanino, G. P., Gallego, P., & Carmo, P. (2006). Live-captures of common bottlenose dolphins, *Tursiops truncatus*, and unassessed bycatch in Cuban waters: evidence of sustainability found wanting. *Latin American Journal of Aquatic Mammals* 5: 39–48.

Weihe, P., Debes, F., White, R. F., Sørensen, N., Budtz-Jørgensen, E., Keiding, N., & Grandjean, P. (2003). Environmental epidemiology research leads to a decrease of the exposure limit for mercury [in Danish]. *Ugeskr Laeger* 165: 107–111.

Weihe, P., Grandjean, P., Debes, F., & White, R. (1996). Health implications for Faroe islanders of heavy metals and PCBs from pilot whales. *Science of the Total Environment* 186: 141–148.

Weihe, P., Grandjean, P., & Jørgensen, P. J. (2005). Application of hair-mercury analysis to determine the impact of a seafood advisory. *Environmental Research* 97: 201–208.

Weinrich. M. (2005). A review of collisions between whales and whale watch boats. Paper presented to the Scientific Committee at the 57th Meeting of the International Whaling Commission, May 30–June 10, 2005, Ulsan, Korea.

Wells, R. S., Tornero, V., Borrell, A., Aguilar, A., Rowles, T. K., Rhinehart, H. L., Hofmann, S., Jarman, W. M., Hohn, A. A., & Sweeney, J. C. (2005). Integrating life-history and reproductive success data to examine potential relationships with organochlorine compounds for bottlenose dolphins (*Tursiops truncatus*) in Sarasota Bay, Florida. *Science of the Total Environment* 349: 106–119.

Wright, A. J., Aguilar Soto, N., Baldwin, A. L., Bateson, M., Beale, C., Clark, C., Deak, T., Edwards, E. F., Fernández, A., Godinho, A., Hatch, L., Kakuschke, A., Lusseau, D., Martineau, D., Romero, L. M., Weilgart, L., Wintle, B., Notarbartolo di Sciara, G., & Martin, V. (2007). Do marine mammals experience stress related to anthropogenic noise? *International Journal of Comparative Psychology* 20(2–3): 274–316.

Yoshitome, R., Kunito, T., Ikemoto, T., Tanabe, S., Zenke, H., Yamauchi, M., & Miyazaki, N. (2003). Global distribution of radionuclides (137Cs and 40K) in marine mammals. *Environmental Science and Technolology* 37: 4597–4602.

Zakharov, V. M., & Yablokov, A. F. (1990). Skull asymmetry in the Baltic grey seal: effects of environmental pollution. *Ambio* 19: 266–269.

CHAPTER 16

Threats to Pinnipeds

CHAPTER OUTLINE

Endangered and Threatened Pinniped Populations

    Mediterranean Monk Seal, *Monachus monachus*
    Hawaiian Monk Seal, *Monachus schauinslandi*
    Steller Sea Lion, *Eumetopias jubatus*
    Exploring the Depths: "Junk Food" Hypothesis
    Exploring the Depths: Orca Predation Hypothesis
    Galápagos Fur Seal, *Arctocephalus galapagoensis*
    Australian Sea Lion, *Neophoca cinerea*

    New Zealand Sea Lion, *Phocarctos hookeri*
    Northern Fur Seal, *Callorhinus ursinus*
    Caspian Seal, *Pusa caspica*
    Exploring the Depths: Northern Elephant Seal, Mirounga angustirostris
    Harbor Seal, *Phoca vitulina*
    Exploring the Depths: The Canadian Harp Seal Harvest

Selected References and Further Reading

Endangered and Threatened Pinniped Populations

There is considerable concern for pinnipeds throughout the world but especially for the four species of threatened pinnipeds that are geographically restricted to tropical regions. A recent review: *Global Threats to Pinnipeds* by an international team of 19 experts reported that "one in three species of pinnipeds is threatened compared to one of five mammals generally" (Kovacs et al. 2011, p. 4). However, this work also importantly highlighted that conservation status at the species level may hide serious issues for pinnipeds at the subspecies or population level. Historically, hunting may have been the most important factor influencing pinniped populations, but today fisheries interactions appear to be the major threat. For 11 of the 13 species/subspecies that are either classified as "critically endangered," "endangered," or "vulnerable" by the International Union for the Conservation of Nature (IUCN): "direct or indirect fisheries interactions are identified as the primary, or an important secondary threat" (Kovacs et al. 2011, p. 12). Pollution, hunting, predation, prey abundance, and disease are some of the other factors influencing declining populations. Threats from global climate change are likely to be important in the future, particularly for ice associated species but it may be difficult to assess changes in these populations as data on population size and distribution are deficient for species living in polar regions.

Mediterranean Monk Seal, *Monachus monachus*

The most threatened pinniped in the world is currently the Mediterranean monk seal (**Figure 16.1**). This pinniped species was formerly found throughout the Mediterranean Sea, Black Sea, and northwest African coast. However, the current population consists of only about 600 animals. For centuries the seals were killed by fishermen who saw them as competitors or accused them of destroying their fishing

FIGURE 16.1 A Mediterranean monk seal.

gear. They were also killed for their skin and body parts, which were used for "medicinal" purposes. Such takes are rare today; however, entanglement in fishing gear is currently a major cause of mortality.

Because of the high intensity of human activities in the Mediterranean, this species is vulnerable to disturbance, pollution, marine litter, and habitat degradation. Moreover, disease has been a problem for this species. In 1997, there was a mass mortality in Cap Blanc Peninsula in which 70% of the population died, and the number of animals in this population dropped from approximately 310 to less than 90. The IUCN lists the Mediterranean monk seal as "critically endangered," and the Convention on International Trade in Endangered Species of Wild Fauna and Flora (CITES) list them on Appendix I, prohibiting the international trade of Mediterranean monk seals or their body parts (see Chapter 17).

Hawaiian Monk Seal, *Monachus schauinslandi*

The second most threatened pinniped is a related species to the Mediterranean monk seal dwelling in the United States: the Hawaiian monk seal (**Figure 16.2**). As their name suggests, Hawaiian monk seals inhabit the Hawaiian Islands, although much of their historical habitat has been lost to development or disturbance by humans. Approximately 1,300 to 1,400 of these animals remain on islands around northwestern Hawaii, with a few on the main islands (**Figure 16.3**).

They breed from December to August, and there is much concern about the large sex imbalance at breeding beaches, with three times as many males as females. This gender imbalance is thought to have contributed to the observed high levels of male aggression.

Hawaiian monk seals have been heavily hunted throughout history, especially during the 19th century when populations were severely depleted by commercial sealers and opportunistic hunters. In the early 20th century their numbers recovered slightly, but animals were also hunted and killed during World War II. The species was reported to have declined from the 1950s through the 1980s. Disturbance, pollution, military activities, algal blooms, depletion of prey species, entanglement in fishing gear, and marine trash and debris have degraded the monk seal's habitat and depleted the population. By the 1980s only an estimated 1,500 animals remained. In 1998 the Chief of Protected Species Investigations of the U.S. National Marine Fisheries Service (the government agency with responsibility for most pinnipeds) stated that any further decline in Hawaiian monk seal numbers "could very seriously compromise" the survival of the species. Hawaiian monk seals are considered to be "endangered" under the U.S. Endangered Species Act and "critically endangered" by the IUCN.

FIGURE 16.2 A Hawaiian monk seal.

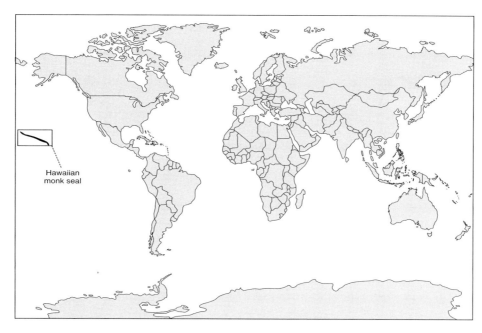

FIGURE 16.3 Map showing the current, known distribution of the Hawaiian monk seal. Data from: IUCN Red List.

Steller Sea Lion, *Eumetopias jubatus*

Another endangered pinniped species in the United States is the Steller sea lion, a species about which there has been much controversy as to the cause of its decline (**Figure 16.4**). Steller sea lions (or perhaps more correctly, Steller's sea lions as they are named after their "discoverer," a naturalist named Steller) are found across the north Pacific from Russia in the western Pacific to Alaska, Canada, and the west coast of continental United States in the eastern Pacific, from mid-California northward (**Figure 16.5**). The species is split into two genetically distinct stocks. The eastern stock extends from California, through Oregon, British Columbia to southeast Alaska, and the western stock extends from the Gulf of Alaska, across the Bering Sea to coastal Russia, and also includes small numbers of sea lions in Japan. The species breeds in the spring and summer, often on isolated islands, and more than 1,000 individuals can crowd a haul-out site. They feed on cephalopods and fish, including schooling and benthic (seabed) species, generally at night. They also often feed around trawlers and other fishing

FIGURE 16.4 A Steller sea lion.

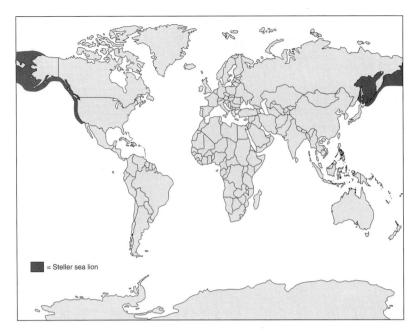

FIGURE 16.5 Map showing the current, known distribution of the Steller sea lion. Data from: IUCN Red List.

boats, taking discards or even taking fish from nets. There is an extensive history of the hunting of this species for oil and pelts first by native people and then by Russians and Europeans. Animals are still taken in subsistence hunts by some native communities. There is an annual subsistence harvest of approximately 100 western Steller sea lions in Alaska.

Steller sea lions were culled in the United States in the 20th century because of perceived conflicts with human fisheries. Culls of this species still continue in Japan to this day. Animals also continue to be killed on occasion by fishermen in Russia, Canada, and the United States, although such killing is technically illegal in most cases.

Other problems facing this species include entanglement in discarded fishing gear and marine litter (which kills 30 or more animals per year), contamination from organic pesticide and heavy metals, oil pollution from the 1989 *Exxon Valdez* and other smaller spills, prey changes caused by El Niño events and adverse climatic conditions, and predation. However, the main concern for Steller sea lions is believed to be linked to changing prey availability.

The eastern stock of Steller sea lions are considered to be "threatened" under the U.S. Endangered Species Act due to a long-term decline during the 20th century, although numbers have been improving since the 1980s and are now thought to be relatively stable. The western stock of Steller sea lions in the Aleutian Islands has decreased substantially since the 1960s. Consequently, the western stock is listed as "endangered" under the U.S. Endangered Species Act. The species as a whole is considered to be "endangered" by the IUCN.

In 2000 there were an estimated 45,000 to 46,000 western Steller sea lions (versus 80,000 eastern sea lions) down from a population estimate of 227,000 for the western stock in 1960. Because of the decline of the western stock of Steller sea lions and a perceived lack of action by the U.S. government, environmental groups brought legal cases against the government in 2000. The court ruled for the environmental groups and against the government, forcing it to take more action. Therefore, in August 2000 a ban of all groundfish trawl fishing within 20 nautical miles of designated Steller sea lion critical habitat rookeries, haul outs, and feeding areas was instigated, because depletion of the sea lion's food supply was thought to be the major cause of its decline. However, in November and December 2000 an Alaskan senator attached a "rider" to a congressional appropriations bill that prevented the full implementation of the management measures and limited the government's ability to restrict the fisheries. This is just one example of where politics has hindered taking effective measures to conserve this species.

Exploring the Depths: "Junk Food" Hypothesis

Many scientists believe the decline of the Steller sea lion is linked to their poor nutritional status. One idea to explain this poor nutritional status is the so-called "junk food" hypothesis. The basic concept is that the abundance of high-fat-content fish that was traditionally the main prey species for Steller sea lions has declined dramatically, forcing the sea lion to seek alternatives. They now feed primarily on pollock (Family: Gadidae), a commercially important fish stock for human fisheries that has a lower fat content than traditional prey species. The sea lions thus have trouble meeting their nutritional needs from pollock and must eat more in the process. The fat content of pollock is especially low during the sea lion's reproductive season, which likely impacts pup production and survival. The low energetic value of their prey places them on a "nutritional knife edge" in that any factor that places additional stress on the sea lions (further prey depletion, disturbance, disease, pollution) could lead to debilitation or mortality.

There is some debate, however, as to whether this is all due to an ecological "regime shift," where one of the dominant animals in an ecosystem switches from one group of fish species to another (perhaps linked to climate change), or if it is instead the result of human overfishing. Perhaps something else is at work, or a combination of factors are involved. However, despite well over $100 million of U.S. government funding having been spent on Steller sea lion research to investigate this issue, the answer remains elusive, and major management actions to protect this marine mammal and to prevent their decline have been limited.

Exploring the Depths: Orca Predation Hypothesis

Another hypothesis as to the cause of the decline of Steller sea lions revolves around increased predation on the species by killer whales. The hypothesis is that commercial whaling severely depleted large whales in the North Pacific, leaving less prey for mammal-eating killer whales. Therefore, killer whales may have turned to feeding on large numbers of Steller sea lions and sea otters (for a discussion on sea otter predation see Chapter 5) as an alternate prey species, thus causing Steller sea lions to decline. This hypothesis is somewhat controversial, however, and several senior scientists have highlighted various important criticisms of this hypothesis that have not yet been addressed. For example, patterns of sea lion decline do not match hypothesized patterns of prey removal for the killer whale. There are also abundant populations of northern minke whales and Dall's porpoise in the area, both of which are likely to be nutritionally more beneficial as prey to killer whales than sea lions. Moreover, killer whales predating baleen whales primarily eat whale calves (and typically eat the calf's tongue and discard much of the rest of the carcass); the initial energetic calculations by scientists assumed that the killer whales ate adult whales and thus the decline in their food availability may not be as much as thought. Therefore, the hypothesis should be treated with caution.

However, many individuals with links to the fishing industry in Alaska (including some policy-makers) are championing this hypothesis as it removes overfishing as being to blame for Steller sea lion collapse. In fact, there have even been calls for killer whale culls to "protect" Steller sea lions.

Galápagos Fur Seal, *Arctocephalus galapagoensis*

As their name suggests, the Galápagos fur seal hauls out and breeds on the Galápagos Islands, situated 972 km off the coast of Ecuador (**Figure 16.6**). These animals have a very restricted distribution (**Figure 16.7**) and spend more of their time on land than many other pinniped species (up to 70%). This species of fur seal can nurse its pups for periods of up to 3 years, which may help to buffer them against the effects of food shortages during El Niño events.

Galápagos fur seals have been hunted for their fur and the oil from their blubber since the 1800s. An estimated 22,000 animals had been killed by the early 20th century, nearly causing the extinction of the species. The Galápagos Islands were designated a National Park by the Government of Ecuador in 1959, bringing protection to the fur seals, partly by prohibiting hunting, although illegal poaching of pinnipeds on the Galápagos Islands still occurs.

The islands were internationally recognized as an UNESCO World Heritage Site in 1978 and as a Biosphere Reserve in 1985. The surrounding waters, encompassing an area of 70,000 square kilometers, were designated a marine reserve in 1986. In 2001 the Galápagos Biosphere Reserve was extended to include this marine area. Despite this designation, however, there are serious problems with uncontrolled and illegal fishing.

The high level of tourism on the islands may also be a cause of disturbance for the fur seals (and Galápagos sea lions that live in the same area) whereas coastal development and increasing human populations bring the threat of coastal habitat degradation. As with many endemic species on the Galápagos Islands, invasive species have been a problem for the fur seals, with reports of feral dogs killing pups. In January 2001 an oil spill may have also had an impact on the species: although no Galápagos fur seals were reported to be oiled, 79 oiled Galápagos sea lions

FIGURE 16.6 A Galápagos fur seal.

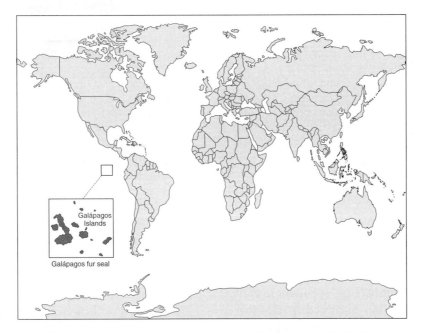

FIGURE 16.7 Map showing the current, known distribution of the Galápagos fur seal. Data from: IUCN Red List.

(*Zalophus wollebaeki*) were observed. Other consequences, such as contamination of the food supply, are likely to have occurred.

In 1983 the extreme El Niño caused such a change in the Galápagos marine ecosystem that almost all the pups born between 1980 and 1983 died, as well as almost all of the large "beachmaster" males and approximately 30% of the remaining adult population. Although there was another El Niño event in 1989, this does not seem to have had such a substantial impact on the fur seal population. As a result of this major depletion, the Galápagos fur seal continues to be listed as "endangered" by the IUCN.

Australian Sea Lion, *Neophoca cinerea*

Australian sea lions can be found in southern and southwestern Australia (**Figures 16.8** and **16.9**). Altogether they breed on at least 73 islands, but 60% of the pups are born at just five sites. They feed in shallow waters on benthic (seabed) prey species, including fish and cephalopods (species such as cuttlefish). Natural sources of mortality include predation by the great white shark (*Carcharodon carcharias*) and killer whale (*Orcinus orca*). However, as with many pinniped populations, there has also been a long history of hunting of the Australian sea lions dating back thousands of years to hunts by Australian aborigines. Major hunts in the 1600s and 1700s by commercial sealers, however, significantly depleted populations and caused the

FIGURE 16.8 Australian sea lions.

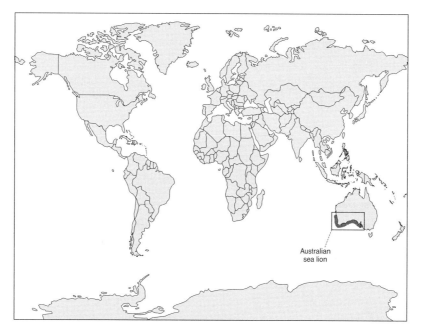

FIGURE 16.9 Map showing the current, known distribution of the Australian sea lion. Data from: IUCN Red List.

species to become locally extinct (extirpated) in some areas, such as Tasmania.

The sea lions are now protected under Australian law. Despite this, entanglement in fishing gear and marine debris is a significant cause of mortality, although a code of conduct for fishermen has been introduced in recent years to try to reduce the incidence of accidental bycatch in fishing gear. Entanglement in the shark nets that skirt many Australian beaches also occurs. There are also growing concerns about the impacts of disturbance as more and more tourists flock to visit the sea lions' haul-out sites.

The first laws to manage and protect Australian sea lions were introduced as early as 1889. At present they are protected under the 1999 Australian Environment Protection and Biodiversity Conservation Act, which lists them as "vulnerable and threatened." Population models estimate that if bycatch and other sources of mortality continue, several local populations of this species will go extinct in the near future. The IUCN acknowledges this and lists this species as "endangered."

New Zealand Sea Lion, *Phocarctos hookeri*

The New Zealand sea lion is restricted to a few breeding sites (**Figures 16.10** and **16.11**). More than 85% of pups are born in three colonies in the Auckland Islands, less than 7 km apart. They feed on benthic squid, octopus, fish, and occasionally other species such as penguins.

Similar to many pinnipeds, there has been a history of exploitation of the New Zealand sea lion, which was extensively hunted for its hide and oil from its blubber in the 19th century (some hunting also occurred in the early 20th century). In 1893 killing this species became illegal in New Zealand. Despite this, the sea lion was ultimately designated as a "threatened" species under New Zealand's Marine Mammals Protection Act in 1997. Additional protections include a 20-km Marine Mammal Sanctuary around the Auckland Islands, which has recently been proposed for

FIGURE 16.10 New Zealand sea lions.

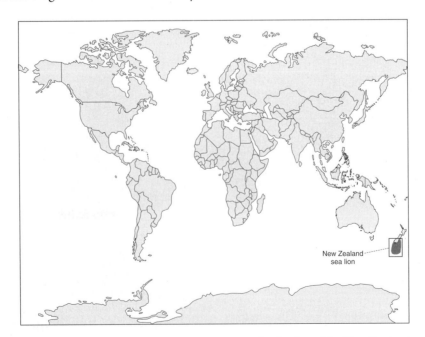

FIGURE 16.11 Map showing the current, known distribution of the New Zealand sea lion. Data from: IUCN Red List.

an extension out to 100 km. In 1998 New Zealand's sub-Antarctic islands, where most of the breeding sites for the species are found, were designated World Heritage Sites by the United Nations.

Since the 1970s the biggest threat to this species has been entanglement in the nets of the squid trawl fishery. To reduce the impacts of the fishery on the sea lions, fishing has been prohibited within 12 nautical miles of the islands since 1982. Additionally, since 1994, whenever sea lion mortality exceeds agreed limits, the fishery is shut down. This is not an empty threat, as it has become a relatively regular occurrence. Moreover, some vessels have voluntarily withdrawn from the squid fishery.

Marine mammal escape devices are being tested, and alternative methods for catching squid are being developed. In the meantime, however, bycatch of the sea lions still continues, albeit limited by the closures.

Disease has also been a problem for this species. In 1998 an unknown illness struck during the breeding season and killed 50% of the pups of that year, as well as approximately 20% of the adult population. There were resurgent outbreaks in 2002 and 2003 that were linked to the bacterium *Klebsiella pneumoniae*, which killed one-third and one-fifth of the pups of those years, respectively.

Finally, the impact of tourism is another cause for concern. Increasing numbers of tourists are visiting the islands, disturbing the sea lions more and more frequently. In combination, these (and probably also other factors) have led to a 30% reduction in pup production over the past decade. The New Zealand sea lion is considered to be "vulnerable" by the IUCN.

Northern Fur Seal, *Callorhinus ursinus*

The northern fur seal's genus means "beautiful nose" (**Figure 16.12**). This species is found in the northern Pacific from California to Alaska in the east and Japan and Russia in the west (**Figure 16.13**). They feed mostly at night on pelagic fish and squid, spending the majority of their time at sea, except when they come to land to breed (June through August). There are approximately 1.3 million northern fur seals. However, three-fourths of these animals breed on the relatively small Pribilof Islands (St. Paul and St. George) in the Bering Sea, providing a very limited land-based distribution.

Because of their warm fur coats and thick fat layers, these animals have been hunted since prehistoric times and by Europeans since 1786. In 1844, because of declining numbers, hunting ceased on St. Paul Island. By 1911 the decline due to overhunting warranted an international agreement: The North Pacific Fur Seal Convention, which included the United States, Japan, Russia, and United Kingdom/Canada as signatories. The Convention

FIGURE 16.12 A northern fur seal.

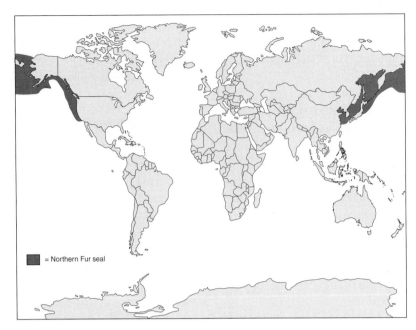

FIGURE 16.13 Map showing the current, known distribution of the northern fur seal. Data from: IUCN Red List.

banned the killing of northern fur seals at sea and limited hunts to immature males on land. Between 1956 and 1968 an experimental hunt on females was introduced in the islands intended to boost growth and birth rates. Sadly, and somewhat predictably, this hunt instead led to a subsequent decline in population by 6–8% between 1975 and 1982. In 1973 hunting largely stopped on St. George Island, again due to reductions in numbers. At present the northern fur seal population is approximately one-half of the population size in the 1950s, but hunting continues.

Subsistence harvests of northern fur seals by Alaskan natives currently remove 400–500 animals (although annual numbers were closer to 1,000 a decade ago). In addition, higher numbers have been taken annually in Russia and Asia in recent times, primarily for their fur but also for meat to feed animals kept in fur farms. Entanglement in marine debris and fishing gear potentially kills more than 5,000 animals a year. Disturbance and pollution related to the oil industry may also be a problem. Finally, there are concerns about the impacts of overfishing and climate change on the ecosystems of the North Pacific and Bering Sea and the effects this could ultimately have on this species (in much the same way as for Steller sea lions). The IUCN lists the northern fur seal as "vulnerable," and the U.S. Marine Mammal Protection Act lists the species as "depleted" (for details on the Marine Mammal Protection Act see Chapter 17).

Caspian Seal, *Pusa caspica*

This seal species is endemic to the Caspian Sea, breeds on winter sea ice, and is thought to have had a historical population size of over 1 million animals (**Figures 16.14** and **16.15**). Current estimates place the population abundance at just over 100,000 animals. Caspian seals have been heavily hunted since the 19th century when the annual harvest was generally over 100,000 animals, reaching up to nearly 300,000 in the 1930s. Hunting declined slightly during and after World War II, when harvests of an average 60,000

FIGURE 16.14 A Caspian seal.

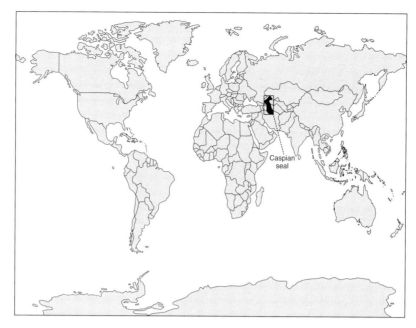

FIGURE 16.15 Map showing the current, known of distribution of the Caspian seal. Data from: IUCN Red List.

animals were reported. Since 1975 takes have been much lower, but 3,000–4,000 pups are still caught every year in commercial and "scientific" takes of this species. Several thousand animals are also killed as the result of entanglement in fishing gear.

In 1997 a seal distemper outbreak resulted in a mass mortality event in Azerbaijan. Seals in this mortality event had high levels of organic pollutants in their body tissues, and it is very possible that this contamination compromised the animal's immune system. These pollutants may have suppressed (and may still be suppressing) reproduction, with 70% of sampled females from this population reported to have fertility problems. In 2000 there was another mass mortality event, again linked to distemper.

Depletion of prey species may be having a synergistic impact on this seal population. The local fishery for the main prey (*Clupeonella* spp.) of the Caspian seal has increased. Many animals in the 1997 and 2000 mortality events were emaciated, and it is entirely possible that lack of food may force the animals to mobilize and metabolize fat-based energy reserves, thereby releasing contaminants stored in the blubber into the blood where they can reach other tissues (for details on pollutants see Chapter 15). The newly released contaminants would then weaken the seal's immune system.

Other factors that may pose a problem for this seal species include habitat loss related to industrial development, in particular activities related to the oil and gas industry, and culls by fishermen because of perceived competition

Exploring the Depths: Northern Elephant Seal, *Mirounga angustirostris*

Elephant seals gain their name from the male's large fleshy nose or "trunk" (**Figures B16.1** and **B16.2**). They are extremely sexually dimorphic; adult males are two-thirds as long as adult females (5 m male vs. 3 m female) and up to five times heavier (up to 2,200 kg male vs. 400–800 kg female). They haul out of the water for prolonged periods only twice a year: once to mate and once to molt. During the mating season larger males defend a "harem" of females from other males. These "beachmasters" aggressively compete against other males by vocalizing, rearing up and displaying, and, ultimately, by fighting. Meanwhile, smaller and often younger males try to sneak their way into the edges of harems unnoticed to mate with females, taking advantage of opportunities when the beach-

masters are distracted, such as when they are fending off other males. They are one of the deepest and longest diving of the pinnipeds, reaching depths of up to 1.5 km and remaining underwater for as long as 80 minutes.

Although northern elephant seals are currently abundant and not threatened, they were so heavily hunted for oil from their blubber they were believed to have gone extinct at the beginning of the 20th century. The northern elephant seals have, however, recovered from this depletion, although they have probably gone through a "genetic bottleneck." Their resulting lack of genetic diversity may be a cause of concern in the future.

Exploring the Depths: Northern Elephant Seal, *Mirounga angustirostris* (continued)

FIGURE B16.1 A northern elephant seal.

FIGURE B16.2 Map showing the current, known distribution of the northern elephant seal. Data from: IUCN Red List.

with fisheries. The Caspian seal is considered to be "endangered" by the IUCN.

Harbor Seal, *Phoca vitulina*

The harbor seal is found in northern coastal waters from temperate to polar regions (**Figures 16.16** and **16.17**). Five subspecies, based on geographical and genetic isolation, are recognized: *Phoca vitulina richardii* in the eastern Pacific; *P. v. vitulina* in the eastern Atlantic; *P. v. stejnegeri* in the western Pacific; *P. v. mellonae* in the Ungava Peninsula, eastern Canada; and *P. v. concolor* in the western Atlantic. The IUCN Red List classifies the population status of the harbor seal as a whole as "least concern" because of its large and generally stable/increasing population (estimated total population size, 350,000–500,000). Despite this favorable conservation status there is increasing concern about population declines in some of the subspecies and in certain regions.

In the past, harbor seals have been hunted by subsistence hunters and killed in efforts to protect fisheries. More recently, concern has switched to the changes in the marine environment that are thought to be affecting the prey on which harbor seals depend. Their coastal distribution makes them susceptible to other anthropogenic factors as well. Isolated harbor seal populations appear to be

FIGURE 16.16 A harbor seal with evidence of entanglement from marine debris around its neck.

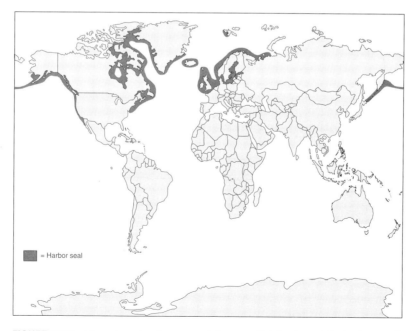

FIGURE 16.17 Map showing the current, known distribution of the harbor seal. Data from: IUCN Red List.

particularly vulnerable to local pressures. The Lake Ontario population went extinct in the 1800s. The Baltic population was almost extinct in the 1970s and may still be as low as a few hundred individuals. The Japanese population of *P. v. stejnegeri* was estimated to be only 350 animals in the 1980s as a result of hunting and mortality associated with salmon trap net fisheries, the latter of which continues to be a problem. The subspecies *P. v. mellonae* comprises no more than 600 individuals and is confined to the Seal Lakes (Lac des Loups Marins) of the Ungava Peninsula, Canada. This population is vulnerable because of its small population size and may be threatened by changes in water level associated with hydroelectric development.

Since the 1970s there have been large-scale population declines (>60%) of *P. v. richardii* in the Gulf of Alaska. No overall cause has been established, but some evidence suggests these declines are related to oceanographic regime changes and/or overfishing on important prey species, as is believed to be the case for the Steller sea lion and northern fur seal. A population decrease in the Prince William Sound area occurred before the *Exxon Valdez* oil spill; however, the dwindling harbor seal population was seriously affected by oiling and has been slow to recover. Elsewhere in British Columbia (Canada) and Washington, Oregon, and California (United States) the *P. v. richardii* population has remained stable or has been increasing since the early 1990s. *P. v. concolor* populations are also slowly recovering from hunting pressure in previous centuries. One local exception to this pattern has been a drop in the population on Sable Island, Nova Scotia, largely as a result of shark predation and competition with an increasing grey seal population. Populations in West Greenland, Iceland, and Norway continue to be low as a result of hunting.

In recent years mass mortalities of harbor seals in Europe have occurred due to viral epidemics. In 1988 more than 20,000 *P. v. vitulina* died from an outbreak of phocine distemper virus (*Morbillivirus*), followed by a second outbreak in 2002 that killed a further 30,000. Pollution may have been a contributing factor to the susceptibility of harbor seals to disease. Harbor seals across Europe accumulate high concentrations of industrial chemicals and pesticides due to their close proximity to large human populations in many parts of their range.

More recently, the numbers of *P. v. vitulina* in the north east of Scotland have almost halved in the past 10 years. As yet there is no clear cause, although a number of likely factors have been suggested. These include decreases in prey species, interactions with fisheries and aquaculture, competition for food with grey seals, and a possible link with increasing orca populations. A recently identified cause of mortality of harbor seals, giving rise to characteristic "cork screw" lacerations, is a specific design of large ducted propeller found on certain types of vessels.

One example of an attempt to resolve historical conflicts between seals and salmon fisheries in Scotland was the establishment of a Seal Management Plan in the Moray Firth. This pilot initiative has brought together local District Salmon Fishery Boards, the Scottish Executive, Scottish Natural Heritage, and local stakeholders to try to manage the number of seals killed by fishermen. In other areas of Scotland Special Areas of Conservation under the European Union Habitats Directive have been set up for the conservation of harbor seals (see Chapter 17).

Exploring the Depths: The Canadian Harp Seal Harvest

The commercial harvest of harp seals (*Pagophilus groenlandicus*) in Canada (**Figure B16.3**) occurs after the spring whelping (pupping) season and is the world's largest hunt of any marine mammal population. Harp seals are found in the North Atlantic and their distribution is associated with the seasonal limits of the pack ice. The largest population of this species occurs in the Northwest Atlantic off eastern Canada, a second breeding population is found in the White Sea (Russia), and the third and smallest breeding population is on the "West Ice" between Jan Mayen and Svalbard. In Canada, harp seals have been hunted for subsistence for 3,000 to 4,000 years and commercially harvested, mainly for oil from its blubber since the 18th century. Today, seal products include leather, handicrafts, meat for human and animal consumption, and recently for seal oil that is high in Omega-3. The current value of a high quality pelt is around $20–$25.

Historically, harp seal populations were severely depleted by commercial sealing. The Northwest Atlantic population, however, has increased from less than 2 million since the 1970s to a current population of around 9 million. In 2008 the gray seal (*Halichoerus grypus*) hunt took a maximum of around 1,500 animals and less than 400 hooded (*Cystophora cristata*) seals have been harvested annually since 1999.

Television in the 1960s featuring the killing of pups on the ice subsequently led to major campaigns against sealing by animal welfare organizations around the world. In 1983, the European Economic Community banned the importation of fur from whitecoats (harp seal pups) and bluebacks (hooded

Exploring the Depths: The Canadian Harp Seal Harvest (continued)

FIGURE B16.3 A harp seal pup killed in Canada during the seal hunt.

seal pups). Recommendations by the Royal Commission on Seals and Sealing in Canada brought about a ban on the commercial harvest of non-molted pups in 1987. It is therefore only legal to kill pups that have molted their first coat and are living independently from their mothers.

The season for the commercial harvest of harp seals is from November 15th to May 15th as set out in the Canadian Marine Mammal Regulations (1993). The majority of sealing occurs in late March in the Gulf of St. Lawrence and in early April on the Front in Newfoundland and Labrador. Records from the government department, Fisheries and Oceans Canada (DFO) show that 67,327 harp seals were harvested in 2010 compared to 217,857 in 2008. In 2010, there were approximately 12,500 commercial licences and 1,600 personal use sealing licences. Personal use licences have been issued to residents adjacent to sealing areas and allows the holder to take up to six seals for personal consumption. The DFO uses population models (based on survey estimates of seal numbers) to set a total allowable catch (TAC) for harp seals, as well as gray and hooded seals. The management objectives for Canada's commercial seal hunts are (i) to allow sealers to maximize their economic benefits without compromising conservation and (ii) to ensure conservation by maintaining the population at a level above 70% of the maximum observed population. The 2011–2015 Integrated Fisheries Management Plan for Atlantic Seals will allow a harvest of up to 400,000 animals over the next 3 years. The setting of these quotas is strongly disputed by anti-sealing organizations. Among several criticisms of population models the TAC is based on percentages of the maximum estimate of population size, rather than an estimate of the carrying capacity of the environment.

The Marine Mammal Regulations state that seals must be killed quickly using only high-powered rifles, shotguns, clubs, or hakapiks. The hakapik is a traditional Norweigian sealing tool that consists of an iron head with curved iron spike on one side and blunt end on the other, mounted on a wooden handle. In 2009, a number of amendments to the Marine Mammal Regulations came into force to improve the humaneness of killing methods. Together with detailed license conditions, this introduced a three-step process to minimize suffering. The three steps are: 1. *Striking*: sealers must shoot or strike animals on the top of the cranium, with either a firearm, a hakapik, or club; 2. *Checking*: the sealer must palpate both the left and right halves of the cranium, following striking (either with a firearm, hakapik, or club), to ensure that the skull has been crushed. This ensures that the seal is irreversibly unconscious or dead; and 3. *Bleeding*: the sealer must bleed the animal by severing the two axillary arteries located beneath the front flippers and must allow a minimum of 1 minute to pass before skinning the animal. Bleeding ensures that the seal is dead.

There is no doubt that the anti-sealing movement has brought about greater regulation of sealing methods and maintained strong pressure on the Canadian government to justify quotas. The issue continues to be highly controversial. The Canadian government strongly defends sealing as a traditional, saying it is a sustainable harvest of a natural resource providing important income and jobs for local communities. Fishers may also be sealers and justify seal culls because they believe declining fish stocks are due to predation by seals. However, many individuals and international animal welfare organizations remain fully committed to ending sealing and each year the harp seal hunt is featured widely in the media.

SELECTED REFERENCES AND FURTHER READING

Alva, J.J., & Salazar, S. (2006). Status and conservation of otariids in Ecuador and the Galápagos Islands. In: *Sea Lions of the World* (Ed. A.W. Trites, S.K. Atkinson, D.P. DeMaster, L.W. Fritz, T.S. Gelatt, L.D. Rea & K.M. Wynne), pp. 495–519. Alaska Sea Grant College Program, Fairbanks, AK.

Alverson, D.L. (1992). A review of commercial fisheries and the Steller sea lion (*Eumetopias jubatus*): the conflict arena. *Reviews in Aquatic Sciences* 6: 203–256.

Bickham, J.W., Loughlin, T.R., Calkins, D.G., Wickliffe, J.K., & Patton, J.C. (1998). Genetic variability and population decline in Steller sea lions from the Gulf of Alaska. *Journal of Mammalogy* 79: 1390–1395.

Bickham, J.W., Patton, J.C., & Loughlin, T.R. (1996). High variability for control-region sequences in a marine mammal: implications for conservation and biogeography of Steller sea lions (*Eumetopias jubatus*). *Journal of Mammalogy* 77: 95–108.

Bowen, W.D., Ellis, S.L., Iverson, S.J., & Boness, D.J. (2003). Maternal and newborn life-history traits during periods of contrasting population trends: implications for explaining the decline of harbour seals (*Phoca vitulina*) on Sable Island. *Journal of Zoology (London)* 261: 155–163.

Burns, J.J. (2008). Harbor seal and spotted seal *Phoca vitulina* and *P. largha*. In: *Encyclopedia of Marine Mammals,* 2nd ed. (Ed. W.F. Perrin, B. Würsig, & J.G.M. Thewissen), pp. 533–542. Academic Press, New York.

Butler, J.R.A., Middlemas, S.J., McKelvey, S.A., McMyn, I., Leyshon, B., Walker, I., Thompson, P.M., Boyd, I.L., Duck, C., Armstrong, J.D., Graham, I., & Baxter, J.M. (2008). The Moray Firth Seal Management Plan: an adaptive framework for balancing the conservation of seals, salmon, fisheries and wildlife tourism in the UK. *Aquatic Conservation* 18: 1025–1038.

Castinel, A., Pomroy, B., & Grinberg, A. (2007). Hookworm infection and *Klebsiella pneumoniae* epidemics in New Zealand sea lion pups. *Veterinary Microbiology* 125: 388–389.

Chilvers, B.L. (2008). New Zealand sea lions: *Phocarctos hookeri* and squid trawl fisheries: bycatch problems and management options. *Endangered Species Research* 5: 193–204.

Chilvers, B.L., Wilkinson, I.S., & Childerhouse, S. (2007). New Zealand sea lion, *Phocarctos hookeri*, pup production—1995 to 2006. *New Zealand Journal of Marine and Freshwater Research* 41: 205–213.

Dalton, R. (2005). Is this any way to save a species? *Nature* 436: 14–16.

Fisheries and Oceans Canada. *Seals and Sealing in Canada.* http://www.dfo-mpo.gc.ca/fm-gp/seal-phoque/index-eng.htm. Accessed: 1/18/12.

Fritz, L.W., & Hinckley, S. (2006). A critical review of the regime shift—"junk food" nutritional stress hypothesis for the decline of the western stock of Steller sea lion. *Marine Mammal Science* 21: 3476–3518.

Frost, K.J., Lowry, L.F., & Ver Hoef, J.M. (1999). Monitoring the trend of harbor seals in Prince William Sound, Alaska, after the *Exxon Valdez* oil spill. *Marine Mammal Science* 15: 494–506.

Goldsworthy, S.D., & Page, B. (2007). A risk-assessment approach to evaluating the significance of seal bycatch in two Australian fisheries. *Biological Conservation* 139: 269–285.

Hall, A.J., Duck, C.D., Law, R.J., Allchin, C.R., Wilson, S., & Eybator, T. (1999). Organochlorine contaminants in Caspian and harbour seal blubber. *Environmental Pollution* 106: 203–212.

Hall, A.J., Law, R.J., Wells, D.E., Harwood, J., Ross, H.M., Kennedy, S., Allchin, C.R., Campbell, L.A., & Pomeroy, P. P. (1992). Organochlorine levels in common seals (*Phoca vitulina*) which were victims and survivors of the 1988 phocine distemper epizootic. *Science of the Total Environment* 115: 145–162.

Hamill, M.O. & Stenson, G.B. (2007). Application of the precautionary approach and conservation reference points to management of Atlantic seals. *ICES Journal of Marine Science* 64:702–706.

Härkönen T., Dietz, R., Reijnders, P. J. H., Teilmann, J., Thompson, P., Harding, K.C., Hall, A., Brasseur, S., Siebert, U., Goodman, S.J., Jepson, P.D., & Rasmussen, T.D. (2006). The 1988 and 2002 phocine distemper virus epidemics in European harbour seals. *Diseases of Aquatic Organisms* 68: 115–130.

Härkönen, T., Jüssi, M., Baimukanov, M., Bignert, A., Dmitrieva, L., Kasimbekov, Y., Verevkin, M., Wilson, S., & Goodman, S.J. (2008). Pup production and breeding distribution of the Caspian Seal (*Phoca caspica*) in relation to human impacts. *Ambio* 37: 356–361.

Hayama, S.I. (1988). Kuril seal-present status in Japan. *Ambio* 17: 75–78.

Heide-Jorgensen, M., & Harkonen, T.J. (1988). Rebuilding seal stocks in the Kattegat-Skagerrak. *Marine Mammal Science* 4: 231–246.

Hoelzel, A.R., Fleischer, R.C., Campagna, C., Le Boeuf, B.J., & Alvord, G. (2002). Impact of a population bottleneck on symmetry and genetic diversity in the northern elephant seal. *Journal of Evolutionary Biology* 15: 567–575.

Kajiwara, N., Watanabe, S.N.M., Ito, Y., Takahashi, S., Tanabe, S., Khuraskin, L.S., & Miyazaki, N. (2002). Organochlorine and organotin compounds in Caspian seals (*Phoca caspica*) collected during an unusual mortality event in the Caspian Sea in 2000. *Environmental Pollution* 117: 391–402.

Kennedy, S., Thijs Kuiken, T., Jepson, P.D., Deaville, R., Forsyth, M., Barrett, T., van de Bildt, M.W.G., Osterhaus, A.D.M.E., Eybatov, T., Duck, C., Kydyrmanov, A., Mitrofanov, I., & Wilson, S. (2000). Mass die-off of Caspian seals caused by canine distemper virus. *Emerging Infectious Diseases* 6: 637–639.

Kitts, D.D., Huynhl, M.D., Hu, C., & Trites, A.W. (2004). Seasonal variation in nutritional composition of Alaskan walleye Pollock. *Canadian Journal of Zoology* 82: 1408–1415.

Kovacs, K.M., Aguilar, A., Aurioles, D., Burkanov, V., Campagna, C., Gales, N., Gelatt, T., Goldsworthy, S.D., Goodman, S.J., Hofmeyr, G.J.G., Härkönen, T., Lowry, L., Lydersen, C., Schipper, J., Sipilä, T., Southwell, C., Stuart, S., Thompson, D., & Trillmich, F. (2011). Global threats to pinnipeds. *Marine Mammal Science:* online June 15, 2011, DOI: 10.1111/j.1748-7692.2011.00479.x.

Krylov, V.I. (1990). Ecology of the Caspian seal. *Finnish Game Research* 47: 32–36.

Kuiken, T., Kennedy, S., Barrett, T., Borgsteede, F., Deaville, R., Duck, C., Eybatov, T., Forsyth, M., Foster, G., Jepson, P.D.,

Kydyrmanov, A., Mitrofanov, I. ,Ward, C.J., Wilson, S., & Osterhaus, A. D. M. E. (2006). The 2000 canine distemper epidemic in Caspian seals (*Phoca caspica*): pathology and analysis of contributory factors. *Veterinary Pathology* 43: 321–338.

Lavigne, D.M. (2009). Harp seal. In: *Encyclopedia of Marine Mammals*, 2nd ed. (Ed. W.F. Perrin, B. Wursig, & J.G.M. Thewissen), pp. 542–546. Academic Press, New York.

Lavigne, D.M. & Kovacs, K.M. (1988). *Harps and hoods. Ice Breeding Seals of the Northwest Atlantic*. University of Waterloo Press, Waterloo, ON.

Lonergan, M., Duck, C.D., Thompson, D., Mackey, B.L., Cunningham, L., & Boyd, I.L. (2007). Using sparse survey data to investigate the declining abundance of British harbour seals. *Journal of Zoology* 271: 261–269.

Lonergan, M., Hall, A.J., Thompson, H., Thompson, P.M., Pomeroy, P.P., & Harwood, J. (2010). A comparison of the 1988 and 2002 phocine distemper epizootics in British harbour seal (*Phoca vitulina*) populations. *Diseases of Aquatic Organisms* 88: 183–188.

Loughlin, T.R., & Nelson R. (1986). Incidental mortality of northern sea lions in the Shelikof Strait, Alaska. *Marine Mammal Science* 2: 14–33.

Lucas, Z., & Stobo, W.T. (2000). Shark-inflicted mortality on a population of harbour seals (*Phoca vitulina*) at Sable Island, Nova Scotia. *Journal of Zoology (London)* 252: 405–414.

Merrick, R.L., Chumbley, M.K., & Byrd, G.V. (1997). Diet diversity of Steller sea lions (*Eumetopias jubatus*) and their population decline in Alaska: a potential relationship. *Canadian Journal of Fisheries and Aquatic Sciences* 54: 1342–1348.

National Marine Fisheries Service (NMFS). (2000). *Endangered Species Act Section 7 Consultation, Biological Opinion and Incidental Take Statement on the Authorization of the Bering Sea/Aleutian Islands and Gulf of Alaska Groundfish Fisheries Based on the Fishery Management Plans*. Protected Resources Division, Alaska Region, NMFS, Juneau AK.

Page, B., McKenzie, J., McIntosh, R., Baylis, A., Morrissey, A., Calvert, N., Haase, T., Berris, M., Dowie, D., Shaughnessy, P.D., & Goldsworthy, S.D. (2004). Entanglement of Australian sea lions and New Zealand fur seals in lost fishing gear and other marine debris before and after Government and industry attempts to reduce the problem. *Marine Pollution Bulletin* 49: 33–42.

Pascual, M.A., & Adkison, M.D. (1994). The decline of the Steller sea lion in the Northeast Pacific: demography, harvest or environment. *Ecological Applications* 4: 383–403.

Reijnders, P.J.H. (1986). Reproductive failure in common seals feeding on fish from polluted coastal waters. *Nature* 324: 456–457.

Rosen, D.A.S., & Trites, A.W. (2000). Pollock and the decline of Steller sea lions: testing the junk-food hypothesis. *Canadian Journal of Zoology* 78: 1243–1250.

Salazar, S. (2003). Impacts of the *Jessica* oil spill on sea lions (*Zalophus wollebaeki*) populations. *Marine Pollution Bulletin* 47: 313–318.

Salazar, S., & Bustamante, R.H. (2003). Effects of the 1997–98 El Niño on population size and diet of the Galápagos sea lions (*Zalophus wollebaeki*). *Noticias de Galápagos* 62: 40–45.

Shaughnessy, P., Kirkwood, R., Cawthorn, M., Kemper, C., & Pemberton, D. (2003). Pinnipeds, cetaceans and fisheries in Australia: a review of operational interactions. In: *Marine Mammals: Fisheries, Tourism and Management Issues* (Ed. N. Gales, M. Hindell, & R. Kirkwood), pp. 136–152. CSIRO Publishing, Collingwood, Australia.

Shaughnessy, P.D., Dennis, T.E., & Seager, P.G. (2005). Status of Australian sea lions, *Neophoca cinerea*, and New Zealand fur seals, *Arctocephalus forsteri*, on Eyre Peninsula and the far west coast of South Australia. *Wildlife Research* 32: 85–101.

Shaughnessy, P.D., McIntosh, R.R., Goldsworthy, S.D., Dennis, T.E., & Berris, M. (2006). Trends in abundance of Australian sea lions, *Neophoca cinerea*, at Seal Bay, Kangaroo Island, South Australia. In: *Sea Lions of the World* (Ed. A.W. Trites, S.K. Atkinson, D.P. DeMaster, L. W. Fritz, T.S. Gelatt, L.D. Rea, & K.M. Wynne), pp. 37–63. Alaska Sea Grant College Program, University of Alaska, Fairbanks, AK.

Shima, M., Hollowed, A.B., & VanBlaricom, G.R. (2002). Response of pinniped populations to directed harvest, climate variability, and commercial fishing activity: a comparative analysis. *Reviews in Fisheries Science* 8: 89–124.

Sinclair, E.H., & Zeppelin, T.K. (2002). Seasonal and spatial differences in diet in the western stock of Steller sea lions (*Eumetopias jubatus*). *Journal of Mammalogy* 83: 973–990.

Smith, R.J. (1997). Status of the Lacs des Loups Marins Harbour seal, *Phoca vitulina mellonae*, in Canada. *Canadian Field-Naturalist* 111: 270–276.

Springer, A.M., Estes, J.A., van Vliet, G.B., Williams, T.M., Doak, D.F., Danner, E.M., Forney, K.A., & Pfister, B. (2003). Sequential megafaunal collapse in the North Pacific Ocean: an ongoing legacy of industrial whaling? *Proceedings of the National Academy of Science* 100: 12223–12228.

Thompson, D., Bexton, S., Brownlow, A., Wood, D., Patterson, T., Pye, K., Lonergan, M., & Milne, R. *Report on recent seal mortalities in UK waters caused by extensive lacerations* October 2010. Sea Mammal Research Unit, University of St. Andrews, Scotland. Accessed at: http://www.smru.st-and.ac.uk/documents/366.pdf.

Towell, R.G., Ream, R.F., & York, A.E. (2006). Decline in northern fur seal (*Callorhinus ursinus*) pup production on the Pribilof islands. *Marine Mammal Science* 22: 486–491.

Trites, A.W. (1992). Northern fur seals: why have they declined? *Aquatic Mammals* 18: 3–18.

Trites, A.W., & Donnelly, C.P. (2003). The decline of Steller sea lions in Alaska: a review of the nutritional stress hypothesis. *Mammal Review* 33: 3–28.

Wada, K., Hayama, S., Nakaoka, T., & Uno, H. (1991). Interactions between Kuril seals and salmon trap net fishery in the coastal waters of southeastern Hokkaido. *Marine Mammal Science* 7: 75–84.

Wade, P.R., Burkanov, V.N., Dahlheim, M.E., Friday, N.A., Fritz, L.W., Loughlin, T.R., Mizroch, S.A., Muto, M.M., Rice, D.W., Barrett-Lennard, L.G., Black, N.A., Burdin, A.M., Calambokidis, J., Cerchio, S., Ford, J.K.B., Jacobsen, J.K., Matkin, C.O., Matkin, D.R., Mehta, A.V., Small, R.J., Straley, J.M., McCluskey, S.M., VanBlaricom, G.R., & Clapham, P.J. (2007). Killer whale and marine mammal trends in the North Pacific—a re-examination of evidence for sequential megafauna collapse and the prey-switching hypothesis. *Marine Mammal Science* 23: 766–802.

Weber, D.S., Stewart, B.S., Garza, J.C., & Lehman, N. (2000). An empirical genetic assessment of the severity of the northern elephant seal population bottleneck. *Current Biology* 10: 1287–1290.

Wilkinson, I., Burgess, J., & Cawthorn, M. (2003). New Zealand sea lions and squid: managing fisheries impacts on a threatened marine mammal. In: *Marine Mammals: Fisheries, Tourism and Management Issues* (Ed. N. Gales, M. Hindell & R. Kirkwood), pp. 192–207. CSIRO Publishing, Collingwood, Australia.

Winship, A.J., & Trites A.W. (2003). Prey consumption of Steller sea lions (*Eumetopias jubatus*) off Alaska: how much prey do they require? *Fishery Bulletin* 101: 147–167.

York, A.E., & Hartley, J.R. (1981). Pup production following harvest of female northern fur seals. *Canadian Journal of Fisheries and Aquatic Sciences* 38: 84–90.

Marine Mammal Protection: Laws and Initiatives

CHAPTER 17

CHAPTER OUTLINE

International Union for Conservation of Nature
- *Exploring the Depths: United Nations Convention on the Law of the Sea*

Convention on the International Trade in Endangered Species

Convention on Migratory Species
- *Exploring the Depths: ASCOBANS*
- *Exploring the Depths: ACCOBAMS*

U.S. Endangered Species Act

U.S. Marine Mammal Protection Act
- *Exploring the Depths: 1966 Fur Seal Act*
- *Exploring the Depths: U.S. Marine Mammal Commission*
- *Exploring the Depths: U.S. Magnuson-Stevens Act*
- *Exploring the Depths: New Zealand Marine Mammals Protection Act*

European Community Habitats Directive
- *Exploring the Depths: European Council Regulation on Bycatch*
- *Exploring the Depths: Pinniped Conservation Law in the United Kingdom*
- *Exploring the Depths: Legal Protections for Marine Mammals in the United Kingdom*
- *Exploring the Depths: Public Opinion on Marine Mammal Protection in the United Kingdom*
- *Exploring the Depths: Marine Protected Areas and Sanctuaries*

Food For Thought: Science-Based Management?

Food For Thought: Enforcement
- *Exploring the Depths: Domestic Laws Around the World*

Selected References and Further Reading

In addition to the International Whaling Commission (see Chapter 14), several treaties and international bodies address the conservation and management of marine mammals. Moreover, there are several specific laws for their protection and conservation. This chapter briefly describes some of these treaties and laws.

International Union for Conservation of Nature

The International Union for Conservation of Nature (IUCN) is an international body composed of numerous domestic governmental bodies and environmental and animal welfare nongovernmental organizations (NGOs). As of 2009 the IUCN comprised 80 sovereign states, 112 government agencies, and 742 NGOs. A main activity of the IUCN is production of the Red Lists, which are scientifically based assessments of the conservation status of plant and animal species. In these lists the IUCN categorizes the status of each population and/or species in various ways (**Table 17.1**). Many nations subsequently use these categories as a basis for prioritizing conservation actions. The three categories of greatest threat are "critically endangered," "endangered," and "vulnerable." Each of these, like the other categories, requires that certain criteria are met before they are assigned to a species or population (**Figure 17.1**).

In addition to assessing status and compiling the Red Lists, the IUCN has a number of committees and working groups that focus on particular issues. The Cetacean Specialist Group is comprised of selected scientists who typically work on threatened cetacean populations. To date, the Cetacean Specialist Group has produced three action plans detailing research and conservation priorities for threatened cetacean species or populations. These

TABLE 17.1 IUCN Categories

EXTINCT (EX) - A taxon is Extinct when there is no reasonable doubt that the last individual has died.

EXTINCT IN THE WILD (EW) - A taxon is Extinct in the Wild when it is known only to survive in cultivation, in captivity, or as a naturalized population (or populations) well outside the past range.

CRITICALLY ENDANGERED (CR) - A taxon is Critically Endangered when it is facing an extremely high risk of Extinction in the Wild in the immediate future.

ENDANGERED - A taxon is Endangered when it is not Critically Endangered but is facing a very high risk of Extinction in the Wild in the near future.

VULNERABLE - A taxon is Vulnerable when it is not Critically Endangered or Endangered but is facing a high risk of Extinction in the Wild in the medium-term future.

LOWER RISK - A taxon is Lower Risk when it has been evaluated, does not satisfy the criteria for any of the categories—Critically Endangered, Endangered or Vulnerable. Taxa included in the Lower Risk category can be separated into three subcategories:

CONSERVATION DEPENDENT - Taxa that are the focus of a continuing species-specific or habitat-specific conservation programs—the cessation of which would result in the taxon qualifying for one of the threatened categories above within a period of 5 years.

NEAR THREATENED - Taxa that do not qualify for Conservation Dependent, but that are close to qualifying for Vulnerable.

LEAST CONCERN - Taxa which do not qualify for Conservation Dependent or Near Threatened status.

DATA DEFICIENT - A taxon is Data Deficient when there is inadequate information to make a direct, or indirect, assessment of its risk of extinction. Listing of taxa in this category indicates that more information is required and acknowledges the possibility that future research will show that threatened classification is appropriate.

NOT EVALUATED - The taxon has not yet been assessed against the criteria.

Data from: IUCN.

plans help to direct conservation funding toward the most threatened populations and focus research in the most needed areas.

Finally, the IUCN also holds major meetings, or congresses, to discuss conservation priorities. In 2008 the IUCN held their 4th Congress. These meetings are composed of two "houses": one made up of governmental organizations and another comprising NGOs. At each congress resolutions (aimed at helping shape international conservation efforts and policy) are discussed. To be passed by the congress, these resolutions have to be voted on and approved by both houses. The 2008 Congress passed a resolution on whale conservation calling for the promotion of the nonlethal use of whales (i.e., through responsible whale watching).

At present, marine mammal populations or species categorized as "critically endangered" are as follows:

- Blue whale (Antarctic population)
- Bowhead whale (Svalbard population)
- Gray whale (northwestern Pacific population)
- North Pacific right whale (northeastern Pacific population)
- Southern right whale (Chilean population)
- Indo-Pacific humpback dolphin (East Taiwan Strait population)
- Vaquita
- Mediterranean monk seal
- Hawaiian monk seal

Species in the "endangered" category currently include the following:

- Blue whale
- Fin whale
- Sei whale
- North Atlantic right whale
- North Pacific right whale
- Indian River dolphin
- Hector's dolphin
- Galápagos fur seal
- Galápagos sea lion
- Steller sea lion
- Australian sea lion
- Caspian seal
- Sea otter
- Marine otter

Finally, the following species are currently listed in the "vulnerable" category:

- Amazonian manatee
- Atlantic humpback dolphin
- Dugong
- Finless porpoise
- Fishing (or fish-eating) bat

CRITICALLY ENDANGERED

(A) Observed, estimated, inferred, or suspected reduction of at least 80% over the last 10 years or 3 generations.
(B) Extent of occurrence estimated to be less than 100 km^2 or area of occupancy estimated to be less than 10 km^2, and two of the following:
   1) Severely fragmented or known to exist at only a single location.
   2) Continuing decline in extent of occurrence and quality/quantity of habitats.
   3) Extreme fluctuations in occurrence/occupied area/number of animals.
(C) Population estimated to number less than 250 mature individuals and either:
   1) An estimated continuing decline of at least 25% within 3 years or 1 generation, or
   2) A continuing decline in numbers of mature individuals and population structure is (a) severely fragmented (i.e. no subpopulation estimated to contain more than 50 mature individuals) (b) all individuals are in a single subpopulation.
(D) Population estimated to number less than 50 mature individuals.
(E) Analysis shows the probability of extinction in the wild is at least 50% within 10 years or 3 generations.

ENDANGERED

(A) Observed, estimated, inferred, or suspected reduction of at least *50%* over the last 10 years or 3 generations.
(B) Extent of occurrence estimated to be less than *5000 km^2* or area of occupancy estimated to be less than *500 km^2*, and two of the following:
   1) Severely fragmented or known to exist in *no more than 5 locations*.
   2) Continuing decline in extent of occurrence and quality/quantity of habitats.
   3) Extreme fluctuations in occurrence/occupied area/number of animals.
(C) Population estimated to number less than *2,500* mature individuals and either:
   1) An estimated continuing decline of at least *20%* within *5 years or 2* generations, or
   2) A continuing decline in numbers of mature individuals and population structure is (a) severely fragmented (i.e. no subpopulation estimated to contain more than *250* mature individuals) (b) all individuals are in a single subpopulation.
(D) Population estimated to number less than *250* mature individuals.
(E) Analysis shows the probability of extinction in the wild is at least *20%* within *20 years or 5 generations*.

VULNERABLE

(A) Observed, estimated, inferred, or suspected reduction of at least *20%* over the last 10 years or 3 generations.
(B) Extent of occurrence estimated to be less than *20,000 km^2* or area of occupancy estimated to be less than *2,000 km^2*, and two of the following:
   1) Severely fragmented or known to exist in *no more than 10 locations*.
   2) Continuing decline in extent of occurrence and quality/quantity of habitats.
   3) Extreme fluctuations in occurrence/occupied area/number of animals.
(C) Population estimated to number less than *10,000* mature individuals and either:
   1) An estimated continuing decline of at least *10%* within *10 years or 3* generations, or
   2) A continuing decline in numbers of mature individuals and population structure is (a) severely fragmented (i.e. no subpopulation estimated to contain more than *1,000* mature individuals) (b) all individuals are in a single subpopulation.
(D) Population estimated to number less than *1,000* mature individuals.
(E) Analysis shows the probability of extinction in the wild is at least *10%* within *100 years*.

FIGURE 17.1 Descriptions of IUCN category thresholds.

Exploring the Depths: United Nations Convention on the Law of the Sea

The United Nations Convention on the Law of the Sea (UNCLOS) came into force on November 16, 1994. The treaty effectively defines marine territorial boundaries and the legal rights and obligations of coastal states to adjacent waters. For example, the "territorial sea" of a country extends from the coastline out to 12 nautical miles, provided this does not overlap with that of another country, when the border is set equidistant from both shorelines. Waters within this strip are governed exclusively by a country's domestic laws. The next 12 nautical miles comprise the Contiguous Zone, where states have limited control over various aspects of the activities of vessels, including pollution. Beyond this and extending out to 200 nautical miles from the coast (unless conflicting with that of another nation, when it is left to the states to delineate the actual boundary) is the Exclusive Economic Zone (EEZ). Countries have no legal control over their EEZs but do still have, according to this convention, exclusive rights to exploit the resources contained within their EEZs, including fisheries (even whale stocks) and submarine minerals. States also maintain certain obligations to manage the environment and the resources within this area (e.g., by controlling pollution). Finally, states may extend their EEZs by declaring their continental shelf area. The continental shelf is defined under UNCLOS as the natural prolongation of the land territory to the continental margin's outer edge, or 200 nautical miles from the coastal state's baseline, whichever is greater. However, a state's continental shelf may never exceed 350 nautical miles from the shore or go more than 100 nautical miles beyond the 2,500-m isobath (i.e., the "contour" for that depth). Beyond the EEZ (however established) is a region known as the "high seas," which may be exploited by anyone, without any specific environmental or conservation obligations, except as noted below.

- Cetaceans are collectively considered to be a "marine living resource" under UNCLOS. As such, member nations are obliged to limit any harvest of these species to sustainable levels, including in EEZs and on the high seas, although UNCLOS does not ban hunting or otherwise exploiting marine mammals entirely. The treaty also explicitly mentions the conservation of migrating marine mammals, requiring member states to cooperate to conserve, manage, and study such marine mammals on the high seas as well as in waters more specifically under their own control. Moreover, Articles 65 and 120 of the treaty state that member states should "co-operate with a view to the conservation of marine mammals and in the case of cetaceans shall in particular work through the appropriate international organizations for their conservation, management and study."

For whales the international organization is the International Whaling Commission, but the appropriate organizations for handling small cetaceans, pinnipeds, sirenians, polar bears, and otters are less clear.

Although a signatory to UNCLOS, the United States is one of the few nations that has not yet ratified the treaty (i.e., formally and legally committed to the treaty). Despite this, Ronald Reagan, through a 1983 executive order, directed U.S. agencies to comply with all the provisions in UNCLOS except for Part XI, which concerns deep-sea mining. Thus, since 1983 the United States has essentially been in voluntary compliance with the Convention in any case. U.S. efforts even led to the official modification of Part XI in 1994, addressing all U.S. concerns. This has led to many calls for full U.S. ratification from groups as disparate as the military, the oil and gas industry, and environmentalists. Ratification would change little in the way of U.S. domestic policy, but it would mean that the extended EEZ (through the continental shelf) could be officially claimed by the United States, ultimately providing unchallenged access to the resources contained within. This is of great importance in the Arctic, where extensive mineral, oil, and gas deposits are thought to exist. Internationally recognized marine borders (and rights of passage elsewhere) would also be beneficial to the Navy. Finally, the Convention would officially commit the United States to sustainable harvests of all marine living resources and oblige them to regulate pollution. Despite all these benefits, many naysayers remain, claiming that UNCLOS will impinge upon U.S. sovereignty.

- Franciscana
- Hooded seal
- Irrawaddy dolphin
- Northern fur seal
- New Zealand sea lion
- Polar bear
- Sperm whale
- West African manatee
- West Indian manatee

Convention on the International Trade in Endangered Species

The international wildlife trade is worth billions of dollars and has been at least partially, if not entirely, responsible for the decline of numerous animal and plant species. In an attempt to control this international trade and stop the decline of impacted species, the Convention on International Trade in Endangered Species of Wild Fauna and Flora (CITES) was signed in 1973 (and it came into effect in 1975). More than 160 nations are signatories of CITES. Signatories are obliged to introduce their own national legislation that controls their trade in wild species. The treaty lists threatened species on two appendices: CITES Appendix I, which lists species that are threatened with extinction or may become threatened by trade and so the international trade in these species (or their parts) for commercial purposes is banned; and CITES Appendix II, which lists species that are not yet threatened with extinction but may become so if there is uncontrolled trade in

TABLE 17.2 Cetacean Species Listed Under CITES Appendix I

Blue whale	*Balaenoptera musculus*
Fin whale	*Balaenoptera physalus*
Bryde's whale	*Balaenoptera edeni*
Sei whale	*Balaenoptera borealis*
Humpback whale	*Megaptera novaeangliae*
Northern minke whale (except West Greenland population—Appendix II)	*Balaenoptera acutorostrata*
Antarctic minke whale	*Balaenoptera bonaerensis*
Gray whale	*Eschrichtius robustus*
Bowhead whale	*Balaena mysticetus*
Right whales	*Eubalaena* spp.
Pygmy right whale	*Caperea marginata*
Yangtze River dolphin	*Lipotes vexillifer*
Indian River dolphins	*Platanista* spp.
Sperm whale	*Physeter macrocephalus*
Arnoux's beaked whale	*Berardius arnuxii*
Baird's beaked whale	*Berardius bairdii*
Bottlenose whales	*Hyperoodon* spp.
Humpback dolphins	*Sousa* spp.
Irrawaddy dolphin	*Orcaella brevirostris*
Tucuxi	*Sotalia fluviatilis*
Finless porpoise	*Neophocaena phocaenoides*
Vaquita	*Phocoena sinus*

Data from: CITES.

them or their parts. In addition, Appendix II contains species that must also be regulated to make the control of the various potentially threatened species effective. Consider occasions when two species may look very similar (e.g., as a result of mimicry) or their bones or other parts might be largely indistinguishable without intensive analyses. As such, CITES Appendix III contains a list of species that are protected by one or more member states that have approached CITES (successfully) for assistance in regulating their international trade.

Any international trade of Appendix I species for noncommercial purposes (e.g., research or captive breeding) must be documented with both an export and import permit. Member states are also obliged to regulate domestic transfers in similar ways. Trade in Appendix II–listed species is allowed because of their less threatened status, but there must be an appropriate permit (mostly with regards to export) and a "nondetriment finding" issued (i.e., a supposedly science-based document that states the level of trade being permitted will not adversely affect wild populations).

Trade in Appendix III–listed species requires a certificate of origin and, if it comes from the state(s) that listed the plant or animal, also a valid export permit.

All species of cetaceans are listed (either individually or collectively) in either CITES Appendix I or II (**Table 17.2**). Most of the "great whale" species (e.g., baleen whales and sperm whales) are listed under Appendix I. However, at the last few meetings of CITES's decision-making body, several pro-whaling nations have proposed "down-listing" (moving a species to a status that reflects less vulnerability and usually offers less protections; in this case moving a species from Appendix I to Appendix II) northern minke whale populations in the North Atlantic.

In Europe CITES is implemented via the European Community Regulation of Trade in Endangered Species. Under this regulation the European community treats all cetaceans as being on Appendix I of the Convention, regardless of which appendix they are actually listed under.

As for the other marine mammal species, Guadalupe fur seals and the two living species of monk seal are listed on CITES Appendix I. The southern elephant seal and the remainder of the fur seals except the northern fur seal (i.e., all *Arctocephalus* species) are currently listed on CITES Appendix II. Canadian populations of the walrus are listed on CITES Appendix III. All living sirenian species are listed on CITES Appendix I, except the West African manatee, which is listed on Appendix II. The marine otter and the southern sea otter subspecies *Enhydra lutris nereis* are listed on CITES Appendix I (with the remainder of the sea otters listed under Appendix II), as is the Eurasian otter (which, as noted previously, could be considered to be a marine mammal in some locations). The polar bear is listed on CITES Appendix II, although in 2010 the US government put a proposal into CITES to have the polar bear up-listed to Appendix I (which was unsuccessful at that time).

Convention on Migratory Species

The 1979 Convention on the Conservation of Migratory Species of Wild Animals (CMS), sometimes referred to as the Bonn Convention, is an international treaty that tries to encourage signatories to develop international agreements for the conservation of species that migrate across the boundaries of different countries. At the time of writing this textbook, the Convention has 116 signatories (including the European Union, but not the United States).

Cetaceans are listed on two of the treaty's appendices: Appendix I (migratory species threatened with extinction) or Appendix II (migratory species that would significantly benefit from international cooperation). Two key cetacean conservation agreements developed as the result of this convention are the 1991 Agreement on the Conservation of Small Cetaceans of the Baltic and North Sea (ASCOBANS) and the 1996 Agreement on the Conservation of Cetaceans of the Black Sea, Mediterranean Sea and Contiguous Atlantic Area (ACCOBAMS). Furthermore, at meetings of CMS member nations, conservation-focused resolutions and recommendations are produced, which member nations are obliged to try to address. For example, in 1999 Resolution 6.2 called for member countries to work to try to reduce fishery bycatch of cetaceans (and marine turtles and seabirds). In 2008 a memorandum of understanding was signed between various West African nations to help promote conservation of cetaceans, and also manatees, in this region (Memorandum of Understanding Concerning the Conservation of the Manatee and Small Cetaceans of Western Africa and Macaronesia). This memorandum of understanding is still relatively new, but hopefully it will generate several conservation initiatives in this region.

For pinnipeds, CMS signatories developed the 1990 Agreement on the Conservation of Seals in the Wadden Sea, a regional agreement similar to ASCOBANS and ACCOBAMS above. In 2007, a Memorandum of Understanding Concerning Conservation Measures for the Eastern Atlantic Populations of the Mediterranean Monk Seal was also signed to help further regional conservation of this threatened species. Few pinniped species are, however, listed under CMS: the South American fur seal and sea lion, and certain populations of the grey and harbor seal are listed under Appendix II, with only the Mediterranean monk seal listed under Appendix I. However, all sirenians are listed under Appendix II of the CMS, and in 2007 a memorandum of understanding was developed specifically to aid dugong conservation (Memorandum of Understanding on the Conservation and Management of Dugongs and their Habitats throughout their Range). Of the other marine mammals, only the marine otter is listed by the CMS (under Appendix I).

Exploring the Depths: ACCOBAMS

ACCOBAMS came into force on June 1, 2001. The agreement is slightly wider in scope than ASCOBANS as it deals with all cetaceans in the region and not just small cetaceans. As of 2011 the agreement had 23 member nations, including countries as diverse as Libya, Tunisia, Algeria, France, Spain, and Monaco. The agreement has developed an action plan of scientific and conservation-oriented projects, such as evaluating bycatch levels and identifying areas of critical habitat for key species. A fund for conservation projects and a scientific advisory panel have also been established. Other projects developed by ACCOBAMS include agreements to restrict drift net use and guidelines for whale watching in the region, scientific specimen collection, dealing with rescues of "animals in distress," and releases of cetaceans into the wild. Members of the agreement have also discussed the issue of underwater noise. The agreement has been slow to move in some areas because it is ultimately reliant on member nations to push activities forward (and some nations are much less active than others), but nonetheless the agreement has been extremely successful in bringing together a very diverse group of nations for the common cause of cetacean conservation.

Exploring the Depths: ASCOBANS

ASCOBANS was signed on March 17, 1992. The agreement was, in part, the result of concern over high levels of fishery bycatch of harbor porpoises and dolphins in the North Sea. The agreement requires member nations to introduce, as far as they are able, "conservation, research and management measures" for small cetaceans. These measures include working toward a reduction in harmful pollutants (including noise), attempting to introduce modifications of fishing gear that would reduce cetacean bycatch, working to reduce the depletion of cetacean food resources, and reducing disturbance of small cetaceans, in particular working to minimize the impacts of underwater noise.

ASCOBANS also calls for coordinated small cetacean research in the region, in particular with respect to identifying important small cetacean habitat. The agreement also requests member parties to introduce national legislation to prohibit the intentional killing of small cetaceans and to require the immediate release of bycaught small cetaceans if they are still alive when found. Additionally, ASCOBANS requires member parties to ensure information is provided to fishermen so that bycatch can be reported accurately and carcasses can be appropriately analyzed by scientists. Although great on paper, the actual effect of ASCOBANS in terms of conservation is debatable. There have been two large-scale surveys, ultimately as a result of this agreement, but little has been done to modify fishing gear, reduce pollutants, or reduce the impact of underwater noise on small cetaceans in the North Sea, particularly in the Baltic regions.

U.S. Endangered Species Act

The 1973 U.S. Endangered Species Act (ESA) offers protection to species threatened to some extent with extinction in the United States. A species (or population) can be listed as an "endangered species," in accordance with the ESA, if it is in danger of extinction throughout all or a significant portion of its range or as a "threatened" species if it is likely to become endangered in the foreseeable future.

The ESA works in two stages. Immediately after a species is listed, the ESA requires the government to introduce measures to stop the species from declining further and to protect it from possible extinction. The ESA then requires the government to take further steps to restore the species' numbers to the point where it is no longer threatened.

The U.S. National Marine Fisheries Service (NMFS; for cetaceans and pinnipeds except the walrus) and the U.S. Fish and Wildlife Service (FWS; for the walrus, polar bear, otters, and sirenians) are the agencies charged with carrying out the appropriate management actions, including creating regulations to protect the species and granting permits, authorizations, and exemptions to those entities or activities entitled to them under the ESA, such as scientists. Accordingly, NMFS and FWS are required to list a species as threatened or endangered if its existence is threatened by any of the following:

- Present or threatened destruction, modification, or curtailment of its habitat or range
- Overutilization for commercial, recreational, scientific, or educational purposes
- Disease or predation
- Inadequacy of existing regulatory mechanisms
- Other natural or artificial factors affecting its continued existence

Marine mammal species currently listed under the ESA include Steller sea lions, West Indian manatees, polar bears, and a number of cetacean species and populations (**Table 17.3**).

Some of the biggest debates that surround the ESA are implications for private landowners on their own land, who can be limited in what they are allowed to do as a result of the presence of a listed species on that land. This has, unfortunately, led some to simply (but illegally) shoot and bury such offending species, in an attempt to avoid these problems.

Other controversy surrounds the declaration of "critical habitat" for the listed species. Largely abandoned by FWS as "useless" (due to the diversion of resources to battle the lawsuits it generates) since the 1980s, the protection of critical habitat is in fact a cornerstone of the Act as mandated by Congress: "to provide a means whereby the ecosystems upon which endangered species and threatened species depend may be conserved." The focus on ecosystems is a crucial recognition of the importance of suitable habitat to the recovery of a species. Once designated, no action (with a few exceptions) can lead to "adverse modification" of the critical habitat. Perhaps most importantly, habitat can be designated beyond the current distribution of the species.

There are two reasons to do this. The first is to provide habitat into which the species can expand as it recovers. The second is to protect an important resource that may always remain outside the distribution of the species. One good example is the rivers in the Pacific Northwest of the United States and Canada that support the salmon runs crucial to resident killer whales in that area. Although not yet listed as critical habitat under the ESA (or the Canadian equivalent, the Species At Risk Act), there have been discussions about this possibility (especially in Canada).

Finally, it is quite possible that, on occasion, provisions in this law will come into conflict with those of other existing laws. Congress displayed unusual foresight here (at least with regards to marine mammals), including within

TABLE 17.3 Marine Mammal Species and Populations Listed Under the United States Endangered Species Act

Common Name	Species Name
Blue whale	*Balaenoptera musculus*
Fin whale	*Balaenoptera physalus*
Sei whale	*Balaenoptera borealis*
Bowhead whale	*Balaena mysticetus*
North Atlantic right whale	*Eubalaena glacialis*
North Pacific right whale	*Eubalaena japonica*
Southern right whale	*Eubalaena australis*
Humpback whale	*Megaptera novaeangliae*
Sperm whale	*Physeter macrocephalus*
Vaquita	*Phocoena sinus*
Indus River dolphin	*Platanista minor*
Baiji	*Lipotes vexillifer*
Guadalupe fur seal	*Arctocephalus townsendi*
Hawaiian monk seal	*Monachus schauinslandi*
Mediterranean monk seal	*Monachus monachus*
Steller sea lion	*Eumetopias jubatus*
Sea otter	*Enhydra lutris*
Marine otter	*Lontra felina*
West Indian manatee	*Trichechus manatus*
Amazonian manatee	*Trichechus inunguis*
West African manatee	*Trichechus senegalensis*
Dugong	*Dugong Dugon*
Polar bear	*Ursus maritimus*
Listed Populations	
Killer whale (southern resident population/stock)	*Orcinus orca*
Gray whale (West Pacific population/stock)	*Eschrichtius robustus*
Northern fur seal (Eastern Pacific [Pribilof Island] stock)	*Callorhinus ursinus*
Saimaa (ringed) seal	*Phoca hispida saimensis*

Data from: NOAA.

the ESA a clause stating that the more restrictive provision would take precedence if any conflicts did indeed occur between the provisions of this Act and those of the Marine Mammal Protection Act.

U.S. Marine Mammal Protection Act

The U.S. Marine Mammal Protection Act (MMPA) was introduced in 1972 partly in response to the depletion of whales by commercial whaling and partly to address the high levels of bycatch experienced by dolphins, particularly in the eastern tropical Pacific tuna fisheries (see Chapter 15). Under the MMPA "taking" marine mammals is prohibited. A "take" was defined not only as the killing or injuring of a marine mammal but also the harassment of these animals. There are currently two legislated levels of harassment. Level A harassment is defined, for the most part, as "any act of pursuit, torment, or annoyance" that "has the potential to injure a marine mammal or marine mammal stock in the wild." In contrast, Level B harassment encompasses similar activities that have the "potential to disturb a marine mammal or marine mammal stock in the wild by causing disruption of behavioral patterns, including, but not limited to, migration, breathing, nursing, breeding, feeding, or sheltering." It should be noted that the MMPA term "stock" is largely interchangeable with the more biologically appropriate term "population."

To comply with the MMPA, those wishing to conduct an activity that could result in one or more takes must apply to the government for a permit or "take authorization." Permits entitle bearers to make directed takes. These may be for qualified activities, such as for scientific research, during the process of taking wildlife photographs, through captures for public display, or for taking animals to enhance their population and conservation status (e.g., captive breeding for conservation purposes). Authorizations are also needed for incidental takes, such disturbance resulting from activities like oil and gas exploration.

In addition to takes, the importing and exporting of marine mammals or marine mammal products is also prohibited under the MMPA. Like takes, permits may be granted for specific activities, in this case scientific research purposes or transferring animals between public display and captive breeding facilities.

Certain exemptions to these prohibitions occur within the Act. Alaskan indigenous populations are exempted, allowing them to hunt marine mammals (while they remain within their subsistence quotas) and to work and sell marine mammal products. Similarly, an amendment to the MMPA in 1994 exempted importing polar bear trophies (e.g., heads and skins taken by sports hunters). Finally, there is an ongoing interim exemption for the bycatch of animals in the due course of fishing activities. However, the MMPA set up a management regime to reduce marine mammal mortalities and injuries in their interactions with fisheries over the longer term. In addition, the MMPA also established basic requirements for public display of captive marine mammals and created a management regime for native subsistence hunting of marine mammals in Alaska.

One major objective of the MMPA is to prevent marine mammal populations from being reduced so they "cease to be a significant functioning element in the ecosystem of which they are a part" (Section 2(2) of the MMPA). This was one of the first laws to specifically note the important role marine mammals play in their marine ecosystems and the importance of maintaining their populations as a result. Consequently, the law requires that marine mammals "should not be permitted to diminish below their optimum sustainable population" (Section 2(2) of the MMPA), with the optimum sustainable population defined as follows (Section 3(9) of the MMPA):

> [T]he number of animals which will result in the maximum productivity of the population or the species, keeping in mind the carrying capacity of the habitat and the health of the ecosystem of which they form a constituent element.

If a marine mammal stock or population is listed as "endangered" or "threatened" under the ESA or it drops below the optimum sustainable population, it is considered to be "depleted." Regulations for the taking or trading of depleted species become somewhat stricter, with permitted exemptions for directed takes only allowed for scientific, public display, or conservation purposes. For example, upon being listed as threatened under the ESA, imports of polar bear trophies became restricted. Incidental takes remain largely unaffected, although the associated requirements for determining that the takes will have only a negligible impact on the population become harder to meet. The government may also develop a conservation plan for depleted populations, but this is not required.

Cetacean stocks currently considered to be depleted include the northeastern Pacific stock of pantropical spotted dolphins (sometimes referred to as *Stenella attenuata attenuata*), the coastal Pacific stock of pantropical spotted dolphins (*S. a. graffmani*), the eastern Pacific spinner dolphin stock (*S. longirostris orientalis*), the Cook Inlet population of beluga whales, and the mid-Atlantic stock of common bottlenose dolphins.

The NMFS is primarily responsible for enforcing the MMPA. (NMFS is part of the National Oceanic and Atmospheric Administration, which is in turn part of the U.S. Department of Commerce.) Under the MMPA, NMFS is responsible for the management and conservation of

Exploring the Depths: 1966 Fur Seal Act

This 1966 statute largely prohibited the taking, transportation, importing, or possession of fur seals and sea otters. The main exceptions to this statute are Native Americans who dwell on the coasts of the North Pacific Ocean and the U.S. Department of the Interior, which is specifically mandated to conduct research on these species. An amendment in 1983 reauthorized the Act, maintaining the exception for native populations provided takes were consistent with the relatively new MMPA. As a result the MMPA has become the dominant Act, and the Fur Seal Act is often overlooked.

Exploring the Depths: U.S. Marine Mammal Commission

Title II of the MMPA of 1972 created the Marine Mammal Commission as an independent agency of the Administration. The Commission consists of three members appointed by the President with the consent of the Senate. It is assisted by a nine-member Committee of Scientific Advisors on Marine Mammals and 14 full-time permanent staff.

The Act also established the major duties of the Commission as follows:

1. Undertake a review and study of the activities of the United States pursuant to existing laws and international conventions relating to marine mammals, including, but not limited to, the International Convention for the Regulation of Whaling, the Whaling Convention Act of 1949, the Interim Convention on the Conservation of North Pacific Fur Seals, and the Fur Seal Act of 1966.
2. Conduct a continuing review of the condition of the stocks of marine mammals, of methods for their protection and conservation, of humane means of taking marine mammals, of research programs conducted or proposed to be conducted under the authority of this Act, and of all applications for permits for scientific research, public display, or enhancing the survival or recovery of a species or stock.
3. Undertake or cause to be undertaken such other studies as it deems necessary or desirable in connection with its assigned duties as to the protection and conservation of marine mammals.
4. Recommend to the Secretary and to other federal officials such steps as it deems necessary or desirable for the protection and conservation of marine mammals.
5. Recommend to the Secretary of State appropriate policies regarding existing international arrangements for the protection and conservation of marine mammals and suggest appropriate international arrangements for the protection and conservation of marine mammals.
6. Recommend to the Secretary such revisions of the endangered species list and threatened species list published pursuant to section 4(c)(1) of the ESA of 1973 as may be appropriate with regard to marine mammals.
7. Recommend to the Secretary, other appropriate federal officials, and Congress such additional measures as it deems necessary or desirable to further the policies of this Act, including provisions for the protection of the Native Americans whose livelihood may be adversely affected by actions taken pursuant to this Act.

To fulfill those duties the Commission reviews and makes recommendations on the domestic and international policies and actions of federal agencies to ensure they are consistent with the MMPA, which seeks to protect and conserve marine mammals as functioning elements of healthy, stable marine ecosystems.

Marine mammals are subject to multiple human-related risk factors, including operational and ecological fishery interactions; the introduction of noise, disease, and contaminants; harmful algal blooms and dead zones; ill-managed coastal development and other forms of habitat modification; collisions with vessels of all sizes; and climate change. The Commission consults with other federal agencies (e.g., NMFS; FWS; U.S. Department of State; U.S. Navy; Bureau of Energy Management, Regulation, and Enforcement), state agencies (e.g., various departments of fish and game), and tribal organizations (e.g., the Indigenous People's Council on Marine Mammals) to characterize those risk factors and identify cost-effective solutions.

The Commission also helps develop and coordinate multi-agency and international research and management initiatives to facilitate marine mammal protection and conservation, as described in the Commission's annual reports to Congress. In all its work the Commission seeks to be a source of useful information, focused and catalytic research funding, and independent and objective oversight.

–Contributing author, Tim Ragen, U.S. Marine Mammal Commission.

Exploring the Depths: U.S. Magnuson-Stevens Act

The U.S. Magnuson-Stevens Act, or, more correctly, the Fisheries Conservation and Management Act of 1976, established governmental control over U.S. marine fish populations. This Act has provisions requiring fish stocks to be monitored and to ostensibly protect fish habitat, so helping to conserve wild fish populations. It sets the U.S. governmental policy on destructive fishing practices, including large-scale drift nets, calling for a ban on these, and allowing sanctions (in terms of fish imports) against countries using large-scale drift nets. The law also introduced the requirement for a conservation plan to reduce bycatches of all sorts and introduced fines of up to $25,000 for infractions. These provisions and other measures should help to protect cetacean prey species and their habitats and reduce mortality through bycatch. Unfortunately, although the Act looks great on paper, fish stocks continue to decline and levels of bycatch remain unsustainably high in many cases.

Exploring the Depths: New Zealand Marine Mammals Protection Act

Alongside the U.S. MMPA, the 1978 New Zealand Marine Mammals Protection Act (NZMMPA) is probably the most comprehensive law specifically protecting marine mammals. The NZMMPA, similar to the U.S. MMPA, prohibits the taking of dolphins and defines "taking" as killing, harming, injuring, attracting, poisoning, herding, or harassing. However, Section 16 of the NZMMPA states that the accidental capture of a cetacean (e.g., bycatch) is not an offense if the instance is reported promptly. This is an important provision as some fishermen in the U.S. realize that killing a cetacean is illegal and when bycatches do occur they may try to dispose of incriminating evidence. Thus, Section 16 allows for the cooperative monitoring of the impact of fishery bycatch in New Zealand while still encouraging fishermen to minimize them.

Whale watching (see Chapter 18) is a major economic activity in New Zealand, and, unlike the U.S. MMPA, the NZMMPA introduces laws governing whale watching (both by boats and also by airplanes) requiring users to "use their best endeavours to operate vessels, vehicles, and aircraft so as to not disrupt the normal movement of any marine mammal" (Clause 18). Other requirements are as follows (Clause 18):

> contact with any marine mammal shall be abandoned at any stage if it becomes or shows signs of becoming disturbed or alarmed
> no person shall cause any marine mammal to be separated from a group of marine mammals or any members of such a group to be scattered
> no rubbish or food shall be thrown near or around any marine mammal
> no sudden or repeated change in the speed or direction of any vessel or aircraft shall be made except in the case of emergency
> where a vessel stops to enable the passengers to watch any marine mammal, the engines shall be placed either in neutral or be switched off within a minute of the vessel stopping
> no person, vehicle or vessel shall cut off the path of a marine mammal or prevent a marine mammal from leaving the vicinity of any person, vehicle or vessel
> the master of any vessel less than 300 meters from any marine mammal shall use their best endeavours to move their vessels at a constant slow speed no faster than the slowest marine mammal in the vicinity, or at idle or "no wake" speed
> vessels departing from the vicinity of any marine mammal shall proceed slowly at idle or "no wake" speed until the vessel is at least 300 m from the nearest marine mammal, except that in the case of dolphins, vessels may exceed idle or "no wake" speed in order to outdistance the dolphins but must increase speed gradually and shall not exceed 10 knots within 300 m of any dolphin

Furthermore, Clause 20 adds as follows:

> no vessel shall proceed through a pod of dolphins; and
> ... no person shall make any loud or disturbing noise near dolphins or seals.

cetaceans and pinnipeds other than the walrus. Walruses, sirenians, sea otters, and polar bears are under the jurisdiction of the FWS, which is part of the U.S. Department of the Interior.

The MMPA also established the Marine Mammal Commission. This is an independent agency for marine mammal conservation and management advice (see Exploring the Depths: the U.S. Marine Mammal Commission).

European Community Habitats Directive

The 1992 European Council Directive on the Conservation of Natural Habitats and Wild Fauna and Flora (more commonly referred to as the "Habitats Directive") is the main piece of legislation that protects wildlife, including marine mammals, in Europe. The Directive has two parts that are relevant to marine mammals. For animal species

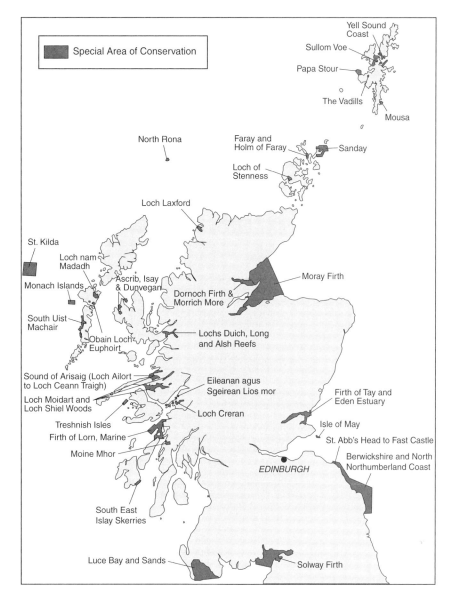

FIGURE 17.2 Map of Special Areas of Conservation for marine mammals in Scotland. Data from: Scottish Heritage (2008).

Exploring the Depths: European Council Regulation on Bycatch

On April 26, 2004 the European Union adopted a new regulation that aimed to reduce levels of cetacean bycatch in European waters (Council Regulation 812/2004). This regulation requires the following:

- Acoustic deterrent devices (also more commonly called "pingers"), which are sound-producing devices used to keep cetaceans away from fishing nets, are attached to fishing gear used by certain fishing vessels (those more than 12 meters in length) in the English Channel, Celtic Sea, and the North Sea.
- Observers attend fishing vessels more than 15 meters in length to accurately monitor levels and distribution of cetacean bycatch.
- Research projects must be established to monitor the effectiveness of acoustic deterrent devices in reducing bycatch and to monitor levels of bycatch in smaller fishing vessels (less than 15 meters).

Exploring the Depths: European Council Regulation on Bycatch (*continued*)

Although this is a good attempt to monitor and reduce bycatch using a legal means, the exception of acoustic deterrent device use on the many smaller vessels (there are approximately 6,000 fishing boats less than 12 meters in the United Kingdom alone), for example, means that bycatch may still occur, especially for species such as harbor porpoises.

The regulation also requires a phasing out of drift net use in the Baltic Sea. Although drift nets have been banned in all other European waters, the Baltic Sea was exempt. High levels of bycatch, particularly of harbor porpoises, have continued in this area as a result, possibly pushing the Baltic Sea harbor porpoise population to extinction.

Exploring the Depths: Pinniped Conservation Law in the United Kingdom

Hot on the heels of the first international treaty for wildlife conservation, the 1911 North Pacific Fur Seal Convention, of which the United Kingdom was a signatory, one of the earliest pieces of seal conservation legislation within the U.K. itself was the Grey Seal Protection Act of 1914. This Act banned the killing of grey seals during the breeding season (October 1 to December 15) and specifically gave protection for grey seals on one of the last known breeding sites on Haskeir, an isolated island in the Outer Hebrides.

Because of the depletion of several seal stocks from overharvesting, the Conservation of Seals Act of 1970 introduced a prohibition on killing of seals in their breeding and molting seasons in the United Kingdom (June 1 to August 31 for harbor seals and September 1 to December 31 for grey seals). As a result of large mortalities of seals, especially harbor seals (**Figure B17.1**), from outbreaks of seal distemper (in 1998 and 2002) in Scotland, the 2002 Conservation of Seals (Scotland) Order extended the closed season for harbor seals to year round. Another order in 2004 gave year-round protection for both harbor and grey seals in the Moray Firth as a result of a major decline of seals in the area, Serious concern regarding a 40% decline in harbor seals on the east coast of Scotland resulted in the extension of protection to a wider area by the Conservation of Seals (Scotland) Order 2007, which also prohibited the killing, injuring, or taking of harbor seals in the Northern Isles (Orkney and Shetland Islands and territorial waters adjacent to them) and in the territorial sea adjacent to the east coast of Scotland.

Most recently, the Marine (Scotland) Act 2010 has substantially increased the protection for seals in Scotland by making it an offense to kill seals at any time, except under license or for animal welfare purposes. Harassing a seal (intentionally or recklessly) at a haul-out site is now also an offense. Fisheries and aquaculture must apply for a seal management license to kill seals, and, for the first time, statutory reporting of seals killed or injured under license is required.

Under the strictures of the Convention on Biodiversity, the United Kingdom has produced conservation action plans for many marine mammals. Both grey and harbor seals are considered conservation priority species under the U.K. Biodiversity Action Plan process, but so far a U.K.-wide action plan has not been produced for seals. However, several local regions have produced conservation action plans, but in some cases these were not adopted because they were deemed too "controversial" (i.e., fishing bodies protested at the idea of conserving seals). Harbor and grey seals are listed on Appendix II of the Habitats Directive; therefore, the United Kingdom is obligated to designate SACs for both species. SACs have been designated in locations such as the Treshnish Isles in Scotland and the coast of southeastern Islay.

FIGURE B17.1 Harbor seals in Scotland.

Exploring the Depths: Legal Protections for Marine Mammals in the United Kingdom

Current legislation in the United Kingdom pertaining to marine mammals is an almost non-navigable set of intertwined provisions arising mainly from the Wildlife and Countryside Act of 1981 and its various amendments (such as, in England, the Marine and Coastal Access Act of 2009) and all the numerous regulations that result from both. Things get more complicated when the equivalent legislation in Scotland and Northern Ireland are taken into account (although, some would say, this is an improvement because Scottish laws are usually even more conservation-oriented). However, in combination, these laws and regulations do, both in their own right and as implementation tools for the European Union Habitats Directive, confer a certain amount of protection to all marine mammals throughout the United Kingdom. Unfortunately, despite the various marine protected areas in place under these laws (be they conservation zones, reserves, or SACs), only two sites are currently designated as "no-take zones," thus affording actual protection.

Of particular note is that all cetaceans are listed in relation to the Wildlife and Countryside Act as "protected animals" and the harbor porpoise and bottlenose dolphin are specifically mentioned as "animals which may not be killed or taken by certain methods." As a result it is illegal for any person to intentionally or recklessly kill, injure, or take any cetacean in the United Kingdom.

Some older laws, although not necessarily taking precedence today, still exist, such as the Whaling Industry Regulation Act of 1934, which prohibited the capture of any whales within British coastal waters and seriously limited allowed takes beyond these boundaries. Perhaps most interestingly, a law dating back to King Edward II (Statute Prerogative Regis, 17 Edward II, AD 1324) declares whales and sturgeon to be "Royal fish", making them the property of the monarch. It is not clear exactly to which cetaceans this applies, although Scottish law limits it to whales over 25 feet (7.6 m). However, this may, in principle, offer them more protection, as any such animals caught in British waters are immediately the property of the crown while Royal fish caught anywhere become the property of the monarch upon landing. This has some far-reaching implications for cetacean bycatch in British waters, which should all be theoretically reported to the Receiver of Wrecks before they can be disposed of. In theory, this should mean British bycatch figures are accurate, but it is likely that many cetaceans are simply cast aside, like elsewhere in the world, and remain unrecorded. Although a fisherman found himself in trouble concerning the catch of a sturgeon in 2004 with regards to this statute, no similar case is known with regards to cetaceans.

Exploring the Depths: Public Opinion on Marine Mammal Protection in the United Kingdom

Only a few studies have assessed public opinion about how well marine mammals are (or are not) being protected, and the majority of these have been conducted in the UK. In particular, several studies in Scotland investigated the issue in both rural coastal communities and in major cities. For those participants that expressed an opinion, most thought whales and dolphins were not protected enough. Moreover, 80% supported the introduction of laws specifically for the conservation of cetaceans in Scotland (e.g., a Cetacean Protection Act similar to the U.S. or New Zealand MMPAs). Those participants who thought cetaceans were sufficiently protected were more likely to be older men than any other demographic group. Also, when asked if it makes them see a politician more favorably if that politician were to introduce such a law, 40% said yes.

Interestingly, although laws and conservation measures to protect seals in Scotland are arguably much weaker than for whales and dolphins (e.g., it is legal to kill seals at certain times of the year in most locations in Scotland; see text box on Pinniped Conservation Law in the United Kingdom, above), a larger percentage of the population (at least for rural areas) thought seals were sufficiently protected by law. However, nearly one-third still thought seals were not protected enough (**Table B17.1**).

TABLE B17.1 Public Opinion on Marine Mammal Protection in Scotland

How Well Are Cetaceans Protected?	Percentage (Southwest)	Percentage (Major Cities)
Don't Know	25.8	60.0
Overprotected	0.4	0.0
Sufficiently protected	28.2	7.0
Not sufficiently protected	45.6	33.0
How Well are Seals Protected?		
Don't Know	20.3	
Overprotected	9.2	
Sufficiently protected	41.0	
Not sufficiently protected	29.5	

From Scott & Parsons (2005a, 2005b).

Exploring the Depths: Marine Protected Areas and Sanctuaries

The practice of area- or site-based protection of the environment, for example, the use of terrestrial wildlife reserves and national parks, has held a prominent place in the long history of conservation. This paradigm has proven that focused protection of a species, community, or ecosystem is achievable in a managed environment or network of ecologically coherent natural habitats. Primarily, this has been successful because the main concept is simple and easy to enforce and understand across a wide range of management aims, conservation demands, and stakeholder desires.

As management and conservation demands have increased in recent years, many agencies have extended their attention to the protection of our marine and oceanic environments, using similar area-based methods developed through the creation of marine protected areas (MPAs). MPA has become a universal yet generic term, first defined by the IUCN to describe "... any area of inter-tidal or submarine terrain, together with its overlying water and associated flora, fauna, historical and cultural features, which has been reserved by law or other effective means to protect part or all of the enclosed environment." The transition, however, of the terrestrial reserve concept to our oceans as MPAs has been a difficult one because of the intrinsic features of the marine environment and its denizens (including, for example, mobility of species, variation in productivity, less obvious geographic borders) and the nature of our impacts (which are typically widespread, difficult to monitor, and challenging in terms of the enforcement of legislation). Despite these problems marine conservationists and managers have worked to develop the MPA model, from early examples in the 1970s (particularly the Great Barrier Reef Marine Park in Australia) to modern high seas and statewide sanctuaries (for example, the Chagos Marine Reserve in the British Indian Ocean Territory). Today, although only around 1% of our oceans have been designated as MPAs, those sites established have proven that area-based protection in the marine environment can promote biodiversity, increase population sizes, and maintain ecosystems services worldwide.

Currently, MPAs have many separate titles and designations depending on the national jurisdiction, legislation, or function of the intended MPA. These include marine parks, marine reserves, wildlife refuges, national sanctuaries, and other specific member state designations (e.g., European Union SACs and special protection areas). Although these types aim to reduce damage to the distribution of a target species, community, or habitat, in many instances the foci, means, or execution of the MPA differs considerably. In certain highly protected types their aim is to prevent all extraction of resource (for example, no-take zones) and disturbance to the sites' conservation features by preventing wide-scale access to the MPA. Others in comparison primarily function as a method by which to ensure the effective management of multiple-use activities (such as fishing, shipping, offshore energy, tourism), ensuring that both industry and conservation objectives can be met. As our understanding of the marine environment increases, the use of MPA networks are now widely acknowledged as important tools in much wider ecosystem management. MPAs are therefore critical to ensure the cohesive management of marine resources, their exploitation and disturbance mitigation, through human activity by spatial management, now widely known as marine spatial planning.

For most MPAs indices such as species richness, ecosystem services, or critical habitat often form the basis of many current and proposed MPAs. Suitable conservation or management metrics need to be implemented that will allow both monitoring across long temporal scales but also quick reaction to rapid climatic or anthropogenic impacts. Therefore, understanding the ways in which marine organisms interact with the surrounding environment has recently become a key tool in their subsequent conservation and management, through the use of ecologically coherent and connected MPA networks.

Depending on the focus of any specific area, sites are often chosen based on either bottom-up or top-down features, including those that either promote productivity (submarine features, reefs) or act as the engineers of the habitat (corals, benthic invertebrates, algae). Conversely, because of their rapid reaction to ecosystem and climatic change, marine megafauna such as sharks, rays, seabirds, and marine mammals are used as featured reasons by which to monitor and designate MPAs that include their critical habitats. These species act both as umbrellas, adding further protection to the wider species community, and as foci for education and public engagement with marine protection and conservation initiatives.

Today, modern MPA tools give special consideration to the connections between the habitats and communities adjacent to the areas they protect. This is often achieved through the use of zonal management and protective buffers. The function of these multiple-site MPAs is that each area will promote the diversity and services of specific parts of the interconnecting community. Thereby this holistic approach would induce a greater diversity and abundance within the reserve to benefit or filter out into neighboring zones within and outside the reserve.

—Contributing author, Mike Tetley, Marine and Ecological Sciences.

listed under Annex IV of the Directive (which includes cetaceans), European member states are required to prohibit the following:

- All forms of deliberate capture or killing
- Deliberate disturbance of cetaceans, particularly during the period of breeding, rearing, hibernation, and migration
- Deterioration and destruction of "breeding sites" or "resting places"

As the result of a court ruling in 1999, the prohibitions of the Habitats Directive extend to 200 nautical miles from the coast of each European country. However, there have been some problems with this legislation because it is often difficult to prove that actions are "deliberate." Moreover, for cetaceans in particular, defining areas that are "breeding sites" and "resting places" is difficult; a term such as "critical habitat" might be more useful.

For species listed on Annex II (which includes harbor porpoises, common bottlenose dolphins, harbor seals, and grey seals), European member nations have an obligation to establish a network of protected areas, or Special Areas of Conservation (SACs), to aid the conservation of these species. Within these protected areas human activities are managed (and potentially prohibited) if they might have a negative impact on the species for which the site was designated.

There have been complaints that, in several marine SACs, activities that have negative impacts on designated species (such as certain types of fishing or oil- and gas-related activities) have either been allowed to continue or have not been prohibited from starting and that management of these areas should be stricter. Also, the size of SACs for marine mammals have been criticized because the areas are relatively small, whereas the ranges of the populations in question are usually much larger (**Figure 17.2**, page 281). As a result, only a small fraction of the population may actually inhabit the SAC at any given time, or the SAC is only used by the population for short periods over the course of a year. Instead, many animals occupy waters outside of the SAC for the majority of the time, where they receive much less protection.

Moreover, the usual small sizes of the SACs do not provide much protection from anthropogenic noise, which can enter the SAC from activities just beyond the boundaries. Similarly, their small sizes do not offer much of a buffer if conditions within the SAC become unfavorable because of anthropogenic, oceanographic, or biological factors, such as a change in prey availability or climate-driven changes. This may, in fact, be the situation in the Moray Firth SAC. Many of the bottlenose dolphins located in Scotland have increased their range, and so the animals that use the Moray Firth are spending far less time within the protected area. Nonetheless, the designation of these areas is an extra layer of protection for many European marine mammal populations.

Food For Thought: Science-Based Management?

Many domestic laws, as well as international agreements and treaties, have various science-based metrics embedded in them, such as determining endangerment, achieving sustainable harvests, limiting negligible impacts, or reducing harmful pollutants. However, the application of science to these standards is not always complete, partly because the information provided by science may be lacking (e.g., how can you determine a negligible impact if the size of the population is not known) and partly because of political pressure to do otherwise. The interpretation of such standards (e.g., what exactly IS a negligible impact, or how bad must a pollutant be before it is considered "harmful") also lies, for the most part, outside the realm of science and firmly within the more whimsical world of policy.

It has also often been argued that managers may delay conservation action while the necessary data are collected, fulfilling, in their minds, obligations such as the ASCOBANS requirement to introduce conservation, research, and management measures through the funding of research alone. One clear example of this has been the delay in implementing bycatch reduction schemes to protect the vaquita in the Gulf of California, until more information on the size of the population and the state of the decline was available. This is unfortunate, because such delays are often in direct conflict with the intent of the laws and agreements to protect the species first and then harvest sustainably if possible. It should also be recognized that delays in implementing management measures pending further research actually constitute a management decision/action. Even when management action is undertaken, the "opinion" of scientists is often tempered by the needs of industry and commerce. This is fine in principle, as policy decisions must ultimately consider such things. However, cynics might note that politicians and managers would rather try to make the available scientific information fit their position so they can claim it is supported by science rather than openly accept the science and choose to ignore it for other political reasons.

Finally, many international agreements and treaties are nonbinding, or still require that member states implement the concepts within their own domestic legislation. Often, the agreements offer little in the way of recourse if member states do not comply with even obligatory conditions.

Exploring the Depths: Domestic Laws Around the World

Although the main focus of this chapter has been on U.S. and European laws, it should be noted that many countries have domestic legislation to protect endangered species. Some laws were enacted to allow the country to become compliant with CITES or other international agreements, whereas others predate these agreements. These national laws include, but are not limited to, Canada's Species At Risk Act, Hong Kong's Wild Animals Protection Ordinance, Australia's Wildlife Conservation and Environment Protection and Biodiversity Conservation Acts, China's Law on the Protection of Wildlife, and India's Wildlife Protection Act.

Some of the above-mentioned acts focus exclusively on animals, whereas others address plants as well. Many include provisions similar to those related to critical habitat in the U.S. ESA. Most offer varying degrees of protection to different categories of species. The effective implementation of these acts also varies around the world. Unfortunately, some are hardly implemented at all.

Even when conflict resolution or noncompliance is addressed adequately, it often requires other member states, and not private citizens or NGOs, to initiate the enforcement proceedings. This can generate international incidents that are often politically unsavory, and thus such proceedings are often considered to be an option of last resort, if considered at all.

In short, science is often ignored by politicians and managers. The handling of climate change is the classic example. Political life spans and terms in office mean that few politicians are brave enough to address long-term issues in favor of their own re-election. Arguably this is especially apparent in the United States, where corporations (and, to some extent, the military) and their allies in government have been slowly eroding protections offered by groundbreaking environmental laws, such as the MMPA and ESA, that were written in the 1970s. All the while, the same people often tout global U.S. environmental leadership, based on the presence of the very same laws. Despite this, many experts believe that if these laws were to be proposed in Congress today, they would almost certainly offer only a small fraction of the protection, if they were passed at all.

Food For Thought: Enforcement

Another problem with international treaties and agreements, and to a lesser extent domestic laws and regulations, is that of enforcement. For example, China and Taiwan have some of the best environmental laws in the world but some of the most degraded ecosystems because these laws are simply ignored. This problem is rampant throughout the world, however, as fishing continues within protected areas, many activities that should be regulated are not, and various corporations and scientists violate the terms of their permits unpunished.

There are various reasons for this, including a lack of political will (e.g., the perception that the value of a new industrial plant merits ignoring the laws restricting it), conflicts of interest (e.g., the fact that research is occurring through our own government scientists means that it is acceptable), a lack of funding for enforcement agencies (e.g., we can only be in one place at a time and/or prosecute one case at a time), and the lack of specific regulations due to the apparently unmanageable scale of a problem (e.g., the noise produced by private boats and their echo sounders). Some of these problems require fairly straightforward solutions, such as additional funding for enforcement agencies, whereas others, such as noise from private boats, might require some creative thinking. Unfortunately, conflicts of interest are harder to solve, although decisions affected in this way can be, and have been (to varying degrees of success), challenged in court in some countries. Institutional and governmental corruption and flagrant disregard for the law is, however, often beyond such reproach.

SELECTED REFERENCES AND FURTHER READING

Bates, L. (2003). *A Critical Evaluation of the Management Measures for Protected Areas, for Coastal Cetaceans in the UK*. M.Sc. thesis, Bangor University, Wales.

Berggren, P., Wade, P.R., Carlström, J., & Read, A.J. (2002). Potential limits to anthropogenic mortality of harbour porpoises in the Baltic Region. *Biological Conservation* 103: 313–322.

Cardigan Bay SAC Relevant Authorities Group. (2003). Cardigan Bay Candidate Special Area for Conservation. Action Plan Review 1. Retrieved from http://www.cardiganbaysac.org.uk/pdf%20files/RAGactionplanrevision2003.pdf.

Commission of the European Communities. (2002). *Incidental Catches of Small Cetaceans*. Report of the Second Meeting of the Subgroup on Fishery and the Environment (SGFEN) of the Scientific, Technical and Economic Committee for Fisheries (STECF).

Hammond, P.S., Benke, H., Berggren, P., Borchers, D.L., Buckland, S.T., Collet, A., Heide-Jørgensen, M.P., Heimlich-Boran, S., Hiby, A.R., Leopold, M.F., & Øien,

N. (1995). *Distribution and Abundance of the Harbour Porpoise and Other Small Cetaceans in the North Sea and Adjacent Waters.* Final Report to the European Commission under contract LIFE 92-2/UK/27. Sea Mammal Research Unit, University of St. Andrews, Scotland.

Hastie, G.D., Barton, T.R., Grellier, K., Hammond, P.S., Swift, R.J., Thompson, P.M., & Wilson, B. (2003). Distribution of small cetaceans within a candidate special area of conservation: implications for management. *Journal of Cetacean Research and Management* 5: 261–266.

Hooker, S.K., & Gerber, L.R. (2004). Marine reserves as a tool for ecosystem-based management: the potential importance of megafauna. *BioScience* 54: 27–39.

Howard, C., & Parsons, E.C.M. (2006). Attitudes of Scottish city inhabitants to cetacean conservation. *Biodiversity and Conservation* 15: 4335–4356.

Hoyt, E. (2004). *Marine Protected Areas for Whales, Dolphins and Porpoises.* Earthscan, London.

Mitchell-Jones, A.J., Marnell, F., Mathews, J.E., & Raynor, R. (2008). Mammals and the law. In: *Mammals of the British Isles: Handbook,* 4th ed. (Ed. S. Harris & D.W. Yalden). pp. 32–53. The Mammal Society, Southampton, UK.

Parsons, E.C.M., Clark, J., & Simmonds, M.P. (2010). The conservation of British cetaceans: a review of the threats and protection afforded to whales, dolphins and porpoises in UK Waters, Part 2. *International Journal of Wildlife Law and Policy* 13: 99–175.

Parsons, E.C.M., Clark, J., Warham, J., & Simmonds, M.P. (2010). The conservation of British cetaceans: a review of the threats and protection afforded to whales, dolphins and porpoises in UK Waters, Part 1. *International Journal of Wildlife Law and Policy* 13: 1–62.

Perrin, W.F. (1989). *Dolphins, Porpoises and Whales. An Action Plan for the Conservation of Biological Diversity: 1988–1992.* International Union for Conservation of Nature and Natural Resources, Gland, Switzerland.

Reeves, R.R., & Leatherwood, S. (1994). *Dolphins, Porpoises and Whales. A 1994–1998 Action Plan for the Conservation of Cetaceans.* International Union for Conservation of Nature and Natural Resources, Gland, Switzerland.

Reeves, R.R., Smith, B.D., Crespo, E.A., & Notarbartolo di Sciara, G. (2003). *Dolphins, Porpoises and Whales. 2002–2010 Conservation Action Plan for the World's Cetaceans.* International Union for Conservation of Nature and Natural Resources, Gland, Switzerland.

Rose, G. (1996). International law and the status of cetaceans. In: *The Conservation of Whales and Dolphins—Science and Practice.* (Ed. M.P. Simmonds & J.D Hutchinson), pp. 23–53. John Wiley & Sons, Chichester, UK.

Scott, N.J., & Parsons, E.C.M. (2005a). A survey of public opinions in Southwest Scotland on cetacean conservation issues. *Aquatic Conservation* 15: 299–312.

Scott, N.J. & Parsons, E.C.M. (2005b). A survey of public opinion on seal management in southwest Scotland. *Aquatic Mammals* 31: 104–109.

Shrimpton, J.H., & Parsons, E.C.M. (2000). *Cetacean Conservation in West Scotland.* Hebridean Whale and Dolphin Trust, Mull, Scotland.

Thompson, P.M., Wilson, B., Grellier, K., & Hammond, P.S. (2000). Combining power analysis and population viability analysis to compare traditional and precautionary approaches to the conservation of coastal cetaceans. *Conservation Biology* 14: 1253–1263.

Tregenza, N.J.C., Berrow, S.D., Leaper, R., & Hammond, P.S. (1997). Harbour porpoise *Phocoena phocoena L.* bycatch in set gill nets in the Celtic Sea. *ICES Journal of Marine Science* 54: 896–904.

Wilson B., Reid, R.J., Grellier, G., Thompson, P.M., & Hammond, P.S. (2004). Considering the temporal when managing the spatial: a population range expansion impacts protected areas-based management for bottlenose dolphins. *Animal Conservation* 7: 331–338.

CHAPTER 18

Marine Mammal Tourism

CHAPTER OUTLINE

Significance of Marine Mammal Tourism

Exploring the Depths: Whale Watching and Ecotourism
Exploring the Depths: Whale Watching Versus Whaling
Exploring the Depths: Pinniped Tourism
Exploring the Depths: Whale Watching Around the World

Who Watches Whales? The Nature of Marine Mammal Tourists

Negative Impacts of Marine Mammal Watching

Exploring the Depths: Negative Impacts of Pinniped Tourism

Managing Marine Mammal Tourism

Exploring the Depths: Sustainability Report Card
Exploring the Depths: Educational Potential of Marine Mammal Tourism
Exploring the Depths: Dolphinaria: Pros and Cons
Exploring the Depths: Keiko, the Whale From Free Willy

Solitary Sociable Dolphin Problem

Selected References and Further Reading

Significance of Marine Mammal Tourism

Marine mammals are a major tourist attraction, bringing visitors from afar into an area. Marine mammal tourism has thus become a huge international industry. In particular, whale watching (here we use the term generically to include the watching of all cetaceans in the wild, not just large or great whales) has become a hugely profitable business. In fact, researcher Erich Hoyt estimated in 1995 that the global whale watching industry was worth a total of $504 million a year, with more than 5.4 million people going whale watching annually (**Figure 18.1**). At the time there were more than 65 countries hosting whale watching operations, and the economic value of the industry was growing at greater than 15% per year. As a result of this substantial expansion, by the year 2000 the whale watching industry as a whole was

- Worth a total of $1,050 million per year
- Growing at a rate of greater than 18% annually
- Attracting more than 9 million people every year
- Present in more than 87 countries

As such, whale watching is one of the fastest growing sections of the global tourism industry and has become the largest economic activity using whales as a resource (see Exploring the Depths: Whale Watching Versus Whaling).

In areas where marine mammal tourism occurs, the activity can have widespread and very substantial local impacts. For example, in a 2000 survey in rural Scotland, an area with a thriving marine mammal watching industry, 47% of tour operators consider whale watching to be important to the local economy and an even greater percentage (75%) consider seal watching also to be important. In fact, in some areas of rural Scotland up to 12% of all tourism income (the number one economic activity) came from whale watching, making it of considerable economic importance to these communities.

There are indications of similar importance elsewhere. For example, in a review in 2000 of whale watching around the world, Erich Hoyt reported that one-fourth of whale watchers in Newfoundland, Canada had only come to the area because of the whales and that for the remainder it was a major reason for their visit. Hoyt also noted that nearly 80% of visitors to North Vancouver had visited the area entirely or mainly to go whale watching. Similar results were found in a survey of whale watchers in Scotland in 2000. Likewise, an article published in 1999 stated that nearly two-thirds of tourists visiting the Pacific island of Tonga

FIGURE 18.1 A whale watching trip.

were visiting primarily because of whales and opportunities to go whale watching. Figures such as these are often used to argue that marine mammal conservation not only benefits the species being conserved but, if managed properly, can also bring substantive economic benefits through increased tourism.

Exploring the Depths: Whale Watching and Ecotourism

Whale watching (and watching of other marine mammals) is often incorrectly referred to as "ecotourism." Ecotourism, however, has a very specific definition and encompasses tourism activities that specifically minimize tourist impacts while using resources in a sustainable way and giving substantive return to the host local community. For example, the International Ecotourism Society defines ecotourism as "responsible travel to natural areas that conserves the environment and sustains the well being of local people." The International Union for Conservation of Nature defines ecotourism a little more narrowly: "environmentally responsible travel and visitation to relatively undisturbed natural areas, in order to enjoy and appreciate nature (and accompanying cultural features, both past and present) that promotes conservation, and provides for beneficially active socioeconomic involvement of local populations."

Many, probably the majority, of marine mammal watching operations do not take steps to ensure their sustainability, specifically promote conservation, or provide benefits to host communities. This means that, at best, most marine mammal tourism businesses should be more accurately described as "wildlife tourism," that is, viewing of key species in their natural habitats (as opposed to zoos or aquariums). However, it should be noted that the sustainability of some whale watching, especially less regulated businesses and those in certain areas with small or particularly vulnerable populations, has been debated recently (see below). Although it is unquestionably preferable to whaling, given the currently depressed nature of many populations, as a means of generating income from a conservation point-of-view (see text box on Whale Watching Versus Whaling, below), such targeted and sometimes aggressive pursuit of marine mammals may have a range of consequences for them, including disturbance and chronic stress.

To help further define ecotourism as it relates to marine mammals, specifically to cetaceans, the International Whaling

Exploring the Depths: Whale Watching and Ecotourism (*continued*)

Commission's Whalewatching Subcommittee produced a definition of "whale ecotourism." This is a commercial operation that has taken major steps to do the following:

- Actively assist with the conservation of cetaceans (for example, assisting local scientists or promoting conservation initiatives)
- Provide accurate educational materials and/or activities about cetaceans and their associated habitats for tourists
- Try to minimize their environmental impact (whether by reducing their carbon footprint, reducing the amount of waste produced by their operation, or introducing other environmentally beneficial practices)
- Abide by a set of whale watching regulations or an appropriate set of guidelines if no specific regulations are available for the area
- Provide benefits to the local host community within which the company operates. Examples of such benefits might include a company policy of preferential employment of local people, selling local handicrafts, or supporting conservation, educational, or social and cultural projects or activities in the local community.

It should be emphasized that although many whale ecotourism operations are boat-based, whale ecotourism could potentially include aerial whale watching (e.g., from a dirigible/airship or aircraft), land-based whale watching platforms, or even visitor centers linked to other whale watching activities.

Exploring the Depths: Whale Watching Versus Whaling

Whale watching is a lucrative industry and is now the major economic use of cetaceans as a resource globally. For example, in 2000 it was estimated that Japan's whaling program (when one excludes direct subsidies from the Japanese government) was worth $31.1 million a year. At the same time whale watching in Japan was estimated to be worth slightly more at $33.0 million (including direct expenditure as well as indirect expenditure through associated spending, e.g., hotel accommodation linked to whale watching tourists).

Similarly, in 2000 Norwegian commercial whaling was estimated to be worth $6 million, whereas whale watching was worth $12.0 million (including both direct and indirect expenditure, as above), thus having an economic value double that of whaling. Furthermore, the value of whale watching in Japan and Norway might be even higher if whaling was to cease. Several surveys have shown that whale watching tourists may actively avoid countries that hunt whales. For example, one survey of whale watchers in the United Kingdom found that 79% of whale watching tourists said they would boycott visiting a country that conducted hunts for cetaceans. A survey of tourists conducted in the Dominican Republic (which at the time was considering joining the International Whaling Commission) found that more than three-fourths stated that if a Caribbean country supported the hunting or capture of whales or dolphins they would be less likely to visit it on holiday. Moreover, more than 80% stated that if a country has a strong commitment to whale and dolphin conservation they would be more likely to visit this country on vacation. However, whether those stating they will or will not boycott a country actually do or whether they are actually aware of a country's stance on whale conservation and whaling may affect the actual impact on tourism revenues. Nonetheless, if only a small fraction of people do indeed boycott a pro-whaling country or choose a pro-conservation country as their holiday destination over others, it could have significant economic repercussions. Therefore, a strong pro-cetacean conservation stance in a country could actually be financially beneficial. Such information may be particularly influential in the Caribbean, because in this region tourism is the most important and influential industry for most nations. Despite this, many Caribbean countries have a pro-whaling stance at the International Whaling Commission.

Such sentiments apply not only to cetaceans but also to other species of marine mammals. For instance, one Scottish study conducted in 2000 found that 17% of all surveyed tourists (not just marine mammal watchers) stated they would avoid visiting Scotland on holiday if they conducted seal culls in the country. In that particular study the projected income losses for Scotland from such a boycott would actually exceed the total value of fisheries income.

The upshot here is that although whale watching appears to coexist with whaling in countries such as Japan and Norway, the economic value of whale watching may be extremely restricted by whaling. It may also be that the total value of whale watching to a nation could increase substantially if the country involved was to cease whaling. An example of this is found in Iceland. After scientific and commercial whaling ceased in the late 1980s, whale watching activities began and the local industry quickly became one of the fastest growing sectors of Icelandic tourism (and of the fast-growing whale watching industries globally). However, on announcement of the resumption of scientific and then commercial whaling (see Chapter 14), whale watching operators reported substantial numbers of cancellations. There have also been conflicts between whale watching operators and whalers and complaints from tourism associations within the country about Iceland's whaling stance and its impact on the economy. It also appears that the industry has not grown as rapidly as projected before the resumption of whaling. In summary, although marine mammal tourism and activities such as whaling can coexist in a country, it is economically detrimental to the whale watching industry, and possibly also to tourism in the country as a whole, when this does occur.

Exploring the Depths: Pinniped Tourism

Pinniped tourism is becoming more popular throughout the world as human attitudes toward seals have transformed from viewing the animals as commodities that provide fur, oil, and meat to a species that has the marketing appeal to generate substantial income in the sector of nature-based tourism (**Figure B18.1**). The growth of pinniped tourism equates to almost 2 million visitors per annum with an economic value of more than U.S.$12.5 million in the southern hemisphere alone. Potential nonfinancial benefits of nature-based tourism include the shift of tourism participants toward biocentric values due to the experiences, education, and interaction attained from these tours that facilitate conscientious environmental behavior. However, the statement that these industries have positive remuneration on the targeted species and their habitat is debatable, particularly due to the plethora of literature that exists on the negative implications nature-based tourism can pose on the targeted species and the environment as a result of industry practices.

Seals make ideal candidates for tourism operations because they are colonial and their presence and location is predictable. Pinniped tourism currently includes observing seals from motorized vessels, viewing platforms, and kayaks and directly walking into their environment and swimming (diving or snorkeling) with seals. These activities occur throughout the southern and northern hemispheres. Guiding principles that regulate seal tourism are variable according to site and can range from no guiding principles to codes of conduct and regulations. Port Phillip Bay, Victoria, Australia provides an example of a seal-swim industry that was implemented with no guiding principles followed by implementation of a code of conduct in 2004 independently by the tour operators. In 2010 regulations were introduced after consultation with all stakeholders.

—Contributing author, Carol Scarpaci, Victoria University.

—Contributing author, Richard Stafford-Bell, Victoria University.

FIGURE B18.1 Tourists watching pinnipeds.

Exploring the Depths: Whale Watching Around the World

Since the humble beginnings of commercial whale watching on the west coast of the United States and in Baja California (Mexico) in the 1950s, whale watching has grown to a multimillion dollar business that covers 119 countries and territories around the globe. It is often described as the fastest growing sector of the tourism industry, and although initial growth rates in North America, Europe, and Australia have been significant, other regions emerged much later. In the past decade annual growth rates of whale watching in Asia (17%), Central America and the Caribbean (13%), South America (10%), and Oceania/Pacific Islands (10%) were well above the general tourism growth rate of 4.2%. The countries with the highest growth rates from 1998 to 2008 were mainland China (107% annually), the Maldives (86%), Cambodia and Laos together (79%), St. Lucia (74%), Madeira (73%), Venezuela (58%), Costa Rica and Nicaragua (both 56%), and Panama (53%). There have been six countries with over half a million annual whale watchers in 2008, representing almost two-thirds of the global whale watching industry (**Table B18.1**).

People watch cetaceans on all continents, but by far the most important region is North America (the United States and Canada). Although Africa and the Middle East were ranked second in 1998, this region was the only one with slightly decreasing numbers in the past decade. The countries in the South Pacific (including Antarctica) grew faster than any other region and are now in second place (**Figure B18.2**).

The most important whale watching countries in Africa and the Middle East are the Canary Islands (although politically part of Spain), Egypt, Mauritius, Namibia, South Africa, and Tanzania. Europe's whale watching industry is dominated by Iceland, Ireland, Portugal, and Scotland, followed by countries such as Gibraltar, Italy, Norway, Spain, and Wales. Some countries in Asia have seen enormous growth rates, with the largest numbers in mainland China, Taiwan, India, and Japan.

Exploring the Depths: Whale Watching Around the World (*continued*)

TABLE B18.1 Countries with More Than 500,000 Annual Whale Watchers (2008)

Country	Whale Watchers in 2008	Percentage of Total Global Whale Watchers
USA	4,899,809	38%
Australia	1,635,374	13%
Canada	1,165,684	9%
Canary Islands (Spain)	611,000	5%
South Africa	567,367	4%
New Zealand	546,445	4%

Data from O'Connor, S., Campbell, R., Cortez, H., & Knowles, T. (2009). Whale Watching Worldwide: Tourism Numbers, Expenditures and Expanding Economic Benefits.

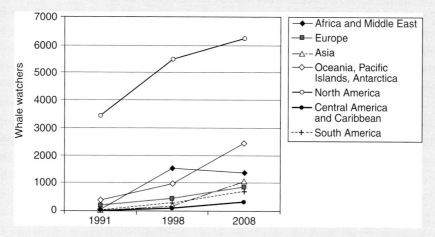

FIGURE B18.2 Whale watching growth rates by region (in thousands). Reproduced from O'Connor, S., Campbell, R., Cortez, H., & Knowles, T. (2009). Whale Watching Worldwide: Tourism Numbers, Expenditures and Expanding Economic Benefits.

Interestingly, Cambodia and the land-locked Laos offer dolphin watching on the Mekong River, with a sizeable 83,000 dolphin watchers in 2008. Australia, Guam, New Zealand, and Tonga dominate whale watching in the South Pacific, with significant numbers of whale watchers in Antarctica, the Cook Islands, French Polynesia, and New Caledonia. In North America, the United States, Canada, and Mexico offer a wide variety of whale watching tours and take more than 6 million visitors on these tours annually. The world's smallest region in terms of whale watchers is Central America, including the islands of the Caribbean. The largest whale watching nations in this region are Costa Rica, the Dominican Republic, and St. Lucia. Finally, most whale watchers in South America follow this pursuit in Argentina and Brazil, followed by Ecuador and Colombia.

—Contributing author, Michael Lück, Auckland University of Technology.

Who Watches Whales? The Nature of Marine Mammal Tourists

Surveys in a variety of countries (such as Australia, Canada, and the United Kingdom) found that marine mammal watchers tend to be slightly older, toward middle age (36–50 years). However, New Zealand whale watching tourists were somewhat unusual in that they were younger (20–34 years). Regardless of age, marine mammal watchers are generally middle class, relatively affluent, and well educated. Approximately three-fourths of whale watchers surveyed in a selection of studies in Argentina, Canada, United Kingdom, United States, and New Zealand had a university degree or equivalent. In many regions of the world (Australia, Canada, Japan, United Kingdom, United States, and New Zealand), whale watchers more often tend to be women.

Demographic information such as this provides useful insights into the whale watching industry for those in the industry, as well as those responsible for managing it. For example, more affluent, slightly older tourists are more likely to pay more for trips. On the other hand, they may also look for a tour with more comfort and luxury, whether on the boat itself (assuming vessel-based watching) or in terms of the hotels and restaurants available in the area where whale watching activities take place. The more educated tourist may also require more in terms of education and interpretation provisions on a marine mammal watching trip. They may perhaps also be interested in more than just the marine mammals, such as the local culture, history, or even geology, as well as wider environmental issues.

Marine mammal tourists may also be more environmentally aware than the average tourist. A study in Scotland in 2001 and 2002 found that about one-half of whale watchers were members of environmental groups and an astonishing one-fourth conducted voluntary work for environmental groups. They tended to engage in environmentally friendly activities, with more than three-fourths regularly recycling, nearly two-thirds using energy-saving light bulbs (with just under one-half using other energy-saving devices), and three-fourths saying they only used cosmetics or toiletries that specifically had not been tested on animals.

Few other studies have looked at environmental attitudes in this sector of tourists, but the information above has several implications. For example, the high level of involvement in environmental groups suggests that whale watching operations could involve environmental groups in targeted marketing (e.g., advertisements in environmental magazines). However, it also means their operations should be as environmentally friendly and conservation-conscious as possible or they run the risk of alienating their customers.

Negative Impacts of Marine Mammal Watching

Although marine mammal tourism can be lucrative, which can be potentially helpful with conservation efforts, marine mammal watching can have some negative impacts that are cause for concern. For example, several studies have recorded changes in cetacean behavior in response to whale watching. These have included changes in surfacing, acoustic and swimming behavior, direction, group size, and coordination. Alterations have also been noted in feeding, including the cessation of feeding, and resting behavior. It is hard to judge the long-term effects of such behavioral changes, but in general they may cause the animals to increase their energy expenditure or decrease energy intake. This would obviously be detrimental and could reduce fitness in the long term, especially in situations where there is prolonged or repeated exposure. Moreover, disruption in resting due to disturbance by tourist activities could also lead to a general deterioration in health. This may be particularly problematic for resting pinnipeds that conserve metabolic heat when hauled out; disturbance can flush whole groups back into the ocean for many hours. This deterioration in health related to energetic costs would be compounded by any issues related to sleep deprivation as well as maladaptive and/or chronic stress responses. Finally, abandonment of an area would likely mean that animals are forced into a less optimal situation.

In addition to altering behavior and inducing indirect detrimental impacts to animals, tourism can also have more direct impacts. For example, many whales have been injured or killed as a result of collisions with whale watching vessels, especially in areas where there is a high intensity of whale watching traffic, such as off the coast of Massachusetts. The increasing number of large, high-speed whale watching vessels is a particular cause of concern because the speed of these vessels limits their ability, as well as that of the whales, to avoid collisions. Moreover, a higher speed means that collisions occur with greater force and are more likely to be lethal. An analysis of cetacean collisions with boat traffic suggested that restricting vessel speeds to 11 knots or lower would greatly reduce the likelihood of lethal collisions. Speed restrictions may thus be a good way of reducing this problem. However, several operators are resisting this restriction as faster speed means more trips to areas in which whales occur in greatest numbers and therefore more tourists and more money earned.

However, the impacts of tourist activity on marine mammals can be, and perhaps more often are, much more subtle. For example, then PhD student Lars Bejder and colleagues conducted a 10-year study investigating the approaches by boats in Shark Bay, Australia and found that dolphin groups became more compact, membership of these groups changed more frequently, and changes in directions of swimming increased. Thus, many of the behavioral changes noted above were observed, but this gave no immediate information on the long-term impact of the disturbance. Therefore, dolphins in two sites were investigated: an area where dolphin watching occurred (site A) and an area undisturbed by dolphin watching some 17 km away (site B). Site B animals were found to be more sensitive to boat approaches. Their behavior changed more dramatically and did not return to normal until the tour boat left. In contrast, the behavior of animals in site A returned to normal fairly quickly. From this you might believe that boat traffic had little impact on animals at site A. Essentially, it appears as if the site A animals became habituated to the disturbance and were no longer affected. Fortunately, the study also observed the change in abundance of the dolphins at the two sites over 10 years in conjunction with whale watching activity. Only one whale watching boat operated at site A until the mid-1990s, but when a second boat started using the area a significant decline in dolphin abundance was observed. The researchers determined that some animals had moved away from the area and suggested that these were the most sensitive to disturbance. Reproductive rates of the remaining disturbed dolphins also declined at this time, which would ultimately affect the abundance and status of the population and could potentially contribute to extirpation (local extinction).

This information led the Australian government to reduce the number of licensed boats for the area from two to only one. This study illustrates that marine mammal tourism can have significant population-level effects with only a small number of operations (in this case just two companies), and the effects can be subtle. It took more than 10 years to gather and analyze the data revealing changes in dolphin behavior, abundance, and reproductive rates. Unfortunately, many studies on the impacts of tourism on marine mammals are short term (a couple of years to even just a couple of weeks). In contrast, more long-term studies may be required to discover the true long-term and population-level impacts of tourism-related activity on marine mammals.

Exploring the Depths: Negative Impacts of Pinniped Tourism

In comparison to cetaceans, the impacts of tourism on pinnipeds has been less studied. However, distinct trends of tourism impact, particularly regarding distance of tourists to seals and attitudes of the tourists, have been demonstrated to elicit behavioral responses in seals. An early study by Terhune and colleagues on the effect of sound produced by human activities on harp seals (*Pagophilus groenlandicus*) examined whether mechanical noises from marine vessels affected seals in the Gulf of St. Lawrence. After the arrival of a vessel a decrease in seal vocalizations was found; however, the researchers could not evaluate if the decrease in vocalization was a result of a change in seals' vocal behavior or the seal moving away from the vicinity. Research by Kovacs and Innes documented that the presence of tourists had an impact on behavior of mother and pup harp seals at a breeding colony in the Gulf of St Lawrence, Canada. In the presence of tourists, females reduced their attendance to pups; the females that maintained pup attendance displayed greater vigilance and spent less time nursing their pups. Pups became more active and alert and engaged in agonistic behavior (competition and/or fighting) more in the presence of tourists, whereas rest behavior decreased. The authors also observed young seals displaying "freeze" behavior when tourists approached within 3 m or touched pups.

At a nonbreeding site in Uruguay, Cassini found that South African fur seals (*Arctocephalus pusillus*) tolerated tourists within close proximity, particularly when tourists were calm and quiet. Seals, however, demonstrated strong behavioral responses, such as threats (directed toward tourists) and physical attacks when tourists crossed a 10-m threshold. Another study demonstrated a response of Australian sea lions (*Neophoca cinerea*) to loud noises emitted by tourists in swim-with-seal programs in South Australia. The seals reacted to loud noises by tourists through vigilant behavior (sitting up, looking at tourist, moving farther up the beach), but the presence of tourists had no significant influence on overall seal behavior. Some studies documented that although seals can become habituated to humans, unusual or unfamiliar sounds can disturb them.

In another tourism impact study, the rate of return of Australian sea lions to the beach was not influenced by the presence of visitors to Carnac Island (Western Australia) at times of low, moderate, and high visitation periods. However, it was documented that sea lion response to human approaches and level of sea lion vigilance were attributable to age (elevated in juveniles rather than adults) and time of day (elevated at beginning of the day). At Montague Island, New South Wales, Australia, tour vessel approach decreased seal rest behavior and the proportion of seals resting further decreased as proximity of vessel to seals decreased.

In Kaikoura, New Zealand various studies have demonstrated that sea-based pinniped tourism resulted in less avoidance responses than land-based tourism. Findings also indicated that colonies of seals exposed to intense tourism visitation exhibited signs of habituation, but approaches still elicited behavioral responses. Tourist numbers did not influence seal reactions, possibly due to habituation of seals to tourists. A study in the area by Boren and colleagues observed seal responses to (1) commercial swim-with programs, (2) independent seal swims (i.e., tourists would simply swim with seals on their own accord without a guide), (3) guided seal tours

> **Exploring the Depths: Negative Impacts of Pinniped Tourism (continued)**
>
> on walking tracks, and (4) nonguided walks. Results indicated that commercial swim-with programs elicited less behavioral responses than independent swims. Similarly, the absence of a guide on the walking tracks resulted in an increase in avoidance and behavioral modifications by seals than in the presence of a guide. In addition, tour group size also influenced seal behavior.
>
> Behavioral responses of Australian fur seals (*Arctocephalus pusillus doriferus*) were documented, at a nonbreeding colony, to vessel traffic and underwater noise emissions. Study results demonstrated that reducing vessel distance significantly increased the rate of aggressive displays (which increased when boats were 39 m away), haul-out events (when the vessel was less than 10 m away), and occurrences of seals entering the water (at 29 m away). Threat postures increased significantly in the presence of three or more tour vessels. However, presence of swimmers and underwater noise had no measurable effect on seal behavior. Simultaneously, data were collected on compliance of swim-with tour operations to conditions in the code of conduct for seal swimming. Results indicated that industry was compliant in their approach speed to the colony and anchoring condition (do not anchor). In most instances (78%) tourists complied with a "no splashing" condition, but swim-with programs were more likely to breach the section of the code regarding swimming with an injured seal. This level of noncompliance was possibly due to a high incidence of injured seals observed during each field trip (90% of field trips involved the presence of an injured seal). Completion of this study indicated that satisfactory compliance does not necessarily yield effective management. It was recommended that regulations be instituted and written in a simple manner with defined quantifiable ranges to develop realistic and enforceable regulations. In 2010 the state government developed regulations for seal swimming in Port Phillip Bay, Victoria, Australia with simple and definable distance ranges.
>
> Research that measures tourism impacts on pinnipeds should document seal behavior in reference to presence and absence of tourists, proximity of tourist or vessel to seals, number of tourists, attitude of tourists (e.g., loud), and underwater noise. Responses of seals have also been attributed to factors such as breeding status of colony age, class composition of targeted animals, mother–pup interactions, and habituation of seals to tourists. Studies that focus on the effectiveness of management should incorporate studies of seal behavior to tourist interactions and compliance of tour operations to their guiding principles.
>
> –Contributing authors, Carol Scarpaci and Richard Stafford-Bell, Victoria University.

Managing Marine Mammal Tourism

To reduce the impacts on marine mammals from tourism, commonly rules or regulations are introduced. To investigate the efficacy of such regulations, in 2004 tourism researchers Brian Garrod and David Fennell published an analysis of whale watching guidelines and codes of conduct worldwide. They found that one-third was regulatory (i.e., legal requirements and nonvoluntary), but two-thirds of guidelines were entirely voluntary, making them completely reliant on the good will of operators to enact. In terms of the content of these guidelines, most whale watching codes of conduct had regulations for minimum approach distances (commonly a distance of 50–100 m, but sometimes farther). However, most did not provide guidance on other major problem areas, with more than two-thirds having no prescriptions on feeding dolphins and three-fourths no prescription on touching dolphins. Their conclusions were that whale watching guidelines were variable and "patchy" in nature and that the sector was growing far more rapidly than the regulations governing the industry, which put doubts on the overall sustainability of the industry.

Even if guidelines, regulations, or laws exist in an area, there is no guarantee that operators will follow these rules. For example, in Doubtful Sound, New Zealand, then-PhD student David Lusseau found that two-thirds of tour boat encounters with common bottlenose dolphins violated the New Zealand Marine Mammal Protection Act (see Chapter 17), with one-third of encounters involving more than one violation. In particular, failure to abide by speed restrictions close to dolphins was a common violation. In fact, Lusseau estimated that dolphins in Doubtful Sound are "at risk" once every 7 minutes when interacting with tour boats. Things were found to be even worse with regards to non–tour boats, as it was estimated that animals were at risk once every 3 minutes when interacting with other vessels. Lusseau also noticed that some dolphin behaviors were observed more frequently when vessels were traveling fast or were not complying with regulations. These included leaps where dolphins landed on their side and diving where the dolphins brought their tail flukes out of the water. Such behaviors might therefore be considered as a good indicator of dolphins being disturbed by boat traffic and possibly also undergoing a stress response.

In Victoria, Australia, Carol Scarpaci and colleagues also studied rates of compliance with whale watching regulations. She found that operators offering swim-with dolphin trips complied with only one of four guidelines that she monitored (specifically the number of swimmers allowed in the water with dolphins). Rules governing boat approaches and time spent with animals were routinely flaunted. After their research was presented to the government and an education program was introduced, the rates of noncompliance with regulations actually increased! Presumably, this was because operators realized that, despite regulators having details of high infraction levels, there was subsequently little in the way of punishment or enforcement, as is a common situation in many parts of the world.

One problem with many marine mammal tourism regulations or guidelines, however, is that they are sometimes difficult to judge, especially by nonspecialists. For example, distances can be difficult to judge accurately at sea, particularly during rough weather, making regulations based on distances problematic. Regulations that rely on operators recognizing specific behaviors can also be a problem, because these may be missed or misinterpreted by nonscientists. The same is true for regulations that require other subjective decisions (such as to avoid "noisy activities"). Complications also arise if marine mammals approach the tour vessels or when tourists walk on the beaches. Efforts to comply with regulations that require maintaining a stand-off distance, especially for boats, may end up causing more disturbances due to the additional maneuvering. Sometimes simple, easily understood, easily monitored, and enforceable guidelines and regulations can be more effective than complicated and detailed rules. Unfortunately, these usually do not cater to all eventualities, leading to issues in the situations that fall through the gaps.

However, refuges are a good and simple method for reducing the impacts of whale watching. These refuges, or "no-go" areas, potentially allow animals to engage in important behaviors (such as feeding, resting, or nursing) away from vessels and without disturbance. Refuges could be purely spatial (e.g., a marine protected area where marine mammal watching vessels cannot enter), but they may also have temporal elements (e.g., at certain times marine mammal tourism is not allowed in a given area or region, or even perhaps out of a particular port). Furthermore, clearly defined refuge areas or time periods would make monitoring for compliance and any subsequent enforcement easier for managers.

Exploring the Depths: Sustainability Report Card

To help assess whether marine mammal tourism in an area is sustainable, whale watching researcher Erich Hoyt suggested a "sustainability report card," which includes such questions as follows:

- Is the marine mammal population growing?
- Are marine mammals moving out of an area?
- Are the marine mammals exhibiting changes in behavior?
- What are the levels of biological and chemical pollutants in coastal waters?
- Are marine mammal tourism operators knowledgeable about marine mammals and local culture?
- Are the operators good education providers?
- Are marine mammal tourism operators concerned about the safety and welfare of their customers?
- Does the marine mammal tourism activity aid or benefit the local community?

A list of questions like these can be used in any area where marine mammal tourism occurs and could help to identify ways to minimize impacts on marine mammals while ensuring a high quality experience for tourists.

Exploring the Depths: Educational Potential of Marine Mammal Tourism

There are many forms of marine mammal tourist excursions, from leisurely trips to physically demanding adventures, that bring interested visitors up close to marine mammals in their natural environments. But these trips do not just bring visitors to see the animal; they can also offer a very influential educational opportunity if programming is done well (**Figure B18.3**).

Reputable marine mammal tourism ventures can offer great educational experiences. Most whale and dolphin watching tours include interpreters or scientists who educate visitors on marine mammal biology and conservation. But do tourists really want to learn anything from these excursions, or do they just want to see the animal? A study by Lück demonstrated that tourists do want to be educated about the animals they observe, not just see them. Also, Foxlee showed that whale watching visitors want to learn about the marine environment as well. The results of these studies support the need to implement structured educational programming into marine mammal tours that addresses the educational needs and wishes of the patron, including education on the marine environment.

If patrons want to be educated about marine mammals and their environment, are they actually retaining this new

Exploring the Depths: Educational Potential of Marine Mammal Tourism (*continued*)

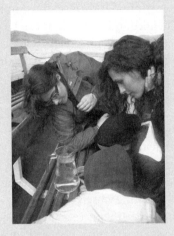

FIGURE B18.3 A presentation on a whale watching boat. The whale watchers are learning about the planktonic prey of right whales.

knowledge? Most visitors of a swim-with dolphin tour that Lück investigated were shown to have increased their knowledge of dolphins and their environment. Similar work by Mayes and colleagues in Australia demonstrated that knowledge of dolphins increased by 81% among patrons of a resort that offered interactive programming. However, most surveys of patrons have been undertaken only on board vessels and at-site with very few assessing the impact the trip has on them over time. Stamation and coauthors, in surveying whale watch visitors in Australia 6 months after their excursion, showed that patrons retained little of the educational content received during their trip and that their conservation behaviors did not change. Assisting long-term retention of information is challenging, just as it is difficult to change a person's attitude or behavior toward marine mammals and/or environmental conservation, but many argue that education is the key.

Marine mammal tourist background knowledge, expectations, and demographics can vary widely from one operation to another and often depends on the type and activity level of the excursion. There is an intrinsic value in wanting to just see a marine mammal, but many tourists want to know more about the animals they are viewing. The challenge to educational programming during these excursions is to create a connection to the patron with the animal and/or environment. This will captivate their interest, hopefully for the long term, which has the potential to incite behavioral and attitudinal change that can assist the conservation of these natural resources and the longevity of the marine tourism industry.

–Contributing author, Jennifer Ambler, George Mason University.

Exploring the Depths: Dolphinaria: Pros and Cons

"Dolphinaria" are facilities where captive cetaceans are held for public display. Bottlenose dolphins are the most commonly kept species, but others, including beluga whales and orcas, are also often found. These facilities can be either nonprofit or for-profit entities and often display a variety of noncetacean marine species as well. The emphases of these institutions vary, with some focusing on education and outreach, whereas others concentrate on performance and exhibition. Either way, there is quite a bit of controversy over their existence (**Figure B18.4**).

Visits to captive facilities such as zoos and aquaria have been shown to provide positive psychological benefits to their visitors. For example, one study demonstrated that visitors reported a wide range of positive feelings, such as relaxation, engagement, and excitement. In addition, trips to these institutions seem to satisfy social interaction needs, as most visitors are accompanied by family and/or friends.

The most commonly cited reason for maintaining dolphinaria is for their conservation outreach and educational benefits (although it should be noted that the type, quantity, and quality of educational materials and programs vary greatly across captive facilities). Because some people do not have the opportunity to see cetaceans in the wild, proponents say, captive facilities provide an opportunity to see and learn about these animals and therefore grow to care about their survival in the wild and the ocean world as a whole—a world that many are not familiar with and that can seem distant. Dolphinaria might also serve to create a sense of caring for and connection with the display animals, which might positively impact conservation-related behaviors. It has been pointed out that awareness of marine mammal conservation issues seemed to coincide with an increase of dolphinaria in the United States; specifically, public knowledge of and attitudes toward orcas changed as this species first commonly appeared in captivity, lending credence to the idea of captive animals acting as "ambassadors" for their species.

However, it appears that visits to captive facilities might serve only to reinforce existing conservation ethics rather than

Exploring the Depths: Dolphinaria: Pros and Cons (*continued*)

A   B

FIGURE B18.4 There are a variety of dolphinariums from (A) bare concrete pools to (B) more "natural looking" pools or sea pens.

to foster positive attitudes in new segments of the population. This also appears to be true for education; in one study, those who arrived at an aquarium wishing to learn more about animals were more likely to come away with lasting knowledge than those who came for other reasons. In addition, the role that cetaceans might play in the educational process is unclear, because very little research has focused on this aspect of aquaria. For example, a study conducted by the Association of Zoos and Aquariums included only one facility that displayed cetaceans, and no questions were asked that related specifically to this exhibit. Finally, there is some evidence to suggest that educational goals are better met outside of captive environments; one study showed that the public believed that watching media documentaries or seeing cetaceans in the wild were better ways to learn about cetaceans' natural habitats, and another project conducted in the Dominican Republic showed that over 90% of tourists participating in the study preferred to see dolphins in the wild rather than in captivity.

Many argue that cetaceans are poorly equipped to deal with captivity, and so dolphinaria are also an animal welfare issue. Offshore bottlenose dolphins are reported to have average movements of 33–89 km a day, a distance that is not replicable in a captive environment. Most commonly held cetacean species have complex social lives; providing the conditions to meet these social needs is difficult if not impossible in captivity. As a result captive dolphins often display stress-related behaviors and disorders, both physically and mentally. Further, captive cetaceans have proven dangerous to humans at times; while this text was being written two trainers in the U.S. and Europe were killed by orcas. It should be remembered that immersive and interactive experiences do not need cetaceans to be effective; Monterey Bay Aquarium's "Jellies" exhibit contains only jellyfish but is presented to the public in such a manner as to be highly effective from a conservation perspective.

Another common reason cited for maintaining captive display populations is for captive breeding purposes, which can theoretically aid conservation efforts. However, only one captive breeding program of an endangered species has been attempted (for the now extinct baiji); most programs are aimed at maintaining a captive population. Although many display cetaceans are now captive-born, live captures continue in many parts of the world, and some dolphinaria (especially those in countries with little to no oversight of these facilities) rely completely on wild animal captures. These cetaceans often come from populations that have been poorly studied, and so the impact of the captures is unknown.

Both sides of the captivity/anticaptivity debate feel very strongly about the subject. As much of the debate over dolphinaria rests on the question of whether or not viewing captive cetaceans strengthens the public's concern over marine conservation issues, future research should directly address this question.

–Contributing author, Megan Draheim, George Mason University.

Exploring the Depths: Keiko, the Whale From *Free Willy*

Keiko, a male killer whale (*Orcinus orca*), was captured off the coast of Iceland in 1979 at approximately 2 years of age for the purpose of public display (**Figure B18.5**). Between 1979 and 1985 he was held by facilities in Iceland and Canada. In 1985 the amusement park Reino Aventura purchased Keiko and moved him to Mexico City. While in Mexico City Keiko lived without other orcas, although bottlenose dolphins were kept in the tank with him for varying periods of time. The small size of the pool combined with high-temperature artificial saltwater contributed to serious health problems as Keiko grew. He was a minimum of 910 kg underweight at maturity and developed a skin condition similar to papilloma warts, which caused lesions over large areas of his body. His muscle tone and cardiovascular condition were also poor. Recognizing that his physical environment was inadequate, Reino Aventura made efforts to sell Keiko to another aquarium beginning in the early 1990s, but these efforts were unsuccessful.

The 1993 movie *Free Willy*, whose plot concerned the release of a captive killer whale into the wild, starred Keiko as the central character. The film brought international attention to his situation in Mexico City. A cover story in *Life Magazine* drew attention to his failing health and the inadequate facilities at Reino Aventura. Efforts to move Keiko to a more suitable facility were pursued, while at the same time the idea to reintroduce him to the wild began to gain momentum. In 1994 Earth Island Institute began discussions with the Oregon Coast Aquarium (OCA) in Newport, Oregon about building a rehabilitation enclosure for Keiko at the facility. The Free Willy/Keiko Foundation (FWKF), a nonprofit organization working to ensure Keiko's reintroduction to the wild, formed in 1994. Substantial donations from the McCaw Foundation, Warner Bros., The Humane Society of the United States, and private individuals were used to construct a unique rehabilitation enclosure at OCA.

In January 1996 FWKF oversaw Keiko's transport from Mexico City to OCA (using a plane donated by United Parcel Service). Once in the natural seawater enclosure, training efforts focused on husbandry and physical conditioning. Keiko did not perform any public shows. Specialized behavioral and physical conditioning programs were implemented to improve his health. During the following 2 years Keiko gained weight and his skin lesions disappeared. His physical condition improved, he became much more active, and he began catching and eating live fish. Based on Keiko's improved health, FWKF prepared to move Keiko to Iceland, consulting with the U.S. National Marine Fisheries Service and the Icelandic government. A contractual dispute that occurred between OCA and FWKF at this time was ultimately resolved in favor of FWKF, making possible the next phase of the reintroduction effort.

In September 1998 FWKF transported Keiko from Oregon to Vestmannaeyjar, Iceland on a U.S. Air Force C-17 transport plane, with the ultimate goal of reintroducing him to the wild. Keiko's new facility in Iceland was a large pen constructed in a bay adjacent to the North Atlantic. Vestmannaeyjar was chosen partly because of its proximity to a herring spawning area and the large number of killer whales present during summer months. Reintroduction efforts during this period focused on continued physical and behavioral conditioning, including live fish training. In March 2000 the bay was netted off and Keiko gained access to the entire area, the size of several football fields. Staff also trained Keiko to follow a specific "walk" boat while working to desensitize him to other vessels in preparation for introduction to the open ocean. To gather data related to free-ranging killer whales in the area, researchers conducted studies involving photo identification, aerial surveys, genetic sampling and analysis, dive data collection, and acoustic data collection and analysis.

In May 2000 open-ocean "walks" began, with Keiko swimming alongside a designated walk boat. Walks continued through the 2000–2002 summer seasons, occasionally lasting several days in length. At times, Keiko left the walk boat and spent periods on his own. Specially designed satellite and VHF (very high frequency) radio tags were fitted to his dorsal fin to remotely track Keiko's movements and gather data related to his diving behavior. During this period Keiko had several interactions with free-ranging killer whales. Between August 5 and August 31, 2002 Keiko left the observation area and swam east to Norway. He arrived in Norway in robust condition, with the same girth measurement as when he departed Iceland; he appeared to have been feeding himself for at least part of the time he was on his own. Keiko spent the remainder of 2002 and 2003 in Taknes Fjord, with caretakers providing food and taking him on "walks". In December 2003, after experiencing symptoms consistent with pneumonia, Keiko died, aged approximately 26 years, and was buried on the shore of the fjord.

–Contributing author, Naomi Rose,
Humane Society International.

FIGURE B18.5 Keiko the whale.

Solitary Sociable Dolphin Problem

Solitary sociable dolphins are those that live separately from their own kind and, instead, interact to an unusually high degree with people. This unusual behavior, which is becoming increasingly common and widespread, is recorded from a range of species, including orcas (the largest of the dolphin species) and beluga whales, although most solitary sociable animals are bottlenose dolphins (**Figure 18.2**).

There is much confusion about this issue. For example, dolphins are generally regarded as friendly and people expect them to exhibit curiosity and friendliness when they meet them in the wild. Indeed, given that dolphins are large, wild carnivores, most interactions between people and dolphins in the wild are remarkably peaceful and the animals are often, albeit usually briefly, willing to spend some time in the company of people. This is something that swim-with tour operators have come to exploit.

However, the solitary sociable dolphin phenomenon is something quite different. For a variety of reasons dolphins can become isolated. Often, the isolated animals are juveniles and disperse naturally away from the school into which they were born. However, in many cases human activities may be involved. For example, dolphins may be scared and scattered by loud noise or a mother may be killed in a fishing net, leaving her calf alone. In addition, for example, in the case of inshore bottlenose dolphins in Europe, the numbers of these animals are depleted, so a lone dolphin venturing out from its original school is now less likely to encounter others of its own kind.

Whatever the reasons for their isolation, these lone dolphins often take up residence within quite a small range close to shore; people then try to interact with them. At first the dolphins are usually relatively wary of people trying to touch them or swim alongside them. Then, over time, the dolphins start to interact with people. Early interactions may include dolphins swimming alongside people. A series of stages in this "habituation" have been suggested by scientists, and, in the later stages, dolphins will eagerly approach people in the water and even allow themselves to be used to tow them by their dorsal fins (something that people try to do having seen it in dolphin shows). Sometimes feeding is involved in the development of the solitary sociable dolphins, but in other instances this has not happened and it appears the dolphins simply enjoy interacting with people. Indeed, it may be that to some extent this replaces the interactions they would otherwise have had with their own kind.

Many of these late-stage solitary sociable dolphins also become increasingly dominant in their play. They may leap on top of swimmers or stop them from leaving the water. They may also play robustly with people on kayaks. Robust treatment of people often leads to calls for the dolphins to be removed or punished.

It is frequently the case that these solitary sociable dolphins are harmed because of their unusual behavior. Hanging around inshore where there are often lots of boat movements and fishing activity makes them vulnerable to ship strikes and entanglements, which can kill them or create significant wounds. Inshore waters may also be polluted by chemicals and pathogens, and these can be further threats to them.

People may even flock in the hundreds to interact with these animals when their presence and friendliness becomes widely known through the media. However, as the habituation process is a gradual one caused by human contact, unknowingly by interacting with dolphins in the wild, people going into the water with dolphins may be making them increasingly vulnerable to harm. Of the four solitary sociable bottlenose dolphins known in U.K. waters in 2006, one was killed by a tug boat propeller; one lost one-third of its tail fluke, probably as a result of entanglement, and disappeared shortly after; one died from an infection likely to have resulted from a combination of human-induced wounds and pathogens in the water; and the fourth has disappeared. The Whale and Dolphin Conservation Society, which has been studying this issue for a number of years, currently strongly recommends that people do not get into the water with these wild isolated animals to try to stop the process of habituation and the harm to the animals that has followed in so many cases.

FIGURE 18.2 A solitary, sociable dolphin.

SELECTED REFERENCES AND FURTHER READING

Adelman, L.M., Falk, J.H., & James, S. (2000). Assessing the National Aquarium in Baltimore's impact on visitors' conservation, knowledge, attitudes, and behaviors. *Curator* 43: 33–61.

Allen, M., & Read, A. (2000). Habitat selection of foraging bottlenose dolphins in relation to boat density near Clearwater, Florida. *Marine Mammal Science* 16: 815–824.

Andersen, M.S., & Miller, M.L. (2006). Onboard marine environmental education: whale watching in the San Juan Islands, Washington. *Tourism in Marine Environments* 2: 111–118.

Asmutis-Silvia, R.A. (1999). *An increased risk to whales due to high-speed whale-watching vessels.* Paper presented to the Scientific Committee at the 51st Meeting of the International Whaling Commission, May 1–15 1999, Grenada.

Barton, K., Booth, K., Simmons, D.G., & Fairweather, J.R. (1998). *Tourist and New Zealand Fur Seal Interactions Along the Kaikoura Coast.* Education Centre Report No. 9. Lincoln University, Lincoln, New Zealand.

Bejder, L., Samuels, A., Whitehead, H., & Gales N. (2006). Interpreting short-term behavioural responses to disturbance within a longitudinal perspective. *Animal Behaviour* 72: 1149–1158.

Bejder, L., Samuels, A., Whitehead, H., Gales, N., Mann, J., Connor, R., Heithaus, M., Watson-Capps, J., Flaherty, C., & Kürtzen, M. (2006). Decline in relative abundance of bottlenose dolphins exposed to long-term disturbance. *Conservation Biology* 20: 1791–1798.

Bonner, N. (1990). *Seals and Sea Lions of the World.* Blandford, London.

Boren, L.J. (2001). *Assessing the Impact of Tourism on New Zealand Fur Seals* (Arctocephalus forsteri). Masters thesis, University of Canterbury, Canterbury, New Zealand.

Boren, L.J., Gemmell, N.G., & Barton, K.J. (2002). Tourist disturbance on New Zealand fur seals *Arctocephalus forsteri. Australian Mammalogy* 24: 85–96.

Boren, L.J., Gemmell, N.G., & Barton, K.J. (2009). The role and presence of a guide: preliminary findings from swim with seal programs and land based seal viewing in New Zealand. *Tourism in Marine Environments* 5: 187–199.

Buckstaff, K. (2004). Effects of watercraft noise on the acoustic behaviour of bottlenose dolphins, *Tursiops truncatus*, in Sarasota Bay, Florida. *Marine Mammal Science* 20: 709–725.

Cassini, M.H. (2001). Behavioural responses of South African fur seals to approach by tourists—a brief report. *Applied Animal Behaviour Science* 71: 341–346.

Clayton, S., & Myers, G. (2009). *Conservation Psychology: Understanding and Promoting Human Care for Nature.* Wiley-Blackwell, West Sussex, UK.

Constantine, R. (2008). Whale Watching. In: *The Encyclopedia of Tourism and Recreation in Marine Environments* (Ed. M. Lück), pp. 527–529. CAB International, Wallingford, UK.

Corbelli, C. (2006). *An Evaluation of the Impact of Commercial Whale Watching on Humpback Whales, Megaptera novaeangliae, in Newfoundland and Labrador, and of the Effectiveness of a Voluntary Code of Conduct as a Management Strategy.* PhD Thesis, Department of Biology, Memorial University of Newfoundland, St. John's, Canada.

Corkeron, P. (2002). Captivity. In: *The Encyclopedia of Marine Mammals* (Ed. W.F. Perrin, B. Würsig, J., & J.G.M. Thewissen), pp. 192–197. Academic Press, New York.

Corkeron, P.J. (2004). Whale watching, iconography, and marine conservation. *Conservation Biology* 18: 847–849.

Corkeron, P.J. (2006). How shall we watch whales? In: *Gaining Ground: In Pursuit of Ecological Sustainability* (Ed. D.M. Lavigne), pp. 161–170. International Fund for Animal Welfare, Guelph, Canada.

Cunningham-Smith, P., Colbert, D.E., Wells, R.S., & Speakman, T. (2006). Evaluation of human interactions with a provisioned wild bottlenose dolphin (*Tursiops truncatus*) near Sarasota Bay, Florida, and efforts to curtail the interactions. *Aquatic Mammals* 32: 346–356.

Curtin, S., & Wilkes, K. (2007). Swimming with captive dolphins: current debates and post-experience dissonance. *International Journal of Tourism Research* 9: 131–146.

Duffus, D.A. (1988). *Non-Consumptive Use and Management of Cetaceans in British Columbia Coastal Waters.* PhD Thesis, University of Victoria, British Columbia, Canada.

Draheim, M., Bonnelly, I., Bloom, T., Rose, N., & Parsons. E.C.M. (in press). Tourist attitudes towards marine mammal tourism: an example from the Dominican Republic. *Tourism in Marine Environments.*

Economist. (2000). The politics of whaling. *The Economist*, September 9, p. 100.

Falk, J.H., Reinhard, E.M., Vernon, C.L., Bronnenkant, K., Heimlich, J., & Deans, N.L. (2007). *Why Zoos and Aquariums Matter: Assessing the Impact of a Visit to a Zoo or Aquarium.* Association of Zoos and Aquariums, Silver Spring, MD.

Finkler, W., & Higham, J. (2004). The human dimensions of whale watching: an analysis based on viewing platforms. *Human Dimensions of Wildlife* 9: 103–117.

Foxlee, J. (2001). Whale watching in Hervey Bay. *Parks and Leisure Australia* 4(3): 17–18.

Garrod, B., & Fennell, D.A. (2004). An analysis of whalewatching codes of conduct. *Annals of Tourism Research* 31: 334–352.

Gordon, J., Leaper, R., Hartley, F.G., & Chappell, O. (1992). *Effects of Whale-Watching Vessels on the Surface and Underwater Acoustic Behaviour of Sperm Whales off*

Kaikoura, New Zealand. Science and Research Services Series No. 52. New Zealand Department of Conservation, Wellington, New Zealand.

Hastie, G.D., Wilson, B., Tufft, L.H., & Thompson, P.M., (2003). Bottlenose dolphins increase breathing synchrony in response to boat traffic. *Marine Mammal Science* 19: 74–84.

Higham J., & Lusseau, D. (2004). Ecological impacts and management of tourist engagements with marine mammals. In: *Environmental Impacts of Ecotourism* (Ed. R. Buckley), pp. 173–188. CABI Publishing Wallingford, U.K.

Higham, J.E.S., & Lusseau, D. (2007). Urgent need for empirical research into whaling and whale watching. *Conservation Biology* 21: 554–558.

Higham, J.E.S., & Lusseau, D. (2008). Slaughtering the goose that lays the golden egg: are whaling and whale-watching mutually exclusive? *Current Issues in Tourism* 11: 63–74.

Hoyt, E. (1995). *The Worldwide Value and Extent of Whale Watching*. Whale and Dolphin Conservation Society, Bath, U.K.

Hoyt, E. (2001). *Whale Watching 2001: Worldwide Tourism Numbers, Expenditures and Expanding Socioeconomic Benefits*. International Fund for Animal Welfare, Yarmouth Port, Massachusetts.

Hoyt, E. (2005a). *Marine Protected Areas for Whales, Dolphins and Porpoises: A World Handbook for Cetacean Habitat Conservation*. Earthscan, London.

Hoyt, E. (2005b). Sustainable ecotourism on Atlantic Islands, with special reference to whale watching, marine protected areas and sanctuaries for cetaceans. *Proceedings of the Royal Irish Academy* 105B: 141–154.

Hoyt, E., & Hvenegaard, G.T. (2002). A review of whale watching and whaling with applications for the Caribbean. *Coastal Management* 30: 381–399.

Hughey, K.F., & Ward. (2002). *Sustainable Management of Natural Assets Used for Tourism in New Zealand*. Lincoln University Report 55. Lincoln University, New Zealand.

Iñiguez, M.A., Tomsin, A., Torlaschi, C., & Prieto, L. (1998). *Aspectos Socio-económicos del Avistaje de Cetáceos en Península Valdés, Puerto San Julián y Puerto Deseado, Patagonia, Argentina*. Fundación Cethus, Buenos Aires.

Janik, V.M., & Thompson, P.M. (1996). Changes in surfacing patterns of bottlenose dolphins in response to boat traffic. *Marine Mammal Science* 12: 597–602.

Kirkwood, R., Boren, L., Shaughnessy, P., Szteren, D., Mawson, P., Huckstadt, L., Hofmeyr, G. Oosthuizen, H., Schiavini, A., Campagna, C., & Berris, M. (2003). Pinniped-focused tourism in the southern hemisphere: a review of the industry. In: *Marine Mammals: Fisheries, Tourism and Management Issues* (Ed. N. Gales, M. Hindell & R. Kirkwood). CSIRO Publishing, Collingwood, Australia.

Kovacs, A., & Innes, S. (1990). The impact of tourism on harp seals (*Phoca groenlandica*) in the Gulf of St. Lawrence, Canada. *Applied Animal Behaviour Science* 26: 25–26.

Laist, D.W., Knowlton, A.R., Mead, J.G., Collet, A.S., & Podesta, M. (2001). Collisions between ships and whales. *Marine Mammal Science* 17: 35–75.

Leaper, R. (2001). *Summary of data on ship strikes of large cetaceans from progress reports (1996–2000)*. Working Paper presented to the Scientific Committee at the 53rd Meeting of the International Whaling Commission, July 3–16, 2001, London.

Lück, M. (2003). Education on marine mammal tours as agent for conservation—but do tourists want to be educated? *Ocean and Coastal Management* 46: 943–956.

Lück, M. (2008). *Encyclopedia of Tourism and Recreation in Marine Environments*. CAB International, Wallingford, UK

Lück, M. (2009). *Environmentalism and Tourists' Experiences on Swim-With-Dolphins Tours. A Case Study in New Zealand*. Verlag Dr. Muller, Saarbrüken, Germany.

Lusseau, D. (2003). Male and female bottlenose dolphins *Tursiops* spp. have different strategies to avoid interactions with tour boats in Doubtful Sound, New Zealand. *Marine Ecology Progress Series* 257: 267–274.

Lusseau, D. (2004). The state of the scenic cruise industry in Doubtful Sound in relation to a key natural resource: bottlenose dolphins. In: *Nature-based Tourism in Peripheral Areas: Development or Disaster?* (Ed. M. Hall & S. Boyd), pp. 246–260. Channel View Publications, Clevedon, UK

Lusseau, D. (2006). Short term behavioral reactions of bottlenose dolphins to interactions with boats in Doubtful Sound, New Zealand. *Marine Mammal Science* 22: 802–815.

Lusseau, D., Bain, D., Williams, R., & Smith, J. (2009). Vessel traffic disrupts foraging behaviour of southern resident killer whales. *Endangered Species Research* 6: 211–221.

Martinez, A. (2003). Swimming with sea lions: friend or foe? Impacts of tourism on Australian sealion *Neophoca cinerea*, at Baird Bay, S. A. Honours thesis, Flinders University of South Australia.

Mattson, M.C., Thomas, J.A., & St. Aubin. D. (2005). The effect of boat activity on the behaviour of bottlenose dolphins (*Tursiops truncatus*) in waters surrounding Hilton Head Island, South Carolina. *Aquatic Mammals* 31: 133–140.

Mayes, G., Dyer, P., & Richins, H. (2004). Dolphin-human interaction: pro-environmental attitudes, beliefs, and intended behaviours and actions of participants in interpretation programs: a pilot study. *Annals of Leisure Research* 7: 34–53.

Muloin, S. 1996. *Whale watching in Hervey Bay: results from Matilda II*. Department of Tourism. James Cook University, Townsville, Australia.

O'Connor, S., Campbell, R., Cortez, H., & Knowles, T. (2009). *Whale Watching Worldwide: Tourism Numbers, Expenditures and Expanding Economic Benefits.* A special report from the International Fund for Animal Welfare (IFAW) and Economists at Large, Yarmouth, MA.

Odell, D.K., & Robeck, T.R. (2002). Captive Breeding. In: *The Encyclopedia of Marine Mammals* (Ed. W.F. Perrin, B. Würsig, & J.G.M. Thewissen), pp. 188–192. Academic Press, New York.

Orams, M.B. (1995). A conceptual model of tourist-wildlife interaction: the case for education as a management strategy. *Australian Geographer* 27: 39–51.

Orams, M.B. (1999). *The Economic Benefits of Whale-Watching in Vava'u, The Kingdom of Tonga.* Centre for Tourism Research, Massey University at Albany, North Shore, New Zealand.

Orams, M.B. (2004). Dolphins, whales and ecotourism in New Zealand: what are the impacts and how should the industry be managed? In: *Nature-based Tourism in Peripheral Areas: Development or Disaster?* (Ed. M. Hall & S. Boyd), pp. 231–245. Channel View Publications, Clevedon, UK.

Orsini, J.P., Shaughnessy, P.D., & Newsome, D. (2006). Impacts of human visitors on Australian sea lions (*Neophoca cinerea*) at Canac Islands, Western Australia: Implications for tourism management. *Tourism in Marine Environments* 3: 101–115.

Parsons, E.C.M. (2003). Seal management in Scotland: tourist perceptions and the possible impacts on the Scottish tourism industry. *Current Issues in Tourism* 6: 540–546.

Parsons, E.C.M., & Draheim, M. (2009). A reason not to support whaling: a case study from the Dominican Republic. *Current Issues in Tourism* 12: 397–403.

Parsons, E.C.M., Fortuna, C.M. Fortuna, Ritter, F., Rose, N.A., Simmonds, M.P., Weinrich, M., Williams, R., & Panigada S. (2006). Glossary of whalewatching terms. *Journal of Cetacean Research and Management* 8 (Suppl.): 249–251.

Parsons, E.C.M., & Gaillard, T. (2003). *Characteristics of high-speed whalewatching vessels in Scotland.* Paper presented to the Scientific Committee at the 55th Meeting of the International Whaling Commission, May 26–June 6, 2003, Berlin, Germany.

Parsons, E.C.M., Lewandowski, J., & Lück, M. (2006). Recent advances in whalewatching research: 2004–2005. *Tourism in Marine Environments* 2: 119–132.

Parsons, E.C.M., Lück, M., & Lewandowski, J. (2006). Recent advances in whalewatching research: 2005–2006. *Tourism in Marine Environments* 3: 179–189.

Parsons, E.C.M., & Rawles, C. (2003). The resumption of whaling by Iceland and the potential negative impact in the Icelandic whale-watching market. *Current Issues in Tourism* 6: 444–448.

Parsons, E.C.M., Warburton, C.A., Woods-Ballard, A., Hughes, A., Johnston, P., Bates, H., & Lück, M. (2003a). Whale-watching tourists in West Scotland. *Journal of Ecotourism* 2: 93–113.

Parsons, E.C.M., Warburton, C.A., Woods-Ballard, A., Hughes, A., & Johnston, P. (2003b). The value of conserving whales: the impacts of cetacean-related tourism on the economy of rural West Scotland. *Aquatic Conservation* 13: 397–415.

Parsons, E.C.M., & Woods-Ballard, A. (2003). Acceptance of voluntary whalewatching codes of conduct in West Scotland: the effectiveness of governmental versus industry-led guidelines. *Current Issues in Tourism* 6: 172–182.

Pearce, D.G., & Wilson, P.M. (1995). Wildlife-viewing tourists in New Zealand. *Journal of Travel Research* 34: 19–26.

Rawles, C.J.G., & Parsons, E.C.M. (2004). Environmental motivation of whale-watching tourists in Scotland. *Tourism in Marine Environments* 1: 129–132.

Richter, C., Dawson, S., & Slooten, E. (2006). Impacts of commercial whale watching on male sperm whales at Kaikoura, New Zealand. *Marine Mammal Science* 22: 46–63.

Rose, N.A., Parsons, E.C.M., & Farinato, R. (2009). *The Case Against Marine Mammals in Captivity,* 4th ed. The Humane Society of the United States and the World Society for the Protection of Animals, Washington DC.

Russell, C.L., & Hodson, D. (2002). Whalewatching as critical science education? *Canadian Journal of Science, Mathematics and Technology Education* 7: 54–66.

Scarpaci, C., Bigger, S.W., Corkeron, P.J., & Nugegoda, D. (2000). Bottlenose dolphins, *Tursiops truncatus,* increase whistling in the presence of "swim-with-dolphin" tour operations. *Journal of Cetacean Research and Management* 2: 183–185.

Scarpaci, C., Lück, M., & Parsons, E.C.M. (2009). Recent advances in whalewatching research: 2008–2009. *Tourism in Marine Environments*: 6: 39–50.

Scarpaci, C., Nugegoda, D., & Corkeron, P.J. (2003). Compliance with regulations by "swim-with-dolphins" operations in Port Philip Bay, Victoria, Australia. *Environmental Management* 31: 342–347.

Scarpaci, C., Nugegoda, D., & Corkeron, P.J. (2004). No detectable improvement in compliance to regulations by "swim-with-dolphin" operators in Port Philip Bay, Victoria, Australia. *Tourism in Marine Environments* 1: 41–48.

Scarpaci, C., Nugeoda, D., & Corkeron, P.J. (2005). Tourists swimming with Australian fur seals (*Arctocephalus pusillus*) in Port Phillip Bay, Victoria: are tourists at risk? *Tourism in Marine Environments* 1: 89–95.

Scarpaci, C., Parsons, E.C.M., & Lück, M. (2008). Recent advances in whalewatching research: 2006–2007. *Tourism in Marine Environments* 5: 55–66.

Scarpaci, C., Parsons, E.C.M., & Lück, M. (2009). Recent advances in whalewatching research: 2007–2008. *Tourism in Marine Environments* 5: 319–336.

Scheidat, M., Castro, C., González, J., & Williams, R. (2004). Behavioural responses of humpback whales (*Megaptera novaeangliae*) to whalewatching boats near Isla de la Plata, Machalilla National Park, Ecuador. *Journal of Cetacean Research & Management* 6: 63–68.

Shaughnessy, P.D., Newsome, D., & Briggs, S.V. (2008). Do tour boats affect seal behaviour at Montague Island, NSW. *Tourism in Marine Environments* 5: 15–27.

Stafford-Bell, R. & Scarpaci, C. (in review). Behaviour of the Australian fur seal *Arctocephalus pusillus doriferus* in response to vessel traffic and underwater noise emissions.

Stamation, K.A., Croft, D.B., Shaughnessy, P.D., Waples, K.A., & Briggs S. V. (2007). Educational and conservation value of whale watching. *Tourism in Marine Environments* 4: 41–55.

Stone, G.S., Katona, S.K., Mainwaring, A., Allen, J.M., & Corbett, H.D. (1992). Respiration and surfacing rates of fin whales, *Balaenoptera physalus*, observed from a lighthouse tower. *Reports of the International Whaling Commission* 42: 739–745.

Terhune, J.M., Stewart, R.E.A., & Ronald, K. (1979). Influence of vessel noises on underwater vocal activity of harp seals. *Canadian Journal of Zoology* 57: 1337–1338.

Tilt, W. C. (1987). From whaling to whalewatching. *Transactions of the North American Wildlife and Natural Resource Conference* 52: 567–585.

Toolis K. (2001). Eat it or save it? *The Guardian* (Weekend Pages), October 27, p. 58.

Wells, R.S., & Scott. M.D. (2002). Bottlenose dolphins. In: *The Encyclopedia of Marine Mammals* (Ed. W.F. Perrin, B. Würsig, & J.G.M. Thewissen), pp. 122–128. Academic Press, New York.

Whitt, A.D., & Read, A.J. (2006). Assessing compliance to guidelines by dolphin-watching operators in Clearwater, Florida, USA. *Tourism in Marine Environments* 3: 117–130.

Woods-Ballard, A., Parsons, E.C.M., Hughes, A.J., Velander, K.A., Ladle, R.J., & Warburton, C.A., (2003). The sustainability of whale-watching in Scotland. *Journal of Sustainable Tourism* 11: 40–55.

Marine Mammal Research Techniques

APPENDIX

CHAPTER OUTLINE

Line Transect Surveys

  Exploring the Depths: Experiment Design
   Hypotheses
   Statistical Errors
   Pseudo-replication
   Endnote

Mark-Recapture Analysis and Photo-Identification

  Exploring the Depths: Photo-Identification
  Exploring the Depths: Taking Marine Mammal Photographs

Land-Based Surveys

  Exploring the Depths: Behavioral Observation Methods for Marine Mammals

Biopsy Darting

  Exploring the Depths: Strandings Analysis
  Exploring the Depths: A Brief Glimpse into Molecular Genetics

Tagging

Crittercams

Acoustic Techniques

  Exploring the Depths: Public Sightings Schemes
  Exploring the Depths: Welfare, Animal Research Ethics, and Invasive Studies

Selected References and Further Reading

Conducting research on marine mammals is an obvious necessity to discover more about the biology, behavior, and ecology of marine mammals. But research is also essential for conservation—policy and management decisions need to be informed by accurate scientific data, for example, whether populations are declining in numbers, or whether habitats are becoming degraded by pollutants. A recent study found that half of the scientific papers published on cetaceans were directly related to conservation, so research and conservation currently go hand-in-hand.

Line Transect Surveys

Line transect surveys are the most commonly used way to calculate how many animals are in a large area at a given time. As pinnipeds can usually be counted on shore, most marine mammal line transect surveys tend to be for cetaceans. The survey estimates the density of animals during the survey in a strip of the ocean. This is then extrapolated to the whole of the area in question. It is called a line transect survey as the survey platform (usually a boat) typically follows a predetermined transect line route (**Figure A.1**). The main assumptions of the basic method are that animals on the transect line are always detected (the probability of detection is 100% or 1) and that there is a diminishing likelihood of detection the further you get from the transect line (i.e., the farther away from the vessel you are).

Two key pieces of information are required: (i) the bearing to the animal from the path of the vessel; and (ii) the estimated distance to the animal. This allows the perpendicular distance of the sighted animal away from the transect line to be calculated with some simple trigonometry. In turn, this can then be used to calculate the effective width of the transect strip (i.e., the width of ocean where animals are highly likely to have been spotted) and, from that, also the density of cetaceans within this strip.

Accuracy in estimating the distance of the cetaceans from the survey vessel is therefore very important. Often simply estimated by observers, this can also be done more

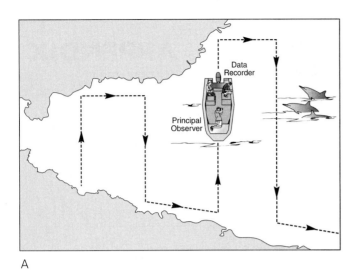 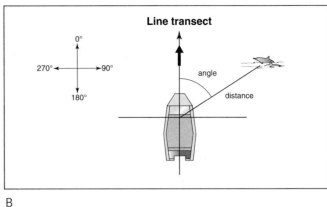

FIGURE A.1 A line transect survey.

accurately with the aid of laser range-finding binoculars to determine the distance to a large cetacean or nearby object. An accurate count of the number of animals in a group is also important.

As noted above, the basic method assumes that animals on the transect line are always sighted, but this is often not the case with cetaceans. Animals submerge underwater, some species such as sperm and beaked whales for extremely long periods of time. Weather conditions can also affect the ability of observers to spot cetaceans. As a result, the likelihood of detecting an animal on the track line has to be determined and incorporated into all later calculations. The probability of detecting an animal on the transect line is commonly referred to as g(0). This is then used to adjust the estimate of the number of animals in an area through the following equation:

$$\hat{N} = A \frac{n\bar{s}}{2\hat{w}L\hat{g}(0)}$$

where $\hat{N}$ is the estimated abundance (the pointy "hat" denotes an estimated value), A is the size of the study area, n is the number of animals seen, $\bar{s}$ is the average group size (the bar above denotes that it is a mean value), $\hat{w}$ is half the effective width of the transect strip (i.e., usually how far away from the boat, on one side, that you could reliably be spotting roughly the same number of animals as you would on the track line), L is the total length of the transect line surveyed and $\hat{g}(0)$ is the probability of detecting an animal on the transect line.

To conduct a line transect survey you need a survey platform (for marine mammals, this is usually a boat, but transect surveys can also be conducted from planes) and at least two experienced observers with binoculars. It is important that these observers are able to estimate distances (accurately, as well as consistently with each other) and numbers of animals in a group and identify species correctly. The two observers will scan the water to the front of the boat. One will be viewing a 90° arc from the track line to the right and the other a 90° arc from the track line to the left. It is also useful to have a third (and maybe a fourth) observer, who is independent and isolated from the others (e.g, if the two primary observers are on deck, the additional observers might be on the ship's bridge, or crow's nest). The sightings of the independent observers can be compared with those of the two primary observers, testing the accuracy of data collection. Any differences in species identification and/or numbers sighted can be used to gauge the accuracy of the sightings data.

Finally, it is very useful to have at least one additional crew member acting as a dedicated data recorder as this allows the observers to concentrate on sighting cetaceans. The data recorder might input the sightings data onto a paper data sheet, or possibly into a hand-held computer or something similar. It is this individual's job to make sure all required information is obtained from the observers. A hand-held GPS receiver or an on-board computer linked to the ship's navigation system means that the locations where marine mammals are sighted can be accurately noted. This also means that the position of the ship can be recorded every few minutes to accurately record the actual track line itself. Other important data to record include environmental factors—such as sea state, swell, amount of sun glare on the water's surface, and relative quantity and type of precipitation—as these can all affect the ability to sight marine mammals. Finally, a computer system with appropriate software is required to eventually analyze the data and produce an estimate of abundance.

Exploring the Depths: Experiment Design

Hypotheses

It is very important to state your hypotheses, what you expect to see, before collecting any data. The null hypothesis, H_0, assumes no difference. For example, there is no difference in the abundance of minke whales at site A as compared to site B. The null hypothesis is tested against the alternative hypothesis, H_1. In this example, one alternative hypothesis could be that there is a difference in abundance between site A and site B (this is called a two-sided alternative). A one-sided alternative could be that there are more minke whales in site A than site B. While a one-sided test is more powerful, a two-sided test is more appropriate when you are not able to predict which site has a higher abundance.

A statistical hypothesis test should be used to decide whether there is enough evidence to reject the H_0 in favor of the H_1, or whether there is insufficient evidence to reject the H_0. When a statistical hypothesis test fails to reject the H_0, the null hypothesis could be true. In our example, we would say that there was no statistically significant difference in the abundance of minke whales at site A as compared to site B.

Statistical Errors

If you reject your H_0, evidence suggests that the H_1 is true, but since this is statistics, you can never be 100% certain that you are making the correct decision. There is always a small possibility that your statistical test rejects H_0 when in fact H_0 is true. This kind of error is called the Type I error and is symbolized by the Greek letter alpha (α).

The other type of error is called the Type II error, symbolized by the Greek letter beta (β). Type II error occurs when you have insufficient evidence to reject H_0, but really H_0 is not true. There is a tradeoff between Type I and Type II errors (when one goes down, the other goes up), so many researchers choose to accept a 5% Type I error rate, setting $\alpha = .05$. This means that if the probability that the H_0 is true is 5% or smaller, you would choose to reject H_0. Ninety-five percent of the time you will be correct, but 5% of the time, the data from your sample will lead you to make an incorrect decision, a Type I error. To decrease both types of errors at the same time, you would need a larger sample size.

Pseudo-replication

The objective of statistics is to draw conclusions about a larger population based on a sample that you, the experimenter, measure. A larger sample will be more representative of the population, so true replication of measurements enhances your ability to generalize the results found from your sample to the wider population. Thus, true replication improves the power $(1 - \beta)$ of your statistical test, improving your statistical decision-making because of a decrease in the Type II error rate.

Pseudo-replication is where a researcher treats each observation as independent for the sake of statistical analysis when really, the observations were not independent at all. This can be taking repeated measurements of the same sample, measuring one sample at multiple time-points, or pooling some samples into one pooled observation. For example, say you want to study the effect of seaweed density on the density of an herbivorous species. You do 100 line transects looking for your herbivorous species in a single low-density seaweed area and 100 in a single high-density seaweed area. Each group of 100 line transect replicates are not true replicates since they took place in the same area—these are pseudoreplicates. You could have identified multiple high seaweed density locations and multiple low seaweed density locations and measured species abundance in each of these areas, accounting for other differences (ocean temperature, proximity to shore, and other factors that could impact your results).

Using pseudoreplicates instead of true replicates underestimates the variance of the data. With an artificially smaller variance, there is a larger possibility of rejecting the null hypothesis when the null hypothesis is actually true (a Type I error).

Endnote

For more detailed advice on your particular study, consult a statistician.

	H_0 is true	H_0 is not true
You reject H_0	Type I error (α)	Correct decision
You fail to reject H_0	Correct decision	Type II error (β)

–Contributing author, Marike Kester, George Mason University.

Mark-Recapture Analysis and Photo-Identification

Mark-recapture studies are the techniques that allow researchers to track the same individual over a period of time (from days to years, depending on the specific method). Such studies underlie much of what is known about marine mammals today. These techniques have also been used in many marine mammal studies to estimate abundance through a mark-recapture analysis. The method hinges on the idea that if you capture and mark a sample at random from a population, the closer the population is to your sample size, the more of the same individuals (as identified through the marks) will be recaptured each time you take an additional sample. Let us consider a working example. On the first day of a project, 10 sea lions are captured on a haul out and marked in some way. Two weeks later, 11 seals are captured, but nine of these were the same individuals as were caught in the previous sample. In all likelihood the total size of the population is unlikely to be much more than 11, as the same individuals are sighted repeatedly. Alternatively, if only two of the original animals were resighted, the population size is likely to be much larger. An estimate of abundance can be calculated using the equation:

$$\hat{N} = \frac{n_1 n_2}{m_2}$$

where n_1 is the numbers of animals in the first sample, n_2 is the number of animals in the second sample, and m_2 is the number of repeat sightings in the second sample. So in the first example above, the number of sea lions in the population would be:

$$\hat{N} = 10 \times 11 / 9 = 12.2$$

and for the second example the number would be:

$$\hat{N} = 10 \times 11 / 2 = 55$$

There are, of course, various considerations when undertaking these types of projects. For example, the idea is that each sample should be a random sample of the population. As a result, each should be independent of each other. This is likely not to be true if you sample on two consecutive days as animals will not have had a chance to move in or out of the area if it incorporates only part of their range. Alternatively, the animals that you handled may leave the area for an extended period of time specifically because they were handled. Instead, you need to wait until you can be fairly sure that the sample is indeed random again and that there is no relation to the previous one. Often this is done by comparing one annual sample with that acquired the following year.

Another issue relates to the capture and marking methods themselves. If you base your analysis on small plastic tags that are clipped to the flippers of sea lions, but some tags are pulled out as the animals climb over rocks, then your resighting rate will be affected. Similarly, it is possible that some handling and marking methods might reduce the survival rates of certain animals. This has ethical considerations (see Exploring the Depths: Welfare, Animal Research Ethics, and Invasive Studies), but also may influence the resighting rate, if an animal dies prior to the resampling.

Various methods exist for marking, ranging from temporary marks (such as dyes, paints, and shaving fur) to more permanent options (such as flipper tags, clipping dorsal fins, and branded markings). Branding is often considered to be the most useful as it produces permanent, highly visible, individually coded marks. Pinnipeds are often marked with hot brands, or brands that have been super-cooled with liquid nitrogen. This leads to scarring and/or loss or discoloration of subsequent hair growth in the skin branded. Branding cetaceans is relatively uncommon, but freeze-branding does occasionally occur, leaving an identifiable scar pattern on the dorsal fin.

Like many aspects of research, there are considerations that need to be balanced here. For example, many studies, including abundance estimates, require that animals be reliably re-identified year after year. Low-impact, temporary marks cannot achieve this and, as mentioned above, flipper tags can be lost. Branding does, for the most part, produce marks that can be spotted and easily identified at a distance. This means that, unlike several other methods, they do not require actually recapturing the animal to resample for abundance estimates and certain other studies. However, concerns have been raised over infections that occasionally occur, as well as the humaneness of the method. Some have pushed for only freeze-branding to be used, as this appears to be less damaging. This method requires the use of more specialized equipment that is not easy to take to more remote areas where some research occurs, so it is not always practical. However, when the results of the work can be influenced by the effects of that work on the animal, this may well be worth the extra effort. It is also likely that branding may be too invasive for use with fur seals that are totally reliant on their fur for insulation. Loss of hair, even temporarily, may result in a huge drop in survival rates that would not only bias the results, but also be unacceptable to most conservation biologists. In summary, such methods should be used by biologists sparingly and only after serious consideration of the following:

- the possible impacts to the animals in the study;
- the status of the population of animals that will be marked;

- the value of the data that will be generated (and that could not be gained through other means) to conservation and management efforts (perhaps to a related endangered species, given consideration of points 1 and 2); and
- the sample size needed to achieve enough statistical power that the work can be effective without over- (or under-) sampling and thus marking animals unnecessarily.

Fortunately, many marine mammals (especially cetaceans) may have individual differences or acquire their own marks as they go through life. Individuals can thus be identified in non-invasive way using these (natural) markings instead. These include:

- tail fluke shape, coloration, or pattern;
- body shape, coloration, or pattern;
- dorsal fin outline, damage, shape, and coloration; or
- scarring patterns.

Sampling and "recapture" of this information is typically done through the use of photography, and so the method is often referred to as photo-identification, or photo-ID for short. Photo-ID has its own set of advantages and disadvantages (see Exploring the Depths: Photo-Identification). Some of the drawbacks can be overcome by combining it with a marking program, which may be especially useful in species that do not generally accumulate their own markings, but then the project becomes subject to the considerations of handling and marking.

As mentioned above, both mark-recapture and photo-ID methods have much wider application than simply as a technique to gain abundance information. As individual animals are identified, it is also possible to gather much additional information; for example, you can determine the size and location of individual home ranges and associations between individuals. It is also possible to get information about reproductive rates, longevity, parental care, alloparental care (i.e., babysitting), individual foraging preferences, and culture. There is also additional value in these methods, as many other techniques, such as genetic analysis, also require information about the individual sampled and its relationships to achieve their full potential (see Exploring the Depths: A Brief Glimpse into Molecular Genetics).

Exploring the Depths: Photo-Identification

With a digital camera, photo-ID is a relatively easy, inexpensive, and fast method of identifying cetaceans. With modern digital cameras, the quality of photographs is effectively as high as film images. Digital photos also have the advantage of being instantly ready to use and easily transferable to the computer where the data can be stored and analyzed using special software. Moreover, digital images are easily exchanged with other researchers, to allow comparison and verification of identification by other scientists.

However, researchers can accumulate huge numbers of images in a photo-ID study (especially now that there is no processing cost for digital images), and comparing photographs from such large data sets can be very time consuming. Fortunately, there are several computer programs (such as the programs Finscan or DARWIN) that can compare and analyze the photographs objectively, speeding up the sorting of photographs (**Figure BA.1**).

The photo-ID data are versatile and can be used to answer various research questions, such as behavioral interactions as noted above. In addition, photo-ID can be used in combination with other research methods, such as genetic sampling (i.e., collecting a biopsy sample from an individual that is simultaneously photographed). The latter method can help, among other things, to estimate the sex and kinship of the individuals in a population. Photo-ID can also provide other valuable information, which can provide insights into the health and causes of injury in the studied population, such as:

- incidences of scarring caused by fisheries and marine litter entanglement;
- proportion of wounds caused by predators, such as shark bites;
- proportion of wounds caused by boat strikes; and
- prevalence of skin disease and ectoparasites.

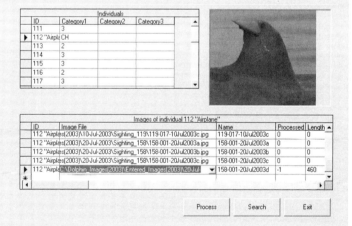

FIGURE BA.1 Screen shot of Finscan photo-ID analysis software. The Finscan program is no longer supported but there are many similar analysis programs available to researchers.

Exploring the Depths: Photo-Identification (*continued*)

Moreover, photo-ID is less invasive for the animal than other methods of identifying marine mammals, such as branding or tagging. As a result of all the above, this method is commonly used in cetacean research. This means that the results of many different studies are directly comparable and also that the various limitations of the method are well known. For example, the individuals of a species being studied using the photo-ID method need to have (or acquire) distinctive markings or patterns that are commonly visible above water. In addition, the species needs to be easily photographed. This does not apply for all species. For instance, the harbor porpoise is very, very difficult to photograph because this species: i) shuns boats, ii) surfaces only for a short time, and iii) has a small dorsal fin. Therefore, the effectiveness of this method will vary between species.

Also, not all individuals within a population may have natural markings to a degree that makes identification possible—some individuals (perhaps males) may have more scars and marks than others. Marks can also change over time. The method is not very effective for the identification of calves and juveniles because their natural markings will often change during their lifetime and they have not had time to acquire other distinctive marks. Injuries from accidents, fighting, or even parasites can change the marks on animals. For example the two images of bottlenose dolphin dorsal fins below (**Figure BA.2**) belong to the same animal (identified by a large scar on the side of the dolphin's body), but within just one year the dorsal fin changes quite substantially.

Additional complications arise in species that are typically found in groups of several hundred individuals, such as offshore populations of common or spinner dolphins. As you might imagine, it can be very difficult to photograph a sufficient proportion of individuals for identification purposes when there are so many animals and you have trouble keeping track of which ones you have already photographed.

Finally, although less invasive than many research methods, photo-ID does require boats to get close enough to marine mammals to take good quality images. The presence of a boat and the noise it produces (see Chapters 6 and 18) could disturb marine mammals, causing their behavior to change and elevating their stress level.

FIGURE BA.2 Two photographs of the dorsal fin of an identified individual dolphin showing how dramatically fin shape can change in just one year.

–Contributing author, Todd McConchie, George Mason University.

Exploring the Depths: Taking Marine Mammal Photographs

Whether for a photo-ID project, or just for fun while on a whale-watching trip, getting good images of whales and dolphins (or other marine mammals at sea) can be tricky. Here we offer a few tips.

A good digital camera (for example, a Canon Rebel or Nikon EOS) body is recommended. A fast write time (i.e., the time gap between pressing the shutter and the picture actually processing) is important—otherwise you will have lots of great photographs of empty ocean where a dolphin used to be! A 35–350 mm lens is probably the most versatile, as it will allow close-up shots if a dolphin is, for example, riding on the bow wave of your boat, but will also allow good photos at a distance. However, these lenses are not necessarily cheap. Good alternatives can be found in the

Exploring the Depths: Taking Marine Mammal Photographs (*continued*)

70–300 mm (75–180, 70–200, 70–300) range, although you will lose out a little on animals very close to the boat. However, you can always keep an inexpensive digital camera close to hand for incidents of bow-riding where the movements of the animals are a little more predictable!

If you can, set the camera to a high shutter speed (1/800 or faster) and high ISO (200 ISO or higher), as this will produce sharper, crisper images of fast-moving marine mammals (and manatees too). However, if you are in an area with consistently poor light, these settings may not allow enough light into the camera. You may need to get a lens that gathers more light. For camera novices, many good cameras have a sports setting (a running man) and this usually has a sufficient shutter speed to get clear photos of many faster moving cetaceans.

If taking pictures for research, it is a good idea to take a photograph of a notepad, slate, or piece of paper, with the date, time, and the number of the group before you start taking photographs of the animals. If you get multiple sightings in the same day it is very easy to quickly lose track of which photographs are from which encounter. This, combined with a *correctly* set camera time and the standard sequential numbering system for images, will resolve any such issues. Finally, when taking photo-ID shots, you should try to fill the whole frame of the camera with the part of the body being used for identification (tail fluke or dorsal fin). Small, blurred images, or images where the edges of, for instance, the dorsal fin are not clearly visible, will be of little use for research. Some examples of images that can and cannot be used for research are shown below (**Figure BA.3**).

–Todd McConchie, Chris Parsons, and Andrew Wright, George Mason University.

FIGURE BA.3 Examples of images of dolphin fins that are (A) useable and (B) unuseable for photo-identification purposes.

Land-Based Surveys

Land-based surveys are less expensive than boat surveys. You effectively just need an elevated view over the ocean to conduct this type of survey. As such it is a common type of study conducted by students or in developing countries where finances may be an issue. Through land-based surveys you can, for example, obtain useful information about daily and seasonal changes in relative dolphin abundance in a specific area. This information can then be analyzed to determine how it is affected by environmental factors (such as tidal state) or anthropogenic activities (such as the presence of boats). Most land-based surveys also use a survey or's theodolite (**Figure A.2**), which is effectively a telescope on a tripod that tells you the vertical and horizontal angle to which the viewfinder of the telescope is pointing. With some basic details about the survey location, theodolites allow you to calculate the exact location of sighted marine

FIGURE A.2 A land based survey for cetaceans, showing a surveyor's theodolite.

mammals using simple trigonometry. In turn, this information can be used in combination with the local depth charts and other oceanographic details to investigate how the animals use their habitat.

To do all this you need to: (i) measure the height of the theodolite lens from the ground; (ii) know the height of the survey site (above the level at which tide height is measured from, which is known as the *chart datum*); (iii) have information on tidal height for the area; (iv) know the exact location of the theodolite site (via a good GPS); and (v) know the direction of north. From adding (i) and (ii) you get information on the height of the theodolite above chart datum. By subtracting the tide height (ii) from this, you know how high the theodolite viewfinder is above the current sea level (h). To get the horizontal distance (as the crow flies) from the survey site to the sighted cetacean (d), you simply take the vertical angle that the theodolite gives when it is centered on a marine mammal group (θ) (assuming that the theodolite is set so that straight down is 0°) and use the following formula (**Figure A.3**):

$$d = h\tan\theta$$

Combining the distance to the marine mammal with the horizontal bearing to the animal recorded from the theodolite allows the position of the marine mammal to be determined and plotted. This process has been made a lot easier thanks to various computer software products that calculate and plot positions for you.

The ability of land-based surveys to accurately plot the location of sighted animals (and thus accurately assess their distance to oceanographic features and human activity) is exceedingly useful and cannot be matched by boat-based surveys. However, land-based surveys are, understandably, very limited in the area that they cover. Although a higher survey vantage point improves the viewing area, the effective distance to detect animals is only a kilometer or so for smaller marine mammals (pinnipeds and porpoises) and perhaps 7 to 10 km for large whales, even during good weather. To get the highest vantage point, cliff tops are frequently used as survey sites. However, structures such as lighthouses are also excellent, as they often have good unobscured views and the height of the platform is often readily known.

To conduct a land-based survey, you ideally need a crew of at least three (four is better), with two observers watching in shifts of approximately 2 hours. As noted above, a surveyor's theodolite is a very useful piece of equipment and older theodolites can often be borrowed from university engineering or geography departments, or companies, for no cost.

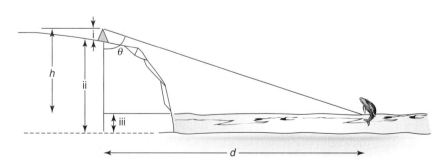

FIGURE A.3 The angle and distances used to calculate the position of a marine mammal using a theodolite (see text for definitions).

Exploring the Depths: Behavioral Observation Methods for Marine Mammals

There are two main methods for collecting behavioral data: (i) time sampling; and (ii) continuous sampling. Time sampling is a common method of making behavioral observations of long duration behaviors, or behavioral states, such as swimming or feeding. These are behaviors that the animal may perform for a long period of time. This type of observation involves establishing a set time interval, such as one minute and then making a single observation of the behavioral state every minute for a set observation period. Time sampling can be done by either *instantaneous* sampling, which measures the behavior at a set point in time, for example at 60, 120, 180 seconds, or by *one-zero* sampling where the observer records the behavior if it occurs at any point within a set interval, for example between 0–60 seconds. Time sampling creates an activity budget for the animal. In order to do this, the data obtained is converted to the proportion of time spent in each state. If one-minute observations are taken for a 15-minute period, then the average for each behavioral state is divided by 15, (or the total number of observations), to determine the proportion of time or percentage of time the animal spent in each state.

There are, however, many behaviors that are significant to marine mammal research that are relatively short in duration or may only occur during specific activities, such as breaching. However, often these may be missed during time sampling and can be the most important to record. These behavioral events can be recorded using continuous sampling. Continuous sampling offers an exact measure of the behavioral event by recording a specific behavior each and every time it occurs during a set observational period. Once these behavioral events are recorded, they can be analyzed to determine the frequency or rate of this behavior. For example, in a 15-minute observational period, the number of times the animal breaches during that period would be multiplied by 4 to obtain the rate per hour that breaching occurs.

Often marine mammals are found in large groups and under these circumstances there are several methods that can be used to study their behavior. One of the main methods is *focal* sampling, which is done by selecting a single individual in the group and observing and recording only the behaviors of that specific individual. *Scan* sampling can also be used with multiple animals and at regular intervals the entire group is scanned and each individual's behavior is recorded. Another method for observations is *behavior* sampling, which focuses on data collection of a specific behavior, for example aggression. Behavior sampling allows the observer to record specific behaviors and which individuals within the group are involved.

There are various ways to record behavioral data when studying marine mammals. It is most important to keep the specific hypothesis in mind to determine which method will provide the most appropriate data to answer the question being analyzed.

When performing behavioral observations it is also important to establish good inter-observer reliability. Often, multiple observers are needed to acquire the necessary data, therefore this reliability check measures the agreement between the observers and helps identify observer bias. Inter-observer reliability should be checked prior to data collection to ensure the accuracy of the data collected.*

—Contributing author, Jillian Fazio, George Mason University.

*Reliability testing can be done using standard correlation statistics, such as Spearman's rank-order correlation; Kendall coefficient of concordance (for more than 2 observers); or the Kappa coefficient, which accounts for random or chance agreements. Most acceptable inter-observer reliability measures are at 70% agreement or better.

Biopsy Darting

As in humans, biopsies involve removing a small piece of tissue from the animal. Often this is a small amount of blubber and skin that can be taken incidentally to other activities (e.g., marking or tagging) while the animal is in hand. However, biopsies of blubber and even muscle can be specifically taken for various analyses. Skin tissue provides, among other things, genetic information that can be used to verify species and stock identity and describe mating patterns. Blubber and muscle can also provide various data, including the trophic level at which the animal is feeding and the contaminant loads that it is carrying. If conducted simultaneously with mark-recapture studies, a great deal of information about the social behavior and ecology of the animals can be gathered. (It should be noted that assorted other samples that are not strictly speaking biopsies may also be (and often are) collected while the animal is "in hand," including fecal material, saliva, nasal/blow hole mucus, blood, milk, pulled teeth, and fur or hair.)

Remote biopsy darting was developed to reduce the need to capture and handle animals to gain biopsies, partly

Exploring the Depths: Strandings Analysis

A wealth of information can be gained from stranded or by-caught marine mammal carcasses through careful autopsies, known as necropsies in animals. Many countries have set up networks for reporting and collecting such carcasses. Visible injuries on carcasses, such as propeller wounds and net marks, can provide immediate information about threats to a population, but carcasses can also be used to investigate many things, including (but not limited to):

- the anatomy and physiology of sometimes rarely seen animals;
- the preferred prey of the species by analysis of stomach contents;
- pollutant loads (heavy metals, PCBs, DDT and derivatives, etc.);
- health of marine mammals via analysis of the histology of tissues and examination for tumors and parasitic, viral, and bacterial infections;
- natural history, breeding cycles, longevity, and the age of sexual maturity by analysis of reproductive tissues and estimation of age from layering patterns in teeth (for those marine mammals that have teeth); and
- species relationships/taxonomy and breeding population structure by comparison of genetic material or by a comparison of cranial and skeletal morphology.

It should, however, be emphasized that many countries have official stranding schemes and that permits may be required to handle dead marine mammals. Moreover, marine mammals can carry a variety of zoonotic diseases (diseases that can be transferred between species) including diseases that can infect and cause serious health problems for humans. So special caution should be taken when encountering a marine mammal carcass (or an animal that has stranded live on the beach).

---

to reduce stress on the animal and partly to allow scientists to get data on more animals. This technique involves shooting marine mammals with a modified rifle dart or crossbow bolt. These dart/arrows have short, sharp-edged, metal tubes on the tips that remove small plugs of blubber when they strike the sides of marine mammals. The width and depth of the plug taken depends on the species, with smaller darts being used for dolphins. A metal disk behind the head of the metal tube generally prevents the tip from sinking any deeper than the animal's blubber layer. Generally biopsy darting is considered not to "hurt" larger animals, but may produce violent reactions in some animals. Occasionally accidents causing serious injury can also occur, so, careful consideration needs to be given to the welfare implications of any program of work using biopsy sampling.

It is often hard to hit animals successfully with a biopsy dart and so the success rate is generally low. It can also be hard to tell if the same animal has been hit twice, and repeat samples can seriously skew studies. Additionally, the biopsy dart can sometimes leave a permanent scar or mark, which might impact photo-ID studies reliant upon scarring patterns for identification (primarily when the photo-ID work is being done by other scientists, perhaps in other areas, who will be unaware of the biopsy). Finally, as noted above, this method does cause, at best, a minor injury to animals, which could become infected, or more serious injuries in the case of an accident. It may also disturb both the target animal and those around it, depending upon the species and often the individual animals involved. Because of these latter concerns, permits and training are often required for biopsy sampling.

Exploring the Depths: A Brief Glimpse into Molecular Genetics

In order to understand populations, geneticists use many tools. Both nuclear DNA (nDNA) and mitochondrial DNA (mtDNA) are useful when examining population structure. Since these two types of DNA mutate at different rates in the cell, they can be used to examine different taxonomic levels in groups of animals. Mitochondrial DNA, a small circular molecule, mutates faster than nuclear DNA and is very helpful in studies of animals that are closely related. The genes that are most helpful for genetic study are the displacement loop (D-loop) and the cytochrome b (cyt b) gene. The D-loop, also called the control region, mutates faster than cyt b because it is a non-coding region of DNA. This gene is useful for studying small populations of the same species to determine genetic diversity. The cyt b gene is useful for studying animals down to the family or genus level of organization. However, since this is a gene that codes for a useful cellular product, it does not mutate as fast as the D-loop.

Although nuclear DNA does not mutate as rapidly as mtDNA, there are sections of nDNA that are useful for taxonomic evaluation and do mutate quickly. Microsatellites are short pieces of non-coding DNA that repeat in the genome a random number of times. These short pieces of DNA, also

Exploring the Depths: A Brief Glimpse into Molecular Genetics (*continued*)

called short tandem repeats, are very useful for population studies and also for determining paternity.

Molecular genetic studies can be conducted using any kind of tissue, including ancient bone. Of course, it is easier to extract DNA from blood or other wet tissues. Genetic analysis is currently underway to study animals as diverse as the woolly mammoth (*Mammuthus primigenius*) and the Neanderthal (*Homo neanderthalensis*) and the DNA used in these studies was extracted from ancient bone. Although this type of study is interesting intellectually, there are other uses for DNA that are much more practical.

The easiest application of molecular genetics is to determine the sex of animals that are not sexually dimorphic and are difficult to sex in the field. This is a simple examination that can be done using skin scrapings or other non-invasive methods. If the sex of individuals is established safely and accurately, scientists can use the data to keep track of animal breeding and add sex to photo-identification data. This type of DNA testing makes positively identifying animals very simple, which allows marine mammalogists to identify individuals as they migrate from one location to another.

Taxonomy is a subject that changes constantly. As new information is gathered, animals are reclassified into an accurate taxonomic tree. A vital consideration when classifying animals is to use as much information as possible. It is very important to take not only DNA but also anatomy into account when classifying animals. One gene can give researchers an incorrect impression of the relationships between animals. Thus it is best to use several genes in order to diagnose relationships.

The differences in DNA can be used to identify animals. Taking advantage of slight differences in the placement of restriction enzyme sites, DNA is cut into variable length pieces and visualized using electrophoresis in an agarose gel. This technique is called restriction fragment length polymorphism (RFLP) and is used to correctly identify criminals of violent crimes from evidence left at the scene. RFLP is equally useful for solving other DNA mysteries.

Some marine mammals are critically endangered and hunting them is illegal based on the Convention on International Trade in Endangered Species of Wild Fauna and Flora (CITES). RFLP is a simple method for identifying retail products of endangered species. Combined with polymerase chain reaction and DNA sequencing, RFLP can identify endangered humpback whale (*Megaptera novaeangliae*) meat found in retail establishments and keep whales from being the target of poachers. In the future, a test will be available for identifying products from extinct animals to discriminate them from products of endangered species.

–Contributing author, Lorelei Crerar,
George Mason University.

Tagging

As mentioned in Chapter 9, tags are often used as a means of tracking individual marine mammals. Typically, this involves attaching a device that either (i) transmits a short-range radio signal for scientists to track; or (ii) records the animal's position and relays that information to scientists via some means while the animal is at the surface. Today satellite tags are most commonly used, although others are also deployed (see Chapter 9). Many modern tags not only can relay information about the position of the marine mammal via satellite (or even by the mobile phone network), but they also contain sensors that record depth of dive, dive duration, water temperature and salinity, light levels, the sound levels animals are exposed to, and even the orientation and acceleration of the tagged animal, among many other things. A variety of tags are used, including instruments that are physically attached to marine mammals, for example by glue (for furred marine mammals such as seals), hooks, or even by bolting them onto the dorsal fin. One common tag attachment method nowadays is the suction cup, which allows tags to stay attached to animals for shorter periods than other methods, but is less invasive. Several cups are used in most tags and the tag unit is generally attached via a pole from a short distance. As a result, animals do not have to be physically captured in order for a tag to be attached, which is an additional consideration. The trade-off here is that a lot of effort can be invested in deploying suction cup tags, success rates can be quite low, and they sometime detach within minutes of deployment. It can also be tricky to precisely determine tag deployment position and orientation, which can have some effects on certain types of tags.

Tags allow a wealth of information to be gathered about the behavior of tagged animals, their environment, and (increasingly) their physiology (e.g., heart rates and stomach temperature). These data could, in many cases, not be obtained in any other way. However, as noted above, many methods of tag attachment are invasive. Tags may also affect hydrodynamic efficiency of marine mammals and can catch on rocks (e.g., in pinnipeds) or floating debris, with undesirable consequences. In smaller animals, attached tags have occasionally been reported to tear through dorsal fins, leaving visible scarring, or altering the shape of the fin itself, which can impact photo-ID studies (either positively, or negatively, depending on the precise situation). It has also been suggested that, due to the electrical nature of tags, it might increase predation, with the tags acting like "beacons" for species that can detect electromagnetic fields (e.g., sharks).

It is hard to determine exactly how "natural" the behavior of tagged animals is compared to untagged animals because the data obtained cannot be collected without the tag, so there is no point of comparison. Furthermore, tagging studies typically only track a small number of individuals at a time and exactly how representative these are of the rest of the population is unknown for the same reason. Thus, scientists must balance the need for the data (perhaps for achieving management and conservation goals) obtained through the use of these devices with the welfare of the animals tagged. They must also attempt to quantify the impacts of the tag(s) on the animals so that the results can be better interpreted in terms of the wider, untagged population.

Crittercams

Crittercams are basically video cameras that are attached to animals. They were developed by Greg Marshall, a National Geographic scientist in the mid-1980's, apparently after observing a remora on a shark. Crittercams can be attached to a wide assortment of animals, including pinnipeds and cetaceans, using a variety of attachment methods. They are often deployed by using an adhesive patch in pinnipeds and generally using a suction cup in cetaceans. Although various versions of the Crittercam exist, the package containing the camera is relatively large for marine species, so adding additional instruments (such as time-depth recorders) is not too difficult and does not increase the size of the unit. As a result, quite a lot of information can be gained about marine mammal behavior underwater using the Crittercam. However, the size also means that they can only be deployed on larger animals at this time. This is particularly true for cetaceans, as the smaller species often twist and turn much more than the larger ones, easily dislodging the suction cups from the more curved body surface (relative to larger animals). Crittercams are thus very good for behavioral research in larger marine mammal species, as they allow for the correlation of various behavioral information with a visual record of what they are actually doing. They are also relatively large and easy to retrieve when they fall off.

However, like all tags, Crittercams have problems. They can be costly and they may affect hydrodynamic efficiency, especially in smaller animals, even more so than other tags. Sometimes the cameras are attached with barbs, which may cause injury or stress an animal. There have been concerns about their effect on predators, as noted above with other tags. Finally, the behavior they are recording may not be completely natural as it may be influenced by the presence of the tag, meaning that the recording must be interpreted carefully, as is the case with any tag.

Acoustic Techniques

Acoustic methods can be categorized in two ways: passive and active. Passive acoustic techniques simply involve listening to and/or recording sounds from marine mammals, while active acoustic techniques also involve some sort of sound production. The most basic passive systems need only a hydrophone (an underwater microphone) that is sensitive at the right frequencies, a computer on which to record the data digitally, and the appropriate software (**Figure A.4**). Active systems also need a transducer (an underwater speaker) to produce sounds as well.

Both techniques can be used in various ways. These include static deployments, where the equipment is fixed to marine structures or on some kind of buoy system, vessel-based deployments, where the hydrophone (and transducer) can be lowered over the side or towed behind the vessel, or even on-animal systems where a miniature hydrophone and electronics are housed in a pressure-safe unit and attached to the animal itself. This last system rarely includes active components.

Static passive systems can be deployed for up to several months (or even a year), depending on the duty cycle (the proportion of time the system is actually recording) and the way it handles the data. Many systems simply record the sounds detected at the hydrophone frequencies digitally on (flash) memory. However, other more complex systems sample the incoming sounds for particular signal (e.g., such as T-PODs do with harbour porpoise clicks) and then record only a few details about the "events" they have detected. This uses much less memory, which is important

FIGURE A.4 Image of a hydrophone being used to record dolphin sound.

Exploring the Depths: Public Sightings Schemes

For some species and/or in certain areas, almost all information we have on marine mammal biology, behavior, and conservation status comes from sightings (or even stranding) data collected opportunistically from members of the general public, such as boaters, fishermen, ferry travellers, or whale-dolphin watching operations. This is a relatively cheap way of getting baseline information on species occurrence and distribution. There are a number of voluntary sightings reporting schemes around the world (**Figure BA.4**). In the UK, for example, the Seawatch Foundation (http://www.seawatchfoundation.org.uk/) collects cetacean sighting information from the general public. The basic information required for any public sighting program is relatively consistent, however, and generally includes:

- the time marine mammals were first observed;
- the date observed;
- the location (preferably the longitude and latitude, but may be simply a mark on a map);
- the species seen and, importantly, how sure the reporter is about their identification—an incorrect report is often much worse than a sighting with a note that the species is unknown;
- the number of animals seen (and again how good the estimate is believed to be);
- the place from where the observer sighted the marine mammals (e.g., a bridge, a commercial ferry, or a kayak); and
- the sea state and visibility (especially if it is raining or foggy).

The time of day can be used, among other things, to calculate tidal states, which very often play an important role in marine mammal movements and behaviors, especially close to shore where most public sightings are made. The position (apart from allowing basic distribution maps to be produced) can be compared with submarine features (such as reefs, gullies, wrecks, etc.) to see if they are important to marine mammal distribution. Sea state and visibility conditions are important as basic factors that affect the ability to sight marine mammals and help to gauge the accuracy of sighting information (numbers and species identification are less likely to be accurate in poor weather conditions). Information such as this (and also, if available, wind speed, visibility distances, etc.) allows scientists to determine the probability of seeing a marine mammal when it surfaces—and therefore how many animals might have been present but not spotted.

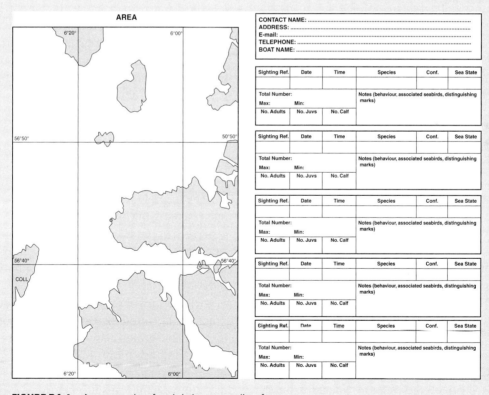

FIGURE BA.4 An example of a sighting recording form.

as recording high frequency acoustic information in its entirety consumes massive amounts of memory, but provides the scientists with much less information upon recovery later. Static systems can include multiple hydrophones to record sounds in different positions to allow for (at least some) localization of the source, or at different frequencies to provide additional information about the various sources, or to help identify the species of animal that made the sounds.

Vessel-based systems can also be very simple. The most basic involve a single hydrophone lowered over the side of the boat that is attached to a recording device, often a laptop, on the boat. While this allows researchers to experience underwater sound, such rudimentary systems are not adequate enough for anything but the most basic research. More often, vessel-based recording systems now involve arrays of two or more hydrophones. Two hydrophones can be used to localize, to some extent, the position of the source of the signal, using the time difference between the arrival of that signal at the two hydrophones. However, this can not discriminate between sources to the left or right of the line of the array. Consequently, more arrays are including three or more hydrophones. Such arrays can then be used to pinpoint an animal making a sound in a social group, for example. Hydrophone arrays may also be towed. On the move, these systems can be used in a similar way to visual line transect surveys for detecting, counting, and, to some extent, tracking marine mammals across wider areas.

Animal-based systems come in various forms. The most common unit is called the D-Tag. Most often deployed on larger cetaceans through use of a long rod to which it is attached, the D-Tag is lowered firmly onto the back of the target animal. At this point it sticks to the animal using time-release suction cups and detaches from the rod. The animal then swims away with the D-Tag attached. In practice, this is not an easy task. However, the D-Tag provides a wealth of information on not only the acoustic behavior of the target animal (in addition to other passing marine mammals), but also the target's diving behavior. We have learned much about the diving and foraging behavior of various animals thanks to the D-Tag, despite the difficulties in deployment.

The other major animal-based acoustic recording system is the A-Tag. Attached on animals in hand, it is less flexible than the D-Tag. Similarly, the housing is less streamlined, at least at present. However, it does record similar data to the D-Tag when combined with other small units. The A-Tag can also be deployed using suction cups, but is also often attached to the fin of an animal using pins. This is more invasive than suction cup deployments, but the tag can remain on the animal for longer periods than the D-Tag to obtain longer data sets. The A-Tag can be combined with a satellite tag to maximize the data collection with minimal impact on the animal, as only one set of attachment pins is needed for the pair of tags. As mentioned in Exploring the Depths: Welfare, Animal Research Ethics, and Invasive Studies that follows, a balance between research impacts on an animal and the information acquired needs to be struck. In this case, much of the data obtained by these tags has management applications and cannot be obtained in any other way at present. However, efforts should always be made to improve these methods and reduce their impacts on the target animals.

Despite all the data that can be obtained through passive acoustic systems, they all suffer from one major failing: the animals can only be detected if they produce sounds. Active acoustic systems can be used to get around this problem. These systems are usually boat-based, although they may also be statically deployed. They come in two forms for achieving specific tasks.

The first is essentially an echo-sonar for finding and tracking marine mammals. Often called whale-finding sonar, these systems have management applications, but can also be used to gather scientific information. Sound produced by a transducer travels through the water, bounces off a target, and returns to the hydrophone in the system. Just like echolocation, information about not only the distance to the target, but also to its likely identity can be gleaned from the response. Receiving hydrophones may also be able to record sounds produced by the target, for additional information. This could be used, again in line-transect style surveys, to gain information about the distribution and abundance of species. Unlike passive surveys, active acoustic surveys can be used to detect marine mammals that produce little sound or even none at all. However, the system does introduce additional sound into the environment, with all the possible consequences that this might have.

The second form of active acoustics may, or may not, involve a receiving hydrophone. These are generally known as playbacks. In these experiments, the scientists play sounds to target animals through the transducer. Scientists then record (visually, acoustically, or both) the responses of the animals to these sounds. The sounds played may be the social sounds of related animals or another group of the same species, sounds of a hunting predator, or even sounds produced by human activities. In any case, however, scientists seek to learn more about what these sounds mean to the animals by assessing their responses. For example, they may approach the transducer curiously, which might be an indication of some sort of recognition, or they might react with hostility, which might be an indication of fear or aggression. In any case, such experiments need to be carefully designed, with control periods (when the transducer is in the water, but not playing sounds) to gauge if the reaction is simply to the foreign object, or actually to the sound. Other considerations are if the animal is responding to that specific sound or that type of sound. By way of example, consider if you wanted to find out if a group of dolphins avoided one specific boat or all boats. In this case, you might need an extended experimental setup with various vessel sounds that could be randomly selected for playing at any time. However, scientists must be careful to avoid exposing any one animal (or group of animals) to the same (or similar) sounds too often as their responses may lessen over time as they become habituated to the sound. Playback experiments can be quite complicated as a consequence of all these difficulties.

Exploring the Depths: Welfare, Animal Research Ethics, and Invasive Studies

Without research, there would be no advancement of our knowledge of the biology and behavioral ecology of marine mammals. Some research, such as monitoring movements of migratory gray whales from the shore with theodolites, may be benign. However, other studies involve invasive procedures and may even result in the death of the animal. Invasive research in particular must address difficult questions. How high a cost must individual animals—or a species—pay to advance our knowledge? Moreover, what responsibility do researchers bear in ensuring that the fewest number of animals are used, the least invasive techniques are chosen, and that effects are properly monitored? It should also go without saying that studies should result in clear benefit to the species.

Endangered Steller sea lions in western Alaska have suffered deep population declines for reasons that are not well understood, although resource limitations have been postulated. For more than a decade, a coalition of government and private institutions has spent tens of millions of dollars to study the decline. Some research has used highly invasive techniques such as gastric lavage, enemas, bio-electrical impedance studies, surgical implantation of life history transmitters into the abdominal cavity that did not transmit information until the death of the animal, and hot branding pups as young as 5 days of age without benefit of anesthesia. Researchers originally requested to undertake abdominal surgery in the field and then release the endangered sea lions with only 10 to 48 hours of post-surgical monitoring. The research program has received criticism by both government and non-government expert panels, including charges that monitoring was deficient, that it lacked replicate studies, and that there was poor impact monitoring. Responding to high levels of Steller sea lion mortality and the research and animal welfare concerns, The Humane Society of the United States in 2006 successfully litigated, resulting in research permits effectively being revoked and a court-mandated review of the environmental impacts of the program. However, this is not the only program that has been controversial.

Satellite telemetry is a common form of distance monitoring of marine mammals. When appropriate technology and diligent monitoring are used, the results can provide information on distribution and behavior that arguably far outweighs the discomfort to individual animals. But it can also go horribly wrong. In 2006, U.S. government researchers captured and bolted non-standard telemetry tags to the dorsal fins of four bottlenose dolphins in North Carolina. Within weeks, members of the public found two of the dolphins stranded and dead. The third had to be recaptured and the tag removed. The fourth was never found. Proper monitoring of the tag transmissions should have indicated that the dolphins had died; instead members of the public found the dead animals. Tagging that risks the lives of endangered cetaceans or that causes visible damage to animals can prompt calls to cease this activity even in cases where the research may have had justifiable goals.

The fact that marine mammals die as a result of research aimed at understanding their lives is no less a cost to the species than dying due to fishery-related entanglement. The lofty goal of a research project should not exempt it from close scrutiny as to the purpose of the research, the appropriateness of the research methodology and a proper understanding of the potential adverse impacts to individuals or species. Guidance to researcher as to legal mandates and ethical responsibilities should be a key part of constructing appropriate research programs.

The European Cetacean Society established a science advisory committee to address concerns and develop guidelines related to ethics and welfare issues of marine mammal studies in Europe. In 2010, the Society for Marine Mammalogy formed an Ethics Advisory Committee to help advise members. These are important steps toward ensuring a responsible approach to studying the animals whose very mystery and magnetism is what drew most researchers to the field.

–Contributing author, Sharon Young, Humane Society of the United States.

SELECTED REFERENCES AND FURTHER READING

Baker, C., Steel, D., Choi, Y., Lee, H., Kim, K., Ma, Y., Hambleton, C., Psihoyos, L., & Brownell, R. (2010). Genetic evidence of illegal trade in protected whales links Japan with the U.S. and South Korea. *Biology Letters* 6: 647–650.

Berman, M. (2008). Endangered species, threatened fisheries: science to the rescue! Evaluating the congressionally designated Steller sea lion research program. *Marine Policy* 32: 580–591.

Calambokidis, J. (2002). Underwater behavior of blue whales using a suction-cup attached CRITTERCAM. Final Technical Report to Office of Naval Research for Grant Number: N00014-00-1-0942. Cascadia Research, Olympia, WA.

Gordon, J., & Tyack, P. Acoustic techniques for studying cetaceans. In: *Marine Mammals: Biology and Conservation* (Ed. P.G.H. Evans & J.A. Raga), pp. 293–324. Kluwer Academic, New York.

Hammond, P.S. (2002). Assessment of marine mammal population size and status. In: *Marine Mammals: Biology and Conservation* (Ed. P.G.H. Evans & J.A. Raga), pp. 269–291. Kluwer Academic, New York.

Hartl, D.L. & Clark, A.G. (2006). *Principles of Population Genetics*. 4th ed. Sinauer Associates, Sunderland, MA.

Irwin, D.M., Kocher, T.D., & Wilson, A.C. (1991). Evolution of the cytochrome *b* gene in mammals. *Journal of Molecular Evolution* 32: 128–144.

Kasamatsu, H., Robberson, D.L., & Vinograd, J. (1971). A novel closed-circular mitochondrial DNA with properties of a replicating intermediate. *Proceedings of the National Academy of Science USA* 68: 2252–2257.

Krützen, M., Barré, L., Möller, L., Haithaus, M., Simms, C., & Sherwin, W. (2002). A biopsy system for small cetaceans: darting success and wound healing in *Tursiops spp*. *Marine Mammal Science* 18: 863–878.

Martin, P. & Bateson, P. (2008). *Measuring Behaviour: an Introductory Guide*. 3rd Ed. Cambridge University Press, Cambridge, UK.

Marshall, G.J. (1998). CRITTERCAM: an animal-borne imaging and data logging system. *Marine Technology Society Journal* 32:11–17.

Mazzoil, M., McCulloch, S.D., Defran, R. H., & Murdoch, M.E. (2004). Use of digital photography and analysis of dorsal fins for photo-identification of bottlenose dolphins. *Aquatic Mammals* 30: 209–219.

McPherson, M. J., & Moller, S.G. (2006). *PCR (The Basics)*. 2nd ed. Taylor & Francis, Oxford, UK.

Rose, N.A., Janiger, D., Parsons E.C.M., & Stachowitsch, M. (2011). Shifting baselines in scientific publications: a case study using cetacean research. *Marine Policy* 35: 477–482.

Scott, M.D., Irvine, A.B., Wells, R.S., & Mate, B.R. (1990). Tagging and marking studies on small cetaceans. In: *The Bottlenose Dolphin* (Ed. S. Leatherwood & R. Reeves), pp. 489–514. Academic Press, San Diego.

Stuart, J.A. (2009). *Mitochondrial DNA: Methods and Protocols*. 2nd ed. Humana Press, New York.

Würsig, B., Cipriano, F., & Würsig, M. (1991). Dolphin movement patterns: information from radio and theodolite tracking studies. In: *Dolphin Societies: Discoveries and Puzzles* (Ed. K. Pryor & K.S. Norris), pp. 79–111. University of California Press, Berkeley.

Index

Note: Page numbers followed by *b* refer to boxed material; those followed by *f* indicate figures; those followed by *n* indicate footnotes; those followed by *t* indicate tables.

A

aboriginal subsistence hunting, 204–205
  of pinnipeds, 207
  of polar bears, 84–85
  of sea otters, 95
aboriginal whaling, 223, 225*t*, 225–227, 226*f*, 227*b*, 227*f*, 231
abyssalpelagic zone, 44*f*
abyssal plains, 44*b*
abyssal zone, 44*f*
ACCOBAMS (Agreement on the Conservation of Cetaceans of the Black Sea, Mediterranean Sea and Contiguous Atlantic Area), 276, 276*b*
acoustic deterrent devices, 236, 236*f*
acoustic research techniques, 318, 318*f*, 320–321
active acoustic techniques, 318
adaptations to marine environments, 43–57, 81, 82*f*, 92–93, 93*f*, 101–102
  diving and, 52*f*, 52–53, 53*f*
  light and, 56*b*, 57*f*
  ocean zones and, 43*b*–44*b*, 44*f*
  osmoregulation and, 55
  pressure effects and, 53–55, 54*f*
  sensory, 55–57, 56*f*, 57*f*
  sleep and, 58*b*
  spermaceti organ and, 47*b*–48*b*, 48*f*
  surface area to volume ratio and, 45, 49*b*, 49*f*
  swimming and, 45*f*, 45–48, 46*f*

  thermoregulation and, 48–52, 49*f*–51*f*
adipose tissue. *See also* blubber
  of polar bears, 81
  thermoregulation and, 50
Aelfrics Colloquy, 208*b*
aerobic respiration, 53
Aetiocetus weltoni, 22–23
African clawless otter (*Aonyx capensis*), 17, 28
Agent Orange, 240
aggression
  in beaked whales, 145*b*–146*b*
  in bottlenose dolphins, 190, 190*t*
  in elephant seals, 122*f*
Agreement on the Conservation and Management of the Alaska-Chukotka Polar Bear Population, 89
Agreement on the Conservation of Cetaceans of the Black Sea, Mediterranean Sea and Contiguous Atlantic Area (ACCOBAMS), 276, 276*b*
Agreement on the Conservation of Polar Bears and their Habitat, 88–89
Agreement on the Conservation of Seals in the Wadden Sea, 276, 285
Agreement on the Conservation of Small Cetaceans of the Baltic and North Sea (ASCOBANS), 286*b*
Ailuropoda melanoleuca (giant panda), 27

alarm calls, 73–74
Alfred of Wessex, King, 207
algal blooms, marine mammals and, 241
Allen, Joel, 49
Allen's rule, 49, 83
Amazonian manatee (*Trichechus inunguis*), 33, 101, 102*f*, 103, 106, 272, 277*t*. *See also* manatees
Amazon River dolphin (*Inia geoffrensis*)
  description of, 149
  legends and, 205
  subspecies of, 37, 149
  thermoregulation in, 51
Amazon River dolphin (*Inia geoffrensis boliviensis*), 149
Amazon River dolphin (*Inia geoffrensis geoffrensis*), 37, 149
Amazon River dolphin (*Inia geoffrensis humboldtiana*), 149
ambergris, 141, 209
Ambler, Jennifer, 298*b*
Ambulocetidae, 20
Ambulocetus natans, 21–22
amphipods, in baleen whale diet, 134, 135*f*
anaerobic respiration, 53
Andrew's beaked whale (*Mesoplodon bowdoini*), 37
animal-based acoustic techniques, 320
Animal Welfare Institute, 212
Antarctic minke whale (*Balaenoptera bonaerensis*), 37, 130, 132
  Japanese hunting of, 221
  status of, 138*t*, 275*t*

Antarctic seal, behavioral adaptation to cold in, 51
Antarctic whaling, 216, 216f
antelopes, 20
Antillean manatee (*Trichechus manatus manatus*), 33, 102, 106
anti-whaling movement, 138b, 212, 212f, 219, 228b
Aonyx spp., 17, 28, 28f
Aonyx capensis (African clawless otter), 17, 28
Aonyx cinerea (Oriental small-clawed otter), 17, 28
aphotic zone, 44, 44b, 44f
appendages
    countercurrent blood system in, 51, 51f
    thermoregulation and, 49
aquatic adaptations
    of dugong, 101
    of manatee, 101–102
    of polar bear, 81, 82f
    of sea otter, 92–93, 93f
archaeocetes (Archaeoceti), 20, 21, 21b, 21f, 35
Architeuthis (giant squid), 142
archosaurs, 3
arctic adaptations, of polar bears, 82, 83f
Arctic cod (*Arctogadus glacialis*), as polar bear prey, 83
Arctic fox (*Vulpes lagopus*), 3, 6–7, 7f, 9
Arctocephalinae, 29
Arctocephalus australis (South American fur seal), 29t
Arctocephalus forsteri (New Zealand fur seal), 29t
Arctocephalus galapagoensis. See Galápagos fur seal
Arctocephalus philippi (Juan Fernández fur seal), 29t, 115
Arctocephalus pusillus doriferus (Australian fur seal), 29t, 296b
Arctocephalus pusillus pusillus (South African fur seal), 29t, 295b
Arctocephalus townsendi (Guadalupe fur seal), 29t, 115, 116b–117b, 116f, 117f, 277t
Arctocephalus tropicalis (subantarctic fur seal), 29t
Arctogadus glacialis (Arctic cod), as polar bear prey, 83
Arctoidea, 27, 29
ARGOS satellite system, 121b

Aristotle, 203
Arnoux's beaked whale (*Berardius arnuxii*), 37, 144, 147–148, 275t
artiodactyls (Artiodactyla), 3, 15, 20
ASCOBANS (Agreement on the Conservation of Small Cetaceans of the Baltic and North Sea), 286b
A-Tag, 320
Atlantic gray whale, 130
    hunting of, 208
Atlantic humpback dolphin (*Sousa teuszii*), 38, 177, 179, 180f, 181, 272
Atlantic spotted dolphin (*Stenella frontalis*), 38, 182, 182f, 190
Atlantic walrus (*Odobenus rosmarus rosmarus*), 30
Atlantic white-sided dolphin (*Lagenorhynchus acutus*), 38, 177, 178f
Australian Environment Protection and Biodiversity Conservation Act, 260
Australian fur seal (*Arctocephalus pusillus doriferus*), 29t, 296b
Australian legends, 205
Australian sea lion (*Neophoca cinerea*), 29t, 126f, 259f
    abundance and status of, 272
    threats to, 259–260
    tourism and, 295b
Australian snubfin dolphin. See snubfin dolphin

B

badgers, 27
Bahamonde's beaked whale (*Mesoplodon bahamondi*), 37, 39b
baiji (*Lipotes vexillifer*), 37, 150f, 150–153, 151f, 275t, 277t
Baikal seal (*Pusa sibirica*), 31t, 118b, 118f, 119f
Baird's beaked whale (*Berardius bairdii*), 37, 144, 147–148
    hunting of, 218, 233
    status of, 275t
Balaena mysticetus. See bowhead whale
Balaenidae, 35, 35t, 36. See bowhead whale; right whale(s); *specific right whales*
Balaenoptera acutorostrata (northern minke whale), 36, 132, 138b

    call of, 71
    hunting of, 208b, 221, 222
    status of, 138t, 275t
Balaenoptera bonaerensis (Antarctic minke whale), 37, 130, 132
    Japanese hunting of, 221
    status of, 138t, 275t
Balaenoptera borealis (sei whale), 36, 37, 130, 134
    abundance and status of, 138t, 272, 275t, 277t
    ban on hunting, 220
    blue whale unit and, 218
    hunting of, 216–217, 221, 222
Balaenoptera brydei (Bryde's whale), 36, 37, 130, 131f, 132, 137t
    Japanese hunting of, 221
    status of, 138t, 275t
Balaenoptera edeni (dwarf Bryde's whale or Eden's whale), 36, 37, 138t
Balaenoptera musculus. (blue whale), 3, 37, 130, 132
    abundance and status of, 3, 136t, 138b, 138t, 272, 275t, 277t
    ban on hunting, 220
    hunting of, 208b, 216
    size of, 129
    sound production by, 66b
Balaenoptera omurai (Omura's whale), 36, 37, 138t
Balaenoptera physalus (fin whale), 37, 130, 132, 134, 137t
    abundance and status of, 138t, 272, 275t, 277t
    blue whale unit and, 218
    hunting of, 216, 221, 222, 226, 227b
    quotas for, 225t
Balaenopteridae. See rorquals
Balaenopterinae, 36–37
baleen whales, 21, 35f, 35–37, 36f, 129f, 129–138, 130f. See also specific baleen whales
    abundance and status of, 136b, 136–138, 136t–138t
    description of, 129
    distribution of, 130, 131f
    feeding behavior and ecology of, 132–135, 132f–135f
    killer whale attacks on, 167, 169
    migration of, 130b, 131f
    reproduction by, 135–136
Baltic ringed seal (*Pusa hispida botnica*), 118b, 119b, 119f
bans, on whaling, 219–220
Barnum, P. T., 154

basilaurids (Basilosauridae), 20, 22
Basilosaurus, 22, 22f, 24b
Basques, whaling by, 207, 208–209
Bass, Claire, 228b
Bastian, Jarvis, 64b
bat(s), inhabiting marine waters, 3
bathyal zone, 44f
bathypelagic zone, 44f
beachmaster seals, 121, 122f
beaked whales (Ziphiidae), 35, 37, 39b, 40f, 144f, 144–148, 145f. *See also specific beaked whales*
   aggression in, 145b–146b
   distribution of, 146, 147f
   diving by, 55, 146–147
   high pressure effects on, 54–55
   status of, 147–148
   strandings of, 75b, 172b
beakless dolphins (*Cephalorhynchus* spp.), 39b, 167
bear(s), 27. *See also specific types of bears*
bearded seal (*Erignathus barbatus*), 31t, 83
Bechshøft, Thea Østergaard, 86b
behavioral adaptations, of polar bears, 83
behavioral observation methods, 315b
Bejder, Lars, 294
beluga whale (*Delphinapterus leucas*), 37, 153f, 153–154
   abundance and status of, 153–154, 278
   Cook Inlet, 154, 154b
   description of, 153
   distribution of, 153
   head shape of, 46, 46f
   hunting of, 231
   as polar bear prey, 83
   pollution and, 241
   sound frequencies used by, 63t
benthic zone, 43b, 44f
Berardius arnuxii (Arnoux's beaked whale), 37, 144, 147–148, 275t
Berardius bairdii. *See* Baird's beaked whale
Bergmann, Christian, 49
Bergmann's rule, 49
Bible, marine mammals in, 201–202, 202f
bioacoustics, 64b–66b, 66f
Biodiversity Action Plan (U.K.), 282b
biopsy darting, 315–316

black-chinned dolphin (*Lagenorhynchus australis*), 38
black dolphin (*Cephalorhynchus eutropia*), 38, 175
Black Sea bottlenose dolphin (*Tursiops truncatus ponticus*), 191, 233b
Blainville's beaked whale (*Mesoplodon densirostris*), 37
blood system
   in blubber, 51
   countercurrent, in appendages, 51, 51f
blowholes
   in sperm whales, 65b, 66f
   swimming and, 45
blubber, 45–46, 47b, 47f
   blood vessels in, 51
   of polar bears, pollutants in, 86
   thermoregulation and, 50
blue whale (*Balaenoptera musculus*), 3, 37, 130, 132
   abundance and status of, 3, 136t, 138b, 138t, 272, 275t, 277t
   ban on hunting, 220
   hunting of, 208b, 216
   size of, 129
   sound production by, 66b
blue whale unit, 218
blunt-headed dolphins. *See* Irrawaddy dolphin; Risso's dolphin; snubfin dolphin
boats
   collisions with. *See* watercraft collisions
   noise from shipping and, 75
   speed, habitat loss and, 246f
   whaling, life on, 210b–211b, 211f
body shape, swimming and, 45
bomb lance, 215
bones, swimming and, 47–48
Bonn Convention, 275–276
bonobo (*Pan paniscus*), 191b
boto. *See* Amazon River dolphin
bottlenose dolphins (*Tursiops* spp.), 167, 189f, 189–192. *See also* common bottlenose dolphin; Indo-Pacific bottlenose dolphin
   abundance and status of, 191
   aggression in, 190, 190t
   dolphinaria and, 298b
   "Flipper" TV series and, 212
   hunting of, 208b, 233, 234b
   life span of, 191
   reproduction by, 190–191
   research on, 321b

bottlenose whale (*Hyperoodon* spp.), 144, 144f
   hunting of, 208b
   status of, 275t
bottling, of pinnipeds, 58b
Bovidae, 20
bowhead whale (*Balaena mysticetus*), 36, 130, 131f, 134, 216
   abundance and status of, 137t, 138b, 272, 275t, 277t
   ban on hunting, 219
   blubber layer of, 47b
   climate change and, 244
   hunting of, 208–209, 218, 227b
   quotas for, 225t
brains, mammalian, 1
branding, 120b, 310
British. *See* United Kingdom
brown bear (*Ursus arctos*), 17, 28, 81
brown fur seal, 29t
Brucella spp., cetacean contamination and, 224b, 224f
Bryde's whale (*Balaenoptera brydei*), 36, 37, 130, 131f, 132, 137t
   Japanese hunting of, 221
   status of, 138t, 275t
bubble netting, by humpback whales, 133f, 133–134, 134f
building, noise from, 75b–76b
bulldog bat (*Noctilio leporinus*), 3, 4, 5, 6f
bull shark (*Carcharhinus leucas*), 179
bunyip, 205
buoyancy, adaptations for, 47
Burmeister's porpoise (*Phocoena spinipinnis*), 38, 159, 159f, 232
burrunan dolphin (*Tursiops australis*), 38n
butylin, 241b
bycatch
   of beaked whales, 147
   of cetaceans, 235–238, 236f
   of common dolphins, 186b,–187b, 187f
   of humpback whales, 199f

C

cadmium, cetaceans and, 239–240
Caldwell, David, 74
Caldwell, Melba, 74
California sea lion (*Zalophus californianus*), 29t, 30b, 30f
Callorhinus ursinus (northern fur seal), 29t, 115, 126f, 261f, 262f

326 INDEX

Calpurnius, on polar bears, 89*b*
camel(s), 20
Camelidae, 20
Caniformia, 27, 28, 29
Canis lupus (gray wolf), 7, 8*b*, 8*f*
cannibalism, among polar bears, 83
Caperea marginata (pygmy right whale), 35, 36, 129, 134, 138*t*, 275*t*
captive breeding, 299*b*
Carcharhinus leucas (bull shark), 179
Carcharodon carcharias (great white shark), 15, 259
Caribbean monk seal *(Monachus tropicalis)*, 32*t*, 33, 125, 126*b*, 126*t*, 153
carnivores (Carnivora), 3, 15, 18. *See also specific animals*
    evolution of, 15, 17–18
    marine, 3, 27–33
    taxonomy of, 27
Carson, Rachel, 240
Caspian seal *(Pusa caspica)*, 31*t*, 118*b*, 119*f*, 126*f*, 262*f*
    abundance and status of, 272
    threats to, 262–263, 263*f*, 265
Center for Biological Diversity, polar bear status and, 89
Central American spinner dolphin *(Stenella longirostris centroamericana)*, 38
Cephalorhynchus spp. (beakless dolphins), 39*b*, 167
Cephalorhynchus commersonii (Commerson's dolphin), 38, 175, 175*f*
Cephalorhynchus eutropia (black dolphin), 38, 175
Cephalorhynchus eutropia (Chilean dolphin), 38, 175
Cephalorhynchus heavisidii (Heaviside's dolphin), 38, 175, 175*f*
Cephalorhynchus hectori hectori (Hector's dolphin), 38, 175, 176*b*–177*b*, 176*f*, 272
Cephalorhynchus hectori maui (Maui's dolphin), 38, 176*b*
Cervidae, 20
Cervus elaphus (red deer), 71–72
cetacean(s) (Cetacea), 3, 4, 21, 35–38. *See also* dolphin(s); porpoises; whale(s); *specific animals*
    air exchange by, 53
    blubber layer of, 47*b*, 47*f*
    breeding of, 50

    climate change and, 243–245
    contamination of, 224*b*, 224*f*
    culls of, 233*b*
    cultural transmission in, 193*b*, 194*f*
    direct takes of, 231–233, 232*f*, 235
    diving depth of, 55
    ears of, 67*b*–68*b*, 68*b*, 68*f*
    echolocation by, 69
    evolution of, 20*b*, 20*f*, 21–23, 22*f*
    eye of, 55, 56*f*
    fisheries and, 235–236, 236*f*
    habitat degradation and, 245, 246*f*
    hearing of, 57
    high pressure effects on, 54–55
    live takes of, 233*b*–234*b*, 234*f*, 235*f*
    mother-calf cohesion in, 73
    noise pollution and, 75*b*
    osmoregulation in, 55
    oxygen storage in, 52, 52*f*, 53
    pollution and, 239–243, 241*b*, 243*b*
    pressure adaptations of, 54
    prey stunning in, 74
    as resources, 201
    rock etchings of, 201, 201*f*
    ship strikes and, 238, 239*f*
    small, direct takes of, 231–232, 232*f*, 235
    smell sense of, 56
    sound production by, 61, 64*b*–65*b*
    sound spectra for, 62, 63*t*
    stress and, 237*b*
    swimming adaptations of, 45, 45*f*, 46
    taste sense of, 57, 57*f*
    thermoregulation in, 51–52
    touch sense of, 56
    tuna fishing and, 236–238, 238*b*
Cetacean Protection Act (U.K.), 283*b*
Cetancodonta, 20
Chagos Marine Reserve, 284*b*
charcharinid sharks, 15
Charles II, King of England, 209
chart datum, 314
Chase, Owen, 211*b*
Chilean dolphin *(Cephalorhynchus eutropia)*, 38, 175
Chilean marine otter, 28, 28*f*
chimpanzee *(Pan troglodytes)*, 191*b*
Chinese finless porpoise *(Neophocaena phocaenoides sunameri)*, 38
Chiroptera, 3
Chlorocebus pygerythrus (vervet monkey), 73
chromium, cetaceans and, 239
Chukchi people, whaling by, 227*b*
CITES. *See* Convention on International Trade in Endangered Species of Wild Fauna and Flora
click production, 65*b*, 65*f*, 66*f*
climate change
    cetaceans and, 243–245
    in Eocene, Miocene, and Oligocene, 23*b*, 23*f*
    Lagenorhynchus species and, 179*b*
    polar bears and, 86*f*, 86–87
    in Quaternary, 33
Clupea (herring), 169*b*
Clupeonella spp., 263
clymene dolphin *(Stenella clymene)*, 38, 182, 183*f*
CMS (Convention on Migratory Species of Wild Animals), 275–276
coastal construction, habitat loss and, 246*f*
cold-blooded organisms, 48
colossal squid *(Mesonychoteuthis hamiltoni)*, 142
Columbus, Christopher, 207
Commerson's dolphin *(Cephalorhynchus commersonii)*, 38, 175, 175*f*
common bottlenose dolphin *(Tursiops truncatus)*, 39*b*, 39*t*, 189*f*, 189–190
    air exchange in, 53
    eye of, 56*f*
    individual recognition in, 74
    intelligence of, 10
    mother-calf cohesion in, 73, 74*f*
    tool use by, 93*b*–94*b*
common dolphins *(Delphinus* spp.), 63*t*, 185*f*, 185–187, 186*f*, 189
common porpoise. *See* harbor porpoise
common seal *(Phoca vitulina)*, 31*t*
communication, 71–74
    danger avoidance and, 73–74
    group cohesion and, 73
    humpback whale songs and, 71, 72, 72*b*–73*b*, 72*f*, 135, 193*b*
    individual recognition and, 74
    intersexual selection and, 71*b*, 71*f*, 71–72
    intrasexual selection and, 71
    mother-calf cohesion and, 73, 74*f*
competition, intrasexual, 71
conservation
    of bottlenose dolphin, 191
    of dugong, 108*b*–109*b*, 109*f*
    of Florida manatee, 106*b*–108*b*, 107*f*, 108*f*

initiatives for, 271–276
IWC Conservation Committee and, 224b
of *Lagenorhynchus* species, 179b
legislation and, 276–285
of polar bear, 88–89
science-based management and, 285–286
whaling and. See International Convention on the Regulation of Whaling; whaling
Conservation of Seals Act, 282b
Conservation of Seals Order (Scotland), 282b
construction, habitat loss and, 246f
continental shelf, 43b–44b
continuous sampling, for behavioral observation, 315b
Convention for the Regulation of Whaling, 218
Convention on Biodiversity, 282b
Convention on Migratory Species of Wild Animals, 275–276
Convention on International Trade in Endangered Species of Wild Fauna and Flora (CITES), 138, 222, 274–275, 317b
  Amazon River dolphin and, 149–150
  Appendix I of, 274, 275t
  Appendix II of, 274–275
  Appendix III of, 275
  bottlenose dolphin and, 191
  dwarf sperm whale and, 144b
  harbor porpoise and, 158b
  humpback dolphin and, 181
  killer whale and, 170
  La Plata dolphin and, 152b
  manatee and, 106
  Mediterranean monk seal and, 254
  pygmy sperm whale and, 144b
  right whale dolphin and, 174
  sperm whale and, 144
  Yangtze River dolphin and, 153
copepods, 35, 36f, 134, 135f, 208
Cornick, Leslie, 154f
cosmopolitan spinner dolphin (*Stenella longirostris longirostris*), 38
costero (*Sotalia guianensis*), 38, 39t, 188b, 188f
countercurrent blood system, in appendages, 51, 51f
cows, 20
crabeater seal (*Lobodon carcinophaga*), 32t, 34f, 121, 125

Craseonycteris thonglongyai (Kitti's hog-nosed bat), 3
Crerar, Lorelei, 9f, 111b, 317b
Cretaceous period, mammalian evolution during, 3
critical habitat, 277
crittercams, 318
crocodilians, 3
cultural transmission, in cetaceans, 193b, 194f
Cuvier's beaked whale (*Ziphius cavirostris*), 37, 144, 146, 147
cylindrical spreading loss, 64
cynodonts, 3, 3f

Cystophora cristata. See hooded seal

D

Dalebout, Merel, 39b
Dall's porpoise (*Phocoenoides dalli*), 38
  description of, 156b
  hunting of, 156b–158b, 157f, 233
  killer whale predation on, 97b, 169
dams, habitat loss and, 246f
danger avoidance, 73–74
Danish unicorn throne, 206b, 206f
darting, biopsy, 315–317
dBs (decibels), 62–63, 63b
DDE (dichlorodiphenyldichloroethylene), cetaceans and, 240
DDT (dichlorodiphenyltrichloroethane), cetaceans and, 240
De Bestiis Marinis (Steller), 9b
decibels (dBs), 62–63, 63b
decompression problems, 53–54, 54f
Deepwater Horizon oil spill, 243, 243f
deer, 20
Delphinapterus leucas. See beluga whale
delphinids (Delphinidae), 35, 37–38, 65b, 167–191. See also specific animals
Delphinus spp. (common dolphins), 63t, 167, 181–189, 185f, 185–187, 186f
Delphinus capensis (long-beaked common dolphin), 38, 39b, 185, 185f, 186f
Delphinus delphis (short-beaked common dolphin), 38, 39b, 185, 185f, 186b, 186f
Delphinus tropicalis, 185, 189
dense beaked whale (*Mesoplodon densirostris*), 37
The Descent of Man and Selection in Relation to Sex (Darwin), 71b

Desmostylia, 18, 19f
DFO (Fisheries and Oceans Canada), 267b
Dicentrarchus labrax (sea bass), 186b
dichlorodiphenyldichloroethylene (DDE), cetaceans and, 240
dichlorodiphenyltrichloroethane (DDT), cetaceans and, 240
Dicrostonyx richardsoni (lemming), 7
dinosaurs, 3
dioxins, cetaceans and, 240
disease
  cetaceans and, 243
  New Zealand sea lion and, 261
disturbances
  manatees and, 105–106, 106b–108b, 107f, 108f
  polar bears and, 87, 87f, 88f
  sperm wales and, 142, 144
diving
  adaptations for, 52f, 52–53, 53f
  by beaked whales, 146–147
  by sea otters, 92–93, 93f
dolphin(s), 37–38. See also specific types of dolphins
  individual recognition by, 74
  intelligence of, 192b–193b, 193f
  new species of, 39b, 39t
  sleep of, 58b
  solitary sociable dolphin problem and, 301
  sounds produced by, 74
dolphinaria, 298b–299b, 299f
dolphin meat, contamination of, 224b
dominance hierarchies, 71
dorsal fins, swimming and, 47
Dorudon, 22, 22f, 24b
Draheim, Megan, 299f
dredging, habitat loss and, 246f
drift nets, 235–236
drive fisheries, 233, 234b, 235, 235f
D-Tag, 320
dugong(s) (Dugongidae), 18, 19, 33. See also dugong (*Dugong dugong*); Steller's sea cow
dugong (*Dugong dugong*), 18, 19, 33
  abundance and status of, 106, 272, 277t
  aquatic adaptations of, 101
  in Australian legend, 205
  conservation of, 108b–109b, 109f
  distribution of, 104, 104f
  feeding behavior and ecology of, 105
  reproduction by, 105
Dugonginae, 33

dusky dolphin *(Lagenorhynchus obscurus)*, 38, 177, 178*f*, 191*b*, 232
Dutch whaling, 209
dwarf Bryde's whale *(Balaenoptera edeni)*, 36, 37, 138*t*
dwarf minke whale, 36
dwarf sperm whale *(Kogia sima)*, 37, 39*b*, 143*b*–144*b*, 143*f*

E

earthquakes, noise from, 75
East Asian finless porpoise *(Neophocaena asiaorientalis sunameri)*, 161*b*
Eastern North Pacific harbor porpoise *(Phocoena phocoena vomerina)*, 38
Eastern Pacific coastal spotted dolphin *(Stenella attenuata graffmani)*, 38, 278
Eastern Pacific spinner dolphin *(Stenella longirostris orientalis)*, 38, 278
echidna (spiny anteater), 1
echolocation, 65*b*, 69*f*, 69–70
echo-sonar, for research, 320
ecotourism, 290*b*–291*b*
ectothermic organisms, 48
Eden's whale *(Balaenoptera edeni)*, 36, 37, 138*t*
Edward II, King of England, 283*b*
Edward III, King of England, 208*b*
EEZs (Exclusive Economic Zones), 274*b*
electronic tags, 120, 120*b*
elephants, 3, 4*f*
    evolution of, 18, 18*f*
    skull of, 18*f*
elephant seal
    molting in, 125*b*, 125*f*
    reproduction by, 121–123, 122*f*
    roar of, 71
    sexual selection in, 71*b*, 71*f*
    thermoregulation in, 52
Elizabeth I, Queen of England, 206*b*, 208*b*
El Niño, Galápagos sea lions and, 258
embolisms, in beaked whales, 75*b*
Enaliarctidae, 18, 29
Enaliarctos, 18
endotherms, 49
endothermy, mammalian, 1
Enhydra lutris. See sea otter
Enhydra lutris kenyoni, 92
Enhydra lutris lutris, 92

Enhydra lutris nereis, 92, 275
Enhydra macrodonta, 17
Enhydriodon, 17
Enhydritherium spp., 17
Enhydritherium lluecai (sea otter, prehistoric), 17
Enhydritherium terraenovae (sea otter, prehistoric), 17
environmental sound, underwater, 75
Environment Protection and Biodiversity Conservation Act (Australia), 286*b*
Eocene period, 15, 16*f*
    cetacean evolution in, 21–22
    climate changes in, 23*b*, 23*f*
    mammalian evolution during, 3
    Pakistan in, 20*b*, 20*f*
    sirenian evolution in, 18–19, 19*f*
Eosiren, 19, 19*f*
epipelagic zone, 44, 44*b*, 44*f*
Erignathus barbatus (bearded seal), 31*t*, 83
ESA. *See* U.S. Endangered Species Act
Eschricht, Daniel, 35
Eschrichtiidae, 35, 36
Eschrichtius robustus. See gray whale
Eskimos. *See also* entries beginning with term Inuit
    whaling by, 227*b*
Essex (ship), 211*b*
estuarine dolphin *(Sotalia guianensis)*, 38, 39*t*, 188*b*, 188*f*
ethical issues, in research, 321*b*
Etruscan shrew *(Suncus etruscus)*, 3, 4*f*
Eubalaena australis. See southern right whale
Eubalaena glacialis (North Atlantic right whale), 35, 35*f*, 36, 138*t*, 272, 277*t*
Eubalaena japonica (North Pacific right whale), 36, 138*t*, 216, 272, 277*t*
Eumetopias jubatus. See Steller sea lion
Eurasian otter *(Lutra lutra)*, 3, 8*b*, 8*f*, 17, 28, 97, 275
European Cetacean Society, 321*b*
European Commission Communication on the Precautionary Principle, 88*b*
European Council Directive on the Conservation of Natural Habitats and Wild Fauna and Flora (Habitats Directive), 280, 281*f*, 282*b*, 283*b*, 285
    harbor porpoise and, 158*b*
European Council Regulation on Bycatch, 281*b*–282*b*

Eutheria, 1
evolution of marine mammals, 15–24
    carnivores, 15, 17–18
    cetaceans, 21–23, 22*f*
    changing oceans and, 15, 16*f*
    fossil record and, 17*b*, 17*f*
    sirenians, 18*f*, 18–19, 19*f*
Exclusive Economic Zones (EEZs), 274*b*
experimental design, 309*b*
Exxon Valdez oil spill
    harbor seals and, 243, 266
    sea otters and, 95, 96*b*, 96*f*
    Steller sea lion and, 256
eyes
    adaptation of, 55–56, 56*f*
    of polar bears, 81
    swimming and, 45

F

factory ships, 216, 217*f*
false killer whale *(Pseudorca crassidens)*, 37, 171, 233
fast Fourier transform (FFT), 62
Fazio, Jillian, 315*b*
feeding behavior
    of baleen whale, 132–135, 132*f*–135*f*
    of dugong, 105
    of killer whale, 169
    of manatee, 104–105
    of marine otter, 98
    of polar bear, 83*f*, 83–84
    of sea otter, 93, 93*f*
feet, of polar bears, 82, 83*f*
Feliformia, 27
Fennell, David, 296
Feresa attenuata (pygmy killer whale), 37, 171
FFT (fast Fourier transform), 62
finless porpoise. *See* Indo-Pacific finless porpoise
fin whale *(Balaenoptera physalus)*, 37, 130, 132, 134, 137*t*
    abundance and status of, 138*t*, 272, 275*t*, 277*t*
    blue whale unit and, 218
    hunting of, 216, 221, 222, 226, 227*b*
    quotas for, 225*t*
fisheries
    cetaceans and, 235–236, 236*f*
    drive, 233, 234*b*, 235, 235*f*
    pinnipeds and, 253
    tuna, purse seine, 238*b*, 238*f*
Fisheries and Oceans Canada (DFO), 267*b*

Fishes Royale decree, 208*b*
fishing (fish-eating) bat *(Myotis vivesi)*, 3, 4, 6, 6*f*, 272
fishing gear. *See also* bycatch
　discarded, 241, 242*f*
　　New Zealand sea lion entanglement in, 261
　　sea otter entanglement in, 95
fitness, demonstration of, 71–72
"Flipper" (TV series), 212
flippers, 45, 45*f*
　swimming and, 47
　thermoregulation and, 52
Florida manatee *(Trichechus manatus latirostris)*, 33, 102
Florida Manatee Sanctuary Act, 107*b*, 107*f*
folklore, marine mammals in, 201–203, 202*f*, 203*f*
forelimbs, swimming and, 45, 45*f*, 47
fossil record, 17*b*, 17*f*
Foyn, Svend, 215
Franciscana. *See* La Plata dolphin
Franks Casket, 208*b*, 208*f*
Fraser's dolphin *(Lagenodelphis hosei)*, 38
Freeman, Milton, 213*b*
Free Willy/Keiko Foundation (FWKF), 300*b*
freeze-branding, 310
Friends of the Earth, 212
Fund for Animals, 212
fur
　lanugo and, 50, 51*f*
　of polar bears, 82
　of sea otters, 92
　thermoregulation and, 49–50, 50*f*
fur lions, evolution of, 18
fur seal(s), 29, 29*t*. *See also specific types of fur seals*
　distribution of, 115, 116*f*
　fur of, 50–51
　killer whale predation on, 169
　status of, 275
　thermoregulation in, 52
Fur Seal Act, 279*b*
FWKF (Free Willy/Keiko Foundation), 300*b*
FWS (U.S. Fish and Wildlife Service), 277

G

Galápagos fur seal *(Arctocephalus galapagoensis)*, 29*t*, 115, 123, 126*f*, 258*f*
　abundance and status of, 272
　threats to, 257–258
Galápagos sea lion *(Zalophus wollebaeki)*, 29*t*, 124*b*, 257–258, 272
Galeocerdo cuvier (tiger shark), 15, 130*b*, 179
Ganges River dolphin *(Platanista gangetica)*, 37, 148*f*, 148–149, 149*f*
Garrod, Brian, 296
Gervais' beaked whale *(Mesoplodon europaeus)*, 37
giant otter *(Pteronura brasiliensis)*, 28
giant panda *(Ailuropoda melanoleuca)*, 27
giant polar bear *(Ursus maritimus tyrannus)*, 17
giant squid *(Architeuthis)*, 142
gill nets, 236
ginkgo-toothed beaked whale *(Mesoplodon ginkgodens)*, 37, 144, 145*f*
giraffes, 20
Giraffidae, 20
Global Positioning System (GPS) tags, 121*b*
global warming. *See also* climate change; temperature, environmental
　Eocene, 15
Globicephala spp. *See* long-finned pilot whale; pilot whales; short-finned pilot whale
Globicephala macrorhyncus. See short-finned pilot whale
Globicephala melas. See long-finned pilot whale
gorilla *(Gorilla gorilla)*, 191*b*
GPS (Global Positioning System) tags, 121*b*
Grampus griseus. See Risso's dolphin
Gray's beaked whale *(Mesoplodon grayi)*, 37
gray (grey) seal *(Halichoerus grypus)*, 31*t*, 32*b*, 266*b*
gray whale *(Eschrichtius robustus)*, 35, 36, 130, 131*f*, 216
　abundance and status of, 137*t*, 137–138, 138*t*, 272, 275*t*, 277*t*
　ban on hunting, 220
　feeding behavior of, 135, 135*f*
　hunting of, 209, 226, 227*b*
　quotas for, 225*t*
gray wolf *(Canis lupus)*, 7, 8*b*, 8*f*
Great Barrier Reef Marine Park, 284*b*
greater bulldog bat *(Noctilio leporinus)*, 3, 4, 5, 6*f*
great white shark *(Carcharodon carcharias)*, 15, 259
Greek mythology, marine mammals in, 202–203
greenhouse gases, 244
Greenland, aboriginal whaling in, 226
Greenpeace, 89, 212
Grey Seal Protection Act (U.K.), 282*b*
grindadráp hunt, 231, 232
grizzly bear *(Ursus arctos horribilis)*, 17, 28
grizzly bear/polar bear hybrid, 28, 28*f*
group cohesion, 73
Guadalupe fur seal *(Arctocephalus townsendi)*, 29*t*, 115, 116*b*–117*b*, 116*f*, 117*f*, 277*t*
Guiana/Guyana dolphin *(Sotalia guianensis)*, 38, 39*t*, 188*b*, 188*f*
Gulf of California harbor porpoise *(Phocoena sinus)*, 38, 160, 160*f*, 272, 275*t*, 277*t*
guyots, 44*b*, 44*f*

H

habitat degradation, cetaceans and, 245, 246*f*
Habitats Directive (European Council Directive on the Conservation of Natural Habitats and Wild Fauna and Flora), 158*b*, 280, 281*f*, 282*b*, 283*b*, 285
hadalpelagic zone, 44*f*
hadal zone, 44*f*
Haida people, whale legend of, 204
hair analysis, for contaminants, in polar bears, 86*b*
hairy-nosed otter *(Lutra sumatrana)*, 28
Halichoerus, 32*b*
Halichoerus grypus (gray [grey] seal), 31*t*, 32*b*, 266*b*
harbor porpoise *(Phocoena phocoena)*, 38, 158*b*, 158*f*, 159*f*, 186*b*
　blubber layer of, 47*b*
　bycatches of, 236
harbor seal *(Phoca vitulina)*, 31*t*, 123, 265*f*, 282*f*
　declining population of, 96*b*
　lanugo of, 50
　oil spills and, 243
　subspecies of, 265, 266
　threats to, 265–266
Harlan, Richard, 24*b*

Harold the Fairhaired, polar bears kept by, 89*b*
harpoons
   explosive, 215, 228*b*, 233
   hand-held, 233
harp seal (*Pagophilus groenlandicus*), 31*t*, 123–124, 266*b*–267*b*, 267*f*
harp seal (*Phoca groenlandica*)
   as polar bear prey, 83
   tourism and, 295*b*
Harvey, George, 65*b*
Hawaiian monk seal (*Monachus schauinslandi*), 32*t*, 33, 34*f*, 126*t*, 254*f*, 255*f*
   abundance and status of, 272, 277*t*
   threats to, 254
hearing, 57, 67, 67*b*–69*b*, 67*f*, 68*f*
heart, mammalian, 1
Heaviside's dolphin (*Cephalorhynchus heavisidii*), 38, 175, 175*f*
heavy metals, cetaceans and, 239–240
Hector's beaked whale (*Mesoplodon hectori*), 37
Hector's dolphin (*Cephalorhynchus hectori hectori*), 38, 175, 176*b*–177*b*, 176*f*, 272
Henry III, King of England, 89*b*
Henry VII, King of England, 208*b*
Henry VIII, King of England, 208*b*
Henry's gas law, 53
Herman, Louis, 192*b*
herring (*Clupea*), 169*b*
hibernation, of polar bears, 83–84
high pressure nervous syndrome, 54
hind limbs, swimming and, 45
Hines, Ellen, 109*b*
Hippopotamidae, 20
hippos, 3, 4*f*, 20
Historia Animalium (Aristotle), 203
Histriophoca fasciata (ribbon seal), 31*t*
homeothermic organisms, 48
Homo neanderthalensis (Neanderthal), 317*b*
hooded seal (*Cystophora cristata*), 31*t*, 126*f*
   abundance and status of, 274
   harvest of, 266*b*
   as polar bear prey, 83, 123*b*, 123*f*
   sexual selection in, 71*b*, 71*f*
hourglass dolphin (*Lagenorhynchus cruciger*), 38, 39*b*, 177, 178*f*
Hoyt, Erich, 289, 297*b*
Hubb's beaked whale (*Mesoplodon carlhubbsi*), 37
Hudson, Henry, 207, 209
human disturbances. *See* disturbances
Humane Society of the United States, 212
humpback dolphins (*Sousa* spp.), 167, 177, 179, 180*f*, 181. *See also* Atlantic humpback dolphin; Indo-Pacific humpback dolphin
   mother-calf cohesion in, 73
   status of, 275*t*
humpback whale (*Megaptera novaeangliae*), 36
   abundance and status of, 137*t*, 138*b*, 138*t*, 275*t*, 277*t*
   ban on hunting, 220
   blue whale unit and, 218
   bycatch of, 199*f*
   distribution of, 130, 130*b*, 131*f*
   feeding behavior of, 132, 133*f*, 133–134, 134*f*
   group cohesion in, 73
   hunting of, 216, 226, 227*b*
   quotas for, 225*t*
   reproduction by, 135
   song of, 71, 72, 72*b*–73*b*, 72*f*, 135, 193*b*
hunting. *See also* aboriginal subsistence hunting; Nordic whaling; whaling
   of Dall's porpoise, 156*b*–158*b*, 157*f*
   of manatee, 106
   Norwegian, 207, 208*b*
   of polar bear, 84–85, 85*f*
   of sea otter, 94, 95
   of whale sharks, 227*b*
hunting behavior. *See also* predation
   of killer whale, 169*b*
   of polar bear, 83, 83*f*
Hussey, Christopher, 209
Hutchinson's rule, 132
Hydrictis maculicollis (spotted-necked otter), 28
Hydrodamalinae, 33
Hydrodamalis gigas. *See* Steller's sea cow
hydrodynamic adaptations, 45–48, 46*f*
hydrophones, 318, 318*f*, 320
Hydrurga leptonyx (leopard seal), 32*t*, 34*f*
hyperexcitability, high pressure causing, 54–55
Hyperoodon spp.. *See* bottlenose whale
Hyperoodon ampullatus (northern bottlenose whale), 37, 218
Hyperoodon planifrons (southern bottlenose whale), 37
Hyperoodontidae, 37
Hyperoodontinae, 144. *See also specific beaked whales*
hypotheses, 309*b*

I

ice, seasonal, melting of, polar bears and, 86*f*, 86–87
Iceland
   whale watching and, 222*b*
   whaling by, 221–222
Icelandic Tourist Industry Association, 222*b*
Indian humpback dolphin (*Sousa chinensis plumbea*), 38, 179, 180*f*
Indian River dolphin (*Platanista* spp.), 37, 272, 275*t*
individual recognition, 74
Indohyus, 21*b*, 21*f*
Indopacetus pacificus (Indo-Pacific beaked whale or Longman's beaked whale), 37, 39*b*
Indo-Pacific beaked whale (*Indopacetus pacificus*), 37, 39*b*
Indo-Pacific bottlenose dolphin (*Tursiops aduncus*), 39*b*, 39*t*, 189, 190
   live takes of, 233*b*
   thermoregulation in, 51
Indo-Pacific finless porpoise (*Neophocaena phocaenoides*), 38, 39*t*, 159–160, 160*b*–161*b*, 161*f*
Indo-Pacific humpback dolphin (*Sousa chinensis*), 38, 39*b*, 39*t*, 177, 179, 180*f*, 181, 181*b*, 272, 275*t*
Indus River dolphin (*Platanista minor*), 37, 277*t*
Inia geoffrensis. *See* Amazon River dolphin
Inia geoffrensis boliviensis (Amazon River dolphin), 149
Inia geoffrensis boliviensis (Rio Madeira dolphin), 37
Inia geoffrensis geoffrensis (Amazon River dolphin), 37, 149
Inia geoffrensis humboldtiana (Amazon River dolphin), 149
Inia geoffrensis humboldtiana (Orinoco River dolphin), 37
Iniidae, 37
intelligence, of dolphins, 10, 192*b*–193*b*, 193*f*

intensity, of sound, 62–63
Inter-American Tropical Tuna Commission, 237
Intergovernmental Panel on Climate Change (IPCC), 244
International Agreement for the Regulation of Whaling, 218
International Convention for the Regulation of Whaling, 218t, 218–223, 219t
    aboriginal whaling under, 223, 225
    Icelandic whaling and, 221–222
    Japanese scientific whaling and, 220–221
    Norwegian whaling and, 220
    quotas and bans and, 218–220
    Revised Management Procedure and Revised Management Scheme and, 222–223
    whaling moratorium and, 220
International Ecotourism Society, 290b
International Otter Survival Fund, 99b
International Union for Conservation of Nature (IUCN), 271–274
    Amazonian manatee and, 272
    Amazon River dolphin and, 149–150
    Antarctic minke whale and, 221
    Atlantic humpback dolphin and, 272
    Australian sea lion and, 260, 272
    beluga whale and, 154
    blue whale and, 272
    bowhead whale and, 272
    Caspian seal and, 265, 272
    categories of, 271, 272, 272t, 273f, 274
    Cetacean Specialist Group of, 271–272
    Conservation Action Plan for the World's Cetaceans of, 234b
    costero and, 188b
    dugong and, 272
    finless porpoise and, 272
    fin whale and, 272
    on fisheries interactions, 253
    fishing bat and, 272
    Franciscana and, 274
    Galápagos fur seal and, 272
    Galápagos sea lion and, 258, 272
    gray whale and, 272
    Guadalupe fur seal and, 117b
    Gulf of California harbor porpoise and, 160
    harbor porpoise and, 158b
    Hawaiian monk seal and, 254, 272
    Hector's dolphin and, 272
    hooded seal and, 274
    humpback dolphin and, 181
    Indian River dolphin and, 272
    Indo-Pacific humpback dolphin and, 272
    Irrawaddy dolphin and, 274
    killer whale and, 170
    Lagenorhynchus dolphin and, 177
    manatee and, 106
    marine otter and, 97–98, 272
    Mediterranean monk seal and, 254, 272
    minke whale and, 137b
    New Zealand sea lion and, 261, 274
    North Atlantic right whale and, 272
    northern fur seal and, 274
    North Pacific right whale and, 272
    polar bear and, 6b, 89, 274
    Red List of Threatened Species of, 6b
    right whale dolphin and, 174
    sea otter and, 95, 272
    sei whale and, 272
    southern right whale and, 272
    sperm whale and, 274
    Steller sea lion and, 256, 272
    tucuxi and, 188b
    vaquita and, 272
    West African manatee and, 274
    West Indian manatee and, 274
    Yangtze River dolphin and, 153
International Whaling Commission (IWC), 147, 231
    aboriginal whaling under, 223, 225
    cetacean bycatch and, 187
    Conservation Committee of, 224b
    Dall's porpoise and, 157
    formation of, 218
    hunting quotas of, 204
    narwhal and, 155
    Scientific Committee of, 220b, 220–221, 221, 222
    viability of, 223
intersexual selection, 71b, 71f, 71–72
intertidal zone, 43b, 44f
intrasexual selection, 71
Inuit hunting, 204–205
    of pinnipeds, 207
    of polar bears, 84
Inuit legends, 205
Inuit whaling, 207, 225, 227b
Inuvialuit-Inupiat Agreements, 89
IPCC (Intergovernmental Panel on Climate Change), 244
Irrawaddy dolphin (*Orcaella brevirostris*), 38, 39t, 172–173, 173f, 274, 275f
IUCN. *See* International Union for Conservation of Nature
IWC. *See* International Whaling Commission

J

Japan
    contamination of cetacean meat in, 224b, 224f
    dolphin fishing in, 207
    scientific whaling in, 220–221
    whaling in, 207–208, 209f, 217b, 226, 226f, 233
Japanese sea lion (*Zalophus japonicus*), 29t, 125, 126b, 126t, 153
JARPN program, 221
Jonah and the whale story, 201–202, 202f
Juan Fernández fur seal (*Arctocephalus philippi*), 29t, 115
junk food hypothesis, 257b
Jurassic period, mammalian evolution during, 3

K

Kaschner, Kristen, 221
Keiko, 300b, 300f
Kellert, Stephen, 213b
kelp, sea otters and, 93, 94, 95f
Kempthorne, Dirk, 89
Kentriodontidae, 23
Kester, Marike, 309b
kidneys, osmoregulation and, 55
killer whale (*Orcinus orca*), 37, 130b, 167–171, 168f
    abundance and status of, 170–171, 171f, 277t
    behavior of, 169b–170b
    in captivity, 300b, 300f
    description of, 168f
    dorsal fins of, 168b, 168f
    ecotypes of, 168–169
    group cohesion in, 73
    hunting of, 208b, 234b
    longevity of, 170b
    in Native American legends, 204
    predation by, 95, 96b–97b, 167, 257b, 259
    sound frequencies used by, 63t
Kitti's hog-nosed bat (*Craseonycteris thonglongyai*), 3

Klebsiella pneumoniae, New Zealand sea lion and, 261
Knossos, Crete, dolphin mural in, 202, 202f
Kogia breviceps (pygmy sperm whale), 37, 143b–144b, 143f
Kogia sima (dwarf sperm whale), 37, 39b, 143b–144b, 143f
Kogiidae, 35, 37
krill, in baleen whale diet, 132–133, 133f

L

lactation. *See* reproduction
lactic acid, diving and, 53
Ladoga seal *(Pusa hispida ladogensis)*, 118b, 119f
Lagenodelphis hosei (Fraser's dolphin), 38
Lagenorhynchus spp., 39b, 177, 178f
 climate change and conservation of, 179b
Lagenorhynchus acutus (Atlantic white-sided dolphin), 38, 177, 178f
Lagenorhynchus albirostris (white-beaked dolphin), 38, 177, 178f, 179b
Lagenorhynchus australis (black-chinned dolphin), 38
Lagenorhynchus australis (Peale's dolphin), 38, 39b, 177, 178f
Lagenorhynchus cruciger (hourglass dolphin), 38, 39b, 177, 178f
Lagenorhynchus obliquidens (Pacific white-sided dolphin), 38, 177, 178f
Lagenorhynchus obscurus (dusky dolphin), 38, 177, 178f, 191b, 232
Lagenorhynchus obscurus (New Zealand dusky dolphin), 38
Lagenorhynchus obscurus fitzroyi (South American dusky dolphin), 38
Lagenorhynchus obscurus obscurus (South African dusky dolphin), 38
lake seals, 118b–119b, 118f, 119f
land-based surveys, 313–314, 314f
lanugo, of seals, 50, 51f
La Plata dolphin *(Pontoporia blainvillei)*, 37, 152b, 152f, 274
Laptev Sea walrus *(Odobenus rosmarus laptevi)*, 30
largha seal *(Phoca largha)*, 31t
law(s). *See* legislation; *specific laws*
Law on the Protection of Wildlife (China), 286b
Layard's beaked whale *(Mesoplodon layardii)*, 37, 144, 145f
lead, cetaceans and, 239
legislation, 276–286. *See also specific laws*
 enforcement of, 286
 European, 280–285
 science-based management and, 285–286
 United States, 276–280
lemming *(Dicrostonyx richardsoni)*, 7
leopard seal *(Hydrurga leptonyx)*, 32t, 34f
Leptonychotes weddellii (Weddell seal), 32t, 51–52, 123
Leptophoca, 18
lesser beaked whale *(Mesoplodon peruvianus)*, 37
light
 in marine environment, 55, 56b, 57f
 in ocean, 44, 44b, 44f
Lillie, Harry, 228b
line transect surveys, 307–308, 308f
Lipotes vexillifer (baiji or Yangtze river dolphin), 37, 150b, 150–153, 151f, 275t, 277t
Lipotidae, 37
Lissodelphis spp. (right whale dolphins), 167
Lissodelphis borealis (northern right whale dolphin), 38, 174
Lissodelphis peronii (southern right whale dolphin), 38, 174, 174f
littoral zone, 43b, 44f
llamas, 20
Llanocetus denticrenatus, 22
Lobodon carcinophaga (crabeater seal), 32t, 34f, 121, 125
loggers, 121b
long-beaked common dolphin *(Delphinus capensis)*, 38, 39b, 185, 185f, 186f
long-beaked dolphins *(Delphinus, Stenella, Steno* spp.), 39b, 167, 181–189
long-finned pilot whale *(Globicephala melas)*, 37, 172
 hunting of, 208b, 231–232, 232f
 sound frequencies used by, 63t
Longman's beaked whale *(Indopacetus pacificus)*, 37, 39b
Lontra canadensis (North American river otter), 4, 28, 97
Lontra felina. *See* marine otter
Lontra longicaudis (neotropical river otter), 28
Lontra provocax (southern river otter), 28
Lück, Michael, 293f
Luksenburg, Jolanda, 185f
lungs
 of sea otters, 92
 size of, 52, 52f
Lusseau, David, 296
Lutra spp., 17, 28
Lutra lutra (Eurasian otter), 3, 8b, 8f, 17, 28, 97, 275
Lutra sumatrana (hairy-nosed otter), 28

M

MacFarquhar, Christine, 181b
MacLeod, Colin, 146b, 179b
Magnuson-Stevens Act, 280b
Makah people, whaling by, 225–226, 227b
mammals (Mammalia), 1
 characteristics of, 1, 1f
 evolution of, 2f–4f, 3
 number of species of, 3
 placental, 1
mammary glands, 1f, 1–10
Mammuthus primigenius (woolly mammoth), 317b
manatees, 18, 33, 101–111, 102f. *See also specific types of manatees*
 abundance and status of, 105–106
 aquatic adaptations of, 101–102
 conservation of, 106b–108b, 107f, 108f
 description of, 101
 distribution of, 102–104, 103f
 evolution of, 19
 eye of, 56f
 feeding behavior and ecology of, 104–105
 hearing of, 67
 mermaid legends and, 205
 migration of, 51
 reproduction by, 105
 skull of, 18f
Maori culture, 205
Marguerite formation, 142
Marine Act (Scotland), 282b
marine litter and debris, cetaceans and, 241–242, 242f
marine mammal(s). *See also specific animals*
 changing attitudes toward, 212f, 212–213

classification and diversity of, 27–40. *See also* carnivores; cetaceans; sirenians; *specific animals*
controversial, 8*b*, 8*f*
definition of, 3–5, 4*f*
extinct, 9
in folklore and history, 201–207, 202*f*–205*f*
quasi-marine, 5–7
special status of, 9–10
Marine Mammal Commission, 279*b*
Marine Mammal Protection Act (MMPA), 220
 beluga whale and, 154
 bottlenose dolphin and, 191
 bowhead whale and, 225
 dolphin bycatch and, 237
 manatee and, 106, 107*b*
 Marine Mammal Commission created by, 279*b*
 marine mammals defined by, 4–5
 northern fur seal and, 262
 polar bear and, 85, 88–89
 sea otter and, 95
Marine Mammal Regulations (Canada), 267*b*
Marine Mammal Sanctuary, New Zealand sea lion and, 260–261
Marine Mammals of the World (Rice), 3
marine mammal tourism, 289–301. *See also* whale watching
 ecotourism and, 290*b*–291*b*
 educational potential of, 297*b*–298*b*, 298*f*
 on Galápagos Islands, 257
 managing, 296–300
 nature of tourists and, 294
 negative impacts of, 294–296
 New Zealand sea lion and, 261
 pinniped, 292*b*, 292*f*, 295*b*–296*b*
 polar bears and, 87–88, 88*f*
 risk posed by, 212
 significance of, 289–293, 290*f*
 solitary sociable dolphin problem and, 301, 301*f*
 sustainability report card and, 297*b*
marine otter *(Lontra felina)*, 3, 4, 28, 28*f*, 91, 97–98, 98*f*
 abundance and status of, 272, 275, 277*t*
 description of, 97
 distribution, abundance, and status of, 97–98, 98*f*
 feeding behavior and ecology of, 98

reproduction by, 98
rescue and rehabilitation of, 99*b*
marine protected areas (MPAs), 284*b*
mark-recapture analysis, 310–311
Marshall, Greg, 318
marsupials, 1
Mary Queen of Scots, 206*b*
mass extinctions, 3
mating. *See* reproduction
Maui's dolphin *(Cephalorhynchus hectori maui)*, 38, 176*b*
McConchie, Todd, 312*b*, 313*b*
Mediterranean monk seal *(Monachus monachus)*, 32*t*, 33, 125, 126*t*, 253*f*
 abundance and status of, 272, 277*t*
 threats to, 253–254
Megaptera novaeangliae. *See* humpback whale
Megapterinae, 36
melon, 65*b*
melon-headed whale *(Peponocephala electra)*, 37, 171
Melville, Herman, 210*b*–211*b*
Memorandum of Understanding Concerning the Conservation and Management of Dugongs and their Habitats throughout their Range, 276
Memorandum of Understanding Concerning the Conservation of the Manatee and Small Cetaceans of Western Africa and Macaronesia, 276
mercury, cetaceans and, 239, 240
mermaid legends, 205, 205*f*, 207
mesonychids, 15, 21
Mesonychoteuthis hamiltoni (colossal squid), 142
mesopelagic zone, 44, 44*b*, 44*f*
Mesoplodon bahamondi (Bahamonde's beaked whale), 37, 39*b*
Mesoplodon bidens (Sowerby's beaked whale), 37
Mesoplodon bowdoini (Andrew's beaked whale), 37
Mesoplodon carlhubbsi (Hubb's beaked whale), 37
Mesoplodon densirostris (densoe or Blainville's beaked whale), 37
Mesoplodon europaeus (Gervais' beaked whale), 37
Mesoplodon ginkgodens (ginkgo-toothed beaked whale), 37, 144, 145*f*
Mesoplodon grayi (Gray's beaked whale), 37

Mesoplodon hectori (Hector's beaked whale), 37
Mesoplodon layardii (strap-toothed or Layard's beaked whale), 37, 144, 145*f*
Mesoplodon mirus (True's beaked whale), 37
Mesoplodon perrini (Perrin's beaked whale), 37, 39*b*, 40*f*
Mesoplodon peruvianus (lesser or pygmy beaked whale), 37
Mesoplodon stejnegeri (Stejneger's beaked whale), 37
Mesoplodon traversii (spade-toothed whale), 37, 39*b*
metabolic rate
 diving and, 53
 of manatee, 105
 of sea otter, 92
Metatheria, 1
migration
 of baleen whales, 130*b*, 131*f*
 of manatees, 51
milk. *See also* reproduction
 of pinnipeds, 124, 125*t*
minke whales, 36, 137*b*. *See also specific types of minke whales*
 abundance of, 138*t*
 climate change and, 244
 hunting of, 217, 218, 221, 222, 226, 227*b*
 quotas for, 225*t*
Miocene period, 16*f*, 17
 cetacean evolution during, 23
 climate changes in, 23*b*, 23*f*
Mirounga angustirostris. *See* northern elephant seal
Mirounga leonina (southern elephant seal), 32*t*, 125*b*, 275
MMPA. *See* Marine Mammal Protection Act
Moby Dick (Melville), 210*b*–211*b*, 211*f*
Møhl, Bertel, 65*b*
molecular genetics, 316*b*–317*b*
molting, in pinnipeds, 125*b*, 125*f*
Monachinae, 29, 32*t*, 121
Monachus monachus. *See* Mediterranean monk seal
Monachus schauinslandi. *See* Hawaiian monk seal
Monachus tropicalis (Caribbean monk seal), 32*t*, 33, 125, 126*b*, 126*t*, 153
monkey lips, 65*b*, 65*f*
Monodon monoceros. *See* narwahl
Monodontidae, 37, 153. *See also* beluga whale; narwahl

monotremes (Monotremata), 1, 1*f*
Monterey Bay Aquarium, 299*b*
Morbillivirus (phocine distemper virus), 266
mother-calf cohesion, 73, 74*f*
MPAs (marine protected areas), 284*b*
museau de singe, 65*b*, 65*f*
Mustelidae, 17–18, 27, 28
Musteloidea, 27, 28
Myotis vivesi (fishing [fish-eating] bat), 3, 4, 6, 6*f*, 272
mysticetes (Mysticeti), 21, 35*f*, 35–37, 36*f*. *See also specific animals*
   bones of, 47
   ears of, 67*b*, 68*b*
   evolution of, 22–23
   group cohesion in, 73
   hearing of, 67
   sound production by, 66*b*, 70

N

Nagati Porou legend, 205
NAMMCO (North Atlantic Marine Mammal Commission), 223*b*
Nantucket, 209, 211*b*
narrow-ridged finless porpoise (*Neophocaena asiaeorientalis*), 39*t*, 161*b*
narwahl (*Monodon monoceros*), 37, 154–156, 155*f*
   diving depth of, 54–55
   hunting of, 231
   tusk of, 155*b*–156*b*, 155*f*
Native Americans. *See also specific groups of Native Americans*
   whaling by, 207
Naturalis Historia (Pliny the Elder), 167, 203, 205
navigation, sound and, 70, 70*f*
Nazca lines, 17*f*
Neanderthal *(Homo neanderthalensis)*, 317*b*
Neobalaenidae, 35, 36
Neophoca cinerea (Australian sea lion), 29*t*, 126*f*, 259*f*
Neophocaena asiaeorientalis (narrow-ridged finless porpoise), 39*t*, 161*b*
Neophocaena asiaorientalis asiaorientalis (Yangtze finless porpoise), 38, 161*b*
Neophocaena asiaorientalis sunameri (East Asian finless porpoise), 161*b*

Neophocaena phocaenoides (Indo-Pacific finless porpoise), 38, 39*t*, 159–160, 160*b*–161*b*, 161*f*
Neophocaena phocaenoides sunameri (Chinese finless porpoise), 38
neotropical river otter *(Lontra longicaudis)*, 28
neritic zone, 43*b*, 44*f*
Nero, Emperor, polar bears in arena of, 89*b*
nerpa. *See* Baikal seal
New Management Procedure, 219
New Zealand dusky dolphin *(Lagenorhynchus obscurus)*, 38
New Zealand fur seal *(Arctocephalus forsteri)*, 29*t*
New Zealand Marine Mammals Protection Act (NZMMPA), 260, 280*b*
New Zealand sea lion *(Phocarctos hookeri)*, 29*t*, 126*f*, 260*f*
   abundance and status of, 274
   threats to, 260–261
nitrogen, decompression problems and, 53–54, 54*f*
nitrogen narcosis, 54
NMFS (U.S. National Marine Fisheries Service), 277, 278, 280
Noctilio leporinus (greater bulldog bat), 3, 4, 5, 6*f*
noise pollution, 75*b*–76*b*
Nordic history, marine mammals in, 204
Nordic whaling, 207, 208*b*, 215
   whaling moratorium and, 220
Norris, Ken, 65*b*
North American river otter *(Lontra canadensis)*, 4, 28, 97
North Atlantic harbor porpoise *(Phocoena phocoena phocoena)*, 38
North Atlantic Marine Mammal Commission (NAMMCO), 223*b*
North Atlantic right whale *(Eubalaena glacialis)*, 35, 35*f*, 36, 138*t*, 272, 277*t*
northern bottlenose whale *(Hyperoodon ampullatus)*, 37, 218
northern elephant seal *(Mirounga angustirostris)*, 32*t*, 33, 34*f*, 120, 121, 263*b*, 264*f*
   diving depth of, 54
   molting in, 125*b*
northern fur seal *(Callorhinus ursinus)*, 29*t*, 115, 126*f*, 261*f*, 262*f*
   abundance and status of, 274, 275, 277*t*

   declining population of, 96*b*
   eye of, 56*f*
   threats to, 261–262
northern minke whale *(Balaenoptera acutorostrata)*, 36, 132, 138*b*
   call of, 71
   hunting of, 208*b*, 221, 222
   status of, 138*t*, 275*t*
northern right whale, 131*f*
   ban on hunting, 219
northern right whale dolphin *(Lissodelphis borealis)*, 38, 174
northern seals, 29, 31*t*, 33*f*
North Pacific bottlenose whale. *See* Baird's beaked whale
North Pacific Dall's porpoise *(Phocoenoides dalli dalli)*, 38
North Pacific Fur Seal Convention, 261–262, 282*b*
North Pacific right whale *(Eubalaena japonica)*, 36, 138*t*, 216, 272, 277*t*
Norwegian marine mammal hunting, 207, 208*b*. *See also* Nordic whaling
nostrils, swimming and, 45
nursing behavior. *See also* reproduction
   in Galápagos fur seal, 124*b*, 124*f*
NZMMPA (New Zealand Marine Mammals Protection Act), 260, 280*b*

O

OCA (Oregon Coast Aquarium), 300*b*
ocean(s), changing, evolution of marine mammals and, 15, 16*f*
oceanic zone, 43*b*, 44*f*
ocean zones, 43*b*–44*b*, 44*f*
Odobenidae, 29, 116. *See also* walrus; *specific species of walruses*
Odobenus rosmarus. *See* walrus
Odobenus rosmarus divergens (Pacific walrus), 30, 116
Odobenus rosmarus laptevi (Laptev Sea walrus), 30
Odobenus rosmarus rosmarus (Atlantic walrus), 30
odontocetes (Odontoceti), 21, 35, 37. *See also* dolphin(s); porpoises; toothed whale(s)
   ears of, 67*b*–68*b*, 68*f*
   echolocation by, 61, 69, 70
   group cohesion in, 73
   hearing of, 67

sound production by, 64b–65b, 69
sound spectra for, 62, 63t
Odontoceti (toothed whales), 21, 35, 37–38, 61, 141–161. *See also* dolphin(s); porpoises; *specific toothed whales*
oil and gas exploration, polar bears and, 87, 87f
oil spills
　cetaceans and, 242–243, 243f
　Deepwater Horizon, 243, 243f
　Galápagos sea lion and, 257–258
　harbor seal and, 243, 266
　sea otter and, 95, 96b, 96f
　Steller sea lion and, 256
Oligocene period, 16f
　cetacean evolution in, 22–23
　climate changes in, 23b, 23f
Ommatophoca rossii (Ross seal), 32t, 117b–118b, 117f, 118f
Omura's whale (*Balaenoptera omurai*), 36, 37, 138t
Oppian, 203
orangutans, 3, 4f
orca. *See* killer whale
Orcaella brevirostris (Irrawaddy dolphin), 38, 39t, 172–173, 173f, 274, 275f
Orcaella heinsohni (snubfin dolphin), 38, 39t, 173, 173f
Orcinus orca. See killer whale
Oregon Coast Aquarium (OCA), 300b
organochlorines, cetaceans and, 240
organohalogens, cetaceans and, 240–241
Oriental small-clawed otter (*Aonyx cinerea*), 17, 28
Orinoco river dolphin (*Inia geoffrensis humboldtiana*), 37
osmoregulation, 55
Otaria flavescens (South American sea lion), 29t
otariids (Otariidae), 29. *See also specific animals*
　ears of, 68b
　evolution of, 18
　individual recognition in, 74
Otariinae, 29
otters, 27, 28, 28f. *See also specific types of otters*
　evolution of, 17
Ovis aries (sheep), living on seaweed, 8b, 8f
oxygen

access and storage for diving, 52f, 52–53, 53f
toxicity of, 54

P

Pacific humpback dolphin (*Sousa chinensis chinensis*), 38, 51, 179, 180f
Pacific walrus (*Odobenus rosmarus divergens*), 30, 116
Pacific white-sided dolphin (*Lagenorhynchus obliquidens*), 38, 177, 178f
Pagophilus groenlandicus (harp seal), 31t, 123–124, 266b–267b, 267f
PAHs (polyaromatic hydrocarbons), 243b
Pakicetidae, 20
Pakicetus, 20b, 20f, 21b
Pakistan
　in Eocene period, 20b, 20f
　fossils found in, 17b, 20
Panama Declaration, 237–238
Pan paniscus (bonobo), 191b
Pan troglodytes (chimpanzee), 191b
pantropical spotted dolphin (*Stenella attenuata*), 38, 182f, 183–184, 278
Paralutra lorteii, 17
Parsons, Chris, 313b
passive acoustic techniques, 318, 320
passive integrated transponder (PIT) tags, 120b
Patterson, Katheryn, 94b
Pauly, Daniel, 221
PCBs, cetaceans and, 240
PCDD (polychlorinated dibenzodioxin), cetaceans and, 240
Peale's dolphin (*Lagenorhynchus australis*), 38, 39b, 177, 178f
peduncle, swimming and, 46–47
pelage, of pinnipeds, 50
Pelagiarctos thomasi, 18
pelagic zone, 43b, 44f
Peponocephala electra (melon-headed whale), 37, 171
perissodactyls (Perissodactyla), 3, 15, 20
Permian period, mammalian evolution during, 3
Perrin, William, 39b
Perrin's beaked whale (*Mesoplodon perrini*), 37, 39b, 40f
Perry, Clare, 158f
Peru, whaling in, 232–233

Pezosiren, 19, 19f
Phoca, 32b
Phoca groenlandica (harp seal), 83, 295b
Phoca hispida (ringed seal), 31t
Phoca largha (spotted or largha seal), 31t
Phoca mimica, 9b
Phocarctos hookeri (New Zealand sea lion), 29t, 126f, 260f
　abundance and status of, 274
　threats to, 260–261
Phoca vitulina (harbor seal), 31t, 123, 265f, 282f
　declining population of, 96b
　lanugo of, 50
　oil spills and, 243
　subspecies of, 265, 266
　threats to, 265–266
Phoca vitulina concolor, 265, 266
Phoca vitulina mellonae, 265, 266
Phoca vitulina richardii, 265, 266
Phoca vitulina stejnegeri, 265, 266
Phoca vitulina vitulina, 265, 266
phocids (Phocidae), 18, 29. *See also* true seals; *specific animals*
　ears of, 68b–69b
　hearing of, 67
Phocinae, 29, 31t, 121
　recent revisions in taxonomy of, 32b
phocine distemper virus (*Morbillivirus*), 266
Phocoena dioptrica (spectacled porpoise), 38, 159, 159f
Phocoena phocoena (harbor porpoise), 38, 158b, 158f, 159f, 186b
　blubber layer of, 47b
　bycatches of, 236
Phocoena phocoena phocoena (North Atlantic harbor porpoise), 38
Phocoena phocoena vomerina (Eastern North Pacific harbor porpoise), 38
Phocoena sinus (Gulf of California harbor porpoise), 38, 160, 160f, 272, 275t, 277t
Phocoena spinipinnis (Burmeister's porpoise), 38, 159, 159f, 232
Phocoenidae, 35, 37, 38
Phocoenoides dalli. See Dall's porpoise
Phocoenoides dalli dalli (North Pacific Dall's porpoise), 38
Phocoenoides dalli truei (West Pacific Dall's porpoise), 38
Phoenician whaling, 202

phonic (phonetic) lips, 65b, 65f
photic zone, 44, 44b, 44f
photo-identification, 311, 311b–313b, 311f–313f
Physeter catadon. See sperm whale
Physeteridae, 35, 37
Physeter macrocephalus. See sperm whale
Pictish beast, 203, 203f
Pilleri, Giorgio, 161b
pilot whales (*Globicephala* spp.), 167. See also long-finned pilot whale; short-finned pilot whale
pingers, 236, 236f
pinniped(s) (Pinnipedia), 3, 4, 15, 27, 28–33, 29. See also sea lions; seal(s); walrus
　abundance and status of, 125, 126t, 253–267
　behavioral adaptation to cold in, 51
　blubber layer of, 47b
　distribution of, 115–116, 116f, 120
　ears of, 68b–69b
　evolution of, 17–18
　eyes of, 55
　fur of, 50
　oxygen storage in, 52, 52f
　pressure adaptations of, 54
　public attitudes toward culling of, 213b
　reproduction by, 121–124, 122f, 125t
　sleep of, 58b
　smell sense of, 56
　sound production by, 64b
　swimming adaptations of, 45, 45f
　thermoregulation in, 50
　touch sense of, 56
Pinniped Conservation law (U.K.), 283b
pinniped tourism, 292b, 292f
　negative impacts of, 295b–296b
pitch, of sound, 62
Pithanotaria starri, 18
PIT (passive integrated transponder) tags, 120b
placental mammals, 1
Platanista spp. (Indian River dolphin), 37, 272, 275t
Platanista gangetica (Ganges River dolphin), 37, 148f, 148–149, 149f
Platanista minor (Indus River dolphin), 37, 277t
Platanistidae, 37
platypus, 1, 1f
playbacks, 320
Pliny the Elder, 167, 203, 205

poikilothermic organisms, 48
polar bear (*Ursus maritimus*), 3, 4, 4f, 15, 27–28, 28f, 81–89
　abundance and status of, 84–88, 274, 275, 277t
　adipose tissue of, 50
　aquatic adaptations of, 81, 82f
　arctic adaptations of, 83, 83f
　in captivity, 89b, 89f
　climate change and, 244
　conservation of, 88–89
　description of, 81
　distribution of, 81–82, 82f
　ears of, 69b
　evolution of, 17
　feeding behavior and ecology of, 83f, 83–84
　fur of, 50
　reproduction by, 84, 84f
　sound production by, 64b
　swimming adaptations of, 45
　taste sense of, 56
polar bear/grizzly bear hybrid, 28, 28f
polar habitats, climate change and, 244, 245f
pollution. See also noise pollution
　beluga whale and, 241
　bottlenose dolphin and, 191–192
　cetaceans and, 224b, 224f, 239–243, 241b, 243b
　killer whale and, 170
　pilot whale and, 172
　polar bear and, 86, 86b
　sea otter and, 95
polyaromatic hydrocarbons (PAHs), 243b
polychlorinated dibenzodioxin (PCDD), cetaceans and, 240
Pompeii, 17f
Pontoporia blainvillei (La Plata dolphin), 37, 152b, 152f, 274
Pontoporiidae, 37
porpoises, 37, 38, 156–161, 159f, 160f. See also specific porpoises
　hunting of, 208b
Potamotherium, 17–18
precautionary principle, 88b
predation. See also hunting behavior
　group, 73
　prey stunning and, 74
pressure effects, 53–55, 54f
　adaptations to, 53–55, 54f
primates, 3
　appearance of, 15
Proboscidea, 3, 18
Project Ahab campaign, 212

Prophoca, 18
Prorastomidae, 33
Prorastomus, 19
Protocetidae, 20, 22
Protosiren, 19
Protosirenidae, 33
Prototheria, 1
Pseudorca crassidens (false killer whale), 37, 171, 233
pseudo-replication, 309b
Pteronura brasiliensis (giant otter), 28
Ptolemy II, polar bear owned by, 89b
public attitudes, toward whaling, 213b
Public Law 99-625, sea otters and, 95
public sightings, research using, 319b, 319f
Puijila darwini, 18
purse seine nets, 236–237
purse seine tuna fisheries, 238b, 238f
Pusa caspica (Caspian seal), 31t, 118b, 119f, 126f, 262f
　abundance and status of, 272
　threats to, 262–263, 263f, 265
Pusa hispida (ringed seal), 32b, 121
　as polar bear prey, 82, 83
　subspecies of, 118b, 119f
Pusa hispida botnica (Baltic ringed seal), 118b, 119b, 119f
Pusa hispida ladogensis (Ladoga seal), 118b, 119f
Pusa hispida saimensis (Saimaa seal), 118b, 119f, 277t
Pusa sibirica (Baikal seal), 31t, 118b, 118f, 119f
pygmy beaked whale (*Mesoplodon peruvianus*), 37
pygmy killer whale (*Feresa attenuata*), 37, 171
pygmy right whale (*Caperea marginata*), 35, 36, 129, 134, 138t, 275t
pygmy sperm whale (*Kogia breviceps*), 37, 143b–144b, 143f

Q

quasi-marine mammals, 5–7
quota system, for whaling, 218–219, 223, 225, 225t

R

raccoons, 27
radioactive discharges, 242b
Ragen, Tim, 279b
rain, noise from, 75

red deer *(Cervus elaphus)*, 71–72
red fox *(Vulpes vulpes)*, as Arctic fox competitor, 7
Red List of Threatened Species, 6*b*
red panda, 27
reflection of sound, 64
refraction of sound, 64
rehabilitation, of marine otters, 99*b*
Reino Aventura, 300*b*
Remingtonocetidae, 20, 22
Renker, Ann, 227*b*
reproduction
    by baleen whale, 135–136
    by bottlenose dolphin, 190–191
    by dugong, 105
    by manatee, 105
    by marine otter, 98
    by pinnipeds, 121–124, 121–125, 122*f*, 125*t*
    by polar bear, 84, 84*f*
    by sea otter, 94
    sperm competition and, 191*b*
    by sperm whale, 142
rescue, of marine otters, 99*b*
research, 307–321
    acoustic techniques for, 318, 318*f*, 320–321
    behavioral observation methods for, 315*b*
    biopsy darting for, 315–316
    crittercams for, 318
    ethical issues and, 321*b*
    experimental design and, 309*b*
    land-based surveys for, 313–314, 314*f*
    line transect surveys for, 307–308, 308*f*
    mark-recapture analysis and, 310–311
    molecular genetics and, 316*b*–317*b*
    photo-identification for, 311, 311*b*–313*b*, 311*f*–313*f*
    public sightings for, 319*b*, 319*f*
    satellite telemetry for, 321*b*
    strandings analysis and, 316*b*
    tagging for, 317–318, 321*b*
retia mirabilia, 52, 53*f*
Revised Management Procedure (RMP), 222–223
Revised Management Scheme (RMS), 223
ribbon seal *(Histriophoca fasciata)*, 31*t*
Rice, Dale, 3, 32*b*
right whale(s) (Balaenidae), 35, 35*f*, 35*t*, 36. *See also specific right whales*
    abundance and status of, 136–137, 137*t*, 138*b*, 275*t*
    feeding behavior of, 35, 35*f*134, 134*f*
    hunting of, 218
    name of, 208
    reproduction by, 135, 136
    sleep of, 58*b*
right whale dolphins *(Lissodelphis* spp.), 167
ringed seal *(Phoca hispida)*, 31*t*
ringed seal *(Pusa hispida)*, 32*b*, 121
    as polar bear prey, 82, 83
    subspecies of, 118*b*, 119*f*
Rio Declaration on Environment and Development, 88*b*
Rio Madeira dolphin *(Inia geoffrensis boliviensis)*, 37
Risso's dolphin *(Grampus griseus)*, 38, 167, 173*f*, 173–174
    description of, 173
    head shape of, 46
    hunting of, 233
    scarring of, 173–174
river dolphins, 37, 148–153. *See also specific river dolphins*
    osmoregulation in, 55
river otter *(Lontra canadensis)*, 4, 28, 97
RMP (Revised Management Procedure), 222–223
RMS (Revised Management Scheme), 223
rock etchings, 201, 201*f*
Rodentia, 3
Rodhocetus spp., 22
Roland, Elly, 66*b*, 69*b*
rorquals (Balaenopteridae), 36, 36*f*
    feeding behavior of, 132
    hunting of, 209
Rose, Naomi, 97*b*, 170*b*, 300*b*
Ross seal *(Ommatophoca rossii)*, 32*t*, 117*b*–118*b*, 117*f*, 118*f*
rough-toothed dolphin *(Steno bredanensis)*, 38, 184, 184*b*–185*b*, 184*f*
Ruminanta, 20

S

SACs. *See* Special Areas of Conservation
Saimaa seal *(Pusa hispida saimensis)*, 118*b*, 119*f*, 277*t*
St. Brendan, 204, 204*f*
St. Kitts declaration, 221
Saint Columba, 204
sanctuaries, for manatees, 107*b*, 108*f*
satellite telemetry, 321*b*
satellite telemetry tags, 120, 120*b*–121*b*, 121*f*
Save the Whales campaign, 138*b*, 212, 212*f*, 219, 228*b*
Scarpaci, Carol, 292*f*, 296, 296*b*
scientific whaling
    Icelandic, 221–222
    Japanese, 220–221
Scotland
    attitudes toward marine mammals in, 212
    folklore of, 206*b*, 206*f*
    harbor seals and salmon fisheries in, 266
    marine mammals in history of, 203*f*, 203–304
scrimshaw, 209, 210*f*
sea ape, 9*b*
sea bass *(Dicentrarchus labrax)*, 186*b*
seabirds, declining population of, 96*b*
seal(s), 15, 27. *See also specific types of seals*
    breeding by, 50
    evolution of, 18
    pressure adaptations of, 54
    in Scottish culture, 203–204
    thermoregulation in, 51–52
sea lions, 15, 27, 29, 29*t*. *See also specific types of sea lions*
    breeding by, 50
    distribution of, 115, 116*f*
    evolution of, 18
    fur of, 50, 51
    killer whale predation on, 169
    thermoregulation in, 52
Seal Management Plan (Scotland), 266
sea monkey, 9*b*
seamounts, 44*b*, 44*f*
sea otter *(Enhydra lutris)*, 3, 4, 5*f*, 15, 28, 28*f*, 91*f*, 91–97
    abundance and status of, 94–95, 272, 275, 277*t*
    aquatic adaptations of, 92–93, 93*f*
    behaviors of, 94, 95*f*
    description of, 91
    distribution of, 92, 92*f*
    ears of, 69*b*
    evolution of, 17
    feeding behavior and ecology of, 93, 93*f*
    fur of, 49–50, 50*f*
    as killer whale prey, 95, 96*b*–97*b*
    lungs of, 52
    metabolic rate of, 51

sea otter *(Enhydra lutris) (continued)*
    oil spills and, 96*b*, 96*f*, 243
    reproduction by, 94
    sound production by, 64*b*
    subspecies of, 92
    swimming adaptations of, 45
    taste sense of, 56
    thermoregulation in, 49
    tool use by, 94*b*
sea otter, prehistoric *(Enhydritherium lluecai)*, 17
sea otter, prehistoric *(Enhydritherium terraenovae)*, 17
seasonal ice, melting of, polar bears and, 86*f*, 86–87
Seawatch Foundation, 319*b*
sei whale *(Balaenoptera borealis)*, 36, 37, 130, 134
    abundance and status of, 138*t*, 272, 275*t*, 277*t*
    ban on hunting, 220
    blue whale unit and, 218
    hunting of, 216–217, 221, 222
self-awareness, in dolphins, 193*b*, 193*f*
sensory adaptations, 55–57, 56*f*, 57*f*
sewage, cetaceans and, 243
sheep *(Ovis aries)*, living on seaweed, 8*b*, 8*f*
Shepherd's beaked whale *(Tasmacetus shepherdi)*, 37, 144, 145*f*, 146
shipping, noise from, 75
ship strikes. *See* watercraft collisions
short-beaked common dolphin *(Delphinus delphis)*, 38, 39*b*, 185, 185*f*, 186*b*, 186*f*
short-finned pilot whale *(Globicephala macrorhyncus)*, 38, 172
    hunting of, 233
    ship strikes and, 238
short tandem repeats, 317*b*
Silent Spring (Carson), 240
silty coastal waters, habitat loss and, 246*f*
sirenians (Sirenia), 3, 4, 18, 33, 101–111, 102*f*. *See also* dugong(s); manatees
    blubber layer of, 47*b*
    evolution of, 18*f*, 18–19, 19*f*
    osmoregulation in, 55
    status of, 275
    swimming adaptations of, 45, 45*f*
    touch sense of, 56
skin
    of cetaceans, 45
    of polar bears, 82

skunks, 27
sleep, of cetaceans, 58*b*
Slooten, Liz, 177*b*
smell sense, 56–57
Smith, John, 207, 209
smooth-coated otter *(Lutrogale perspicallata)*, 28
snubfin dolphin *(Orcaella heinsohni)*, 38, 39*t*, 173, 173*f*
SOFAR (SOund Fixing And Ranging) channel, 70
solitary sociable dolphin problem, 301
sonar, noise from, 75*b*
Soricomorpha, 3
Sotalia fluviatilis (tucuxi), 38, 39*t*, 188*b*, 188*f*, 275*t*
Sotalia guianensis (costero, estuarine dolphin, Guiana/Guyana dolphin, or tucuxi), 38, 39*t*, 188*b*, 188*f*
sound, underwater, 61–76
    bioacoustics and, 64*b*–66*b*, 65*f*, 66*f*
    communication and, 71–74
    echolocation and, 69*f*, 69–70
    environmental, 75
    hearing and, 67, 67*b*–69*b*, 67*f*, 68*f*
    navigation and, 70, 70*f*
    noise pollution and, 75*b*–76*b*
    physics of, 61–64, 62*f*, 63*t*
    prey stunning and, 74*b*
SOund Fixing And Ranging (SOFAR) channel, 70
Sousa spp., *See* Atlantic humpback dolphin; humpback dolphins; Indo-Pacific humpback dolphin
Sousa chinensis (Indo-Pacific humpback dolphin), 38, 39*b*, 39*t*, 177, 179, 180*f*, 181, 181*b*, 272, 275*t*
Sousa chinensis chinensis (Pacific humpback dolphin), 38, 51, 179, 180*f*
Sousa chinensis plumbea (Indian humpback dolphin), 38, 179, 180*f*
Sousa teuszii (Atlantic humpback dolphin), 38, 177, 179, 180*f*, 181, 272
South African dusky dolphin *(Lagenorhynchus obscurus obscurus)*, 38
South African fur seal *(Arctocephalus pusillus pusillus)*, 29*t*, 295*b*
South American dusky dolphin *(Lagenorhynchus obscurus fitzroyi)*, 38

South American fur seal *(Arctocephalus australis)*, 29*t*
South American sea lion *(Otaria flavescens)*, 29*t*
southern bottlenose whale *(Hyperoodon planifrons)*, 37
southern elephant seal *(Mirounga leonina)*, 32*t*, 125*b*, 275
southern right whale *(Eubalaena australis)*, 36, 130, 131*f*
    abundance and status of, 138*t*, 272, 277*t*
    ban on hunting, 219
southern right whale dolphin *(Lissodelphis peronii)*, 38, 174, 174*f*
southern river otter *(Lontra provocax)*, 28
southern seals, 29, 32*t*, 34*f*
Sowerby's beaked whale *(Mesoplodon bidens)*, 37
spade-toothed whale *(Mesoplodon traversii)*, 37, 39*b*
Special Areas of Conservation (SACs), 285
    bottlenose dolphins and, 191
    under Habitats Directive, 282*b*, 283*b*
Species At Risk Act (Canada), 277, 286*b*
species loss, 244
species-specific quota system, whaling and, 219
spectacled porpoise *(Phocoena dioptrica)*, 38, 159, 159*f*
Speculum Regale (The Mirror of Royalty), 204, 205
speed boats, habitat loss and, 246*f*
spermaceti oil, 209
spermaceti organ, 47*b*–48*b*, 48*f*, 141, 209
sperm competition, 191*b*
sperm whale *(Physeter macrocephalus)*, 35, 37, 141–144, 142*f*
    abundance and status of, 142, 144, 274, 275*t*, 277*t*
    ban on hunting, 220
    blowholes in, 65*b*, 66*f*
    distribution of, 142
    diving depth of, 55
    feeding behavior and ecology of, 142
    hunting of, 208*b*, 209, 211*b*, 227*b*
    Japanese hunting of, 221
    in New Zealand legend, 205
    Norse hunting of, 208*b*
    reproduction by, 142

sleep of, 58b
sound production by, 65b, 66f
spermaceti organ of, 47b–48b, 48f, 141, 209
Spes et Fides (boat), 215
spinner dolphin *(Stenella longirostris)*, 38, 182, 183f
   head shape of, 46, 46f
   tuna fishing and, 236
spiny anteater (echidna), 1
spotted dolphin *(Stenella attenuata)*. *See also* pantropical spotted dolphin
   hunting of, 233
   tuna fishing and, 236
spotted-necked otter *(Hydrictis maculicollis)*, 28
spotted seal *(Phoca largha)*, 31t, 121
spreading loss, 64
Squalodon, 23
Stafford-Bell, Richard, 292f, 296b
static passive acoustic systems, 318, 320
statistical errors, 309b
Statute Prerogative Regis, 283b
Stejneger, Leonhard, 9b
Stejneger's beaked whale *(Mesoplodon stejnegeri)*, 37
Steller, Georg Wilhelm, 9b
Steller sea lion *(Eumetopias jubatus)*, 9b, 29t, 126f, 255f, 256f
   abundance and status of, 272, 277t
   research on declining population of, 321b
   threats to, 255–256, 257b
Steller's sea cow *(Hydrodamalis gigas)*, 9b, 18, 19, 33, 101
   declining population of, 96b
   extinction of, 109b–111b, 110f
Stenella spp. (long-beaked dolphins), 39b, 167, 181–189
Stenella attenuata. *See* pantropical spotted dolphin; spotted dolphin
Stenella attenuata attenuata, 278
Stenella attenuata graffmani (Eastern Pacific coastal spotted dolphin), 38, 278
Stenella clymene (clymene dolphin), 38, 182, 183f
Stenella coeruleoalba (striped dolphin), 38, 183f, 183–184
Stenella frontalis (Atlantic spotted dolphin), 38, 182, 182f, 190
Stenella longirostris. *See* spinner dolphin

Stenella longirostris centroamericana (Central American spinner dolphin), 38
Stenella longirostris longirostris (cosmopolitan spinner dolphin), 38
Stenella longirostris orientalis (Eastern or tropical Pacific spinner dolphin), 38, 278
Steno spp. (long-beaked dolphins), 167, 181–189
Steno bredanensis (rough-toothed dolphin), 38, 184, 184b–185b, 184f
strandings
   of beaked whales, 172b
   of common dolphins, 186b–187b
   of pilot whales, 172, 172b
strandings analysis, 316b
strap-toothed beaked whale *(Mesoplodon layardii)*, 37, 144, 145f
stress, 237b
striped dolphin *(Stenella coeruleoalba)*, 38, 183f, 183–184
subantarctic fur seal *(Arctocephalus tropicalis)*, 29t
sublittoral zone, 43b, 44f
subtidal zone, 43b, 44f
Suncus etruscus (Etruscan shrew), 3, 4f
surface area, ratio to volume
   of polar bears, 83
   swimming and, 45, 49b, 49f
   thermoregulation and, 49
sustainability report card, 297b
swimming adaptations, 45f, 45–48, 46f
   of polar bears, 81, 82f

T

tagging, 317–318, 321b
   of pinnipeds, 120b–121b, 120f, 121f
   satellite telemetry tags and, 120, 120b–121b, 121f
tail flukes, swimming and, 45, 46
tapetum lucidum, 55–56
Tasmacetus shepherdi (Tasman or Shepherd's beaked whale), 37, 144, 145f, 146
Tasman beaked whale *(Tasmacetus shepherdi)*, 37, 144, 145f, 146
taste sense, 56, 57f
taxonomy, 317b
teeth, mammalian, 1

temperature, body, thermoregulation and, 48–52, 49f–51f
temperature, environmental, increase in. *See also* climate change
   during Eocene, 15
   during Quaternary, 33
territories, establishing, 71
Tethys Sea, 18, 23
   fossils found in, 20b, 20f
Tethytheria, 18
Tetley, Mike, 284b
2,3,7,8-tetrachlorodibenzodioxin (TCDD), cetaceans and, 240
Theria, 1
thermoregulation, 48–52, 49f–51f
Three Gorges Dam, 151, 151f
tiger shark *(Galeocerdo cuvier)*, 15, 130b, 179
time sampling, for behavioral observation, 315b
tool use
   by dolphins, 93b–94b, 192b
   by sea otters, 94b
toothed whales (Odontoceti), 21, 35, 37–38, 61, 141–161. *See also* dolphin(s); porpoises; *specific toothed whales*
touch sense, 56
Tougaard, Jakob, 63b, 63f
tourism. *See* marine mammal tourism; whale watching
trace elements, cetaceans and, 239–240
transmission loss, 64
Triassic period, mammalian evolution during, 3
Trichechidae, 18, 33
Trichechus inunguis (Amazonian manatee), 33, 101, 102f, 103, 106, 272, 277t. *See also* manatees
Trichechus manatus. (West Indian manatee), 33, 101, 102f
   abundance and status of, 274, 277t
   distribution of, 102–103, 103f
   subspecies of, 102
Trichechus manatus latirostris (Florida manatee), 33, 102
Trichechus manatus manatus (Antillean manatee), 33, 102, 106
Trichechus senegalensis (West African manatee), 33, 101, 102f, 103, 103f, 274, 275, 277t. *See also* manatees
tropical Pacific spinner dolphin *(Stenella longirostris orientalis)*, 38, 278

True's beaked whale (*Mesoplodon mirus*), 37
true seals, 29, 30–33, 31*t*, 32*f*–34*f*, 32*t*
  distribution of, 116, 120
  fur of, 50
tucuxi (*Sotalia fluviatilis*), 38, 39*t*, 188*b*, 188*f*, 275*t*
tuna fishing
  cetaceans and, 236–237, 238*b*
  dolphins and, 212, 236–238, 237, 238*b*, 238*f*
Tursiops spp.
  See bottlenose dolphins; common bottlenose dolphin; Indo-Pacific bottlenose dolphin
Tursiops aduncus. See Indo-Pacific bottlenose dolphin
Tursiops australis (burrunan dolphin), 38*n*
Tursiops truncatus. See common bottlenose dolphin
Tursiops truncatus ponticus (Black Sea bottlenose dolphin), 191, 233*b*
tusk, of narwahl, 155*b*–156*b*, 155*f*
tympanic bullae, 67*b*–68*b*

U

Uhen, Mark, 24*f*
UNCLOS (United Nations Convention on the Law of the Sea), 274*b*
ungulates, 15
United Kingdom
  legal protections for marine mammals in, 283*b*
  public opinion on marine mammal protection in, 283*b*, 283*t*
  whaling in, 208*b*, 208*f*, 209
United Nations Conference on the Human Environment, 220
United Nations Convention on the Law of the Sea (UNCLOS), 274*b*
United Nations Food and Agriculture Organization, 235
United Nations World Charter for Nature, 88*b*
U.S. Dolphin Conservation Act, 238
U.S. Endangered Species Act (ESA), 276–278, 277*t*
  Amazonian manatee and, 277*t*
  baiji and, 277*t*
  blue whale and, 277*t*
  bowhead whale and, 225, 277*t*
  dugong and, 277*t*
  fin whale and, 277*t*
  gray whale and, 277*t*
  Guadalupe fur seal and, 277*t*
  Hawaiian monk seal and, 254, 277*t*
  humpback whale and, 277*t*
  Indus River dolphin and, 277*t*
  killer whale and, 277*t*
  manatee and, 106, 107*b*
  marine otter and, 277*t*
  Mediterranean monk seal and, 277*t*
  North Atlantic right whale and, 277*t*
U.S. Endangered Species Act (ESA), northern fur seal and, 277*t*
  North Pacific right whale and, 277*t*
  polar bear and, 85, 88–89, 277*t*
  polar bear status and, 89
  saimaa (ringed) seal and, 277*t*
  sea otter and, 95, 277*t*
  sei whale and, 277*t*
  southern right whale and, 277*t*
  sperm whale and, 277*t*
  Steller sea lion and, 256, 277*t*
  vaquita and, 277*t*
  West African manatee and, 277*t*
  West Indian manatee and, 277*t*
  Yangtze River dolphin and, 153
U.S. Fish and Wildlife Service (FWS), 277
U.S. Marine Mammal Protection Act, 278, 280
U.S. National Marine Fisheries Service (NMFS), 277, 278, 280
Ursidae, 18, 27
Ursoidea, 27
Ursus actos horribilis (grizzly bear), 17, 28
Ursus arctos (brown bear), 17, 28, 81
Ursus maritimus. See polar bear
Ursus maritimus marinus, 27
Ursus maritimus maritimus, 27
Ursus maritimus tyrannus (giant polar bear), 17

V

vaquita. See Gulf of California harbor porpoise
ventral vestibular air sac, 65*b*
vertebrae, swimming and, 47
vervet monkey (*Chlorocebus pygerythrus*), 73
vessel-based acoustic techniques, 320
video cameras, 318
Vietnamese legend, 205
vision, 55–56, 56*f*
Vulpes lagopus (Arctic fox), 3, 6–7, 7*f*, 9
Vulpes vulpes (red fox), as Arctic fox competitor, 7

W

walrus (*Odobenus rosmarus*), 15, 18, 29, 30, 31*b*, 31*f*
  distribution of, 116
  Norse hunting of, 207
  as polar bear prey, 83
  status of, 275
  thermoregulation in, 51
Wang, John, 161*b*
warm-blooded organisms, 48
warming, global. See climate change; temperature, environmental
water, thermal conductivity of, 48–49
watercraft collisions
  cetaceans and, 238, 239*f*
  killer whales and, 170
  manatees and, 106, 106*b*–108*b*, 107*f*
wave(s), noise from, 75
wavelength, of sound, 62
Waymouth, George, 207
weaning. See also reproduction
  of polar bears, 84
weasels, 27, 28
webbed feet, for swimming, 45
Weddell seal (*Leptonychotes weddellii*), 32*t*, 51–52, 123
West African manatee (*Trichechus senegalensis*), 33, 101, 102*f*, 103, 103*f*, 274, 275, 277*t*. See also manatees
Western North Pacific harbor porpoise (*Phocoena phocoena*), 38
West Indian manatee (*Trichechus manatus*), 33, 101, 102*f*. See also manatees
  abundance and status of, 274, 277*t*
  distribution of, 102–103, 103*f*
  subspecies of, 102
West Pacific Dall's porpoise (*Phocoenoides dalli truei*), 38
whale(s). See also specific types of whales
  ancient, 21
  fossil, 24*b*
  taxonomy of, 211
Whale and Dolphin Conservation Society, 301
whale meat, contamination of, 224*b*, 224*f*, 232

whale oil, demand for, 216, 217t
whale sharks, hunting of, 227b
whale watching. *See also* marine mammal tourism
    ecotourism and, 290b–291b
    global, 292b–293b, 293f, 293t
    Icelandic, 222b
    killer whales and, 170, 171f
    nature of tourists and, 294
    negative impacts of, 294
    risk posed by, 212
    ship strikes and, 238
    significance of, 289–290
    whaling versus, 291b
whaling
    aboriginal, 223, 225t, 225–227, 226f, 227b, 227f, 231
    in America, early, 209
    Antarctic, 216, 216f
    anti-whaling movement and, 138b, 212, 212f, 219, 228b
    bans on, 219–220
    Basque, 207
    beluga whales and, 154
    commercial, early, 208–209, 210f, 211
    factory ships and, 216, 217f
    by Inuits, 207
    Japanese, 207–208, 209f, 217b
    life on whaling boats and, 210b–211b, 211f
    modern, history of, 215–218, 216f, 217f, 217t
    narwhal and, 155
    by Native Americans, 207
    Nordic, 207, 208b, 215, 220
    by Phoenicians, 202
    pilot whales and, 172
    public attitudes toward, 213b
    quota system for, 218–219, 223, 225, 225t
    scientific, 220–221, 221–222
    sperm wales and, 142
    in United Kingdom, 208b, 208f
    welfare and, 228b
    whale watching versus, 291b
Whaling Industry Regulation Act (U.K.), 283b
whistle production, 65b–66b
white-beaked dolphin *(Lagenorhynchus albirostris)*, 38, 177, 178f, 179b
white-sided dolphins, 177. *See also* Atlantic white-sided dolphin; Pacific white-sided dolphin
white whale. *See* beluga whale
Wild Animals Protection Ordinance (Hong Kong), 286b
Wildlife and Countryside Act (U.K.), 283b
Wildlife Conservation Act (Australia), 286b
Wildlife Protection Act (India), 286b
wind farms, marine, building, noise from, 75b–76b
wolf *(Canis lupus)*, 7, 8b, 8f
wolverines, 27
woolly mammoth *(Mammuthus primigenius)*, 317b
World Society for Protection of Animals, 226
Wright, Andrew, 66f, 69b, 313b

Y

Yangtze finless porpoise *(Neophocaena asiaorientalis asiaorientalis)*, 38, 161b
Yangtze River dolphin *(Lipotes vexillifer)*, 37, 150f, 150–153, 151f, 275t, 277t
Young, Sharon, 321b
Yoxton, Grace, 99b
Yupik people, whaling by, 225, 227b

Z

Zalophus californianus (Californian sea lion), 29t, 30b, 30f
Zalophus japonicus (Japanese sea lion), 29t, 125, 126b, 126t, 153
Zalophus wollebaeki (Galápagos sea lion), 29t, 124b, 257–258, 272
Ziphiidae. *See* beaked whales; *specific beaked whales*
Ziphiinae, 37, 144. *See also specific beaked whales*
Ziphius cavirostris (Cuvier's beaked whale), 37, 144, 146, 147
zoos, polar bears in, 89b, 89f
zygapophyses, 45, 45f
Zygorhiza, 24b

Photo Credits

Section Openers

Section Opener 1 © Beth Morley/ShutterStock, Inc.; **Section Opener 2** © Rich Lindie/ShutterStock, Inc.; **Section Opener 3** Courtesy of Great Whales/NOAA 200th Photo Contest/NOAA.

Chapter 1

1.1 © Susan Flashman/ShutterStock, Inc.; **1.4A** © WILDLIFE GmbH/Alamy; **1.4B** © iStockphoto/Thinkstock; **1.4C** © Hemera/Thinkstock; **1.4D** © Digital Vision/Thinkstock; **1.4E** © Matej Hudovernik/ShutterStock, Inc.; **1.5** © Nik Niklz/ShutterStock, Inc.; **1.6** © Kirsten Wahlquist/ShutterStock, Inc.; **1.7A** © Dr. Merlin D. Tuttle/Bat Conservation International/Photo Researchers, Inc.; **1.7B** © Merlin Tuttle/BCI/Photo Researchers, Inc. **1.9** © JG Photo/ShutterStock, Inc.; **B1.1** © Jupiterimages/Photos.com/Thinkstock; **B1.2** © Paul Glendell/Alamy; **B1.3** © John A Cameron/ShutterStock, Inc.; **author photo 1** Courtesy of Lorelei Crerar, George Mason University.

Chapter 2

B2.1A © iStockphoto/Thinkstock; **B2.1B** © JeniFoto/ShutterStock, Inc.; **B2.2** Courtesy of Carl Buell; **B2.4** Courtesy of Carl Buell; **author photo 1** Courtesy of Mark Uhen.

Chapter 3

3.1 © Steven J. Kazlowski/Alamy; **3.2A** © Kevin Schafer/Alamy; **3.2B** © neelsky/ShutterStock, Inc.; **3.3A** Courtesy of U. S. Fish and Wildlife Service; **3.3B** © iStockphoto/Thinkstock; **3.3C** © Jupiterimages/Photos.com/Thinkstock; **3.3D** Courtesy of U. S. Fish and Wildlife Service; **3.5A** © iStockphoto/Thinkstock; **3.5B** Courtesy of Dr. James P. McVey, NOAA Sea Grant Program; **3.5C** © hallam creations/ShutterStock, Inc.; **3.5D** © iStockphoto/Thinkstock; **3.7** © Tim Robinson/ShutterStock, Inc.; **3.8** © Geoffrey Morgan/Alamy **3.9** © digitalbalance/ShutterStock, Inc.; **3.10** © Jose Gil/ShutterStock, Inc.; **B3.1** © iStockphoto/Thinkstock; **B3.3** © Morten Hilmer/ShutterStock, Inc.

Chapter 4

4.3A © Hemera/Thinkstock; **4.3B** © iStockphoto/Thinkstock; **4.3C** © Anna segeren/ShutterStock, Inc.; **4.3D** © iStockphoto/Thinkstock; **4.6** Courtesy of Dominic McAfferty; **4.13** © Bochkarev Photography/ShutterStock, Inc.; **B4.2** Courtesy of Ocean Explorer/NOAA; **B4.3** Courtesy of Dr. Dave Casper.

Chapter 5

5.4 Screen shot from Simrad EK60, courtesy of Tom Stevenson; **5.5** © Yaniv Eliash/ShutterStock, Inc.; **B5.1** Courtesy of Jakob Tougaard, Aarhus University, Denmark; **B5.2A** Courtesy of Wayne Hoggard NOAA/NMFS/SEFSC; **B5.2B** © Hemera/Thinkstock; **B5.3** Image courtesy of Sound in the Sea 2001/NOAA; **B5.5A** © worldswildlifewonders/ShutterStock, Inc.; **B5.5B** © WILDLIFE GmbH/Alamy; **B5.6** © Jan Kratochvila/ShutterStock, Inc.; **author photo 1** Courtesy of Jakob Tougaard, Aarhus University, Denmark; **author photo 2** Courtesy of Adele Roland, George Mason University, photo taken by Chris Parsons.

Chapter 6

6.1 © Christian Musat/ShtterStock, Inc.; **6.3** © Vlad Ghiea/ShutterStock, Inc.; **6.4** © Thomas Barrat/ShutterStock, Inc.; **6.5** © Uryadnikov Sergey/ShutterStock, Inc.; **6.6** © louise murray/Alamy; **6.9** © iStockphoto/Thinkstock; **6.10** © Kevin Schafer/Alamy; **B6.1** © Hemera/Thinkstock; **author photo 1** Courtesy of Thea Østergaard Bechshøft, Aarhus University.

Chapter 7

7.1 © iStockphoto/Thinkstock; **7.3** © Francois Gohier/Photo Researchers, Inc.; **7.4** © iStockphoto/Thinkstock; **7.5** © iStockphoto/Thinkstock; **7.6** © Kevin Schafer/Alamy; **B7.1** Courtesy of U. S. Fish and Wildlife Service; **author photo 1** Courtesy of Katheryn Patterson; **author photo 2** Courtesy of Naomi Rose, Humane Society International; **author photo 3** Courtesy of Grace Yoxon, International Otter Survival Fund.

Chapter 8

8.1A © Tsuneo Nakamura/Volvox Inc/Alamy; **8.1B** © Comstock Images/Comstock/Thinkstock; **8.1C** © Andre Seale/Alamy; **8.1D** © Ant Clausen/ShutterStock, Inc.; **B8.1** © Tom Stack/Alamy; **B8.2** © Bull's-Eye Arts/ShutterStock, Inc.; **B8.4** © Dray van Beeck/ShutterStock, Inc.; **author photo 1** Courtesy of Ellen Hines San Francisco State University.

Chapter 9

9.1A © John Foxx/Stockbyte/Thinkstock; **9.1B** © Vladimir Melnik/ShutterStock, Inc.; **9.2** © iStockphoto/Thinkstock; **9.3** © worldswildlifewonders/ShutterStock, Inc.; **B9.1** © Michael Patrick O'Neill/Alamy; **B9.3** Courtesy of NOAA; **B9.5A** © iStockphoto/Thinkstock; **B9.5B** © withGod/ShutterStock, Inc.; **B9.5C** © Uryadnikov Sergey/ShutterStock, Inc.; **B9.7** Courtesy of Dr. Bernie McConnell, Scottish Oceans Institute, University of St. Andrews; **B9.9** © Norman Lightfoot/Photo Researchers, Inc.; **B9.10** © Bryan and Cherry Alexander/Photo Researchers, Inc.; **B9.11A** © iStockphoto/Thinkstock; **B9.11B** © iStockphoto/Thinkstock

Chapter 10

10.1A © Sebastien Burel/ShutterStock, Inc.; **10.1B** © iStockphoto/Thinkstock; **10.5A** Courtesy of Matt Wilson/Jay Clark, NOAA NMFS AFSC; **10.5B** Courtesy of Jamie Hall/NOAA; **10.6** Courtesy of Christine Gleason; **10.9** © Eduardo Rivero/ShutterStock, Inc.; **10.10A** Courtesy of Matt Wilson/Jay Clark, NOAA NMFS AFSC; **10.10B** Courtesy of NOAA-OE/HBOI; **10.11** © Minden Pictures/SuperStock

Chapter 11

11.1 © Doug Perrine/Alamy; **11.2** Courtesy of Chris Parsons; **11.7** © Universal Images Group Limited/Alamy; **11.9** © guentermanaus/ShutterStock, Inc.; **11.10** © Mark Carwardine/Visuals Unlimited, Inc.; **11.12** © JingAiping/ShutterStock, Inc.; **11.13** © iStockphoto/Thinkstock; **11.14** © Martin Tiller/ShutterStock, Inc.; **11.15** © David Fleetham/Visuals Unlimited; **B11.1B** © Dorling Kindersley RF/Thinkstock; **B11.4** © Bryan & Cherry Alexander/Photo Researchers, Inc.

B11.5 Courtesy of Clare Perry, Environmental Investigation Agency; **B11.6** Courtesy of Clare Perry, Environmental Investigation Agency; **B11.7** © Hemere/Thinkstock; **B11.9** © Volvox Inc/Alamy; **author photo 1** Courtesy of Dr. Colin D. MacLeod, GIS In Ecology (www.GISinEcology.com); **author photo 2** Courtesy of Leslie Cornick; **author photo 3** Courtesy of Clare Perry; **author photo 4** Courtesy of John Wang.

Chapter 12

12.1 © Carl Dawson/ShutterStock, Inc.; **12.2** © Jupiterimages/Photos.com/Thinkstock; **12.4** © Mike Price/ShutterStock, Inc.; **12.5A** Courtesy of Lieutenant Elizabeth Crapo, NOAA Corps; **12.5B** Courtesy of NOAA/NOS/NMS/MBNMS; **12.8A** © blickwinkel/Alamy; **12.8B** © Eduardo Rivero/ShutterStock, Inc.; **12.8C** © Nickolay Stanev/ShutterStock, Inc.; **12.10A** Courtesy of A.Reckendorf/Whale and Dolphin Conservation Society; **12.10B** Courtesy of NOAA/MBARI 2006; **12.10C** © Eirik Gronningsaeter/Alamy; **12.10D** © Steven Vancoillie/ShutterStock, Inc.; **12.10E** © National Geographic Image Collection/Alamy; **12.10F** Courtesy of Lieutenant Elizabeth Crapo, NOAA Corps; **12.11A** © Dorling Kindersley RF/Thinkstock; **12.11B** © Brett Atkins/ShutterStock, Inc.; **12.11C** Courtesy of Samuel Hung, Hong Kong Dolphin Conservation Society; **12.11D** Courtesy of Samuel Hung, Hong Kong Dolphin Conservation Society; **12.14A** Courtesy of NOAA FIsheries service; **12.14B** © iStockphoto/Thinkstock; **12.14C** Courtesy of Wayne Hoggard NOAA/NMFS/SEFSC; **12.14D** Courtesy of Wayne Hoggard NOAA/NMFS/SEFSC; **12.14E** © Anna segeren/ShutterStock, Inc.; **12.15** © Jose Gil/ShutterStock, Inc.; **12.16** Courtesy of Wayne Hoggard NOAA/NMFS/SEFSC; **12.17A** Courtesy of NOAA Fisheries Service; **12.17B** © Picture Hooked/Malcolm Schuyl/Alamy; **12.19** © Tom Brakefield/Stockbyte/Thinkstock; **B12.1A** © Mogens Trolle/ShutterStock, Inc.; **B12.1B** © slava296/ShutterStock, Inc.; **B12.2** Courtesy of Dr. Mridula Srinivasan, NOAA/NMFS/OST/AMD; **B12.3** Courtesy of Wayne Hoggard NOAA/NMFS/SEFSC; **B12.5** © Digital Vision/age fotostock; **B12.6** © Visual&Written SL/Alamy; **B12.8** Courtesy of Diana Reiss, Ph.D.; **B12.9** © melissaf84/ShutterStock, Inc.; **author photo 1** Courtesy of Naomi Rose, Humane Society International; **author photo 2** Courtesy of Liz Slooten, University of Otago; **author photo 3** Courtesy of Dr. Colin D. MacLeod, GIS In Ecology (www.GISinEcology.com).; **author photo 4** Courtesy of Christina MacFarquhar; **author photo 5** Courtesy of Jolanda Luksenburg, George Mason University.

Chapter 13

13.2 © World History Archive/Alamy; **13.3** © Karel Gallas/ShutterStock, Inc.; **13.5** © World History Archive/Alamy; **13.6** © iStockphoto/Thinkstock; **13.7** Courtesy of Library of Congress, Prints & Photographs Division [LC-DIG-jpd-01790]; **13.8** © Jim Barber/ShutterStock, Inc.; **13.9** © Cliff Hide/Alamy; **B13.2** © INTERFOTO/Alamy; **B13.3** © The Protected Art Archive/Alamy.

Chapter 14

14.1 © Rich Lindie/ShutterStock, Inc.; **14.2** © wdeon/Shutterstock.com; **B14.1** © Peter Gordon/ShutterStock, Inc.; **B14.2** © Steven J. Kazlowski/Alamy; **author photo 1** Courtesy of Ann Marie Renker.

Chapter 15

15.1 © Nick Haslam/Alamy; **15.2** © Digital Vision/age fotostock; **15.3** Courtesy of John Higgins/NOAA; **15.4** Courtesy of R.Cesar/Whale and Dolphin Conservation Society; **15.5** Courtesy of Christine Gleason; **15.6** © Cheryl Casey/Shutterstock.com; **15.7** © Jupiterimages/Photos.com/Thinkstock; **15.8A** © fotosunkid/ShutterStock, Inc.; **15.8B** © luigi nifosi'/ShutterStock, Inc.; **15.8C** © eAlisa/ShutterStock, Inc.; **15.8D** © Hemera/Thinkstock; **15.8E** © iStockphoto/Thinkstock; **B15.1** Courtesy of Naomi Rose; **B15.2** © The Yomiuri Shimbun/AP Images ; **B15.3** © Jones & Bartlett Learning. Photographed by Lauren Miller.

Chapter 16

16.1 © Images & Stories/Alamy; **16.2** © iStockphoto/Thinkstock; **16.4** © iStockphoto/Thinkstock; **16.6** © Stacy Funderburke/ShutterStock, Inc.; **16.8** © iStockphoto/Thinkstock; **16.10** Courtesy of Joel Orkin-Ramey; **16.12** © DPS/ShutterStock, Inc.; **16.14** Courtesy of Simon Goodman, University of Leeds/Caspian International Seal Survey; **16.16** © Steffen Foerster Photography/ShutterStock, Inc.; **B16.1** © iStockphoto/Thinkstock; **B16.3** © canadabrian/Alamy.

Chapter 17

B17.1 © Paula Fisher/ShutterStock, Inc.; **author photo 1** Courtesy of Tim Ragen.

Chapter 18

18.1 © Sam Chadwick/ShutterStock, Inc.; **18.2** © Comstock Images/Thinkstock; **B18.1** © Keith Levit Photography/Thinkstock; **B18.3** Courtesy of Chris Parsons; **B18.4A** © Hannu Liivaar/ShutterStock, Inc.; **B18.4B** © Hemera/Thinkstock; **B18.5** Courtesy of Naomi Rose, Human Society International; **author photo 1** Courtesy of Carol Scarpaci, University of Victoria; **author photo 2** Courtesy of Michael Lück, Auckland University of Technology, New Zealand; **author photo 3** Courtesy of Carol Scarpaci, University of Victoria; **author photo 4** Courtesy of Jennifer Ambler, Ph.D., George Mason University; **author photo 5** Courtesy of Megan Draheim, George Mason University

Appendix

A.4 © SuperStock; **BA.2A** Courtesy of Amigos de los Delfines; **BA.3B** Courtesy of Chris Parsons; **author photo 1** Courtesy of Marieke Kester, George Mason University; **author photo 2** Courtesy of Jill Fazio; **author photo 3** Courtesy of Sharon Young, The Humane Society of the United States.

Unless otherwise indicated, all photographs and illustrations are under copyright of Jones & Bartlett Learning, or have been provided by the authors.